Physical Optics of Dynamic Phenomena and Processes in Macromolecular Systems

Physical Optics of Dynamic Phenomena and Processes in Macromolecular Systems

Proceedings
27th Microsymposium on Macromolecules
Prague, Czechoslovakia, July 16-19, 1984

Editor
Blahoslav Sedláček

W
DE
G

Walter de Gruyter · Berlin · New York 1985

Editor
Blahoslav Sedláček, PhD., DSc.
Institute of Macromolecular Chemistry
Czechoslovak Academy of Sciences
Heyrovský sq. 2

CS-162 06 Prague 616
Czechoslovakia

Library of Congress Cataloging in Publication Data

Prague IMPAC Microsymposium on Macromolecules (27th : 1984)
Physical optics of dynamic phenomena and processes in macromolecular systems.
„Under the sponsorship of the International Union of Pure and Applied Chemistry (IUPAC)"--Pref.
Includes bibliographies and index.
1. Macromolecules--Analysis--Congresses. 2. Spectrum analysis--Congresses. I. Sedláček, B. (Blahoslav) II. International Union of Pure and Applied Chemistry. III. Title.
QD380.P73 1984 547.7'046 84-28793

CIP-Kurztitelaufnahme der Deutschen Bibliothek

Physical optics of dynamic phenomena and processes in macromolecular systems : proceedings / 27th Microsymposium on Macromolecules, Prague, Czechoslovakia, July 16–19, 1984. Ed. Blahoslav Sedláček. – Berlin ; New York : de Gruyter, 1985.
ISBN 3-11-010234-X (Berlin)
ISBN 0-89925-011-4 (New York)
NE: Sedláček, Blahoslav [Hrsg.]; Microsymposium on Macromolecules <27, 1984, Praha>

3 11 010234 X Walter de Gruyter · Berlin · New York
0-89925-011-4 Walter de Gruyter, Inc., New York

 Printing: Gerike GmbH, Berlin. Binding: Dieter Mikolai, Berlin. – Printed in Germany.

PREFACE

This volume includes the majority of the short special lectures and about half of the poster contributions presented at the 27th IUPAC Microsymposium "Physical Optics of Dynamic Phenomena and Processes in Macromolecular Systems", held in Prague (Czechoslovakia) from July 16 to 19, 1984.

The Prague Meetings on Macromolecules (including Microsymposia, Discussion Conferences and Summer Schools) have been organized, almost from their very beginning in 1967, under the sponsorship of the International Union of Pure and Applied Chemistry (IUPAC). Due to their specific, usually monothematic character, these meetings would seem to be in contrast to the regular IUPAC Symposia and Congresses, which have a broad, polythematic program: in fact, however, the micro-Symposia supplement the macro-Symposia by bringing together specialists in the respective field(s) for an effective exchange of ideas and experience. Thus, the volume also offers the reader a good opportunity to inform himself about recent progress in the respective disciplines.

This book deals with a problem area which could also be called "Dynamics of Polymers and Biopolymers as Studied by Physico-Optical and Related Methods", acting in solutions, colloid dispersions, gels and solids. The subject (macromolecules) and a tool for their study (method) are 'percolated' to an extent which makes the papers rather complex to read, but 'holographically' illustrative nonetheless; at the same time, however, this makes any systematic classification of papers into sections difficult or arbitrary. Therefore, only methods (and theories) not directly connected with the application to a given polymer are grouped together in a special section.

The contents are divided into five sections: Section A deals with the (hydro)dynamic behaviour of (preferably model) polymers in dilute and semidilute solutions, followed by light scattering and other physico-optical methods. In Section B similar problems connected with concentrated systems (such as dense solutions and gels) are discussed; also, some new or less well-known methods for their research are introduced. Sections C (Intramolecular Phenomena and Processes) and D (Intermolecular Phenomena and Processes) stress processes which occur in macromolecular systems either individually or, in most cases, simultaneously. For several papers, no strict borderline can be drawn between the individual sections: the inclusion in one or another section is made, for the reader's convenience, with respect to the main emphasis and similarity to subsequent papers. Section E deals with theoretical and/or experimental methods outlined or used for a general or, sometimes, special purpose. Occasionally, here and in other sections of this volume, a theory, technique or application may depart somewhat from the scope of the book, but it has been retained as it might be of interest for comparison, application or as additional information.

Due to the very limited time available, pressing demands were sometimes made upon the more than 140 authors, whose efficient cooperation in preparing nearly seventy camera-ready papers is highly appreciated. This form of publishing has the indisputable advantage of an early publication of papers, which is of great value, especially in the case of conference proceedings. With regard to the editing, compromises are in some cases unavoidable in view of the 'least-change principle' and the 'language barrier': I hope that the papers will be found at least understandable, if not well formulated.

My thanks are also due to de Gruyter Publishers and their leading coworkers for a very efficient and helpful approach to all problems which arose during our cooperation. I hope that these fruitful contacts in the Prague Meetings on Macromolecules will also continue in the future.

Blahoslav Sedláček
Chairman and Editor

CONTENTS

Preface V

A. Dilute and Semidilute Solutions

Hydrodynamics of Polymer Chain in Dilute Solution 1
C.J.C. Edwards and R.F.T. Stepto

QELS from Very Long and Semiflexible Filaments in Solution 17
S. Fujima and T. Maeda

Theoretical and Experimental Studies of Anisotropy in Translational Diffusion with Special Reference to Tobacco Mosaic Virus 33
T. Maeda, S. Fujime, K. Kubota, H. Urabe and Y. Tominaga

Molecular Weight and Second Virial Coefficient Influences on the Concentration Dependence of Diffusion Coefficient of Polystyrene Molecules in Dilute Solution 37
K. Witkowski, L. Wolinski, and B. Kołodziej

The Concentration Dependence of Dimensions of Flexible Polymer Molecules in Solution 43
J.S. King, W. Boyer, G.D. Wignall, and R. Ullman

Diffusion and Relaxation in Polymer-Solvent Systems by Photon Correlation Spectroscopy 49
C. Cohen and D.H. Hwang

Quasi-Elastic Light Scattering in Semi-Dilute Solutions at the Θ-Temperature 59
M. Adam and M. Delsanti

Light Scattering Observations of the Plateau Modulus in Semi-Dilute Theta Solutions 67
A.-M. Hecht, H. Bohidar, and E. Geissler

Dynamic Light Scattering on Semidilute Solutions of Polystyrene as a Function of Solvent Quality 71
W. Brown

Dynamic Properties of Entangled Polymer Solutions by Forced Rayleigh Light Scattering 83
M.F. Millet, H. Hervet, and L. Léger

Dynamics of Polyelectrolytes in Aqueous Solutions Probed by Electro-Optical Relaxation 87
S. Wijmenga, F. van der Touw, and M. Mandel

Dynamic Light Scattering from a Rod-Like Polymer in Semidilute Solution 107
K.H. Langley and P.S. Russo

Laser Light Scattering of Dilute and Semi-Dilute Hydrosoluble Biopolymer and Synthetic Polyelectrolyte Solutions 117
G. Muller

B. Concentrated Systems: Gels and Solids

Statistical Theory of Streaming Birefringence in Temporary Polymer Networks 131
E. Kröner and R. Takserman-Krozer

Quasi-Elastic Light Scattering of an Irreversible Sol-Gel Transition 145
S.J. Candau, M. Ankrim, J.P. Munch, P. Rempp, G. Hild, and R. Okasha

Microphase Separation in Poly(acrylamide-bisacrylamide) Copolymerized Gels 157
E. Geissler and A.-M. Hecht

Rayleigh and Brillouin Scattering from Linear Polymer During Gelation 165
T. Igarashi and S. Kondo

Quasi-Elastic Light Scattering Studies of the Diffusion of Compact Macromolecules through Gels 177
D. B. Sellen

Slow Diffusion of Labeled Macromolecules Studied by a Holographic Grating Technique 191
M. Antonietti, J. Coutandin, D. Ehlich, and H. Sillescu

Mechanical Stability of Gels Undergoing Large Swelling 197
Y. Hirokawa, J. Kucera, S.-T. Sun, and T. Tanaka

Photon Correlation Spectra of Amorphous Low Molecular Weight Polyethyl Methacrylate above the Glass Transition Temperature 205
G. Fytas

Rayleigh-Brillouin Scattering of Amorphous Polymers 217
C.H. Wang and B. Stuhn

Brillouin Scattering Study of Segmental Motions of Polyalkyl Methacrylates (PMMA, PEMA and PnBMA) at Low Temperature 229
E. Kato and Y. Saji

Time Resolved Small-Angle X-Ray Scattering Studies on Kinetics and Molecular Dynamics of Order-Disorder Transition of Block Polymers 233
T. Hashimoto

Thermostimulated Luminescence in the Relaxation Regions of Polymers 245
J. Pospíšil, A. Havránek and I.A. Tale

C. Intramolecular Phenomena and Process

Dynamic Phenomena, Interactions, Conformational Changes and Chemical Processes in Macromolecular Systems as Studied by Polarized Luminescence 251
E. Anufrieva and M. Krakovyak

Intramolecular Excimer Formation and Polymer Dynamics: The Coil-Globule Transition 267
C. Cuniberti and A. Perico

Excimer Formation in Polystyrene Solutions: Effect of the Coil-Globule Transition 271
P. Štěpánek, Č. Koňák, and B. Sedláček

The Influence of Chromophoric Groups Rotation on the Decay of Fluorescence of Polystyrene in Solution 275
K. Sienicki and C. Bojarski

Small-Angle Scattering of Polyelectrolyte Solutions 279
R. Koyama

Light-Induced Conformational Changes of Macromolecules in Solution as Detected by Flash Photolysis in Conjunction with Light-Scattering Measurements 287
M. Irie and W. Schnabel

Photon Correlation Spectroscopy of Polymethacrylic Acid 301
P. Brak and A. Persoons

Photon Correlation Spectroscopy of Macromolecular Solutions under High Pressure 305
B. Nyström and J. Roots

Intramolecular Motility in Pig Immunoglobulin G Studied by Neutron Spin Echo Technique 317
S. Borbély, Y.M. Ostanevich, L. Cser, B. Faragó, F. Mezei, and F. Franěk

The Conformation of a DNA-Protein Complex 321
M.A. Scheerhagen, H. van Amerongen, M.E. Kuil, R. van Grondelle, and J. Blok

Electro-Optical Kerr Effect Studies of Hemoglobin 325
Z. Błaszczak, M. Witrh, and B. Nordén

Molecular Mobility in Liquid Polymers and Polymerization Process Investigated by Acoustical Methods 329
P. Hauptmann

D. Intermolecular Phenomena and Processes

Theory of Light Scattering from Systems of Interacting Spherical Macromolecules in Solution 335
R. Klein, G. Nägele and W. Hess

Cooperative Growth of Molecular Aggregation in Semiconcentrated PBLG Solutions Investigated by the Dynamic Electro-Optical Method 349
H. Watanabe, T. Nakano, and Y. Fukuda

Light Scattering Anomaly Observed in Dilute Solutions of Poly(vinyl alcohol) Aged at High Concentration 353
L. Mrkvičková, Č. Koňák, and B. Sedláček

Light Scattering from Carrageenan Solutions 359
C. De Jonghe, H. Reynaers,
K. Bloys van Treslong, and F. Varkevisser

Studies of Protein-Nucleic Acid Interactions by Photon Correlation Spectroscopy. II. tRNA-BSA Interactions at Intermediate and High Ionic Strength 363
A. Patkowski and B. Chu

Solvation of Protein Molecules Studied by Combined QELS and SAXS Investigations 367
K. Gast, D. Zirwer, P. Plietz, J.J. Müller,
G. Damaschun, and H. Welfle

Light Scattering Studies of Bovine Eye Lens α_L-Crystallin at Higher Concentrations: Short and Long-Range Order 371
C. Andries, M. Van Laethem, and J. Clauwert

Magneto-Optics of Colloids 377
R.V. Mehta

Characterization of Inverse Polyacrylamide Latices by Quasi-Elastic Light Scattering 397
F. Candau, C. Holtzcherer, and S. Candau

Interaction between Micelles of Poly(styrene-b-hydrogenated butadiene-b-styrene) in 1,4-Dioxane/n-Heptane Mixtures 401
Č. Koňák, P. Štěpánek, and Z. Tuzar

Properties of Block Copolymer Micelles Near the C.M.C. and C.M.T. 405
Z. Tuzar, P. Štěpánek, and Č. Koňák

Lateral Diffusion of Micelles Measured by Fluorescence Recovery after Photobleaching 409
W. Van de Sande and A. Persoons

E. Theoretical and Experimental Methods

Molecular Dynamics Simulation of the n-Alkanes Rotator Phase 413
M.A. Mazo, E.F. Oleinik, N.K. Balabaev, L.V. Lunevskaya, and A.G. Grivtsov

The Kerr Effect Relaxation in High Electric Fields 427
H. Watanabe and A. Morita

Dielectric Relaxation of a Flexible Molecule 443
A. Morita and H. Watanabe

Photon Statistics in Light Scattering 447
J. Peřina

Attenuation of a Coherent Field in a Dense Dispersion of Particles 457
Č. Koňák, P. Štěpánek, J. Křepelka, and J. Peřina

Determination of Polydispersity Index by Quasi-Elastic Light Scattering 461
P. Štěpánek, Z. Tuzar, and Č. Koňák

Application of Picosecond Laser Pulses in Light Scattering Studies 465
B. Van Wonterghem and A. Persoons

Application of Experimental Techniques for Obtaining Photon Correlation Functions at High Pressure 469
J. Roots and B. Nyström

Bimodal Analysis of QELS Data 473
R. Johnsen

Application of Speckle Techniques in Macromolecular Physics 477
J. Holoubek, J. Mikeš, and B. Sedláček

Light Scattering Speckle Photography: Determination of Slow Correlation Times 481
J. Holoubek

Study of Polymer Solution/Solid Interfaces by Evanescent Wave Spectroscopy 485
H. Hervet, D. Ausserré, and F. Rondelez

Singular Value Decomposition in the Analysis of Spectra 493
W. Curtis Johnson, Jr.

Flow Birefringence of Macromolecular Solutions in Various Elongational Flows 507
R. Cressely, R. Hockquart, J.-P. Decruppe, and T. Wydro

Use of Light Scattering Methods for the Study of Dynamic Processes of Macromolecules Coagulation in Hydrocarbon Medium 511
G.F. Bolshakov

Association of Alcohols Investigated by Rayleigh Light Scattering 517
B.M. Fechner

Refractometric Study of Pressure Effect on Preferential Sorption of Mixed Liquids on Gels under Dynamic Conditions 521
T. Macko, D. Berek, and M. Chalányová

Compatibilization through Hydrogen Bonding of Pairs of Copolymers 525
O. Aouadj, A. Lassoued, and S. Djadoun

Contributors 529

Index 541

HYDRODYNAMICS OF POLYMER CHAINS IN DILUTE SOLUTION

Christopher J C Edwards and Robert F T Stepto

Department of Polymer Science and Technology, University of Manchester Institute of Science and Technology, Manchester M60 1QD, UK

Introduction

The translational diffusion coefficient, D, of a molecule is related to its effective hydrodynamic radius, r_D, by the Stokes-Einstein equation:

$$D = kT/f = kT/6\pi\eta_0 r_D \tag{1}$$

where f is the molecular friction coefficient and η_0 is the solvent viscosity. The relation between D and the detailed molecular structure of a polymer chain is complicated, mainly because of hydrodynamic interactions between the centres of friction in the chain. Currently, the equation due to Kirkwood (1,2)

$$D = \frac{kT}{x\zeta} + \frac{kT}{6\pi\eta_0}\frac{[R^{-1}]}{x^2} \tag{2}$$

represents the best analytical solution to the problem. Here $\zeta = 6\pi\eta_0 a$ is the segmental friction coefficient with a the radius of a segment and x the number of segments in the chain $[R^{-1}] = \Sigma\Sigma\langle r_{ij}^{-1}\rangle$ where $\langle r_{ij}^{-1}\rangle$ is the mean reciprocal separation of segments i and j and the angled brackets denote a configurational average. Eq(2) can be written in a form which is independent of solvent viscosity (3), namely:

Physical Optics of Dynamic Phenomena and Processes in Macromolecular Systems

$$\frac{1}{r_D} = \frac{1}{x.a} + \frac{[R^{-1}]}{x^2} \qquad (3)$$

The first term on the righthand side (rhs) of Eq(3) is the so-called free-draining term associated with the flow of solvent through the molecular domain and this term becomes insignificant as $x\rightarrow\infty$. The second term on the rhs of Eq(3) takes account of hydrodynamic interactions between segments and relates to the effective impermeable diffusion radius of the chain.

The derivation of Eq(2) makes several important assumptions which have been discussed recently in detail(4). For example, the chain is assumed to diffuse as a rigid body in a solvent continuum. In addition, hydrodynamic interactions between segments are assumed to be weak and the interactions are assumed not to perturb the equilibrium configurations of the chain.

In the present paper, previous interpretations in terms of Eq(2) of r_D for short polymethylene (PM) (3,5), polyoxyethylene (POE) (5,6) and linear and cyclic poly(dimethylsiloxane) (PDMS) (7,8) chains are brought together and compared. For a given chain structure ζ is found to vary with solvent, temperature and chain length, and possible reasons for these variations are discussed. In addition, for systems with strong specific interactions Eq(2) is shown not to apply with free-draining being of negligible consequence. Finally, the predicted errors (4) in Eq(2) compared with the full Oseen evaluation of D for high molar-mass PM chains are discussed. A key point in this comparison and in the correlations of r_D for short chains is the use of realistic chain models and Monte Carlo calculations (4,9) to evaluate $\langle r_{ij}^{-1}\rangle$.

Interpretation of experimental data

Multiplication of each term in Eq(3) by $x^{\frac{1}{2}}$ yields:

$$\frac{x^{\frac{1}{2}}}{r_D} = \frac{x^{\frac{1}{2}}}{r_F} + \frac{x^{\frac{1}{2}}}{r_E} \tag{4}$$

where r_F, the effective free-draining radius, is equal to $x.a$ and r_E, the effective impermeable radius, is equal to $x^2/[R^{-1}]$. For unperturbed chains, $x^{\frac{1}{2}}/r_D$ and $x^{\frac{1}{2}}/r_E$ tend to the same constant value and $x^{\frac{1}{2}}/r_F$ tends to zero as $x \to \infty$. Thus the effective diffusion radius, r_D, is the sum-in-parallel of the free-draining radius, r_F, and the effective impermeable radius r_E. Eq(4) has been used previously (3,10,11) to illustrate graphically the relative magnitudes of the contributions of r_F and r_E to r_D. Values of $r_E = x^2/[R^{-1}]$ have been evaluated for PM, POE and cyclic and linear PDMS using a Monte Carlo technique (4,9) which employs Metropolis sampling (12) and rotational-isomeric-state models of the polymer chains given by Flory *et al* (13).

1. Short polymethylene chains. In Figure 1 $x^{\frac{1}{2}}/r_D$, $x^{\frac{1}{2}}/r_F$ and $x^{\frac{1}{2}}/r_E$ are plotted versus x for short PM chains (3). Also shown in Figure 1 is the Gaussian curve, calculated by assuming that $r_E = 3\pi^{\frac{1}{2}}\langle s^2\rangle/8$. The free-draining curve $x^{\frac{1}{2}}/r_F$ has been calculated for a = 0.08 nm which is a reasonable estimate of the size of a $-CH_2-$ segment.

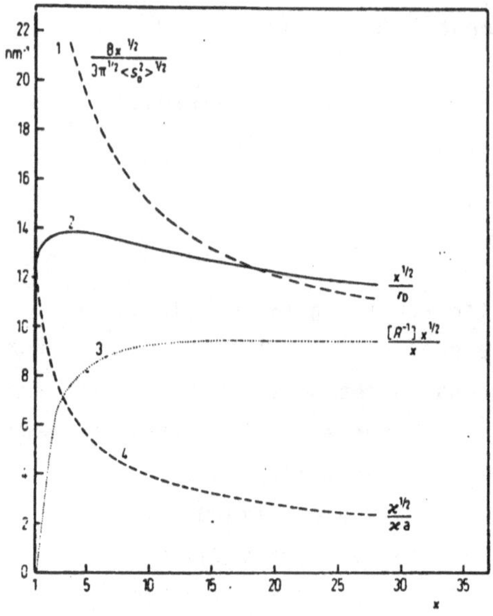

Figure 1

Variation of reciprocal diffusion radii of polymethylene at 298K with chain length (x). $[R^{-1}]$ and $<s^2{}_0>^{\frac{1}{2}}$ calculated using Hill-Stepto model (5) . *a* = 0.08 nm. Curves: 1. Gaussian approximation to Stokes-Einstein radius, 2. Stokes-Einstein radius, 3. impermeable radius and 4. free-draining radius.

Inspection of Figure 1 shows that the effects of free-draining are still significant for x = 50. The Gaussian approximation to $x^{\frac{1}{2}}/r_E$ is poor at low x but tends with increasing chain length to the same value as the Monte Carlo curve rather faster than the effects of free-draining disappear. Figure 2 shows the same plots with experimental data for short PM chains in benzene and quinoline also included (10). It is apparent from Figure 2 that a constant value of the segment radius, a, will not explain the experimental data and that, furthermore, the value of a depends on solvent.

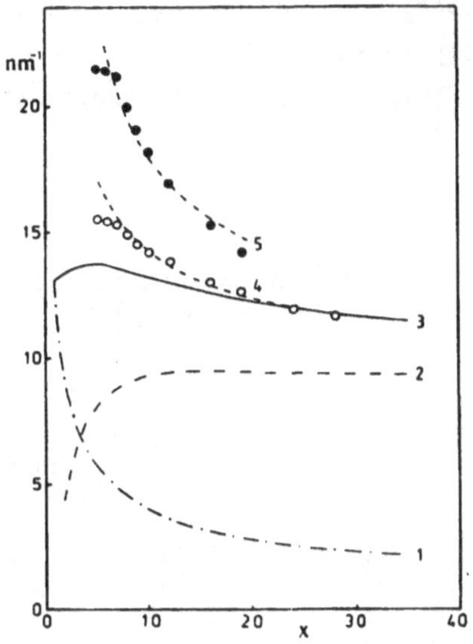

Figure 2

Reciprocal diffusion radii versus chain length (x) for (●) polymethylene/quinoline/298K and (O) polymethylene/benzene/298K. Curves: 1. Free-draining radius, $x^{\frac{1}{2}}/r_F$ with $r_F = x.a$ and $a = 0.08$ nm, 2. impermeable radius, $x^{\frac{1}{2}}/r_E$ with $r_E = x^{\frac{1}{2}}/[R^{-1}]$, 3. Kirkwood-Riseman diffusion radius, $x^{\frac{1}{2}}/r_D = x^{\frac{1}{2}}/r_F + x^{\frac{1}{2}}/r_E$, 4. and 5. calculated diffusion radius with $a = \langle c_b \rangle . b$ and (4) $\langle c_b \rangle = 0.216$ and (5) $\langle c_b \rangle = 0.121$.

Previously, this increase in the apparent segment radius with increasing x has been rationalised by allowing a to vary in proportion to the effective bond length of the chain $b=(\langle r^2 \rangle /n)^{\frac{1}{2}}$, such that

$$\zeta = 6\pi\eta_0 c_b . b \qquad (5)$$

where c_b is a constant which depends on polymer, solvent and temperature.

c_b can be thought of as characterising the "slippage" which occurs at the segment-solvent interface, although an exact molecular interpretation of c_b is not possible. Values of c_b required to reproduce the experimental data for PM in a variety of solvents are given in Table 1 together with the values obtained by averaging over chain lengths, $\langle c_b \rangle$. As has been noted previously (3,10) the values of $\langle c_b \rangle$ generally tend to decrease as the molar volume of the solvent and/or the solvent viscosity increase.

Table 1

Values of c_b required to give agreement between calculated and experimental Stokes-Einstein radii for polymethylene chains in various solvents

x	b/nm	benzene 25°C	carbon tetrachloride 25°C	tetralin 22.2°C	quinoline 25°C	decalin 22.2°C
		c_b	c_b	c_b	c_b	c_b
5	0.228	0.267	0.287	0.157	0.148	0.134
6	0.247	0.230	0.229	0.139	0.121	0.118
7	0.263	0.221	0.225	0.127	0.115	0.104
8	0.277	0.218	0.204	-	0.116	0.098
9	0.289	0.217	-	0.127	0.116	0.099
10	0.300	0.215	0.193	-	0.118	0.092
12	0.318	0.204	0.182	0.105	0.120	0.092
14	0.331	-	-	0.112	-	0.091
16	0.343	0.203	0.196	-	0.125	-
18	0.352	-	0.213	0.108	-	0.096
19	0.356	0.203	-	-	0.136	-
20	0.359	-	0.185	-	-	-
24	0.371	0.220	-	-	-	-
28	0.380	0.226	0.178	0.134	-	0.109
$\langle c_b \rangle$		0.216	0.201	0.122	0.121	0.100
η_o cP		0.608	0.938	2.1	3.44	2.64
V_o $cm^3 mol^{-1}$		88.9	96.5	136	118	159

The values of $<c_b>$ in benzene and quinoline have been used to recalculate $x^{\frac{1}{2}}/r_F$, with $r_F = x<c_b>.b$, and the curves obtained in this way are those shown dashed in Figure 2. The agreement between the calculated curves and the experimental data is now good except at the shortest chain lengths, where end-effects become important. The general decrease in $<c_b>$ with increasing molar volume and/or viscosity of solvent may be thought of as resulting from the assumption of a solvent continuum. $<c_b>$ is a measure of the effective friction experienced by a "hydrodynamic segment" and this decreases as slippage due to voids in the solvent structure (or molar volume) increases and as the perturbation of solvent flow for a given shear gradient decreases (i.e. viscosity increases). The constancy of $<c_b>$, or the concomitant increase in a with chain length is not so easily rationalised. It may result from the assumption of rigid-body hydrodynamics in the derivation of the Kirkwood equation[(3)]. Given that segmental velocities depend less on chain length than the centre-of-mass velocity, the difference between these velocities will increase with chain length giving an effective increase in ζ. The proportionality of ζ to b remains unexplained.

2. Polyoxyethylene chains. Figure 3 shows plots according to Eq(4) for POE in water and quinoline with $x^{\frac{1}{2}}/r_F$ calculated for a = 0.1 nm (10). In this case the experimental data for POE in quinoline agree well with the calculated curve, $x^{\frac{1}{2}}/r_D$, suggesting that a is independent of x. However, the reason for this behaviour, which contrasts with that found previously for PM, becomes apparent when data for methoxy terminated POE chains are considered (points shown as "+" in Figure 3). Similar behaviour is now observed to that found previously for PM and it is clear that quinoline molecules hydrogen-bond to the terminal -OH groups of the POE chain and thereby increase the effective chain length. It is probable that a doubly end-blocked POE chain would show an even larger increase in $x^{\frac{1}{2}}/R_D$ as x decreases similarly to that found previously for PM.

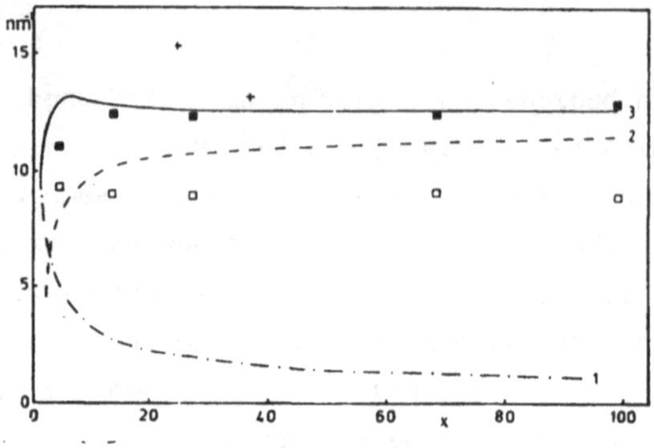

Figure 3

Reciprocal diffusion radii versus chain length (x) for (■) POE/quinoline/298K and (□) POE/water/298K. The points marked (+) refer to POE terminated with one -OH and one -OMe group in quinoline at 298K. Curves: 1. Free-draining radius $x^{\frac{1}{2}}/r_F$, a = 0.1 nm, 2. Impermeable radius $x^{\frac{1}{2}}/r_E$, 3. Kirkwood-Riseman diffusion radius $x^{\frac{1}{2}}/r_D$.

The data for POE in water (5,6) lie well below the calculated curve for $x^{\frac{1}{2}}/r_D$ at all values of x. In this case water molecules are hydrogen-bonded to the skeletal oxygen atoms as well as the terminal -OH groups, thereby increasing the effective chain cross-section. Thus POE chains in quinoline and water provide examples of systems which show non-Kirkwood behaviour due to strong specific interactions between polymer and solvent.

3. <u>Linear and cyclic poly(dimethylsiloxane) (l-PDMS and r-PDMS)</u>. In Figure 4 an analysis is presented for l-PDMS in toluene and bromocyclohexane (10). In this case the data can only be reproduced by considering the friction centres to be $[Si(CH_3)_2O]$ units with a = 0.25 nm.

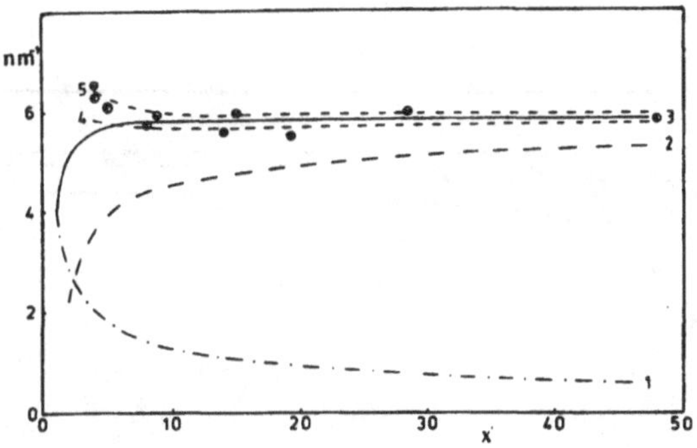

Figure 4

Reciprocal diffusion radii versus chain length (x) for (●) l-PDMS/toluene/298K and (◐) l-PDMS/bromocyclohexane/301K. Curves: 1. Free-draining radius, $x^{\frac{1}{2}}/r_F$, a = 0.25 nm, 2. Impermeable diffusion radius $x^{\frac{1}{2}}/r_E$, 3. Kirkwood-Riseman diffusion radius $x^{\frac{1}{2}}/r_D$, 4. and 5. calculated with $a = \langle c_b \rangle b$ and (4) $\langle c_b \rangle = 0.509$ and (5) $\langle c_b \rangle = 0.405$.

Again at low x the effective segment size varies with chain length, although this effect is less marked for l-PDMS than for PM, possibly due to the larger segment radius of l-PDMS. The values of $\langle c_b \rangle$ required to reproduce the experimental data are 0.509 in toluene and 0.405 in bromocyclohexane. Curves recalculated using those values of $\langle c_b \rangle$ are shown dashed in Figure 4 and as for PM the agreement with the experimental data is good. An equivalent analysis (11) for data on r-PDMS is shown in Figure 5. Since l-PDMS and r-PDMS differ only in topology it would be expected that the same value of a would reproduce the data for both species. However, Figure 5 shows that this is not the case and that for r-PDMS $a \simeq 0.4$ nm as compared with the value of 0.25 nm found for l-PDMS. The difference between the values of _ca_ 0.4 nm and _ca_ 0.25 nm for the segment radius of r-PDMS compared with l-PDMS is consistent with less free-draining flow in the cyclic molecules as would be expected from their higher segment densities.

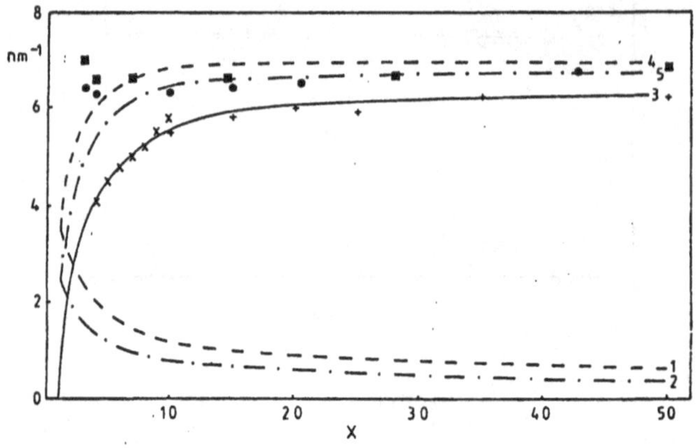

Figure 5

Reciprocal diffusion radii versus chain length (x) for cyclic PDMS in (●) toluene at 298K and (■) bromocyclohexane at 301K. Curves: 1. and 2. free draining radius $x^{\frac{1}{2}}/r_F$ with $r_F = x.a$ and $a = 0.25$ nm and $a = 0.4$ nm respectively, 3. impermeable diffusion radius $x^{\frac{1}{2}}/r_E$, 4. and 5. Kirkwood-Riseman diffusion radii with (4) the sum of curves (1) and (3) (a=0.25 nm) and (5) the sum of curves (2) and (3) (a=0.4 nm). (x) complete enumeration (+) Monte Carlo calculations.

Again agreement with the experimental data is improved (11) if the segment radius is allowed to vary in proportion to an effective bond length. For cyclic species the effective bond length may be redefined as $b_r = (6\langle s_r^2\rangle/x)^{\frac{1}{2}}$ because the mean-square end separation $\langle r^2\rangle$ is undefined for such molecules. Values of b_r have been used in Eq(5) to define values of $\langle c_b\rangle$ for the data in toluene and bromocyclohexane. The improvement in agreement with the experimental data at short chain lengths using a constant value of $\langle c_b\rangle$ is not so marked as for l-PDMS, but the difference in values of $\langle c_b\rangle$ between the two solvents is still significant.

4. Characteristic values of segment radii at infinite chain length. The use of Eq(5) to define the segmental friction coefficient means that at infinite chain length the effective segment radius tends to a constant value as the effective bond length b reaches its limiting value for the unperturbed chain, viz

$$a_\infty = \langle c_b \rangle . b_\infty \qquad (6)$$

with $\langle c_b \rangle$ independent of chain length, and evaluated as described from data on short chains. The parameter a_∞ is useful for summarising differences in segmental friction for different polymer/solvent systems at given temperatures. Table 2 gives the resulting values of a_∞ for PM, 1-PDMS and r-PDMS. Dependences of a_∞ on solvent are clearly seen and relate to solvent viscosity and molar volume as previously discussed. The values of a_∞ increase in the order PM < 1-PDMS < r-PDMS showing that free-draining decreases in the same order.

Table 2

Values of $\langle c_b \rangle$ and a_∞ for 'Kirkwood-Riseman' systems. $\langle c_b \rangle$ evaluated from experimental data on finite chains.

		$\langle c_b \rangle$	b_∞/nm	a_∞/nm
PPM/CCl_4/298K		0.225	0.435	0.098
PM/benzene/298K		0.213	"	0.093
PM/tetralin/295K		0.122	"	0.053
PM/quinoline/298K		0.121	"	0.052
PM/decalin/298K		0.100	"	0.044
1-PDMS/toluene/298K	$\langle r_o^2 \rangle$	0.509	0.566	0.288
	$\langle s_o^2 \rangle$	0.559	"	0.316
r-PDMS/toluene/298K		1.31	0.402	0.527
1-PDMS/BrC_6H_{11}/301K	$\langle r_o^2 \rangle$	0.405	0.566	0.229
	$\langle s_o^2 \rangle$	0.449	"	0.254
r-PDMS/BrC_6H_{11}/301K		0.923	0.402	0.371

Oseen analysis and r_D at infinite chain length

Notwithstanding the rigid body approximation, the Kirkwood equation is known to be approximate. For this reason Edwards, Kaye and Stepto (4) have recently examined for PM chains the magnitudes of the errors introduced by the assumptions of pre-averaging and weak hydrodynamic interactions. Using a development of the equations for the frictional forces on a polymer chain which is related to that originally used by Kirkwood and Riseman (1), it has been shown that the Kirkwood equation relies only on the assumption of weak hydrodynamic interactions and not on the use of a pre-averaged Oseen tensor. In addition, ensembles of rigid-body configurations of rotational-isomeric-state model PM chains were generated using the Monte Carlo procedure described previously. The diffusion tensor equations were then solved numerically for two different cases. Firstly, using the pre-averaged form of the Oseen tensor:

$$\langle \zeta \underline{\underline{T}}_{ij} \rangle_{\text{Oseen}} = \frac{\zeta}{6\pi\eta_0 \ell \tilde{r}_{ij}} \underline{\underline{I}} \qquad (7)$$

when ℓ is the bond length, $\tilde{r}_{ij}$ is the magnitude of the distance between friction centres i and j and $\underline{\underline{I}}$ denotes the unit matrix. In this case the approximation of weak hydrodynamic interactions is not present, but the interactions are pre-averaged. Secondly, the full form of the Oseen tensor

$$\zeta \underline{\underline{T}}_{ij} = \frac{\zeta}{8\pi\eta_0 \ell \tilde{r}_{ij}} (\underline{\underline{I}} + \underline{n}_{ij}\underline{n}_{ij}) \qquad (8)$$

has been used, where $\underline{n}_{ij}\underline{n}_{ij}$ denotes the direct product of the unit vector connecting segments i and j. This approach gives the exact Oseen solution for the case of a rigid body moving with constant velocity and random orientation (i.e. Brownian motion is assumed to be overwhelming) in a solvent continuum. For ease of comparison of results from the two cases with those from the Kirkwood equation, Eq(2) can be expressed as:

$$\frac{\zeta}{f} = \frac{1}{x} + \frac{\kappa[\tilde{R}^{-1}]}{x^2} \qquad (9)$$

where $[\tilde{R}^{-1}]$ represents the sum of reciprocal separations of segments normalised to unit bond length and $\kappa = \zeta/6\pi\eta_0 \ell$ characterises the strength of the hydrodynamic interactions between segments. The previous observation that ζ is related to b implies that κ increases to a limiting value of $x \to \infty$. The lower limit of $\kappa = 0$ represents complete free-draining. Figure 6 shows a plot for PM in benzene at 298 (K) of the percentage reduction in ζ/f relative to the Kirkwood result, for the pre-averaged Oseen solution, as a function of $x^{-\frac{1}{2}}$.

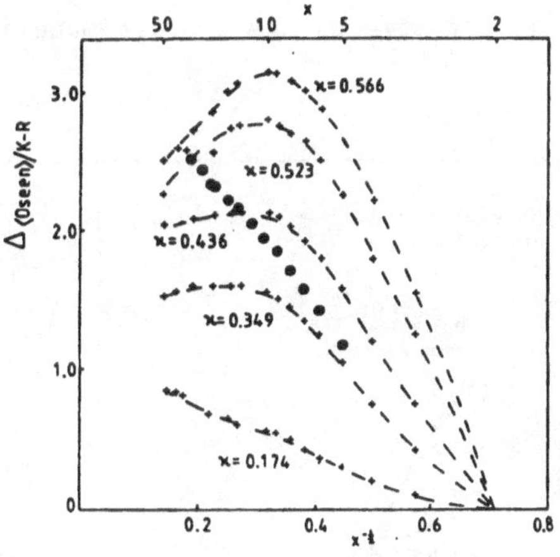

Figure 6

Percentage difference between values of ζ/f for PM chains from <Oseen> and K-R treatments as a function of $x^{-\frac{1}{2}}$. (+) percentage differences at percentage constant κ, (●) percent differences derived from experimental diffusion coefficients of PM in benzene at 298K.

Several values of κ are employed and the points marked (O) correspond to the values of κ derived by interpreting the experimental data using Eq(9). It can be seen that the limiting reduction in ζ/f at high molar-mass lies in the range 2-3%. Corresponding plots (4) are shown in Figure 7 for the results from the full Oseen treatment compared with those from the Kirkwood equation. Here the limiting reduction in ζ/f as $x \to \infty$ is of the order of 8%. The overestimate of 8% in ζ/f from Eq(9) compared with the full Oseen treatment is characteristic of the PM chain in benzene at 298K. Since κ (or ζ) depends on chain structure, solvent and temperature, the overestimate in ζ/f from Eq(9) will also depend on these variables. In addition, it can be seen from Figure 7 (and 6) that actual polymer/solvent systems must be used in order to determine the values of κ to be considered. There is no way at present of determining κ independently. The overestimate in ζ/f for a given value of κ is the same as the overestimate in the parameter (3,10).

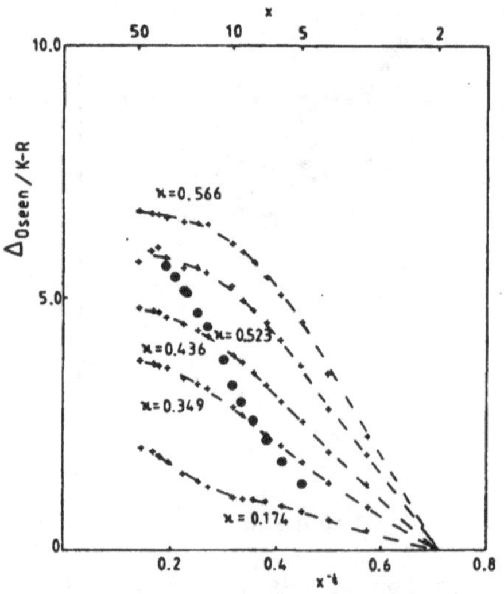

Figure 7

Percentage difference between values of ζ/f for PM chains from Oseen and K-R treatments as a function of $x^{-\frac{1}{2}}$. Symbols as in Figure 6.

$$\psi(x) = \langle s_0^2 \rangle^{\frac{1}{2}} / r_D \qquad (10)$$

The calculations and data indicate that the value of $\psi(\infty)$ for PM in benzene at 298K, as given by the Kirkwood equation (2), namely $8/3\pi^{\frac{1}{2}}$ = 1.505 is about 8% too high. Thus the predicted value of $\psi(\infty)$ in this case is 1.385. Similar results have been deduced by Burchard and Schmidt (14) and ter Meer, Burchard and Wunderlich (15) from experimental diffusion data for high molar-mass polystyrene (14) and poly(methyl acrylate) (15) in θ-solvents for which the Kirkwood-Riseman theory apparently overestimates $\psi(\infty)$ by about 15% and 27% respectively, corresponding to $\psi_{PS}(\infty)$ = 1.27 and $\psi_{PMMA}(\infty)$ = 1.16. The non-universality of $\psi(\infty)$ means that the diffusion radius is not a fixed proportion of the equilibrium radius and that the fractional penetration of the molecular domain by solvent depends on polymer, solvent and temperature.

Conclusions

Analyses of experimental diffusion data for short PM, POE and l-PDMS and r-PDMS chains in terms of the Kirkwood equation show that ζ increases with chain length in proportion to an effective bond length. ζ also varies with solvent and this variation can be correlated approximately with the molar volume and viscosity of the solvent. The same segment radius, *a*, will not explain the experimental data for both l-PDMS and r-PDMS. In addition, POE in quinoline and water provide examples of systems where hydrogen-bonding between polymer and solvent gives rise to deviations from "Kirkwood-type" behaviour.

Exact numerical solutions of the diffusion equations in the rigid body approximation using the Oseen tensor confirm that the Kirkwood equation underestimates the diffusion radius and that the ratio of equilibrium to diffusion radius is not a universal constant.

References

1. Kirkwood, J.G. and Riseman, J.: J. Chem. Phys. 16, 565, (1948).
2. Kirkwood, J.G.: J. Polym. Sci. 12, 1, (1954).
3. Mokrys, I.J., Rigby, D. and Stepto, R.F.T.: Ber. Bunsenges. Phys. Chem. 83, 446, (1979).
4. Edwards. C.J.C., Kaye, A. and Stepto, R.F.T.: Macromolecules, 17, 773, (1984).
5. Hill, J.L. and Stepto, R.F.T.: Proc. IUPAC Int. Symp. Macromol. Helsinki, 3, 325, (1972).
6. Couper, A. and Stepto, R.F.T.: Trans. Faraday Soc., 65, 2486, (1969).
7. Edwards, C.J.C., Stepto, R.F.T. and Semlyen, J.A.: Polymer, 21, 781, (1980).
8. Edwards, C.J.C., Stepto, R.F.T. and Semlyen, J.A.: Polymer, 23, 865, (1982).
9. Edwards, C.J.C., Rigby, D., Stepto, R.F.T., Dodgson, K. and Semlyen, J.A.: Polymer, 24, 391, (1983).
10. Edwards, C.J.C., Rigby, D. and Stepto, R.F.T.: Macromolecules, 14, 1808, (1981).
11. Dodgson, K., Edwards, C.J.C. and Stepto, R.F.T.: British Polymer Journal, in press.
12. Wood, W.W.: 'Physics of Simple Liquids' (Eds. Temperley, H.N.V., Rowlinson, J.S. and Rushbrooke, G.C.) N. Holland Publishing Company, Amsterdam, 1968.
13. Flory, P.J.: 'Statistical Mechanics of Chain Molecules' Interscience, London, 1969.
14. Burchard, W. and Schmidt, M.: Macromolecules, 14, 210, (1981).
15. ter Meer, H.-U, Burchard, W. and Wunderlich, W.: Colloid Polym. Sci. 258, 675, (1980).

QELS FROM VERY LONG AND SEMIFLEXIBLE FILAMENTS IN SOLUTION

Satoru Fujime, Tadakazu Maeda
Mitsubishi-Kasei Institute of Life Sciences, Machida, Tokyo 194, Japan

Introduction

We have studied a quite large effect of anisotropy in translational diffusion on the intensity correlation function $G^2(\tau)$ of light scattered from solutions of long and rigid rods. The theoretical formulation (1) has been successfully examined with special reference to tobacco mosaic virus (2). A concise summary of these studies will be found in this volume (3). A simple extension of this formulation makes it possible to account for the case of a long and semiflexible (or slightly bendable) filament. We will discuss here mainly our theoretical model and shortly an application of the model to the analysis of experimental data.

Theoretical Section

Let define the conformation of a long, thin and semiflexible (or slightly bendable) rod by a space curve $\mathbf{r}(s,t)$, where s is the coordinate of a segment measured from the center of the rod and t is time (Fig. 1);

$$\mathbf{r}(s,t) = \mathbf{R}(t) + s\mathbf{t}(t) + \sum_m'' \mathbf{q}(m,t)Q(m,s) \tag{1}$$

Here, $\mathbf{R}(t)$ is the position vector of the center-of-resistance of, and $\mathbf{t}(t)$ is the unit vector parallel to the 'mean' axis of the rod undergoing small bending. The terms under the summation sign represent the bending motion of the semiflexible rod and the double prime (") means to exclude m = 0 (translational) and m = 1 (rotational) modes of motion. The unnormalized field correlation function $G^1(\tau)$ of polarized light scattered from a solution of a semiflexible filament is given by (4,5):

Physical Optics of Dynamic Phenomena and Processes in Macromolecular Systems

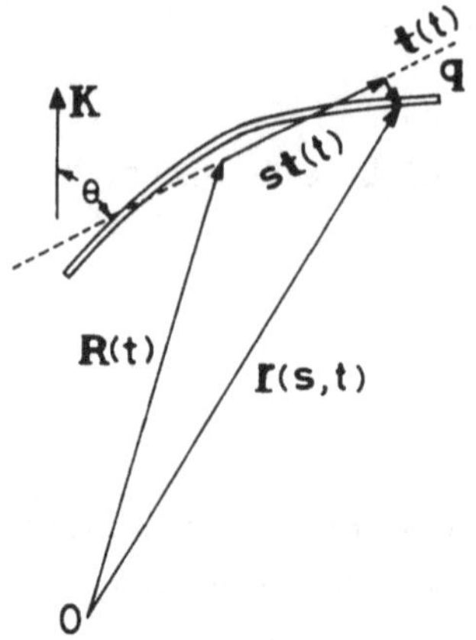

Figure 1. Geometrical relationship among various quantities (see text).

$$G^1(\tau) = G_D(\tau) \times (1/L^2) \iint J(s,s',\tau)dsds' \tag{2}$$

$$J(s,s',\tau) = \langle \exp[iK(s\xi - s'\xi')] \times \Pi_m'' \exp[-\Phi(m,s,s',\xi,\xi',\tau)] \rangle \tag{3}$$

$$G_D(\tau) \equiv \exp[-\{D_0 - \tfrac{1}{3}(D_3 - D_1)\}K^2\tau] \equiv \exp(-D_1K^2\tau) \tag{4}$$

where L is the contour length of the filament, D_1, D_3 and $D_0 = (2D_1 + D_3)/3$ are, respectively, the sideways, the lengthways and the overall translational diffusion coefficients, integrations over s and s' are carried out in the range [-L/2, L/2], and

$$\langle \ldots \rangle = \tfrac{1}{2} \iint (\ldots) g_K(\xi,\xi';\tau)d\xi d\xi' \tag{5}$$

where $\xi = \cos\theta(t)$, $\theta(t)$ being the instantaneous angle between **t**(t) and **K**, and integrations over ξ and ξ' are carried out in the range [-1, 1]. Since the relaxation time τ_m of the internal bending motions is much shorter than the rotational relaxation time of the rod, we can put $\xi = \xi'$ in $\Phi(m,s,s',\xi,\xi',\tau)$ with a very high accuracy and we have (5)

$$\Phi(m,s,s',\xi,\xi',\tau) = (K^2/4)(1 - \xi^2)\langle q_m^2 \rangle \times [Q(m,s)^2 + Q(m,s')^2 - 2Q(m,s)Q(m,s')\exp(-\tau/\tau_m)] \tag{6}$$

where $\langle q_m^2 \rangle$ is the expectation value of the square of q(m,t). The function $g_K(\xi,\xi';\tau)$ is the conditional probability that if the rod is found at orientation ξ' at time t', it will be found at orientation ξ at time t ($\tau \equiv |t - t'|$), and satisfies the following diffusion equation;

$$[\partial/\partial\tau - \Theta(\nabla^2 - \mu^2\xi^2)]g_K(\xi,\xi';\tau) = \delta(\xi - \xi')\delta(\tau) \tag{7}$$

$$\mu^2 \equiv (D_3 - D_1)K^2/\Theta \qquad \text{(coupling constant)} \tag{8}$$

where Θ is the rotational diffusion coefficient, μ^2 is the coupling constant between translational and rotational modes of motion and ∇^2 is the Laplace operator. By expanding $g_K(\xi,\xi';\tau)$ in terms of the Legendre polynomials $P_n(\xi)$, substituting it into eq 7, multiplying both sides by $P_n(\xi)$ and integrating over ξ, we have simultaneous first order differential equations for expansion coefficients $A_n(K,\xi';\tau)$, which can be written in a matrix form as

$$\partial\mathbf{A}/\partial\tau = -\Theta\,\mathbf{M}\,\mathbf{A} \tag{9}$$

where $\mathbf{A} = [A_0, A_1, \ldots, A_N]^T$ (A_n was truncated at n = N) and $\mathbf{M} = \{a_{nm}\}$ with

$$a_{n,n} = n(n+1) + \mu^2 L_0(n), \qquad L_0(n) = (2n^2+2n-1)/(2n-1)(2n+3) \tag{10a}$$

$$a_{n,n-2} = \mu^2 L_1(n), \qquad L_1(n) = n(n-1)/(2n-1)(2n-3) \tag{10b}$$

$$a_{n,n+2} = \mu^2 L_2(n), \qquad L_2(n) = (n+1)(n+2)/(2n+3)(2n+5) \tag{10c}$$

and other elements are all zero. Once the matrix $\mathbf{M}$ is diagonalized, i.e.,

$$\mathbf{M}\,\mathbf{U} = \mathbf{U}\,\mathbf{\Lambda} \qquad \text{with} \qquad \mathbf{\Lambda} = \{\lambda_p \delta_{p,p'}\} \tag{11}$$

we have the solutions of eq 9 and 7 (note $A_\ell(K,\xi';0) = (2\ell+1)P_\ell(\xi')/2$)

$$\mathbf{A}(\tau) = [\mathbf{U}\, e^{-\Theta\mathbf{\Lambda}\tau}\, \mathbf{U}^{-1}]\, \mathbf{A}(0) \tag{12a}$$

$$g_K(\xi,\xi';\tau) = \sum_n \sum_\ell \frac{2\ell + 1}{2} [\mathbf{U}\, e^{-\Theta\mathbf{\Lambda}\tau}\, \mathbf{U}^{-1}]_{n\ell}\, P_n(\xi)P_\ell(\xi') \tag{12b}$$

where n and ℓ run over both even and odd integers (but exclude mixed combination). By use of eq 12b, we have from eq 1 and 2

$$G^1(\tau) = G_D(\tau) \times (1/L^2) \iint dsds' \ \tfrac{1}{2} \int d\xi \ \Pi_m'' \exp[-\Phi(m,s,s',\xi,\xi,\tau)]$$
$$\times \ \sum_n \sum_\ell (2\ell + 1)(-i)^\ell [\mathbf{U} \ e^{-\Theta\Lambda\tau} \ \mathbf{U}^{-1}]_{n\ell} \ j_\ell(Ks')P_n(\xi)\exp(iKs\xi) \qquad (13)$$

where $j_\ell(z)$ is the ℓth order spherical Bessel function. When $\langle q_m{}^2 \rangle = 0$, i.e., for a rigid rod, eq 13 recovers eq 24 in (1) and eq 1 in (3). In the case of a rigid rod, the first cumulant $\bar{\Gamma}$ of $G^2(\tau)$ has been derived (1,3):

$$\bar{\Gamma}/K^2 = D_0 + (L^2/12)\Theta f_1(k) - (D_3 - D_1)[\tfrac{1}{3} - f_2(k)] \qquad (14)$$

where both $f_1(k)$ and $f_2(k)$ are functions depending only on $k \equiv KL/2$, whose values are tabulated in (1). A quite similar procedure to that for eq 14 gives the first cumulant of $G^1(\tau)$ in eq 13 (5);

$$\bar{\Gamma}/K^2 = D_0 + (L^2/12)\Theta f_1^\star(k) - (D_3 - D_1)[\tfrac{1}{3} - f_2^\star(k)] + (k_BT/\zeta L)\textstyle\sum_m'' a_m(k) \qquad (15)$$

where ζ is the friction constant per unit length of the filament. In eq 15, both $f_1^\star(k)$ and $f_2^\star(k)$ depend not only on k but also on the flexibility parameter γL of the filament. The quantity $a_m(k)$ gives the contribution to $\bar{\Gamma}/K^2$ from the *m*th mode of bending motion. In the stiff limit, $\gamma L \to 0$, we have $f_1^\star(k) \to f_1(k)$ and $f_2^\star(k) \to f_2(k)$. In this limit, $a_m(k)$ should be zero (for this, see Appendix C in (4) and also (6)). When $\gamma L \neq 0$, we have the limiting values of $f_1^\star(k) \to 1$, $f_2^\star(k) \to 0$ and $a_m(k) \to 1$ at $k \gg 1$, and thus

$$\bar{\Gamma}/K^2 \to [D_0 - \tfrac{1}{3}(D_3 - D_1)] + (L^2/12)\Theta + (k_BT/\zeta L)\textstyle\sum_m'' 1 \qquad (k \gg 1) \qquad (16)$$

where the factor Σ" 1 means the number of bending modes of motion involved in the scattering process. We have $[D_0 - \tfrac{1}{3}(D_3 - D_1)] = D_1$, and in the long rod limit, $(L^2/12)\Theta = D_1$ and $(k_BT/\zeta L) = D_1$. From eq 16, therefore, each mode of motion is said to contribute by D_1 to $\bar{\Gamma}/K^2$ at large KL values. In an experimentally accessible range of KL values, the m = 2, 3 and possibly 4 modes of bending motion of a long and semiflexible filament are expected to contribute to $\bar{\Gamma}/K^2$ in addition to the m = 0 and 1 modes. Experimentally speaking, when observed $\bar{\Gamma}/K^2$ values exceed those expected from eq 14, one has to consider possible contributions from internal bending motions.

The above consideration deals with the situation of *dilute* solutions. However, the above formulas after simple replacement of D_1, D_3, D_0 and Θ by $\bar{D}_1$, $\bar{D}_3$, $\bar{D}_0$ and $\bar{\Theta}$, respectively, will hold for *semidilute* solutions, where bars over the diffusion coefficients mean that these coefficients are appropriate ones in semidilute solution. They may be written as

$$\bar{D}_1 = \alpha D_1, \quad \bar{D}_3 = D_3, \quad \bar{D}_0 = (2\alpha D_1 + D_3)/3 \quad \text{and} \quad \bar{\Theta} = \beta\Theta \tag{17}$$

Doi and Edwards (7) assumed $\alpha = 0$ and $\beta = 1/(cL^3)^2$, where c is the number of rods in unit volume. This assumption seems to be inadequate for a very long rod (see later).

Now, we consider the motion of a long and semiflexible filament in solution. The elastic potential energy V of a slightly bendable filament is assumed to be given by

$$V = (\varepsilon/2)\int[\partial^2\mathbf{r}(s,t)/\partial s^2]^2 ds \qquad \text{with} \qquad \varepsilon = 2k_BT/(\gamma L) \tag{18}$$

where ε is the flexural rigidity and γL is the flexibility parameter.

1) Free-draining limit. From eq 18, the Langevin equation is derived as

$$\rho(\partial^2\mathbf{r}/\partial t^2) + \zeta(\partial\mathbf{r}/\partial t) + \varepsilon(\partial^4\mathbf{r}/\partial s^4) = \mathbf{A}(s,t) \tag{19}$$

where ρ is the linear mass density and $\mathbf{A}$ is the fluctuating force (8). By normal mode expansion

$$\mathbf{r}(s,t) = \textstyle\sum_m \mathbf{q}(m,t)Q(m,s) \qquad \text{and} \qquad \mathbf{A}(s,t) = \textstyle\sum_m \mathbf{B}(m,t)Q(m,s) \tag{20}$$

eq 19 can be written as

$$\rho\mathbf{q}'' + \zeta\mathbf{q}' + \lambda_m\mathbf{q}(m,t) = \mathbf{B}(m,t) \quad \text{and} \quad \varepsilon(d^4Q/ds^4) = \lambda_m Q(m,s) \tag{21}$$

The second equation in eq 21 with free-end boundary conditions gives

$$\lambda_0 = \lambda_1 = 0 \quad \text{and} \quad \lambda_m = \varepsilon\pi^4(m - \tfrac{1}{2})^4/L^4 \qquad (m \geq 2) \tag{22}$$

$$Q(0,s) = \sqrt{1/L}, \quad Q(1,s) = \sqrt{12/L^3}\, s \quad \text{and} \quad Q(m,s) \quad (m \geq 2, \text{ see } (4)) \tag{23}$$

From eq 18 and 20 we have

$$V = \tfrac{1}{2} \textstyle\sum_m \lambda_m \mathbf{q}(m,t)^2 \tag{24}$$

which determines the expectation values $\langle \mathbf{q}(m,t)^2 \rangle$ for $m \geq 2$ (for which $\lambda_m \neq 0$). From eq 20 and 23, we have transformation formulas from *molecule-fixed coordinates* to *laboratory-fixed coordinates* as

$$\mathbf{q}(0,t) = \sqrt{L}\, \mathbf{R}(t) \qquad \text{and} \qquad \mathbf{q}(1,t) = \sqrt{12/L^3}\, \mathbf{t}(t) \tag{25}$$

which give eq 1. Each $\mathbf{q}(m,t)$ for $m \geq 2$ has mutually perpendicular components $[q_1(m,t), q_2(m,t), q_3(m,t)]$. Putting $q_3(m,t)$ to be parallel to $\mathbf{t}(t)$, the orientation relation between $\mathbf{q}(m,t)$ and $\mathbf{t}(t)$ can be defined. Because $q_1(m,t) \times Q(m,s)$ and $q_2(m,t)Q(m,s)$ represent the bending motion of, and $q_3(m,t)Q(m,s)$ represents the stretch/contraction motion of the filament, we have to put $q_3(m,t) = 0$ for $m \geq 2$. The force $\mathbf{A}(s,t)$ is also written as

$$\mathbf{A}(s,t) = \mathbf{F}(t)/L + s\mathbf{T}(t) \times (12/L^3) + \textstyle\sum_m'' \mathbf{B}(m,t)Q(m,s) \tag{26}$$

where $\mathbf{F}(t) = \int \mathbf{A}(s,t)ds$ is the random force acting on $\mathbf{R}(t)$ and $\mathbf{T}(t) = \int \mathbf{A}(s,t)\, s\, ds$ is the random torque acting on $\mathbf{t}(t)$. Then, the first equation in eq 21 is written for $m = 0$ and 1, respectively, as

$$M\mathbf{R}'' + \Xi_t \mathbf{R}' = \mathbf{F}(t) \qquad \text{and} \qquad I\mathbf{t}'' + \Xi_r \mathbf{t}' = \mathbf{T}(t) \tag{27}$$

where $M = \rho L$, $\Xi_t = \zeta L$, $I = \rho L^3/12$ and $\Xi_r = \zeta L^3/12$. Equation 27 shows translational and rotational Brownian motions of the rod as a whole. Thus, we used $g_K(\xi,\xi';\tau)$ in eq 5 for $m = 0$ and 1 modes. From eq 20 and 24 with the equipartition theorem of thermal energy, we have eq 28 and 29. The factor $2k_BT$ in eq 18 and 29 came from the assumption of $q_3(m,t) = 0$.

$$\langle \mathbf{q}(m,t)\mathbf{q}(m',t')\rangle = \langle q_m{}^2\rangle \exp(-\tau/\tau_m)\delta_{m\,m'} \qquad \text{with} \qquad \tau_m = \zeta/\lambda_m \tag{28}$$

$$\langle q_m{}^2\rangle \equiv \langle \mathbf{q}(m,t)^2\rangle = \langle q_1(m,t)^2\rangle + \langle q_2(m,t)^2\rangle = 2k_BT/\lambda_m \tag{29}$$

2) Non free-draining limit. In order to take account of the hydrodynamic interaction, we follow the method of Hearst, Beals and Harris (HBH). HBH treated the hydrodynamic interaction between beads in its spherically averaged form (9). The interaction tensor has been introduced as

$$H(s,s') = \delta(s - s') + (\zeta/6\pi\eta)f(|s - s'|) \tag{30}$$

The quantity ζ is the friction constant per unit length of the chain, which is assigned the value $3\pi\eta$ throughout by analogy to a string of touching spherical beads. The solvent viscosity is η. The Langevin equation corresponding to eq 19 now has the form

$$\rho(\partial^2\mathbf{r}/\partial t^2) + \boldsymbol{\zeta}(\partial\mathbf{r}/\partial t) + \varepsilon\int H(s,s')[\partial^4\mathbf{r}(s',t)/\partial s'^4]ds' = \mathbf{A}(s,t) \tag{31}$$

where $\boldsymbol{\zeta}$ is the friction factor tensor, which has the diagonal elements ζ_1, $\zeta_2 = \zeta_1$ and ζ_3. For the overall friction factor $\zeta = 3\pi\eta$, we assigned the value $\zeta_1 = 4\pi\eta$ and $\zeta_3 = 2\pi\eta$ by analogy to $D_0 = (2D_1 + D_3)/3$ or $\zeta^{-1} = (2\zeta_1^{-1} + \zeta_3^{-1})/3$. By mode expansion similar to eq 20;

$$\mathbf{r}(s,t) = \textstyle\sum_m \mathbf{q}(m,t)\Psi(m,s) \qquad \text{and} \qquad \mathbf{A}(s,t) = \textstyle\sum_m \mathbf{B}(m,t)\Psi(m,s) \tag{32}$$

eq 31 is written as

$$\varepsilon(d^4\Psi(m,s)/ds^4) + \varepsilon/2\int f(|s - s'|)[d^4\Psi(m,s')/ds'^4]ds' = \sigma_m\Psi(m,s) \tag{33a}$$

$$\rho\mathbf{q}'' + \boldsymbol{\zeta}\mathbf{q}' + \sigma_m\mathbf{q}(m,t) = \mathbf{B}(m,t) \tag{33b}$$

From eq 33b, we have (cf. eq 28)

$$\langle \mathbf{q}(m,t)\mathbf{q}(m',t')\rangle = \langle q_m{}^2\rangle \exp(-\tau/\tau_m)\delta_{m\,m'} \qquad \text{with} \qquad \tau_m = \zeta_1/\sigma_m \tag{34}$$

The friction factor ζ_1 is assumed in eq 34, because we are considering the lateral bending motion. Now, we expand $\Psi(m,s)$ in terms of $Q(m,s)$ in eq 23;

$$\Psi(m,s) = \sum_j a_j(m)Q(j,s) \tag{35}$$

By substituting eq 35 into eq 33a, multiplying both sides by $Q(k,s)$ and integrating over s, we have a set of homogeneous linear simultaneous equations for the expansion coefficients $a_j(m)$;

$$\mathbf{M}\,\mathbf{a}(m) = \sigma_m\,\mathbf{a}(m) \tag{36}$$

with $M_{kj} = G_{kj} + \lambda_k\delta_{kj}$, $\mathbf{a}(m) = [a_0(m), \ldots, a_j(m), \ldots, a_m(m), \ldots, a_N(m)]^T$ and

$$G_{kj} = (\lambda_j/2)\iint Q(k,s)f(|s - s'|)Q(j,s')dsds'$$

In terms of the new variables $p = (s + s')/2L$ and $q = (s - s')/2L$, we have

$$G_{kj} = (2\lambda_j)L^2\int_0^{0.5} f(2Lq)G(k,j,q)dq \tag{37}$$

where $G(k,j,q)$ is analytically given in (10). Since we are interested in the case of $\gamma L \leq 0.5$, $f(|s - s'|)$ in eq 30 is given by (11)

$$f(|s - s'|) = |s - s'|^{-1}(1 + 1/3\,\gamma|s - s'| + \ldots) \tag{38}$$

To avoid divergence of the integral in eq 37, a lower limit of $|s - s'| = b$ or $q = b/2L$ is chosen, where b is a measure of the hydrodynamic diameter of the polymer backbone. The first 11 x 11 portion of **M** is shown in Table I. It is seen that the matrix is almost diagonal, so that perturbation methods may be used to find the eigenvalues. The formula including the second-order perturbation is;

$$\sigma_m = M_{mm} - \sum_{j\neq m} M_{mj}M_{jm}/(M_{jj} - M_{mm}) \tag{39}$$

The numerical values in Table I show that contributions to σ_m from the second-order terms are negligible (at most 2% of M_{mm} as a whole).
The first-order formula for the expansion coefficients $a_j(m)$ for $j \neq m$ is

Table I. The 11 x 11 portion of **M** for L/b = 200 ($10^{-3}M_{kj}$ in units of ε/L^4)

0	0	0.1889	0	-1.971	0	7.271	0	-17.60	0	34.18
0	0	0	1.488	0	-7.925	0	22.25	0	-46.83	0
0	0	1.748	0	4.711	0	-17.37	0	41.29	0	-78.82
0	0	0	12.47	0	9.771	0	-28.86	0	61.34	0
0	0	0.1589	0	44.70	0	17.12	0	-43.42	0	85.13
0	0	0	0.9308	0	114.7	0	27.17	0	-61.48	0
0	0	-0.0961	0.	2.805	0	241.7	0	40.23	0	-83.20
0	0	0	-0.6316	0	6.241	0	447.1	0	56.60	0
0	0	0.0661	0	-2.059	0	11.64	0	755.5	0	76.48
0	0	0	0.4590	0	-4.829	0	19.36	0	1191.	0
0	0	-0.0490	0	1.568	0	-9.347	0	29.71	0	1781.

$$a_j(m) = M_{jm}\, a_m(m)/(\sigma_m - M_{jj}) \tag{40}$$

The value of $a_m(m)$ is to be determined by the normalization requirement that the integral of $\Psi(m,s)^2$ is to be unity; in fact the numerical values in Table I give $a_m(m)$ to be at most 1% smaller than unity for all cases except $a_0(0)$ and $a_1(1)$, both of which are unity.

From eq 18, 32, 34 and 35, we have

$$\langle V\rangle = \tfrac{1}{2} \textstyle\sum_m \lambda_m \langle q_m{}^2\rangle a_m(m)^2[1 + \Phi(m)]$$

where $\Phi(m) = \Sigma_{j\neq m}(\lambda_j/\lambda_m)[a_j(m)/a_m(m)]^2$. The numerical values in Table I give $\Phi(m) < 0.005$. Then, the equipartition theorem of thermal energy gives

$$\langle q_m{}^2\rangle a_m(m)^2 = 2k_BT/\lambda_m \tag{41}$$

Since $a_m(m)^2 \simeq 1$, eq 41 is essentially the same as eq 29 in the free-draining limit. In order to have the expression for $\Phi(m,s,s',\xi,\xi',\tau)$ in eq 6, we need an explicit form of $C(\tau) = \langle\Sigma\mathbf{q}(m,t)\Psi(m,s)\Sigma\mathbf{q}(m',t')\Psi(m',s')\rangle$. From eq 34 and 35, we have

$$C(\tau) = \sum_m'' \langle q_m{}^2\rangle a_m(m)^2 \exp(-\tau/\tau_m)$$
$$\times [Q(m,s)Q(m,s') + \sum_{j\neq m} \sum_{j'\neq m} \Phi(j,j')Q(j,s)Q(j',s')]$$

The factor $\Phi(j,j') = a_j(m)a_{j'}(m)/a_m(m)^2$ is smaller than 0.015 for $j = j' = m - 2$ with $m = 2$, 3 and 4 and negligibly small for other combination of j and j'. Since $a_m(m)^2 \simeq 1$, we have

$$C(t) \simeq \sum_m'' \langle q_m{}^2\rangle \exp(-\tau/\tau_m) Q(m,s)Q(m,s') \tag{42}$$

which is again very close to that in the free-draining limit except that τ_m is given by eq 34 and not by eq 28. Then, the spectrum can be computed by use of eq 13, if we adopt unperturbed eigenfunctions $Q(m,s)$ and average amplitudes $\langle q_m{}^2\rangle$, and the first-order eigenvalues $\sigma_m = \lambda_m + G_{mm}$ for evaluation of τ_m. (Schmidt and Stockmayer (12) have adopted the same procedure as above without proofing eq 41 and 42.)

3) <u>Numerical simulation</u>. The first-order eigenvalues are given by

$$\sigma_m = \lambda_m(1 + f_m) \qquad \text{with} \qquad f_m = G_{mm}/\lambda_m \tag{43}$$

The effect of the hydrodynamic interaction on the normal modes m = 0 and 1 can not be discussed on the basis of eq 43, because λ_0 and λ_1 are zero in the present model. We consider first the translational diffusion coefficient D (11);

$$D_1 = D_1^\circ(1 + f_0) \quad \text{with} \quad f_0 = \ln(L/b) - 1 + \gamma L/6 + b/L + O(\gamma b) \tag{44}$$

where $D_1^\circ = k_B T/\zeta_1 L$ with $\zeta_1 = 4\pi\eta$ is the sideways translational diffusion coefficient in the free-draining limit. The rotational diffusion coefficient Θ is given by (10)

$$\Theta = \Theta^\circ(1 + f_1) \quad \text{with} \quad f_1 = \ln(L/b) - 7/3 + 3b/L + O(\gamma b) \tag{45}$$

where $\Theta^\circ = 12k_B T/\zeta_1 L^3$ is the rotational diffusion coefficient in the free-draining limit (4). By comparing eq 44 and 45 with the corresponding ones of, for example, Broersma's formulas (13), we find $b = d/2$, where d is the diameter of the rod. The cylinder model gives this relationship naturally (14). Figure 2 shows numerical values of $(1 + f_m)$ for $L/b = 2L/d = 200$. From eq 18, 22, 34 and 43, we have the following relationships for the relaxation time τ_m of the *m*th normal mode ($m \geq 2$);

$$\tau_m^{-1} = \sigma_m/\zeta_1 = (\lambda_m/\zeta_1)(1 + f_m) \tag{46}$$

$$= (\pi^4/2\gamma L)(m - \tfrac{1}{2})^4(k_BT/\zeta_1 L^3)(1 + f_m)$$

$$= (\pi^4/24\gamma L)(m - \tfrac{1}{2})^4(12D_1^\circ/L^2)(1 + f_m)$$

$$= (\pi^4/24\gamma L)(m - \tfrac{1}{2})^4\Theta^\circ(1 + f_m)$$

In our previous papers, we assumed $D_1^\circ(1 + f_m) = D_1$ or $\Theta^\circ(1 + f_m) = \Theta$, i.e., $k_BT/(\zeta L)_m = D_1$ or $12k_BT/(\zeta L^3)_m = \Theta$, independently of m in eq 46 in order to obtain an appropriate value of ζ for the normal modes $m \geq 2$. From the result in Figure 2, this assumption seems not to be good. From the above results, we have to modify eq 16; the last term should read $(k_BT/\zeta_1 L)\Sigma''\ (1 + f_m)$.

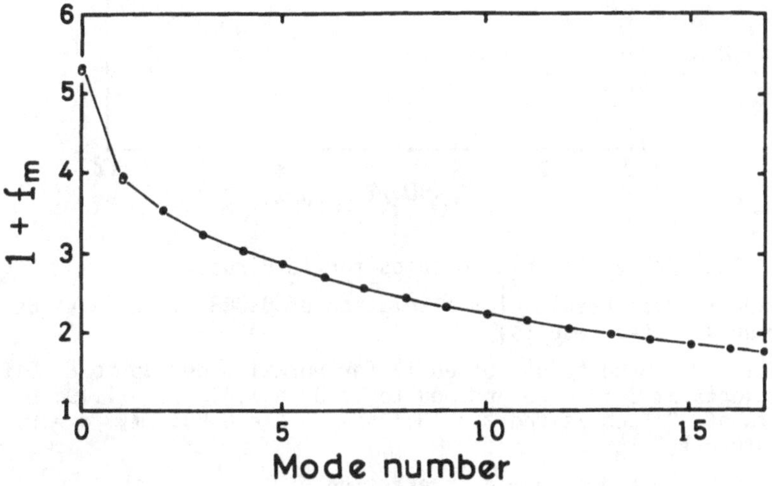

Figure 2. The hydrodynamic interaction factor f_m.
$1 + f_m$ for $L/b = 2L/d = 200$ (•). Corresponding values for m = 0 and 1 from Broersma's formulas are also shown (o).

By use of the above formulation, we computed the field correlation function for dimensions appropriate to fd virus. Figure 3 shows $\bar{\Gamma}/K^2$ vs K^2 relationships obtained by the least-squares fitting of the simulated correlation functions to a second-order cumulant expansion formula. For a value of the flexibility parameter as small as $\gamma L = 0.05$, the internal modes strongly contribute to $\bar{\Gamma}/K^2$ values at $K^2 \gtrsim 2 \times 10^{10}$ cm^{-2}.

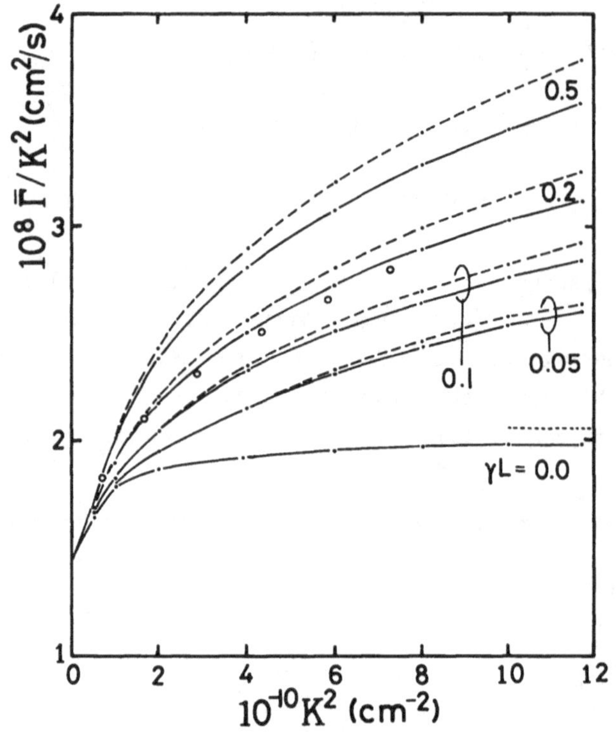

Figure 3. The $\bar{\Gamma}/K^2$ vs K^2 relationships for fd virus.

(o): Experimental results for a solution of 0.088 mg/mL fd virus (parts of our unpublished results).

Simulation was made by use of eq 13 for normal modes up to 8. Diffusion coefficients at 5 °C were assumed to be D_0 = 1.41, D_1 = 1.19, D_3 = 1.87 in units of 10^{-8} cm^2/s and Θ = 13.1 s^{-1}. L = 0.895 μm, d = 9.0 nm and temperature = 5 °C.

(——): With the hydrodynamic interaction.

(---): Without the hydrodynamic interaction, where ζL value was determined from $k_BT/(\zeta L)_m = L^2\Theta/12$ for any $m \geq 2$.

(...): The asymptote for a rigid rod, $D_1 + L^2\Theta/12$.

Experimental Section and Discussion

By use of a 128-channel (*plus* 16 far-point channels for the baseline level) digital correlator with a 488 nm beam from an Ar^+ laser, intensity correlation functions $Y(\tau)$ of light scattered from solutions of fd virus (0.088 - 6 mg/mL or $2.3 \leq cL^3 \leq 160$ in 150 mM KCl and 15 mM Na_2HPO_4 at pH 7.0) were measured over the K^2 range (0.6 - 11) x 10^{10} cm^{-2}. The measured $Y(\tau)$'s were least-squares fitted to a second-order cumulant expansion formula and the apparent diffusion coefficient $\bar{\Gamma}/K^2$ and the normalized dispersion $\mu_2/\bar{\Gamma}^2$ were obtained. Since this study is in progress, we will show only a limited aspect of the experimental study.

1) <u>At dilute regime</u>. Some of experimental $\bar{\Gamma}/K^2$ values for fd virus at 0.088 mg/mL are shown in Fig. 3. This concentration corresponds to $cL^3 = 2.3$, if the molecular weight of 1.64 x 10^7 is assumed for this virus (15). Thus, the data at this concentration are not for a dilute solution ($cL^3 << 1$). According to transient electric birefringnece (TEB) measurements (15,16), however, no sign of the onset of the semidilute region is observed in the experimental Θ values of fd virus up to almost $cL^3 = 5$. Thus, the experimental data in Fig. 3 can be compared with the theoretical results for dilute solutions. Figure 3 suggests that the value of the flexibility parameter γL of fd virus is close to 0.2. To estimate the γL value in two digits, the effect of the hydrodynamic interaction must be taken into account even in a crude model as we have discussed in the Theoretical Section.

2) <u>At semidilute regime</u>. Figure 4 shows some examples of the concentration dependence of the apparent diffusion coefficient $\bar{\Gamma}/K^2$ and of the normalized dispersion $\mu_2/\bar{\Gamma}^2$ at two scattering angles, 30° and 90°. According to TEB measurements (16), almost complete c^{-2} (and fairly good L^{-9}) dependence of observed $\bar{\Theta}$ values has been confirmed for $c_p \geq 0.2$ mg/mL (cf. eq 17; $\bar{\Theta} = \beta\Theta \propto 1/(cL^3)^2$. Thus, the onset of the semidilute region might occur in our range of cL^3 in Fig. 4. However, our $\bar{\Gamma}/K^2$ values at 30° (o) decrease to a little smaller value than D_0 even at $cL^3 = 160$. (Although $cL^3 = 160$ violates the condition $cL^3 << L/d = 100$ for fd virus, no sign of the appearance of a liquid-crystalline phase was observed.) They are very much larger than $\bar{D}_0 = D_3/3$ in the Doi-Edwards model (7) even when the filament flexibility is

taken into account. This means that the decrease in the $\bar{\Gamma}/K^2$ values with increasing cL^3 is primarily due to the decrease in the $\bar{\Theta}$ values and that $\bar{D}_0$ and the contribution to $\bar{\Gamma}/K^2$ from internal bending motions change very little with increasing cL^3. That is, α in eq 17 is nearly unity and bending motions are almost free from surrounding filaments. The latter presumption is further verified from the results at 90° (•), that is, the difference at each cL^3 between the experimental value (•) and the hypothetical one estimated from eq 14 with D_i values at dilute regime and experimental $\bar{\Theta}$ values (16) is almost independent of cL^3. The dotted curve in Fig. 4 roughly sketches this hypothetical situation. These results are qualitatively understood as follows.

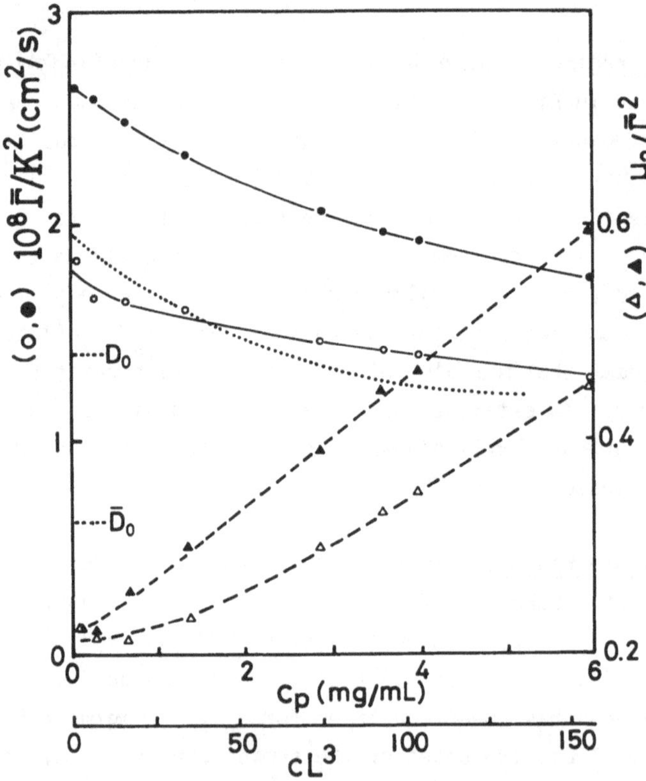

Figure 4. Concentration dependences of the apparent diffusion coefficient $\bar{\Gamma}/K^2$ and the normalized dispersion $\mu_2/\bar{\Gamma}^2$ for solutions of fd virus. Only data at two scattering angles are shown; open symbols at 30° and closed ones at 90°. The abscissa is scaled in c_p (mg/mL virus) and in cL^3 (the number of virus in volume L^3). Corrected to 5°C in water. The dotted curve shows $\bar{\Gamma}/K^2 = D_0 + (L^2/12)\bar{\Theta}f_1(k) - (D_3-D_1)[1/3 - f_2(k)]$ at 90°.

For a very long filament, $L = 1$ μm and $cL^3 = 64$, for example, correspond to the average interfilament separation of 250 nm. This means that a filament is confined in a cage with a diameter of about 250 nm. The sideways translation of the filament in the cage could be detected as if it were free, because we have $1/K = 100$ nm even for such a low angle as $K^2 = 1 \times 10^{10}$ cm^{-2}. Similarly, bending motion in the cage is almost free, because the mean amplitude of lateral bending is estimated to be 60 nm for the $m = 2$ mode at $\gamma L = 0.5$. A more detailed discussion on this point will be found in (5).

Our experimental results are almost the same as those in (17,18) at dilute regime, but quite different from those in (18) at semidilute regime. The intensity correlation function at $cL^3 = 55$ in (18) has a very long tail, whereas our results do not, as seen from a relatively small dispersion $\mu_2/\bar{\Gamma}^2$ in Fig. 4. In our second-order cumulant analysis, however, systematic deviations between experimental and computed correlation functions began to appear when $\mu_2/\bar{\Gamma}^2$ became larger than 0.3. A more detailed analysis will be required. If the experimental correlation functions were fitted to $Y(\tau) = [A_1\exp(-\Gamma_1\tau) + A_2\exp(-\Gamma_2\tau)]^2 + B$ and $\bar{\Gamma} = (A_1\Gamma_1 + A_2\Gamma_2)/(A_1 + A_2)$ were computed, the resultant $\bar{\Gamma}/K^2$ values became larger than those shown in Fig. 4 at larger cL^3 values.

For a systematic analysis of QELS data, especially at semidilute regime, there are too many unknown parameters, $\bar{D}_0$, $\bar{D}_1$, $\bar{D}_3$, $\bar{\Theta}$ and γL. In the framework of the Doi-Edwards model (7), we have $\bar{\Theta} = \kappa\Theta/(cL^3)^2$, where the proportionality constant κ is expected to be within an order of 10. Experimental studies so far made (16,19,20), however, have shown that κ is several orders of magnitude larger than predicted, although almost complete c^{-2} and fairly good L^{-9} dependences have been observed. Also, our results suggest $\bar{D}_1 = \alpha D_1$ with $\alpha \simeq 1$ even at the scattering angle of 30°. Even if we assume $\bar{D}_3 = D_3$ and the γL value estimated at dilute regime, it is desirable to know at least the $\bar{\Theta}$ values by some other methods than QELS. There is a literature values for $\bar{\Theta}$ (16). Unfortunately, however, the origin of the discrepancy between our results and those in (18) has not been elucidated yet, and we have to make TEB measurements for our own samples. Also, we have to make QELS measurements at very low angles in order to know the $\bar{D}_0$ or $\bar{D}_1$ values at the low-angle limit. In such a case, strong multiple scattering has to be taken into account.

[A preliminary result showed that, at $K \simeq 3 \times 10^4$ cm^{-1} and $cL^3 \simeq 160$, $\bar{D}_0 \simeq D_0(1 + 0.29)$ in the absence of, and $\bar{D}_0 \simeq D_0(1 - 0.07)$ in the presence of an analyzer in its parallel setting with a polarizer. Removal of only a depolarized component of scattered light did reduce the $\bar{D}_0$ value by $0.36D_0$.] Therefore, we have to postpone our final conclusion to a future publication. At present, however, it is safe to say that QELS data of fd virus both at dilute and semidilute regimes could be reasonably analyzed *only when the effect of the filament flexibility were properly taken into account* in addition to anisotropic translational diffusion ($D_3 \neq D_1$ or $\bar{D}_3 \neq \bar{D}_1$) and rotational diffusion (Θ or $\bar{\Theta}$).

(840703)

References

1. Maeda, T., Fujime, S.: Macromolecules **17**, 1157-1167 (1984).
2. Kubota, K., Urabe, H., Tominaga, Y., Fujime, S.: Macromolecules (Sept/Oct issue, 1984).
3. Maeda, T., Fujime, S., Kubota, K., Urabe, H., Tominaga, Y.: This volume.
4. Maeda, T., Fujime, S.: Macromolecules **14**, 809-818 (1981).
5. Maeda, T., Fujime, S.: Macromolecules (Sept/Oct issue, 1984).
6. Stockmayer, W.H., Burchard, W.: J. Chem. Phys. **70**, 3138-3139 (1979).
7. Doi, M., Edwards, S.F.: J. Chem. Soc. Faraday II **74**, 560-570 (1978).
8. Harris, R.J., Hearst, J.E.: J. Chem. Phys. **44**, 2595-2602 (1966).
9. Hearst, J.E., Beals, E., Harris, R.J.: J. Chem. Phys. **48**, 5371-5377 (1968).
10. Fujime, S., Maeda, T.: Macromolecules (Feb/Mar issue, 1985).
11. Hearst, J.E., Stockmayer, W.H.: J. Chem. Phys. **37**, 1425-1433 (1962).
12. Schmidt, M., Stockmayer, W.H.: Macromolecules **17**, 509-513 (1984).
13. Broersma, S.J.: J. Chem. Phys. **32**, 1625-1631, 1632-1635 (1960).
14. Yamakawa, H., Fujii, M.: Macromolecules **6**, 407-415 (1973).
15. Newman, J., Swinney, H.L., Day, L.A.: J. Mol. Biol. **116**, 593-606 (1977).
16. Maguire, J.F., McTague, J.P., Rondelez, F.: Phys. Rev. Let. **45**, 1891-1894 (1980); **47**, 148 (1981)..
17. Newman, J., Carlson, F.D.: Biophys. J. **29**, 37-47 (1980).
18. Maguire, J.F.: J. Chem. Soc. Faraday II **77**, 513-518 (1981).
19. Zero, K.M., Pecora, R.: Macromolecules **15**, 87-93 (1982).
20. Mori, Y., Ookubo, N., Hayakawa, R., Wada, Y.: J. Polymer Sci. Phys. **20**, 2111-2124 (1982).

THEORETICAL AND EXPERIMENTAL STUDIES OF ANISOTROPY IN TRANSLATIONAL DIFFUSION WITH SPECIAL REFERENCE TO TOBACCO MOSAIC VIRUS*

Tadakazu Maeda, Satoru Fujime
Mitsubishi-Kasei Institute of Life Sciences, Machida, Tokyo 194, Japan

Kenji Kubota, Hisako Urabe, Yasunori Tominaga
Department of Physics, Ochanomizu University, Bunkyo-ku, Tokyo 112, Japan

Introduction

Let D_1 and D_3 be respectively the sideways and the lengthways translational diffusion coefficients of a rigid rod in solution. The anisotropy ratio is defined as $\Delta \equiv (D_3-D_1)/D_0$, where $D_0 \equiv (2D_1+D_3)/3$ denotes the overall translational diffusion coefficient. For a long rod, the ratio Δ is not small and the anisotropy would affect the correlation function $G^2(\tau)$ of light scattered from a solution of a rigid rod. This was first discussed theoretically by Maeda and Saito (1) and experimentally by Fujime (2). Except for recent works (3,4), however, the large effect of anisotropy on $G^2(\tau)$ or on the power spectrum has been questioned (5-7). Our recent results (4,8) will be given here, which clearly indicate the large effect of anisotropy.

Theoretical Section

Equation 13 in (9) for a rigid rod can be written as

$$G^1(\tau) = G_D(\tau)\Sigma_n\Sigma_\ell\ (2\ell + 1)b_n(k)b_\ell(k)[\mathbf{U}\ e^{-\Theta\mathbf{\Lambda}\tau}\ \mathbf{U}^{-1}]_{n\ell}\,(i)^{n+\ell} \qquad (1)$$

$$b_n(k) = (1/k)\int_0^k j_n(z)dz \quad \text{(even } n\text{);} \quad \text{zero for odd } n \quad (k \equiv KL/2) \qquad (2)$$

where $G_D(\tau) = \exp[-\{D_0 - \tfrac{1}{3}(D_3 - D_1)\}K^2\tau] = \exp(-D_1K^2\tau)$, Θ is the rotational

*The contributors to the theoretical section were Maeda and Fujime (8) and to the experimental section were Kubota, Urabe, Tominaga and Fujime (4).

Physical Optics of Dynamic Phenomena and Processes in Macromolecular Systems

diffusion coefficient and $j_n(z)$ is the *n*th order spherical Bessel function. The matrices **U** and Λ satisfy

$$\mathbf{M}\,\mathbf{U} = \mathbf{U}\,\Lambda \qquad \text{with} \qquad \Lambda = \{\lambda_p \delta_{p\,p'}\} \tag{3}$$

where elements of **M** are given explicitly in (4,8,9). The time derivative of eq 1 gives the first cumulant $\bar{\Gamma}$ of $G^1(\tau)$, after some algebra,

$$\bar{\Gamma}/K^2 = D_0 + (L^2/12)\Theta f_1(k) - (D_3 - D_1)[\tfrac{1}{3} - f_2(k)] \tag{4}$$

where both $f_1(k)$ and $f_2(k)$ are functions depending only on k, whose values are tabulated in (8), L is the rod length and use was made of $\mathbf{U}\,\Lambda\,\mathbf{U}^{-1} = \mathbf{M}$, that is, to obtain eq 4, we do not need to solve eq 3. Equation 4 has been derived via a quite different route (3). To obtain **U** and Λ, machine computation is required. However, for TMV (L = 300 nm), a perturbation technique can be applied with a high accuracy. The results are given by

$$G^1(\tau) = \Sigma_n P_n(k) \exp(-\Gamma_n \tau) \tag{5}$$

$$P_n(k) = S_n(k) - [B_{n-2}(k) - B_n(k)]\delta/\Theta + \ldots \tag{6}$$

$$\Gamma_n = D_0K^2 - \delta/3 + [n(n+1)\Theta + L_0(n)\delta] + [C_{n-2} - C_n]\delta^2/\Theta + \ldots \tag{7}$$

$$S_n(k) = (2n+1)b_n(k)^2, \quad B_n(k) = 2(2n+5)b_n(k)b_{n+2}(k)L_2(n)/(\lambda_{n+2} - \lambda_n) \tag{8}$$

$$C_n = \frac{(n+1)^2(n+2)^2(2n-1)(2n+7)}{2(2n+1)(2n+3)(2n+5)[(2n-1)(2n+3)^2(2n+7) - 2\delta/\Theta]} \tag{9}$$

where n = 0, 2, 4, ..., $\delta = (D_3 - D_1)K^2$, $\mu^2 = \delta/\Theta$ and $\lambda_n = n(n+1) + \mu^2 L_0(n)$ is the *zero*th order eigenvalue. For $L_0(n)$ and $L_2(n)$, see (4,8,9). In case of $\delta = 0$, the above results recover the Pecora formula $G^1(\tau) = \Sigma_n S_n(k) \times \exp[-\{D_0K^2 + n(n+1)\Theta\}\tau]$.

Experimental Section

By use of a 240-channel (*plus* 12 far-point channels for the baseline level) digital correlator with a 488 nm beam from an Ar^+ laser, intensity correlation functions $Y(\tau)$ of light scattered from dilute solutions of TMV (0.02 - 0.06 mg/mL) were measured over the K^2 range (0.6 - 11) x 10^{10} cm^{-2}. The monodispersity of samples was confirmed by electron microscopy. From static intensity measurements, we obtained L = 297 nm, again confirming the sample quality. $Y(\tau)$'s at different K^2 were least-squares fitted to a second-order cumulant expansion formula, and we obtained the $\bar{\Gamma}/K^2$ vs K^2 relationship as shown in Fig. 1a. From a low angle data, we obtained $D_0 = (4.30 \pm 0.10) \times 10^{-8}$ cm^2/s. For this D_0 value and L = 300 nm, the least-squares fitting of $\bar{\Gamma}/K^2$ vs K^2 relationship to eq 4 gave $\Theta = 295$ s^{-1} and $\Delta = 0.412$. The fitting is shown by the dashed line in Fig. 1a. For $\Theta = 320$ s^{-1} from two-exponential analysis, the least-squares fitting gave $D_0 = 4.29 \times 10^{-8}$ cm^2/s and $\Delta = 0.500$. The fitting is shown by the solid line in Fig. 1a. Next, $Y(\tau)$'s at $K^2 \geq 3 \times 10^{10}$ cm^{-2} were least-squares fitted to eq 10, and we obtained the results

$$Y(\tau) = [A_0 \exp(-\Gamma_0\tau) + A_2 \exp(-\Gamma_2\tau)]^2 + B \qquad (10)$$

shown in Figs 1b-d. By using the experimental values of D_0, Θ and Δ, and

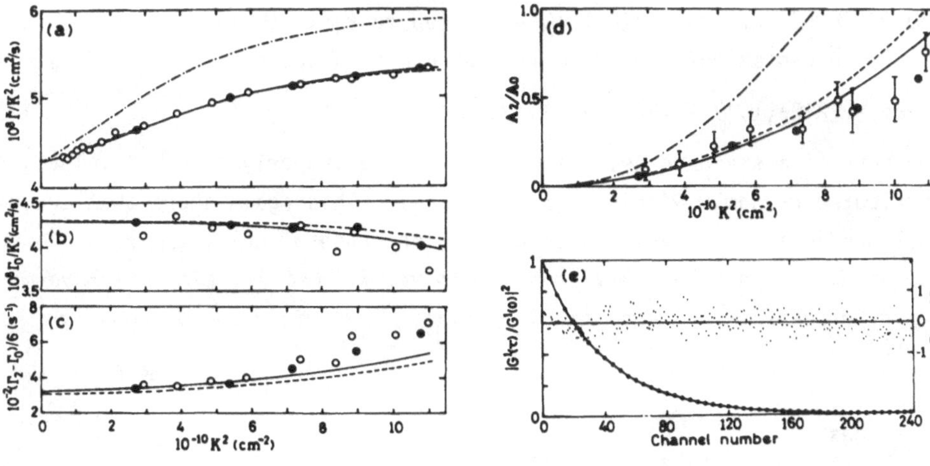

Figure 1. Experimental results for TMV.

(a): The $\bar{\Gamma}/K^2$ vs K^2 relationship corrected to 20°C in water. The theoretical curves are for $D_0 = 4.29 \times 10^{-8}$ cm^2/s, $\Theta = 320$ s^{-1} and $\Delta = 0.500$ (——), and for $D_0 = 4.30 \times 10^{-8}$ cm^2/s, $\Theta = 295$ s^{-1} and $\Delta = 0.412$ (---). The dash-dot line denotes the theoretical curve when $D_3 = D_1$. Filled circles show the results from two-exponential fit for theoretically generated $G^1(\tau)$'s with the former parameter set (see text).
(b-d): The results of two-exponential fit; Γ_0/K^2 in (b), $(\Gamma_2 - \Gamma_0)/6$ in (c) and A_2/A_0 in (d). The filled circles show the results for the theoretically generated $G^1(\tau)$'s. The solid and dashed lines show the theoretical predictions based on the parameter values in (a). The dash-dot line in (d) shows the theoretical prediction when $D_3 = D_1$.
(e): The net correlation function at a scattering angle of 120°. Sampling time = 2.3 μs. The solid line shows the theoretical $|G^1(\tau)|^2$. For definition of Dev, see eq 11.

L = 300 nm, the theoretical $G^1(\tau)$'s were generated according to eq 5-9. The generated $G^1(\tau)$'s (with 0.1% random noise) were least-squares fitted to eq 10, and we obtained the results shown by filled circles in Figs 1a-d, where $\bar{\Gamma} = (A_0\Gamma_0 + A_2\Gamma_2)/(A_0 + A_2)$ was assumed in Fig. 1a. A very good agreement between experimental (o) and theoretical (•) points in Figs 1a-d shows that the theoretical predictions are valid not only for the initial decay rate $\bar{\Gamma}$ of, but also for the entire time behavior of the correlation function. The solid and dashed lines in Figs 1b-d show the theoretical predictions, where renormalization, $[\Gamma_2P_2(k) + \Gamma_4P_4(k)]/[P_2(k) + P_4(k)]$ for Γ_2 in Fig. 1c and $[P_2(k) + P_4(k)]/P_0(k)$ for A_2/A_0 in Fig. 1d, was assumed in order to keep consistency with the least-squares fitting to eq 10. Deviations of the theoretical curves from experimental (o) and theoretical (•) points are observed at large K^2 values. This might come from 'force' fitting to eq 10. Deviation (Dev) of the theoretical $|G^1(\tau)|^2$ (up to $P_4(k)$ terms) from the experimental one can be defined by eq 11, where A is a scale factor and $Y(\tau)^{1/2}$ is

$$\mathrm{Dev} = [A|G^1(\tau)|^2 - (Y(\tau) - B)]/Y(\tau)^{1/2} \qquad (11)$$

a measure of a standard deviation (SD). For a properly chosen A values, Dev distributed randomly within ±1 SD (Fig. 1e). This again shows that theoretical predictions are valid over the entire range of delay times.

This study not only supported the recent result (3) but also afforded much deeper insights into the scattering process by a rigid rod in solution.

(840626)

References

1. Maeda, H., Saito, N.: J. Phys. Soc. Jpn. 27, 894-991 (1969).
2. Fujime, S.: J. Phys. Soc. Jpn. 27, 1370 (1969); 29, 416-430 (1970).
3. Wilcoxon, J., Schurr, J.M.: Biopolymers 22, 849-867 (1983).
4. Kubota, K., Urabe, H., Tominaga, Y., Fujime, S.: Macromolecules (Sept/Oct issue, 1984).
5. Schaefer, W.D., Benedek, G.B., Schofield, P., Bradford, E.: J. Chem. Phys. 55, 3884-3895 (1971).
6. Rallison, J.M., Liel, L.G.: J. Chem. Phys. 74, 4819-4825 (1981).
7. Hartford, S.L.: Ph.D. Thesis (University of Illinois, 1975).
8. Maeda, T., Fujime, S.: Macromolecules 17, 1157-1167 (1984).
9. Fujime, S., Maeda, T.: This volume.

MOLECULAR WEIGHT AND SECOND VIRIAL COEFFICIENT INFLUENCES ON THE CONCENTRATION DEPENDENCE OF DIFFUSION COEFFICIENT OF POLYSTYRENE MOLECULES IN DILUTE SOLUTIONS

Konrad Witkowski, Leszek Woliński, Barbara Kołodziej
Institute of Experimental Physics, University of Gdańsk, ul.Wita Stwosza 57, 80-952 Gdańsk, POLAND

Introduction

The concentration dependence of macromolecular translational diffusion coefficient D(c) is usually expressed (in dilute solution case) as:

$$D(c) = D_o(1 + k_D c_n) \qquad (1)$$

or

$$D(c) = D_o + k_c c \qquad (2)$$

where D_o is the diffusion coefficient extrapolated to infinite dilution, c_n - number density of polymer particles, c- concentration in g/cm^3. Equation (1) is more convenient for theoretical discussion, while constant k_c in equation (2) can be measured directly, for example in dynamic light scattering experiments. Temperature and molecular weight dependence of k_D was theoretically investigated - for monodisperse polymer fractions - by Akcasu (1) . Theory developed in this paper predicts that k_D is an univocal function of the dimensionless parameter $X = R_S/R_H$, where R_S is the effective interaction range (i.e. the radius of the hard sphere with the same second virial coefficient as the investigated polymer molecules) and R_H is the hydrodynamic radius of macromolecules. The shape of $k_D(X)$ function depends on the details of assumed model of inter- and intramolecular interactions.

We present in this work experimental results concerned with $k_D(X)$ dependence for mono- and polydisperse polystyrene fractions.

Physical Optics of Dynamic Phenomena and Processes in Macromolecular Systems

Experimental

The instrument used is a self-contained equipment for the simultaneous measurements of integral and dynamic light scattering. Dynamic light scattering was measured in the homodyne regime, by means of the 64-channel digital correlator built in our laboratory.

Materials. Three sets of atactic polystyrene samples were investigated:

a) eight monodisperse fractions ($5x10^4 \leq M_w \leq 10^7$, $u < 0.15$) in 9 solvents - transdecaline, 2-butanone, 1.4-dioxane, toluene, benzene, ethylbenzene, tetrahydrofuran, carbon tetrachloride and chloroform;

b) two-component blends prepared from fractions a);

c) polydisperse fractions ($0.65 \leq u \leq 2.55$) with known molecular weight distribution functions.

In the light scattering experiments, dilute solutions - $1 \leq c \leq 10$ g/dm^3 - were measured at the scattering angles $25^\circ \leq \theta \leq 155^\circ$.

Results and discussion

a) Monodisperse samples. Fig.1 presents k_D values obtained from our experiments as a function of parameter X for 22°C.

1. It can be concluded (considering the experimental error level) that the molecular mass and the second virial coefficient influence on the k_D values only through parameter X are in accordance with theoretical predictions (1). Some departures from the "universal" $k_D(X)$ dependence were found for chloroform solutions, but they still were within experimental error limits.

2. By fitting a polynomial to experimental points, we have determined an empirical function describing the $k_D(X)$ dependence:

$$k_D(X) = 1.001X^3 + 0.524X^2 + 2.204X - 2.200 \ . \qquad (3)$$

3. Theoretical dependences $k_D(X)$ reviewed in (1) were

compared with the empirical function (3) (Fig.2) .

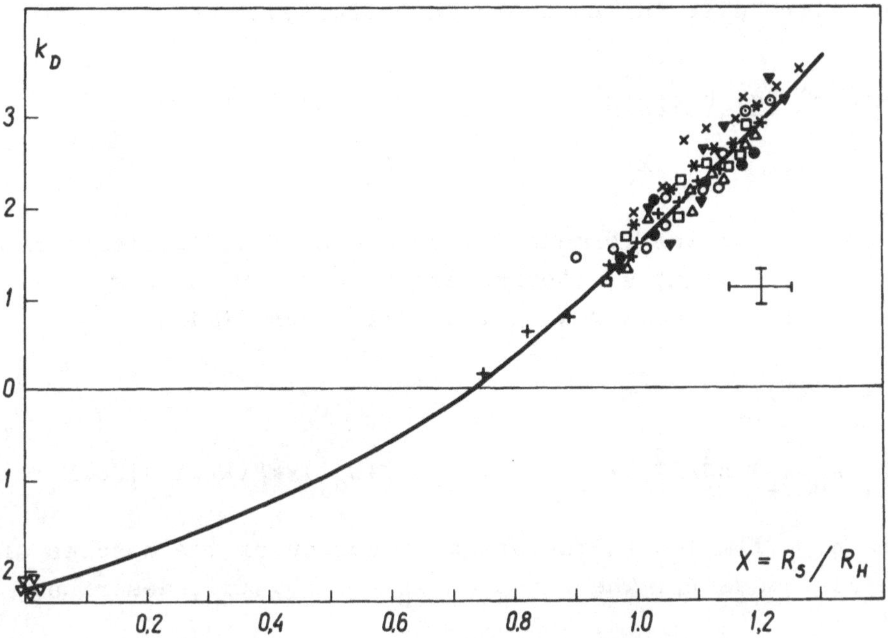

Fig.1 The experimentally obtained $k_D(X)$ dependence for monodisperse fractions. Experimental points for solutions in: ∇- decaline, + 2-butanone, o 1.4-dioxan, ● toluene, ▼ benzene, Δ- ethylbenzene, * tetrahydrofuran, □ carbon tetrachloride, × chloroform; —— empirically fitted function- see formula (3).

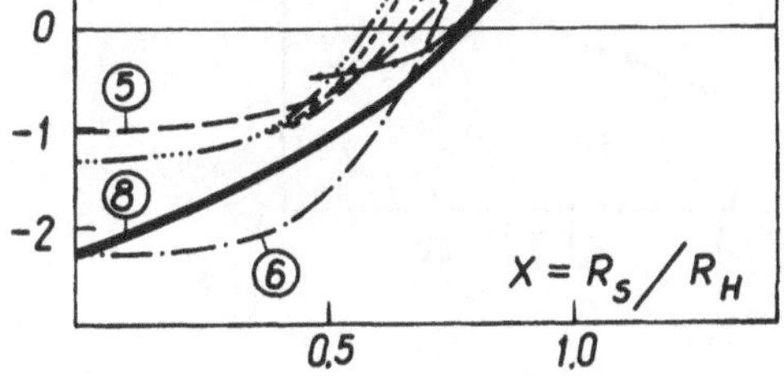

Fig.2 Comparison between theoretical dependences, $k_D(X)$, and our empirical curve 8 ; 1 - 6 taken from (1).

No one of the theoretical models seems to fit well the experimental results in the whole investigated range of X values.

b) and c) Polydisperse samples. Assuming that (2)

$$k_c = 4\pi R_H^3 D_o k_D N_A/3M \tag{4}$$

one can derive the formula for the mean of k_c constants measured in the light scattering experiment for polydisperse samples with molecular weight distribution f(M):

$$\bar{k}_c = (kT)^3 N_A/54\pi^2\eta^2 \times$$
$$\times [\sum_i [k'_D(X_i)\, c_i^2/D_{oi}^2] + \sum_i \sum_{i>j} c_i c_j (D_{oi}^{-2}+D_{oj}^{-2}) \times k''_D(X_i,X_j)]/c\sum M_i c_i \tag{5}$$

where $c_i = f(M_i)\, c$ is the weight fraction of the species of molecular mass M_i. The function $k'_D(X)$, eq. (3), describes

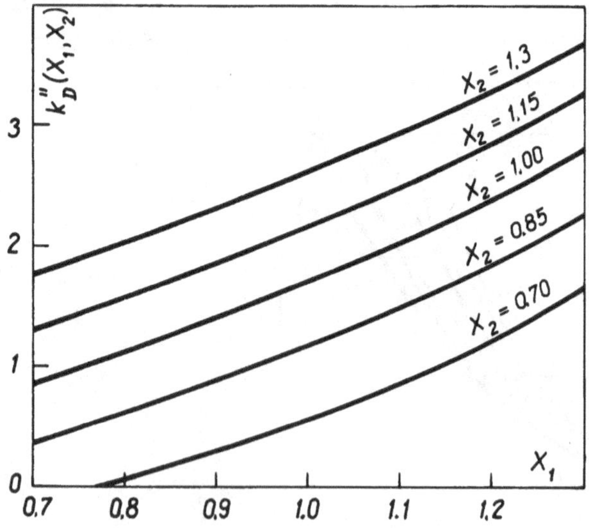

Fig. 3. Function $k''_D(X_i,X_j)$.

interactions of the monodisperse molecules, and the symmetrical function $k''(X_i, X_j)$ involves all terms dependent on the interactions of two molecules differing in their molecular masses. We have determined k''_D empirically from eq. (5) using the measured values of $\bar{k}_c$ for two-component blends of monodisperse fractions b) (Fig. 3). By means of the empirically determined k'_D and k''_D dependences, $\bar{k}_c$ values for some polydisperse PS samples with known, unimodal f(M) distribution c) were calculated. Departures of computed values from that measured for solutions of the samples in 2-butanone by light scattering were less than 15%.

References

1. Akcasu, A.Z.: Polymer 22, 1169 (1981).
2. Han, C.C., Akcasu, A.Z.: Polymer 22, 1019 (1981).

THE CONCENTRATION DEPENDENCE OF DIMENSIONS OF FLEXIBLE POLYMER MOLECULES IN SOLUTION

J. S. King, W. Boyer

Department of Nuclear Engineering, University of Michigan, Ann Arbor, Michigan 48109

G. D. Wignall

Oak Ridge National Laboratory, P.O. Box X, Oak Ridge, Tennessee 37831

R. Ullman

Research Staff, Ford Motor Co., Dearborn, Michigan 48121 U.S.A.

Abstract

Small angle neutron scattering (SANS) experiments were performed on polystyrene of molecular weight 10^5 in deuterotoluene. Determinations of the radius of gyration (R_g) and screening length (ξ) were carried out over a wide range of concentration (c). It was found that $R_g^2 \sim c^{-0.16}$ and $\xi \sim c^{-0.70}$.

Introduction

Flexible polymer molecules expand in good solvents owing to the excluded volume effect. These interactions are screened by the presence of other polymer molecules as concentration increases and the excluded volume vanishes in the bulk polymer.

Small angle neutron scattering provides a direct opportunity to measure how molecular dimensions and screening lengths vary with concentration. Since the variation of R_g with concentration is modest, considerable care must be taken to achieve

Physical Optics of Dynamic Phenomena and Processes in Macromolecular Systems

high enough accuracy for results to be meaningful. Recent techniques for extracting information about single chains at high concentration of the scattering material (1,2) have made this problem a great deal easier than it had originally seemed. The reliability of our results depends strongly on the application of these techniques.

Much of the interpretation of neutron scattering experiments has been based on the analysis of a polymer molecule as a critical object. The strict application of the theory is limited to polymer molecules of infinite chain length, in fact, the theory is applied to polymer molecules which are "sufficiently long". Discrepancy between theory and experiment provides some indication of how long is sufficiently long.

Experimental

Polystyrene (PSH) and its deuterated homolog (PSD) with molecular weights of 1.14×10^5 respectively and $M/M \approx 1.05$ were used in the study. The solvent was deuterotoluene. Measurements were made between concentrations of 0.02 g/ml and bulk polymer.

At each concentration the PSD/PSH molar ratio took on the values 0, 0.5 and 0.8. The measurements were performed on the 30 meter SANS spectrometer at the Oak Ridge National Laboratory, the neutron wave length was 4.75 Å, the distance between source and sample, 8 meters, and the distance between sample and detector, 7 meters. Solutions were contained in quartz cells, solid samples were not enclosed.

Results

Single chain scattering functions were extracted from experiments on which corrections for transmission and incoherent scattering had been made. Values of the radius of gyration as a function of concentration are shown in a log-log plot

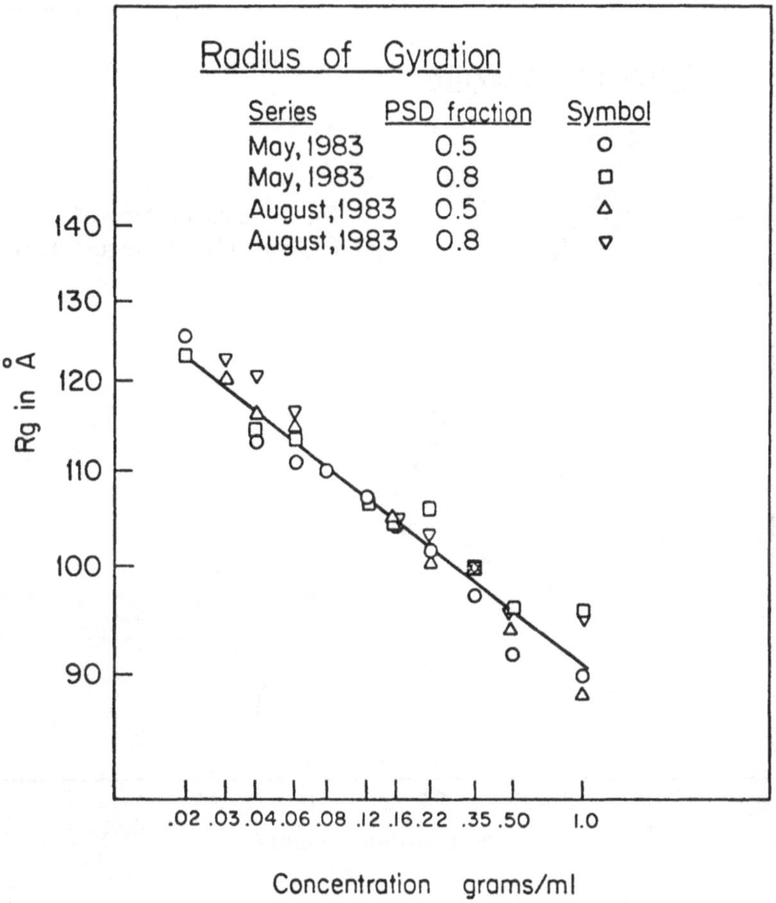

Figure 1: Radii of gyration of polystyrene in toluene as a function of concentration

in Figure 1. These results were obtained from a plot of $S_s(q)^{-1}$ versus q^2 where $S_s(q)$ is a single chain scattering function. Corrections to the scattering experiments for missing data at low q were introduced in calculating these results as well as corrections for polydispersity. A best straight line obeys the equation $R_g^2 = 8.3\times10^3c^{-0.16}$.

Screening lengths were calculated from SANS measurements of PSH in deuterotoluene, and are plotted in Figure 2 versus con-

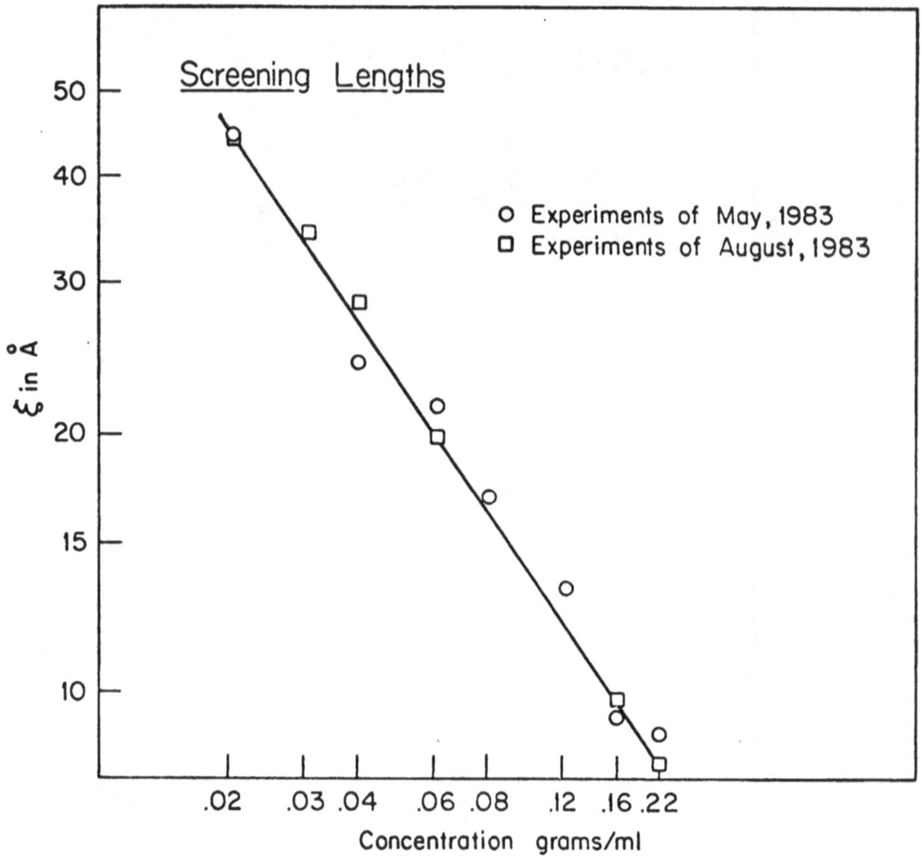

Figure 2. Screening lengths of polystyrene in toluene as a function of concentration

centration. Scattering intensities decreased rapidly with increasing concentration, and reasonably precise numbers were not obtained for c greater than 0.22 g/ml. The best straight line (log-log plot) satisfies the equation $\xi = 2.89\, c^{-0.70}$.

Our measurements of R_g as a function of concentration deviate substantially from that obtained previously and from that calculated using scaling theory (3). Our results of ξ versus c are in good accord with earlier studies (4), and deviate modestly from theoretical predictions.

A theoretical analysis by Muthukumar and Edwards (5) which does not make the assumption of infinite chain lengths provides a set of equations which link ξ and R_g, yield a value of an excluded volume interaction parameter, w, which is the interaction volume per unit length of each of two neighboring polymer chains. Consistent values were obtained from calculations made at different polymer concentrations. Our best estimate of w is 1.8×10^{-8}cm.

Acknowledgement

This work was supported in part by the National Science Foundation under GRANT DMR82-17460.

References

1. Akcasu, A.Z. et al.: J. Polymer Sci., Polymer Phys. 18, 863 (1980).
2. Benoit, H., Koberstein, J., Leibler, L.: Makromol. Chem. Suppl. 4, 85 (1981).
3. Daoud, M. et al.: Macromolecules 8, 804 (1975).
4. Cotton, J.P., Farnoux, B. and Jannink, G.: J. Chem. Phys. 37, 290 (1972).
5. Muthukumar, M. and Edwards, S.F.: J. Chem. Phys. 76, 2720 (1982).

DIFFUSION AND RELAXATION IN POLYMER-SOLVENT SYSTEMS BY PHOTON CORRELATION SPECTROSCOPY

C. Cohen and D.H. Hwang
School of Chemical Engineering, Cornell University
Ithaca, New York 14853, U.S.A.

Introduction

In the area of polymer solutions, photon correlation spectroscopy has been used extensively so far to study dilute polymer solutions on one hand and a regime called "semi-dilute" on the other hand. In the first case, the isotropic translation diffusion coefficient of the macromolecule may be obtained if the molecule is not too large such that the internal motions of the molecule (assumed here to be a chain-like molecule) do not couple and interfere with the measurement of the diffusion coefficient by photon correlation. This requires that $qR<1$, where q is the wave vector and R is the radius of gyration of the molecule (1). However, even for larger macromolecules in dilute solutions ($qR>1$), theory and experiments seem to have reached some concensus on the situation (2). For the case of "semi-dilute" solutions, we need first to define what is meant by "semi-dilute". The term has been associated, in general, with a concentration region just past the concentration c^* at which macromolecular coils begin to overlap. A more restrictive definition of "semi-dilute" solutions would be that advanced by the scaling theory (3) and which is based on an analogy between "semi-dilute" solutions and gels. This analogy requires the mo-molecular weight of the chains to be high enough such that the transition between the dilute regime of independent macromolecules and the "semi-dilute" regime of an entangled gel-like network would occur in a rather very narrow concentration regime past c^*. In the theoretical "semi-dilute" solution, the translation motion of the whole chain in the network is neglected in comparison to the much faster motion of portions of the chain "stuck" between "entanglement" points. The only concentration fluctuations of importance become then those of segment concentration, pretty

Physical Optics of Dynamic Phenomena and Processes in Macromolecular Systems

much like those in the thermal swelling and deswelling fluctuations in a gel. Under these conditions the osmotic pressure of this solution, like that of a gel, would be molecular weight independent and the mutual cooperative diffusion coefficient defined in a frame of reference relative to the solvent velocity is predicted to be molecular weight independent and to follow a power-law in concentration with a power index of 0.75 as indeed it has been shown to do in a number of properly chosen "semi-dilute" solutions.

This short paper summarizes our results of photon correlations from polymer solutions of a very wide range of polymer concentrations from the very dilute to the very concentrated. We do not observe in our systems the scaling law of 0.75 above c*; we suspect that we do not achieve the condition of molecular weight independent osmotic pressure required by the scaling theory because the molecular weights of our polymers are relatively low. By increasing the concentration further, we enter into the concentrated regime before ever attaining the scaling law condition. A new mode of concentration fluctuations, ignored in the theoretical "semi-dilute" solutions, is however very evident in our systems that are not "gel-like" at time scales that are available to the photon correlation experiment. It is natural to associate this new diffusive mode to concentration fluctuations caused by the center of mass motion of entangled polymer chains. Despite the lack of theoretical predictions for this case and for very concentrated polymer solutions, we show that the photon correlation technique provides a wealth of information that is worthy of further investigations. The systems studied here are poly-n-butyl methacrylate and polymethyl methacrylate in methyl ethyl ketone.

Experimental

Solutions with 40% or less by weight of polymer were prepared from linear methacrylates obtained by suspension polymerization at 70°C. The molecular weight and polydispersity of the polymers were determined by gel permeation chromatography; for PMMA, they were 9.6×10^4 and 1.85 respectively; and for PBMA, they were 4.5 x 10^5 and 3. These solutions were freed from dust by filtration and centrifugation. The highly concentrated solutions ($\geq$55%) were prepared directly in the scattering cells from vacuum distilled and degassed monomer (MMA or BMA) and solvent (MEK). The monomer was then thermally polymerized. The set-up and details of the preparation procedures are described in a forthcoming paper (4).

The light scattering spectrometer consisted of a Lexel argon ion laser in conjunction with an EMI 9789 B photomultiplier tube and a Birnboim correlator. The light scattering experiments were performed at 19°C. Pulsed field-gradient spin-echo measurements of solvent self-diffusion were obtained using the Carr-Purcell pulse sequence on a CFT-20 Varian NMR spectrometer.

In dilute solutions below c*, a single exponential fit of the light intensity correlation function gave the isotropic center of mass diffusion of the macromolecule. Above c*, two diffusive processes were observed. Both relaxation modes were clearly q^2-dependent where q is the scattering wave vector. At very high concentrations ($\geq$ 55% PMMA), the "slow diffusion" was too slow to be detected and the fast diffusion was coupled to internal mode motions leading to deviations from q^2-dependence.

Results and Discussion

Figure 1 shows the variation in the nature of the photon correlation function with polymer concentration for a fixed short time delay experiment. At concentration above c*, C(q,t) becomes bimodal, reflecting both the collective and the center of mass modes. Only at fairly high concentrations can the two modes be unequivocally separated. For example, at 30% PBMA in MEK, the time delays required to monitor the two modes are separated by four orders of magnitude (4). Although the separation of the two modes is less clear in the semi-dilute regime, the results obtained with a bimodal correlation lead to a consistent picture for the behavior of the cooperative diffusion D_c and of the center of mass dif-

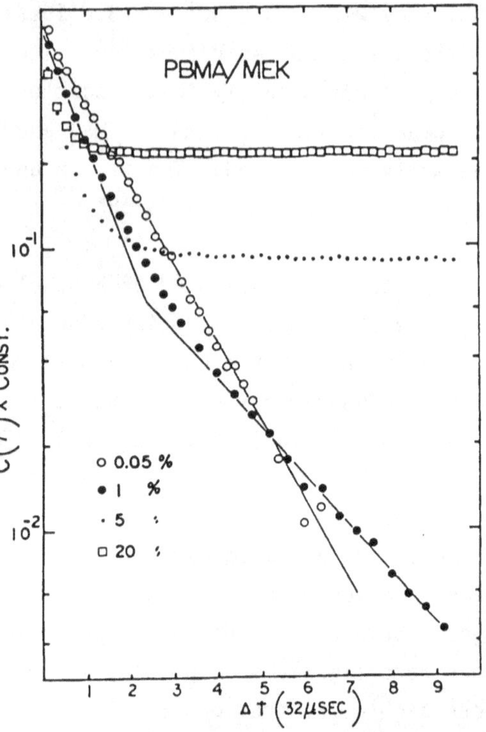

Figure 1. Variation of the correlation function with polymer concentration. The curve for 0.05% solution is in the dilute regime.

fusion D_{CM} with polymer concentrations. The results are shown in Figures 2 and 3 for the PBMA/MEK and PMMA/MEK systems respectively. D_s^* is the self-diffusion of MEK obtained by NMR; D_c and D_{CM} are the binary diffusion coefficients extracted from the light scattering photon correlation experiments. In Figure 2 we show D_c and D_{CM} for $c>c^*$; in Figure 3, for the PMMA/MEK system, we show D_c and D_{CM} merging into the single isotropic diffusion coefficient observed in dilute solutions below c^*. The values of D_c and D_s^* are lower in the PMMA/MEK system than in the PBMA/MEK system; they suggest that the "softer" PBMA, with a glass transition temperature $T_g \simeq 20°C$, provides more "free volume" for the motion of its own segments and of the solvent molecules compared to the "harder" PMMA with its glass transition temperature of $T_g \simeq 105°C$. The power-law exponent of concen-

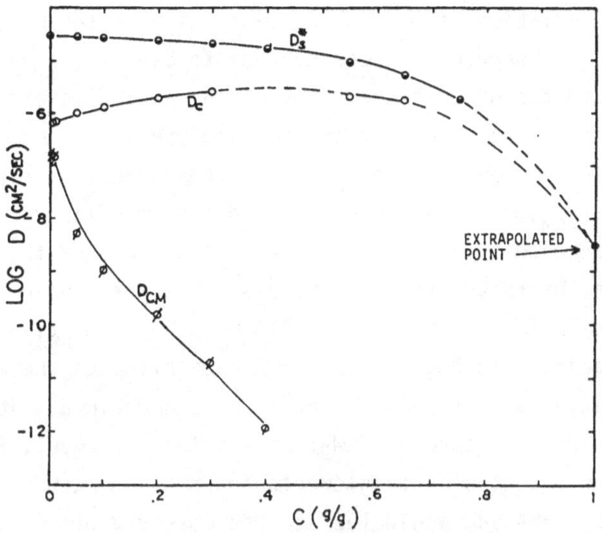

Figure 2. Diffusion coefficients versus concentration for PBMA/MEK solutions. D_s^* is the solvent self-diffusion; D_c and D_{CM} are the cooperative diffusion and the center of mass diffusion respectively.

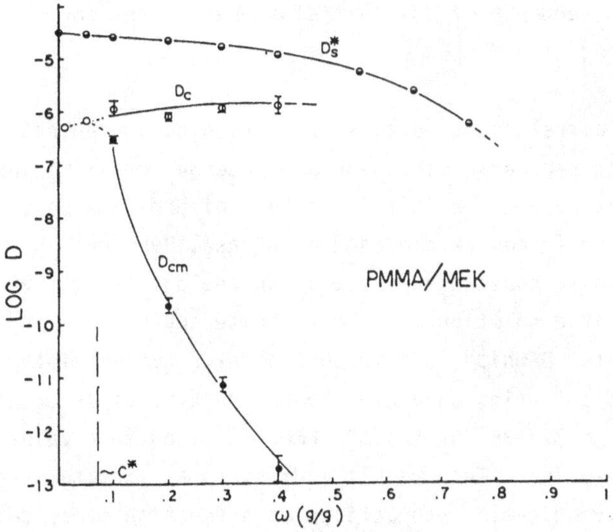

Figure 3. Diffusion coefficients versus concentration for PMMA/MEK solutions. Same notation as in Figure 2.

tration for the variation of D_c with concentration in the semi-dilute regime above c* is observed in these systems to be ~0.4, a value which has also been reported for other systems and which may indicate that the asymptotic conditions required for the scaling prediction of an exponent equal to 0.75 have not been achieved. With regard to the "slow" diffusion coefficient D_{CM}, we observe on one hand the very fast drop of D_{CM} as a function of concentration at high concentrations and on the other hand its merging with the fast diffusion process below the overlap concentration in dilute solutions. The drop of D_{CM} with concentration in the PMMA system is more pronounced than in the PBMA case although the molecular weight of the latter was in fact higher. One can again attribute this effect to the "harder" nature of PMMA; on a molecular level, Ferry observed that the extent of entanglements in PMMA was much higher (~double) than that of PBMA and would lead to the observed effect (5). At concentrations higher than 40% polymer the center of mass diffusion becomes too slow to monitor with the correlator. For 55% and 65% PBMA in MEK the photon correlation is still initially exponential and q^2-dependent and therefore a cooperative diffusion coefficient may be extracted at these very high concentrations in this case. For the PMMA case, deviations from q^2-dependence of the initial slopes of the correlation were observed starting at 55%.

To analyze the correlations obtained at very high concentration when a coupling between segmental diffusion and internal chain motions occur, we examine the behavior of the initial slope Ω of the time correlation function. We use the framework proposed by Akcasu, Benmouna, and Han (2) for the single infinite chain assuming that an analogy exists between the highly concentrated solution and the infinite chain. We find that the draining parameter B which is a measure of hydrodynamic interactions and which we use as a fitting parameter to our results of ln C(q,t) versus Ωt decreases from B>0.38 for the 55% PMMA solution to a value of B≅0.1 for the 65% PMMA solution. This result indicates, as expected, an increase in screening of hydrodynamic interactions as a function of concentration and a trend from a Zimm-like dynamic behavior for the 55% polymer solution toward a Rouse-like behavior around 65% polymer solution. Figure 4 shows

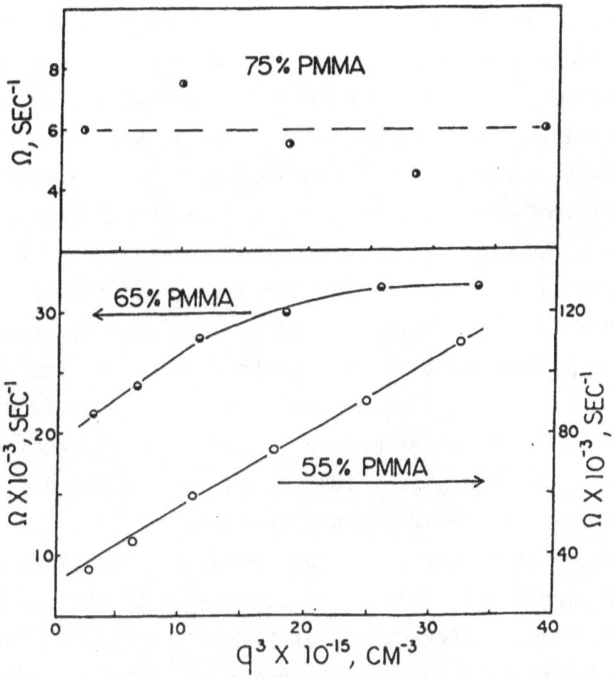

Figure 4. The initial slopes, Ω, of the correlation functions versus q^3 for solutions of 55%, 65%, and 75% PMMA in MEK.

that the q^3-dependence of Ω predicted from the Zimm dynamics of the chain (2) is indeed well obeyed for the 55% PMMA. In light scattering, the Rouse limit leads to q-independent correlation functions as observed in polymer melts. The q^3-dependence observed for the 55% solution will go therefore into a q-independent correlation at higher concentrations. This trend is verified for the results of the 65% and 75% PMMA solutions shown in Figure 4.

Conclusion

For molecular weights not high enough to achieve the asymptotic conditions required by scaling laws in semi-dilute polymer solutions, the isotropic binary diffusion coefficient which is observed by photon correlation in dilute solutions below c* splits into two components above c*. This phenomenon is inherently tied with the anisotropic nature of macromolecular diffusion above the overlap concentration and is caused by the two types of concentration fluctuations that exist in entangled polymer solutions. The diffusion coefficients measured by light scattering are mutual diffusion coefficients that are affected by the thermodynamic interactions between molecules and are different from self-diffusion coefficients measured by techniques such as nuclear magnetic resonance or forced Rayleigh spectroscopy. For very concentrated polymer solutions, an interpretation of the photon correlation results in terms of the theoretical framework developed for the single chain was shown to be fairly instructive. Several important issues may be raised and will require further work: polydispersity, molecular weight dependence of the diffusion coefficients, amplitudes of the fast and slow diffusion as a function of molecular weight, further clarifications between self-diffusion and mutual diffusion coefficients, the surprising appearance of what would be considered as hydrodynamic interactions at polymer concentrations as high as 55% and 65%, and a reverse wave vector behavior (Figure 4, for 65% PMMA) in concentrated polymer solutions from that observed and predicted for dilute solutions. The last two observations have been recently corroborated by neutron spin echo measurements on polydimethylsiloxane solutions presented in this symposium by Dr. Ewen (6).

Acknowledgment

The partial support of the National Science Foundation through grant DMR81-11114 of the Polymers Program is gratefully acknowledged.

References

1. Berne, B.J.; Pecora, R.: Dynamic Light Scattering with Applications to Chemistry, Biology, and Physics, Wiley, New York 1976.

2. Akcasu, A.Z.; Benmouna, M.; Han, C.C.: Polymer 21, 866 (1980).

3. De Gennes, P.G.: Scaling Concepts in Polymer Physics, Cornell University Press, Ithaca, New York 1979.

4. Hwang, D.H.; Cohen, C.: Macromolecules, in press.

5. Ferry, J.D.: private communication. See Ferry, J.D.: Viscoelastic Properties of Polymers, Third edition, Wiley, New York 1980.

6. See Ewen, B.; Stühn, B.; Binder, K.; Richter, D.; Hayter, J.B.: Polymer Communications 25, 133 (1984).

QUASI-ELASTIC LIGHT SCATTERING IN SEMI-DILUTE SOLUTIONS AT THE θ-TEMPERATURE

M. Adam and M. Delsanti
Service de Physique du Solide et de Résonance Magnétique,
CEN-SACLAY, 91191 Gif-sur-Yvette, Cedex, France

Introduction

In order to test the validity of the pseudo-gel model introduced by de Gennes (1-3), many experiments have been performed on semi-dilute polymer solutions in good solvent. But unfortunately in these systems the elastic modulus of the pseudo-gel is negligible in comparison to the osmotic modulus (4) and quasi-elastic light scattering experiments do not demonstrate clearly the pseudo-gel behaviour which is evidenced by viscoelastic measurements (4). Our aim here is to show that quasi-elastic light scattering experiments in a θ semi-dilute polymer solution are useful for understanding the macroscopic dynamical behaviour of the semi-dilute solutions.

Experimental

Using a heterodyne quasi-elastic light scattering spectrometer, we study the spectrum of the light scattered by semi-dilute polystyrene-cyclohexane solutions at 35°C as a function of molecular weight and concentration. The system studied is polystyrene of high molecular weights (M_w = 20.6×10^6, 6.77×10^6, 3.84×10^6, 2.89×10^6, 1.26×10^6, 4.22×10^5) in a concentration range C lower than 0.1 g/cm^3 but larger than the overlap concentration C^*(g/cm^3) = $40/\sqrt{M}$. To prevent formation of an anomalous state, we avoided freezing the samples (6).

Physical Optics of Dynamic Phenomena and Processes in Macromolecular Systems

Results

1) In the case of samples having a "long" viscoelastic relaxation time T_R ($\gtrsim 0.05$ s), the profile of the correlation function of the light scattered is nonexponential . However it is possible to determine two characteristic times : the shortest time τ and the longest time τ_L. Varying the sample time Δt of the 128 channels correlator, one observes that the correlation function is :

- a linearly decreasing line if $\Delta t < \tau/128$
- and a flat base line if $\Delta t > 10\ \tau_L/128$.

The long time τ_L is independent of the momentum transfer q ($\tau_L \sim q^o$) and is equal to the longest relaxation time T_R measured from viscoelastic experiments (7,8) : $\tau_L/T_R = 1.17 \mp 0.2$.

The short time τ shows a

- q^{-2} dependence at low momentum transfer ,
- $q^{-\alpha}$ dependence, $\alpha > 2$, at high momentum transfer .

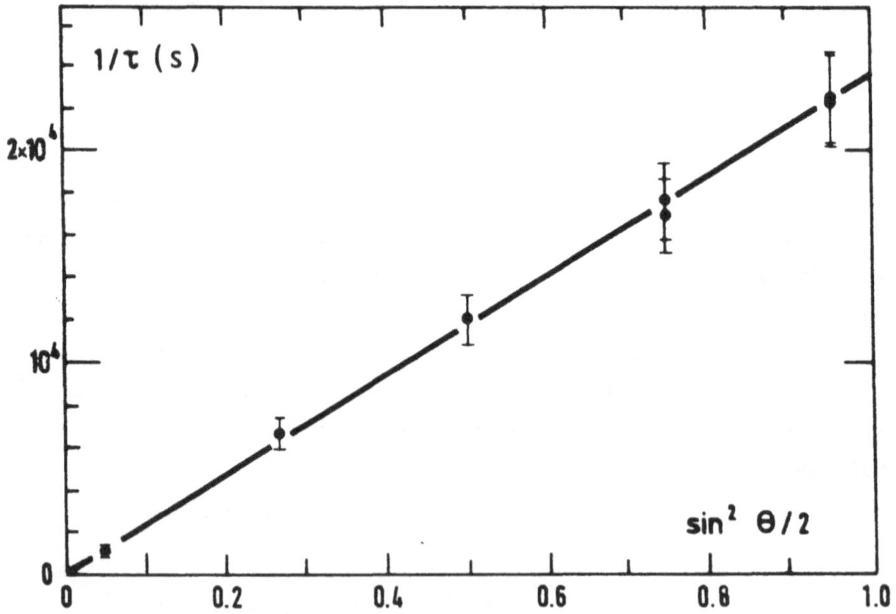

Figure 1 : Example of the variation of the inverse of the characteristic time τ^{-1} as a function of $\sin^2 \frac{\theta}{2}$ with $q\xi < \frac{1}{3}$. θ is the scattering angle ; $q \sim \sin\frac{\theta}{2}$. The sample characteristics are : $M_w = 20.6 \times 10^6$, $C = 7.64 \times 10^{-2}$ g/cm^3.

The transition from q^{-2} to $q^{-\alpha}$ regime occurs when $q\xi \approx 1/3$ where ξ is the concentration correlation length measured by neutron scattering experiments (9).

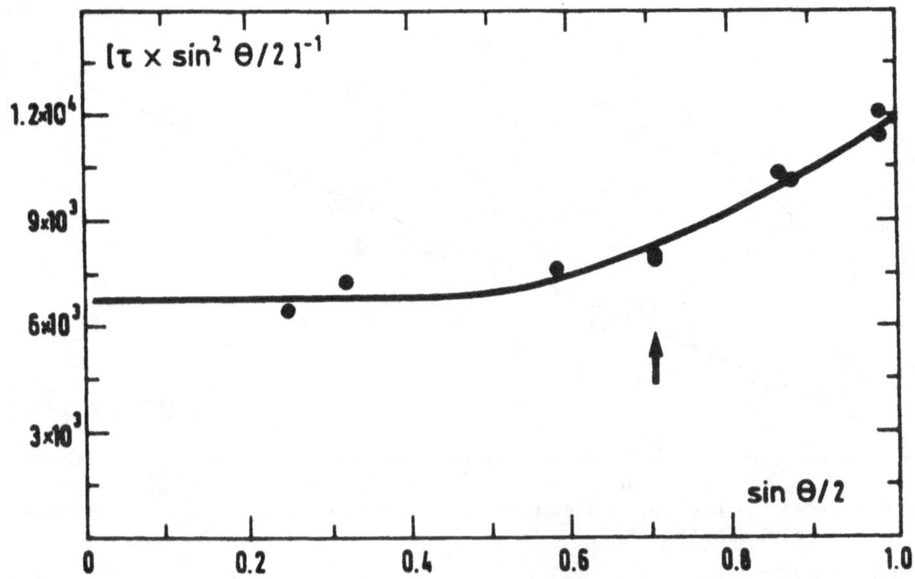

Figure 2 : Variation of the inverse characteristic time τ^{-1} divided by $\sin^2(\frac{\theta}{2})$, as a function of $\sin\frac{\theta}{2}$. The arrow indicates the value of $\sin\frac{\theta}{2}$ for which $q\xi = \frac{1}{3}$ and the solid line is a guide for eye. The sample characteristics are : $M_w = 6.77\times10^6$, $C = 2.47\times10^{-2}$ g/cm^3.

2) In the case of samples having a "short" viscoelastic relaxation time ($T_R \lesssim 0.05$ s) , by varying the momentum transfer we can measure a characteristic time either larger or smaller than T_R. In the first case one observes that the correlation function is described by a single exponential; its characteristic time τ ($\tau > T_R$) is q^{-2} dependent when $q\xi < 1/3$. When $\tau < T_R$ a nonexponential behaviour corresponding to the case 1) is observed.

In all cases the diffusion coefficients D measured are independent of the molecular weight (see Fig.3).

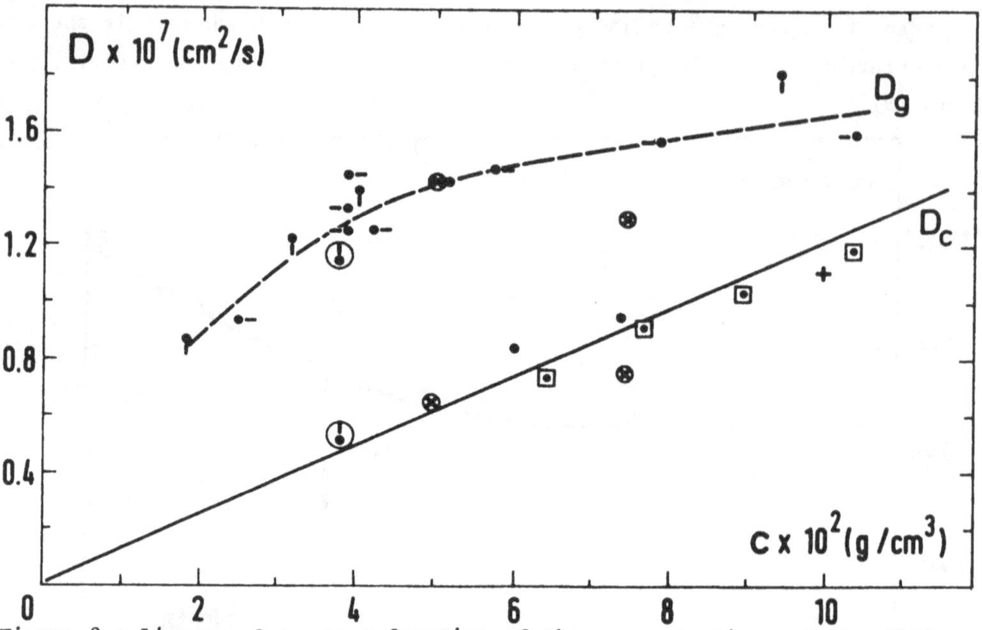

Figure 3 : Linear plot as a function of the concentration, of the diffusion coefficients obtained :

- (D_c) at $\tau/T_R > 1$,
- (D_g) at $\tau/T_R < 1$.

The symbols have the following meaning : . $M_w = 4.22\times10^5$, ⊗ $M_w = 1.26\times10^6$, ↓ $M_w = 2.89\times10^6$, —• $M_w = 3.84\times10^6$, •— $M_w = 6.77\times10^6$, ↑ $M_w = 20.6\times10^6$. + $M_w = 3.9\times10^5$ and ⊡ $M_w = 8.6\times10^5$ diffusion coefficients obtained by gradient concentration technique (10).

The variation of D with concentration depends on the value of the ratio τ/T_R.

At $\tau/T_R > 1$, the diffusion coefficient $D_c = \tau^{-1}/q^2$ follows the same law as that measured by gradient concentration (10) :

$$D_c(cm^2/s) = (1.25 \pm 0.10)\times10^{-6}\ C;$$

it is proportional to the concentration.

At $\tau/T_R < 1$, the diffusion coefficient $D_g = \tau^{-1}/q^2$ is larger than D_c.

The ratio of the two diffusion coefficients follows the relation :

$$D_g/D_c = 1 + 5.22\times10^{-2}\ C^{-1.04+0.2}$$

Discussion

The observations can be explained as follows. The diffusion coefficient D_c is proportional to the osmotic bulk modulus $K = C\ \partial\pi/\partial C$ and inversely proportional to the friction coefficient f per unit volume of the solvent through the polymer : $D_c = \frac{K}{f}$. This relation can be directly demonstrated by comparison with intensity light scattering and sedimentation measurements from which it was found that (11-12) :

$$K\ (\mathrm{dyn/cm^2}) = (2.93 \pm 0.14)\times 10^7\ C^3$$

and

$$s = 1.15\times 10^{-14}\ C^{-0.96\pm 0.04}$$

The sedimentation coefficients, s, is related to the friction coefficient f per unit volume by $s = (1-\bar{v}\rho)\ C/f$ ($\rho\bar{v} = 0.703$ for polystyrene in cyclohexane at 35°C).

Following scaling laws (1-3) the reduced diffusion coefficient D_c/D_o, where D_o is the diffusion coefficient of an isolated chain, is a function only of C/C^*. In Fig.4 we can see that the scaling law is fully respected.

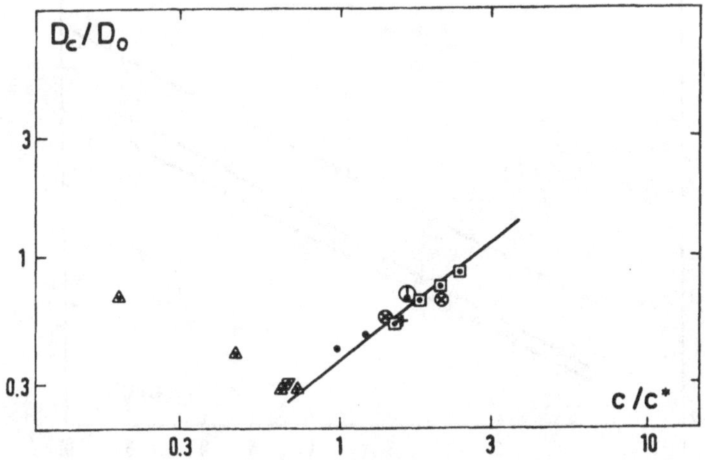

Figure 4 : Triangles are values obtained from ref.(13) : △$M_w = 1.3\times 10^5$, ▽$M_w = 2\times 10^6$. For the meaning of the other symbols see Fig.3, D_c/D_o values have been calculated using : $D_o(\mathrm{cm^2/s}) = 1.30\times 10^{-4}\ \sqrt{M_w}$ (14).

The diffusion coefficient D_g is the diffusion coefficient of the polymer pseudo-gel (7) : $D_g = (K + M)/f$, where M is the longitudinal elastic modulus of the pseudo-gel. Following this hypothesis $D_g/D_c - 1$ must be proportional to M/K. From the present measurements and the measurements done in ref.(11), we deduce that M (dyn/cm^2) $= 1.53 \times 10^6\ C^{1.96}$. The elastic longitudinal modulus of the pseudo-gel is proportional to the volume number density of entanglements ($\sim C^2$). This point is in agreement with the theoretical predictions (1-3) and shear elastic modulus behaviour deduced from viscoelastic measurements (5).

The observation by quasi-elastic light scattering, in a semi-dilute θ polymer solution, of either a liquid or pseudo-gel behaviour, depending on the experimental time scale (2,3), is possible because : M and K are of the same order of magnitude.

The time τ_L, which is independent of the momentum transfer , is a structural relaxation time and is equal to the stress viscoelastic relaxation time T_R. τ_L as well as T_R is related to the disentanglement of one chain. Figure 5 shows that the reduced relaxation time $\frac{T_R}{T_1}$ does not obey a scaling law in C/C^* as is the case in a good solvent (4).

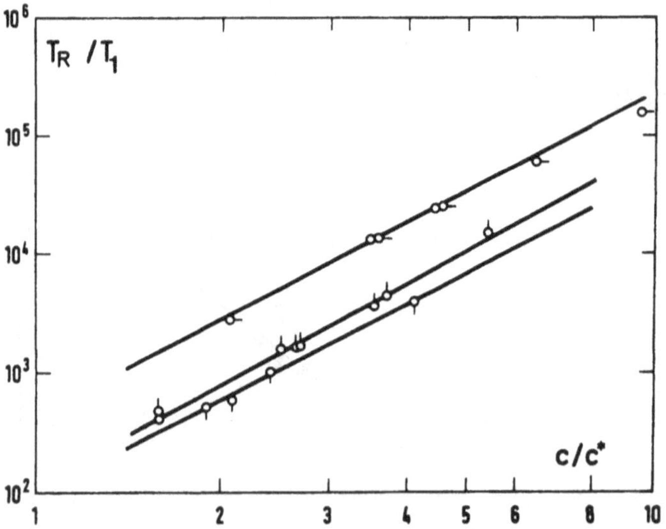

Figure 5: T_R/T_1 as a function of C/C^* where T_1 is the characteristic time of the first mode of a single chain in the dilute regime ($C<C^*$)(5). The symbols have the following meaning : o- $M_w=20.6\times10^6$, ♂ $M_w=6.77\times10^6$, ♀ $M_w=3.84\times10^6$.

In a θ solvent, T_R/T_1 is not just a function of C/C^* because the friction involved in the disentanglement of one chain, is not purely hydrodynamic. Actually hydrodynamic and thermodynamic quantities follow scaling law in C/C^* (see Fig. 4 and ref. 11). In order to explain this observation we must invoke an additional friction which is proportional to the number of entanglements per chain. This number does not obey a scaling law in a θ solvent but does so in a good solvent.

References

1. de Gennes, P.G., Macromolecules 9, 587 (1976).
2. Brochard, F., de Gennes, P.G., Macromolecules 10, 1157 (1977).
3. Brochard, F., J. Physique 44, 39 (1983).
4. Adam, M., Delsanti, M., J. Physique 44, 1185 (1983).
5. Adam, M., Delsanti, M., to appear in J. Physique (1984).
6. He, M.J. et al. Macromolecules 15, 673 (1982).
7. Adam, M., Delsanti, M., submitted to Macromolecules
8. Adam, M., Delsanti, M., J. Physique Lett. 45, L219 (1984).
9. Cotton, J.P. et al., J. Chem. Phys. 65, 1101 (1976).
10. Roots, J., Nyström, B., Macromolecules 13, 1595 (1980).
11. Stepanek, P., Persynski, R., Delsanti, M., Adam, M., Macromolecules, to appear in (1984).
12. Vidakovic, P., Allain, C., Rondelez, F., J. Physique 42, L323 (1981).
13. Munch, J.P., Hild, G., Candau, S., Macromolecules 16, 71 (1983).
14. Schmidt, M., Burchard, W., Macromolecules, 14, 210 (1981).

LIGHT SCATTERING OBSERVATIONS OF THE PLATEAU MODULUS IN SEMI-DILUTE THETA SOLUTIONS

Anne-Marie Hecht, Himadri Bohidar§ and Erik Geissler

Laboratoire de Spectrométrie Physique*, Université de Grenoble I, B.P. 68, 38402 St Martin d'Hères Cedex, France.

Abstract

Dynamic light scattering measurements are described for semi-dilute solutions of polystyrene ($M_w = 2.6 \times 10^7$ Daltons) in cyclohexane at the theta temperature. Observations as a function of scattering angle allow the effects of the shear modulus to be distinguished from those of the osmotic modulus.

Introduction

Theta solutions of polymers of sufficiently high molecular weight are characterized by two distinct cross-over concentrations that separate the dilute from the semi-dilute régime (Figure 1): the usual overlap concentration (proportional to $M^{-0.5}$, where M is the degree of polymerization), and a critical entanglement concentration c_e (proportional to M^{-1}).

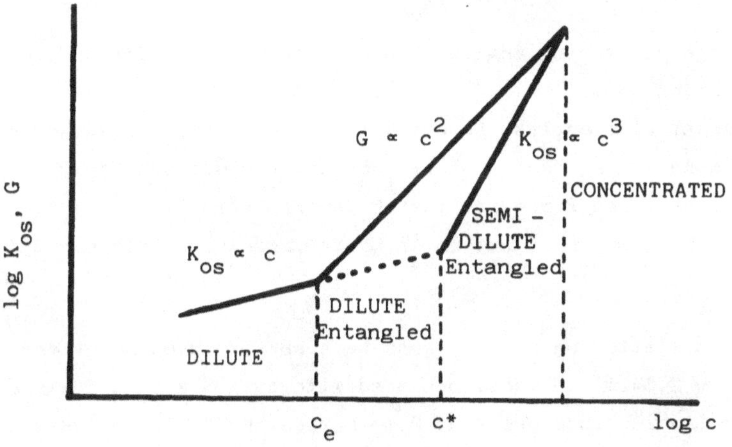

Figure 1
Schematic diagram of the behaviour of the shear (G) and osmotic (K_{os}) moduli as a function of concentration in theta solutions of high molecular weight polymers.

---

§ Present address: Department of Physics, Indian Institute of Technology, New Delhi 110016, India.

* CNRS associate laboratory

Physical Optics of Dynamic Phenomena and Processes in Macromolecular Systems

Brochard (1) has noted that in theta solutions the osmotic modulus is

$$K_{os} = K^{o}_{os} c^{3}, \tag{1}$$

where K^{o}_{os} is a constant. The polymer-solvent friction coefficient f varies as

$$f = f_{o} c^{2}. \tag{2}$$

The shear modulus, G, however, is determined by the probability that two polymer chains encounter one another, and hence

$$G = G_{o} c^{2}. \tag{3}$$

Brochard's suggestion is that in semi-dilute theta solutions ($c^* < c << 1$) $G >> K_{os}$, and hence concentration fluctuations of short time constant should be dominated by the shear modulus. By varying the scattering vector $\underline{Q}$, one can select an appropriate range for the relaxation rate of the concentration fluctuations

$$\Gamma = DQ^{2} \tag{4}$$

where the collective diffusion coefficient D is given by

$$D = (K_{os} + \frac{4}{3} G)/f \tag{5}$$

Thus for large scattering angles, G should be equal to G/f and independent of concentration.

For small scattering angles, however, Γ is small, and the entanglements have time to relax, so that G should no longer contribute to D. In this case, Equations 1 and 2 predict that D is proportional to c.

The purpose of this experiment is to check these predictions.

Experimental Section

Dynamic light scattering measurements were performed on solutions of polystyrene ($M_w = 2.6\times10^{7}$ Daltons) prepared with cyclohexane and sealed in rectangular fluorescence cells of 2 mm optical path. The polystyrene, supplied by Interchim, France, had a stated polydispersity M_w/M_n of less than 1.3. The solutions were kept in an oven at 45° C for more than two years prior to use, in order to ensure homogeneity. The solutions nonetheless contained residual inhomogeneities consisting of insoluble dust particles present in the original polystyrene powder. Such dust particles ensure that the scattered light signal is in the heterodyne mode: this point was checked using a split beam arrangement (2). Their presence, however, prevents meaningful observations of the slowest part of the correlation functions.

In all cases reported here, the photon correlation signals were found to be non-exponential, consisting of an initial well defined decay (itself non-exponential), superimposed on a much slower relaxation. Two-cumulant fits were used for the fast part, the variance of which, $\mu_2/\bar{\Gamma}^2$, was in the range 0.15 - 0.2. This method of analysis gave consistent results for scattering angles (in air) of 15°, 20°, 90° and 160°. For angles in the range 25° - 40°, a weak signal of slightly faster relaxation rate was found to be superimposed on the main correlation signal, and this could not be satisfactorily analysed with the two-cumulant technique. The quoted results are therefore confined to the two ranges: small angles, 15° and 20° in air (λ = 632.8 nm, Q = 2.58×10^4 cm^{-1} and 3.42×10^4 cm^{-1}), and large angles, 90° and 160° in air (λ = 632.8 and 488 nm, Q between 1.99×10^5 and 3.63×10^5 cm^{-1}).

Results

For small scattering angles, D is found to be proportional to $c^{0.94\pm0.08}$ where c is the concentration of polymer in the solution. This result is in agreement with the linear relation described above for simple scaling (1). For large scattering angles, on the other hand, it is found that

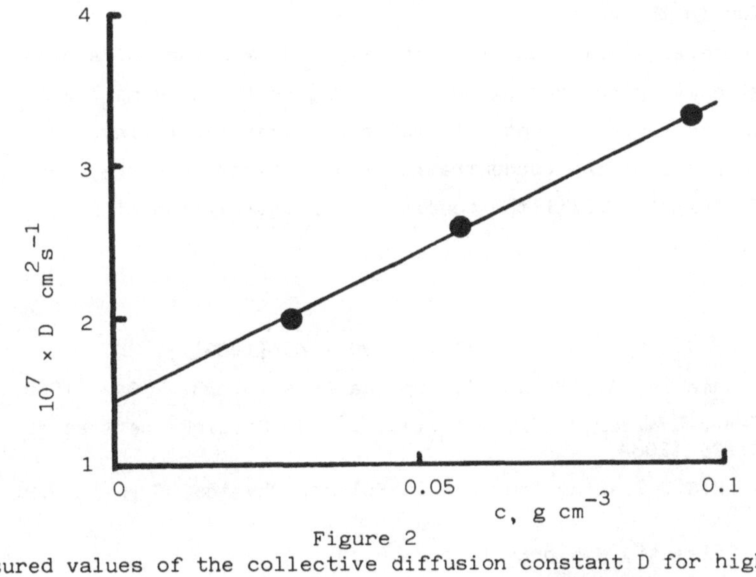

Figure 2
Measured values of the collective diffusion constant D for high transfer wavevectors Q as a function of concentration in polystyrene-cyclohexane.

$$D = A + Bc \tag{6}$$

(see Figure 2), where, from Equation 5, the constant A is equal to 4G/3f, and D is, within experimental error, the same as the proportionality constant found for the small angle scattering. The effects of the plateau modulus G are visible for large scattering angles because the concentration fluctuations decay faster than the transient network formed by the entanglements. In the present case, the apparent network lifetime is of the order of a millisecond (3). Since the characteristic length probed by these dynamic light scattering experiments is the polymer-polymer correlation length ξ, and not the much larger distance L (the length of the extended polymer), the expected lifetime of the constraints due to the local network will be shorter than the chain renewal time T_R (4). A possible alternative reason for the short network lifetime observed here is the polydispersity of our samples (5).

Comparison with earlier dynamic light scattering measurements from the system polystyrene-cyclohexane (6) provide a value for the coefficient K_{os}^{o} of Equation 1, and hence, from the coefficients A and B of Equation 6, the value of G can be extracted.

It is concluded that G varies as c^2 throughout the whole of the entangled régime; moreover, its extrapolated value at $c = 1$ g cm^{-3} coincides with the plateau modulus in the melt.

A second, more paradoxical, result is the implied existence of a dilute entangled régime lying between c_e and c^* in Figure 1. It seems likely, however, that this régime is not detectable by light scattering, since the correlation length ξ of the concentration fluctuations is large, and it is therefore difficult to fulfil the condition for observation of collective modes, $\xi Q < 1$.

References

1. Brochard, F: J. Physique (Paris) 44 , 39 - 43 (1983).
2. Adam, M., Hamelin, A., Bergé, P.: Optica Acta 16, 337 - 342 (1969).
3. Hecht, A.-M., Bohidar, H.B., Geissler, E.: J. Physique Lettres 45, L 121 - L 126 (1984).
4. de Gennes, P.G.: Scaling Concepts in Polymer Physics, Cornell U.P. Ithaca 1979.
5. Adam, M., Delsanti, M.: private communication.
6. Geissler, E., Hecht, A.-M.: J. Chem. Phys. 65, 103 - 110 (1976).

DYNAMIC LIGHT SCATTERING ON SEMIDILUTE SOLUTIONS OF POLYSTYRENE AS A FUNCTION OF SOLVENT QUALITY

Wyn Brown
Institute of Physical Chemistry, University of Uppsala, Uppsala, Sweden

Introduction

The theories of de Gennes[1] for semidilute polymer solutions constituted a major advance in our understanding of the dynamics of these solutions. However, the accumulated data since 1976 tend to show that the original approach cannot explain a number of features, of which one is the trends in the spectral non-exponentiality which is particularly pronounced in poor solvents[2-5], but is also a distinct feature of good solvents[6,7] in the semidilute regime. Thus, for example, using the cumulant method,[8] the normalized second cumulant, $(\mu_2/\bar{\Gamma}^2)$, increases systematically with both concentration and angle.[2-4] As demonstrated by Chu and Nose[2], these trends in $(\mu_2/\bar{\Gamma}^2)$ reflect changes in the relative intensity amplitudes of the contributing modes of relaxation. The latter are observable in dynamic light scattering experiments depending on the choice of probing length, q^{-1}, and reduced concentration C/C^*. However, there will not be a clearly-defined demarcation between the regimes in which the various modes are observable and most experimental arrangements will lead to the detection of a composite of such modes at intermediate molecular weights. Of course, the correlation function may be non-exponential for a more trivial reason such as the presence of dust or other source of local inhomogeneity. This has caused some uncertainty in the literature[4] but should not obscure the fact that non-exponentiality is an intrinsic feature of autocorrelation functions[2,6,7] for semidilute solutions of intermediate molecular weights. The present contribution describes dynamic light scattering experiments on polystyrene systems using solvents differing in quality. The results are treated both by the cumulant method and also using a rapid, on-line, bimodal analysis.

Physical Optics of Dynamic Phenomena and Processes in Macromolecular Systems

Experimental

Narrow distribution polystyrenes were purchased from Pressure Chemical Co, Pittsburg, Pa. U.S.A.

PS 100 ($\bar{M}_w$ = 93,000; $\bar{M}_w/\bar{M}_n$ = 1.06)

PS 300 ($\bar{M}_w$ = 281,000; $\bar{M}_w/\bar{M}_n$ = 1.06)

PS 950 ($\bar{M}_w$ = 928,000; $\bar{M}_w/\bar{M}_n$ = 1.06)

Solvents were of spectroscopic grade from Merck, Darmstadt, FRG. These were exhaustively dried over freshly treated 3Å molecular sieves and filtered through 0.2 µm Millipore filters (Fluoropore).

The dynamic light scattering arrangement has been described briefly in an earlier communication.[11] All experiments were performed in the homodyne mode. The measurements were made as a function of angle but confined to the region corresponding to $q\ R_G < 1$ for each sample.

Results and Discussion

The autocorrelation functions have been analyzed using:

(a) the cumulant method[8] ($\bar{D}_{cum}$) and

(b) bimodal analysis according to an expression of the form:

$$g^{(2)}(\tau) - 1 = \beta\ [A_1\exp(-\Gamma_1\tau)+A_2\exp(-\Gamma_2\tau)]^2 \qquad (1)$$

The point of agreement between approaches (a) and (b) is that $\bar{D}_{cum} = \langle D \rangle$ where the latter is the weighted average of the fast and slow components separable using (b). An estimation of the reliability of the bimodal analyses was established by making a number of model simulations.[17] Normally distributed pseudo-random numbers, typical of the residuals of the fit in a real experiment, were added to a simulated experiment using parameters typical of a real experiment. It was found that if the characteristic relaxation times differ by at least a factor of two, they could be consistently separated even when the minor component constitutes as little as 5% of the total intensity amplitude. The present data establish that, in general, there are two contributing modes ranging over solvent quality from good (THF) to a theta solvent (cyclopentane, 20·8°C).

Good solvent (THF)

The results for PS 950 (M=950,000) in THF are depicted in Fig.1. The data for the main exponent using bimodal analysis coincide with those obtained by the classical gradient technique. Its relative amplitude decreases from a value close to unity to about 0·8 at the highest concentration as entanglement constraints increase (Fig.1 (insert)). The cumulant values, $\bar{D}_{cum}$, on the other hand diverge with increasing C and at the same time there is a progressive increase in the value of the normalized second cumulant (Fig. 2b). It is noted that a single exponent fit according to:

$$g^{(2)}(\tau)-1 = A\exp(-2\Gamma\tau)+B \qquad (2)$$

where B is a baseline term, gives the same value for the relaxation time as the main exponent obtained from bimodal analysis.

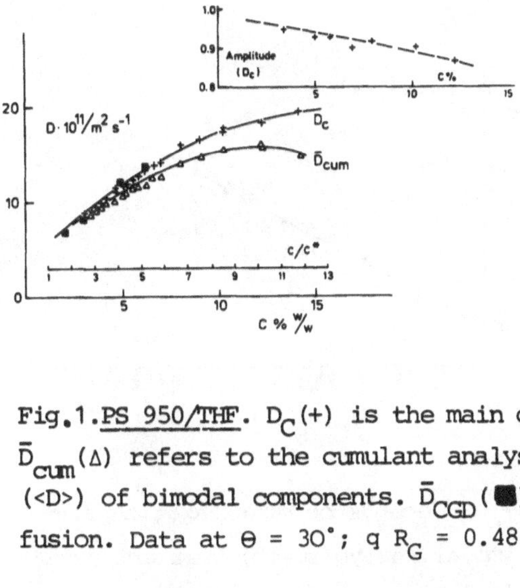

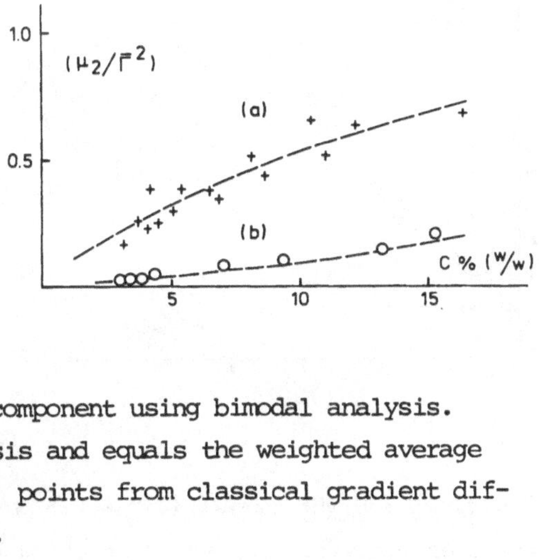

Fig.1.PS 950/THF. D_C(+) is the main component using bimodal analysis. $\bar{D}_{cum}$(Δ) refers to the cumulant analysis and equals the weighted average (<D>) of bimodal components. $\bar{D}_{CGD}$(■) points from classical gradient diffusion. Data at Θ = 30°; q R_G = 0.48.

Fig.2. Dependence of the normalized second cumulant on concentration for
(a) PS 950/cyclopentane (20.8°C).
(b) PS 950/THF (25°C).

However, the statistical fit of the 3-parameter fit was much poorer, i.e.

gave a low value of the quality factor Q :

$$Q = 1 - \sum_{i=1}^{n-1} \varepsilon_i \varepsilon_{i+1}/(n-1) \;/\; \sum_{i=1}^{n} \varepsilon_i \varepsilon_i/n \qquad (3)$$

whereas the bimodal fit gave a Q-value of unity. We thus consider the slow component of small relative intensity amplitude to be distinct from a baseline term. From angular measurements (Fig.3) we provide justification for its interpretation as a slow diffusional mode. It is observed that Yu et al[6] and Amis and Han[9] have also provided experimental evidence for the presence of a slow diffusional component in PS/THF systems.

We make a tentative identification of the slow hydrodynamic mode isolated here with the slow component recently described by Hwang and Cohen[10] and which is readily observed at comparatively low concentration with the more extended polymer chains in aqueous systems[11].

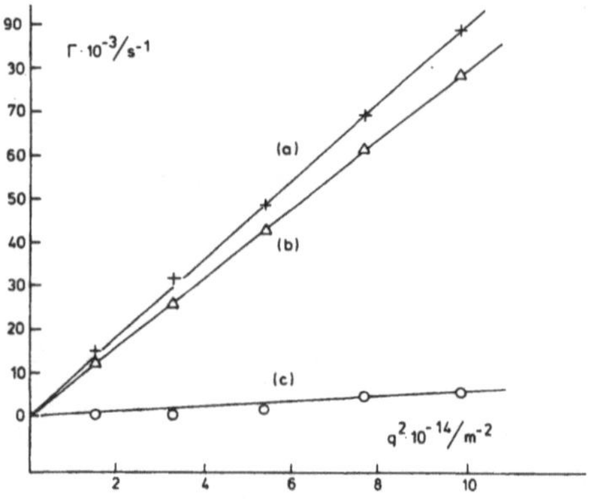

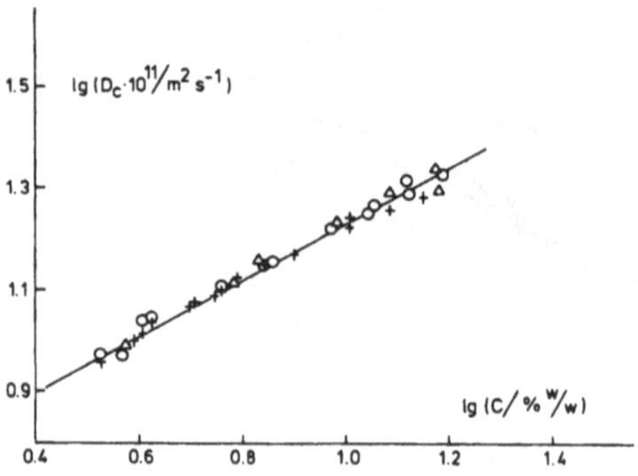

Fig.3. PS 950/THF (25°C) (C = 4.02% w/w). q^2-dependence of relaxation frequencies for: (a) main exponent using bimodal analysis; (b) cumulant data; (c) minor exponent using bimodal analysis.

Fig.4. Double logarithmic plot for data for three polystyrene fractions in THF, where D_C is the main component evaluated using bimodal analysis.
(+ 950,000; O 300,000; Δ 100,000)

The main exponent, on the other hand, describes the cooperative diffusion coefficient, D_C. Taken together with D_C-values for M=93,000 and M=281,000, a common relationship is found for $C/C^* > 1$ (Fig.1) except for a divergence at the highest concentrations which is more pronounced the greater the value of M. The data in Fig.4 correspond to:

$$D_C \sim C^{0.59}$$

where the D_C-values are measured values (D) corrected for solvent backflow according to:

$$D = D_C(1-\emptyset)$$

and $\emptyset$ is the volume fraction of polystyrene. This exponent is substantially lower than the value $\gamma = 0.75$ predicted by de Gennes[1]. Furthermore, there is no tendency for γ to increase with M. This inability of the model to describe the observed variation of D_C with C should serve as a reminder of the extent to which entanglements influence the concentration dependence and which is an aspect ignored in the simple model.

<u>"Marginal" solvent (ethyl acetate)</u>

Results for PS 950 in ethyl acetate are shown in Fig.5. The main feature is the presence of two distinct modes of relaxation which are separable using bimodal analysis. Thus, at all but the lowest concentrations, a single exponent plus baseline gave a very poor fit (at the highest concentration it was appropriate to add a baseline term to the bimodal expression for a marginally improved fit). The slow relaxation has a slightly negative dependence on C whereas the fast has a positive dependence given by:

$$D_C \sim C^{0.47}$$

This exponent is much lower than predicted for cooperative motions of the pseudogel, and there is evidence[15] that this may be the case in poorer solvents. However, it is not suggested that a complete separation of the two modes has been achieved (or can be using the present method) and the points at low concentration in particular scatter considerably making the slope uncertain.

The relative intensity amplitude of the fast mode is strongly concentration dependent (Fig.5 a), the correlation function being close to single exponential at the lowest concentration and rapidly becoming a bimodal distribution as C increases. Marginal and theta solvents reveal a complexity which is not found in good solvents and bimodal autocorrelation functions in ethyl acetate have been noted previously[12,13].

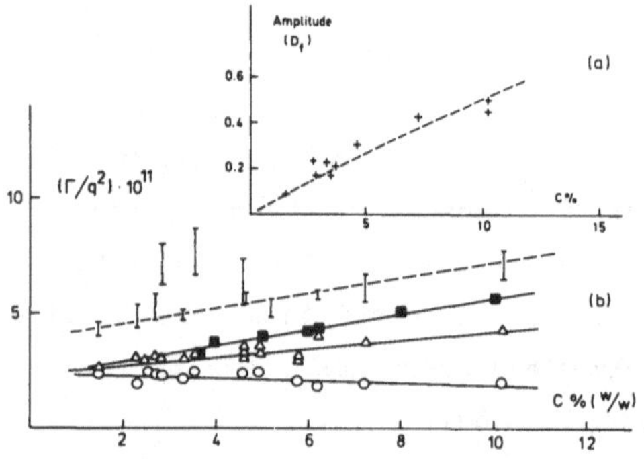

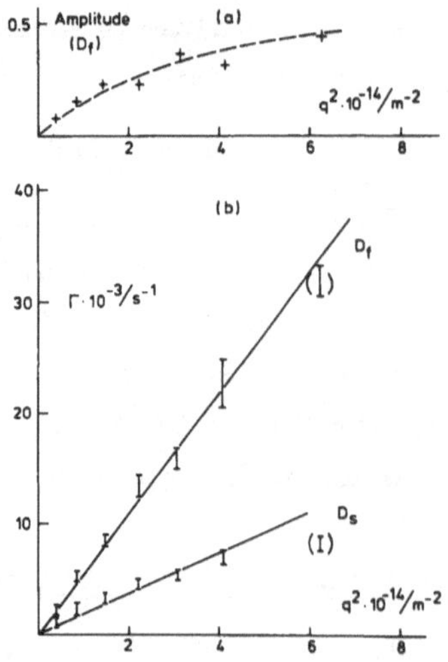

Fig.5. PS 950/ethyl acetate (25°C): (a) Relative intensity amplitude of the fast mode. (b) Fast (I) and slow (O) modes derived using bimodal analysis. Cumulant data (Δ) (=weighted average <D> of fast and slow modes) and $\bar{D}_{CGD}$ (■) data from classical gradient diffusion.

Fig.6. PS 950/ethyl acetate (C = 3.29% w/w): (a) The relative intensity amplitude of the fast mode depicted in Fig. 5. (b) (q^2-dependence of) fast and slow components in Fig. 5.

We observe that the slow relaxation decreases strongly in intensity amplitude with increasing concentration, a finding which contrasts with the exponential increase in the amplitude of a slow relaxation described by Mathiez et al[12]. This suggests, instead, the presence of local concentration inhomogeneities in their solutions particularly since their slow mode amplitude was found to decrease on standing. The amplitude of the fast mode should eventually decrease at even higher concentrations as constraints increase due to the greater entanglement density as shown by Hwang and Cohen[10]. A slow mode as found for good solvents should then become evident.

The angular dependence of the relaxation frequency is linearly dependent on q^2 (Fig.6) suggesting that both modes are diffusional processes. The intensity amplitude of the fast mode increases strongly with increasing angle (Fig.6 a). It is presumed here that the slow relaxation describes the centre of mass translational motions while the fast corresponds to the cooperative motions of the developing pseudogel. At this intermediate molecular weight (M = 950,000) both processes are features of the autocorrelation function. This situation persists over a wide range in concentration, well in excess of C^* where it is usually assumed that the transient gel motions are the only ones of importance.

Values of $\bar{D}_{cum}$ (= <D>) have an intermediate concentration dependence, reflecting the relative weighting of the two modes.

Data for the low molecular weight (M = 10^5) sample are typified by single exponential (<5% of a fast component) correlation functions up to a concentration of ≃ 15%, and the derived diffusion coefficients are concentration independent. A single exponential is also found for this sample in the theta solvent (see below) showing that a minimum chain length is necessary for formation of the transient gel.

Data from classical gradient measurements are included in Fig.6 and have an intermediate concentration dependence. For PS 950 in good solvents $\bar{D}_{CGD}$ falls in line with D_C (Fig.1) while in the theta solvent (Fig.7) agrees with $\bar{D}_{cum}$. Thus the usually assumed identity between $\bar{D}_{CGD}$ and D_C in semidilute solutions does not appear to hold in poorer solvents.

$\bar{D}_{CGD}$ is found to be molecular weight dependent as noted by others[14] in marginal solvents and this is seen to be a consequence of the contributions from the two relaxational modes. For the molecular weight used here $\bar{D}_{cum}$ is also dependent on molecular weight a finding which disagrees with that of Munch et al[13] and Schaefer et al[15] who find a unique diffusion coefficient depending only on concentration. This may well be true at sufficiently high M where the hydrodynamic mode has become too slow to form a major component of the correlation function and the latter will again be approximated by a single exponential plus baseline term.

Theta solvent (cyclopentane ; 20.8°C)

Under theta conditions the spectra from semidilute polymer solutions are known to be strongly non-exponential. This is illustrated by the values of the normalized second cumulant shown in Fig. 2 a for the PS 950 cyclo-pentane (20.8°C) system.

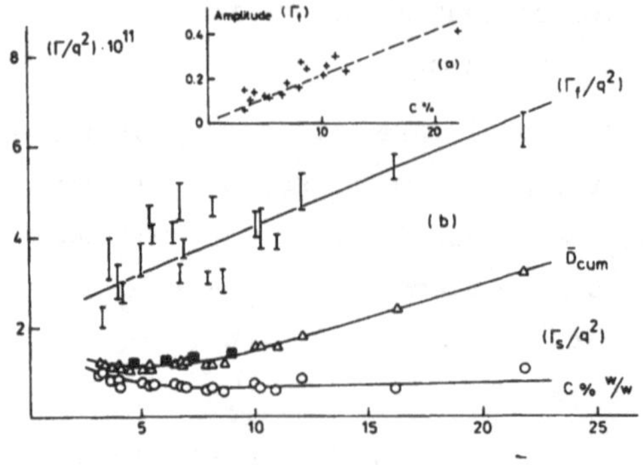

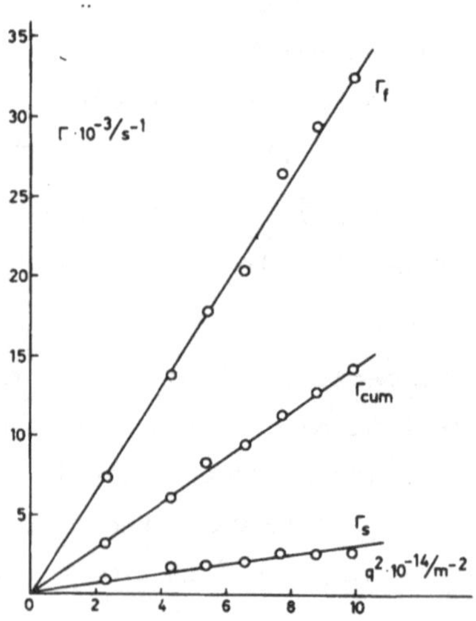

Fig.7. PS 950/cyclopentane (20.8°C) Fast (I) and slow (+) components obtained using bimodal analysis. $\bar{D}_{cum}$ (Δ) refers to cumulant analysis; values from classical gradient diffusion ($\bar{D}_{CGD}$, ■). Measurements at $\Theta=40°$; $q \cdot R_G = 0.63$. Insert: Relative intensity amplitude of the fast component obtained using Eq. (1).

Fig.8. PS 950/cyclopentane, 20.8°C (C=7.8% w/w) q^2-dependence of fast and slow modes using bimodal analysis and cumulant analysis.

$\bar{D}_{cum}$ initially decreases and then increases monotonically with concentration as usually observed[4] at <u>theta</u> conditions over extended ranges of concentration. The shallow minimum observed in Fig.7 b is thus viewed as arising from the increased weighting of the cooperative motions compared to the centre of mass motions above C^*.

The correlation function may, however, be readily resolved using bimodal analysis into two components as shown in Fig.7. One may ascribe the fast component to the developing pseudogel. It follows the relationship:

$$D_C \sim C^{0.58}$$

contrasting with a limiting scaling-law prediction[1] of unity for the exponent. The slow component, however, has a concentration dependence with the same qualitative character as that found for the polymer self-diffusion coefficient (i.e. proportional to the inverse friction coefficient). It will be related to the translational motions of entangled polymer coils in the semidilute region. The resolution into the two exponents is, in general, poor as reflected in the scatter in Fig.7, although the statistical fit for the individual runs as given by the Q-factor was good. It may be noted that for a low molecular weight sample ($M=1\times10^5$) the autocorrelation function is single-exponential; this is assumed to also be the case at very high M. Only for intermediate M is it possible to make observations of both components on the time scale of the usual experiment, the slow mode becoming inaccessible at extreme values of M. To a first approximation, both modes are linearly q^2-dependent (Fig.8). Brochard[16] has predicted bimodal spectra in Θ-solvents but with a q-independent relaxation for the slow mode, and there is some support[5] for this with very high molecular weight samples.

The relative amplitude of the pseudogel component increases approximately linearly with C (Fig.7 a, insert) but the hydrodynamic mode remains the dominant feature at all concentrations for this molecular weight (950,000) at the <u>theta</u> temperature. With increasing temperature the amplitude of the fast mode increases monotonically at the expense of the slow; see Fig.9.

Classical gradient results ($\bar{D}_{CGD}$) fall into line with the $\bar{D}_{cum}/\langle D\rangle$ -

average values from correlation function analysis, as shown in Fig.7. Values of $\bar{D}_{CGD}$ were strongly molecular weight dependent, as they must be if the hydrodynamic mode is an important component.

Conclusions

The main point presented is that for a given system there will be a window of molecular weight in which dynamic light scattering measurements detect contributions to the dynamic structure factor from two complementary modes in semidilute solution. The transition from a good, via marginal to a theta solvent involves a progressive change in the weighting of the two components, which reflect on the one hand collective motions of the partly-formed pseudogel and on the other the centre of mass translational motions. In good solvents the former dominate almost exclusively while the translational mode dominates in a theta solvent. Even at concentrations well in excess of C^*, both modes are detectable and the pseudogel develops much more gradually as a function of concentration under theta conditions.

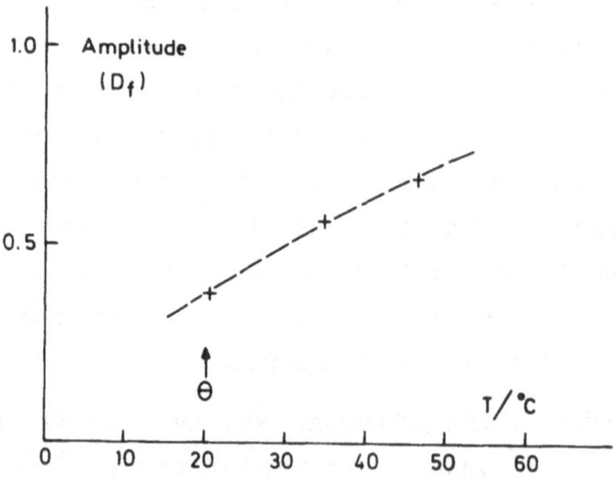

Fig.9. Relative amplitude of the fast mode as a function of temperature (PS 950/cyclopentane; angle θ=40°).

Thus a too rigid view of the C^* -value above which one can only observe the transient gel is unwarranted at intermediate molecular weights.

Marginal solvents show features akin to those of theta solvents in that both are typified by bimodal correlation functions. The present data suggest that there is limited value in the concept of the "marginal" solvent, the differences being of degree rather than kind.

The bimodal character of the correlation function explains why both $\bar{D}_{CGD}$ and $\bar{D}_{cum}$ are molecular weight dependent in poor solvents. Furthermore, the non-identity of D_C and $\bar{D}_{cum}$ for such intermediate molecular weights makes the latter a less suitable quantity for testing theoretical models.

1. de Gennes, P.G.: Macromolecules, 9, 587, 594 (1976).
2. Chu, B., Nose, I.: Macromolecules 13, 122 (1980).
3. Nose, I., Chu, B.: ibid, 12, 590, 599 (1979).
4. Munch, J.-P., Hild, G., Candau, S.J.: Macromolecules, 16, 71 (1983).
5. Adam, M., Delsanti, M., J. Phys. (Paris) Lett. 45, L-279 (1984).
6. Yu, T.L., Reihanian, H., Jamieson, A.M.: Macromolecules, 13, 1590 (1980).
7. Selser, J.C.: J. Chem. Phys. 79(2), 1044 (1983).
8. Koppel, D.E.: J. Chem. Phys. 57, 4814 (1972).
9. Amis, E., Han, C.C.: Polymer 23, 1042 (1982).
10. Hwang, D.-H., Cohen, C.: Macromolecules (in press).
11. Brown, W.: Macromolecules, 17, 66 (1984).
12. Mathiez, P., Mouttet, C., Weisbuch, G.: J. Phys. (Paris) 41, 519 (1980).
13. Munch, J.P., Candau, S., Herz, J., Hild, G.: J. Phys. (Paris) 38, 971 (1977).
14. Carlfors, J., Rymdén, R., Nyström, B.: Polymer, 24, 263 (1983).
15. Schaefer, D.W., Joanny, J.F., Pincus, P.: Macromolecules 13, 1280 (1980)
16. Brochard, F.: J. Phys. (Paris), 44, 39 (1983).
17. see:"Bimodal Analysis of QELS Data" Johnsen, R.; this volume.

DYNAMIC PROPERTIES OF ENTANGLED POLYMER SOLUTIONS BY FORCED RAYLEIGH LIGHT SCATTERING

Marie Françoise Millet, Hubert Hervet, Liliane Léger
Laboratoire de Physique de la Matière Condensée, Collège de France
11 place Marcelin-Berthelot, 75231 Paris Cedex 05, France

Introduction

Forced Rayleigh light scattering (FRS) is a now well-known technique to follow self diffusion of entangled polymer chains (1, 2, 3, 4). Previous measurements, using high-molecular-weight polystyrene samples, have shown good agreement between measurements (2) and predictions from the reptation model (5). However, this model assumes that each chain is confined by its neighbors in a virtual tube that is essentially fixed in space and only slightly affected by the motion of all the surrounding chains themselves (6, 7). We have used the particular flexibility of the FRS technique to test this point in detail, measuring the self-diffusion coefficient of polystyrene chains as a function of their polymerization index N, the polymerization index of the unlabeled surrounding chains, P, and the overall polymer concentration C.

Results

In Fig. 1, we have reported, in log scales, the variations of the self-diffusion coefficient, D_{self}, as a function of the concentration for different values of N and P. We do observe a decrease of D_{self} when P is increased, even far from the cross-over concentration from dilute to entangled solutions. For each N and P value, our data are compatible with a law $D_s \sim c \times 2^{-1.75 \pm 0.1}$. We have reported in Fig.2 the variation of D_{self}, at a given concentration, with the ratio P/N. D_{self} decreases and then saturates for P/N > 5. In Fig. 3, we have reported the saturated value $D_{selfsat}$ as a function of N. Our results are compatible with $D_{selfsat} \sim N^{-2 \pm 0.1}$.

Physical Optics of Dynamic Phenomena and Processes in Macromolecular Systems

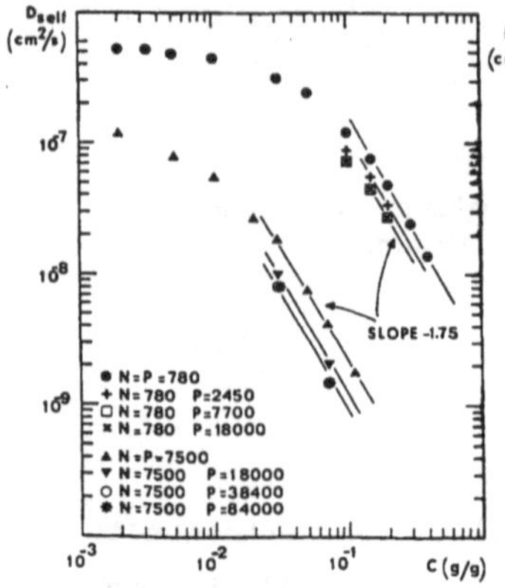

Figure 1

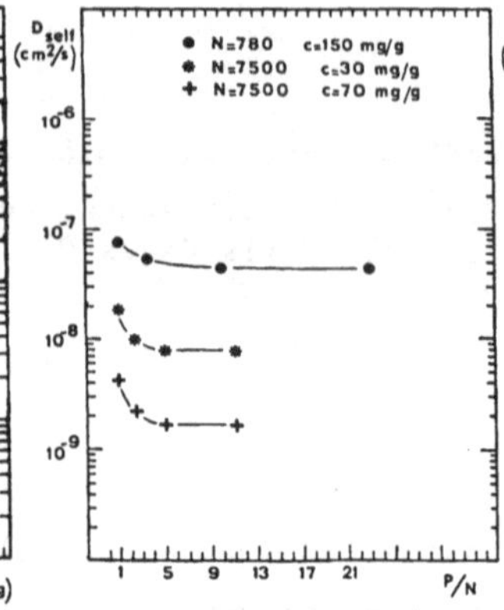

Figure 2

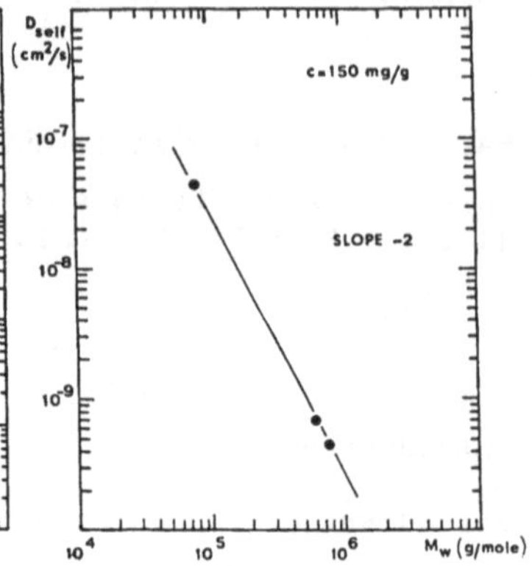

Figure 3

Figure 1 : Variations of D_{self} as a function of the concentration for different values of N and P

Figure 2 : Variations of D_{self}, at given concentrations, with the ratio P/N

Figure 3 : Variations of $D_{self\ (saturation)}$ as a function of the molecular weight of the labeled chain

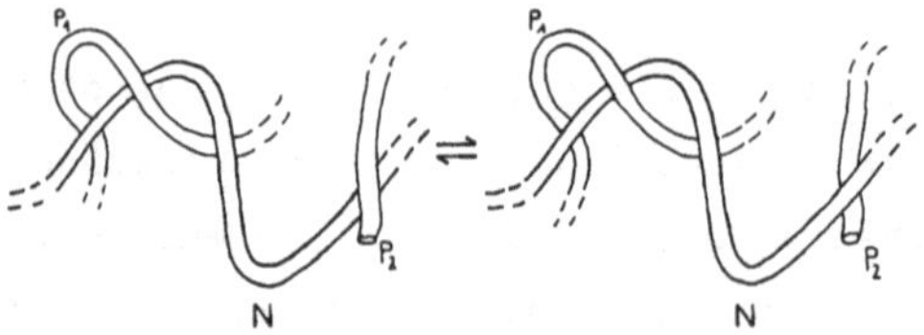

Figure 4

Figure 4 : The chain end P_2 forms a "switching entanglement", by jumping to a new position relative to the N-mer chain. The chain P_1 does not participate in this process.

Discussion

It is tempting to interpret our results in the frame of the reptation model (5). In its former derivation, one long polymer chain is assumed to move among fixed obstacles. The effect of those topological constraints on the Brownian motion of the chain is accounted for by confining the test chain in a virtual tube, enveloping all the obstacles that directly surround the chain. In a real entangled solution, however, all the chains can move. Daoud and de Gennes (6) and Klein (7) have examined this point for polymer melts assuming that all the chains are moving by reptation. They conclude that the motion of the surrounding chains should have only a negligible effect on the test chain. This can be understood qualitatively from the fact that the only way for a surrounding chain to relax a topological constraint is to have its extremity form a "switching entanglement" (see Fig. 4). In a melt of long enough chains, only a negligible amount of such events occur during one reptation time of the test chain. In a semi-dilute solution, however, as the number of entanglements of one chain is $(c/c^*)^{5/4}$ (with c^* the cross-over concentration from dilute to semi-dilute) and may not be very large compared to 1, a single switching entanglement may have an important effect on D_{self}. The arguments of Daoud and de Gennes can be transposed to entangled solutions (9), and lead to a self-diffusion coefficient :

$$D_{self} = D_{rept} \; (1 + g^2 N/P^3) \qquad (1)$$

where g is the number of momomers in one correlation length of the semi-dilute solution (8) ($g \sim c^{-5/4}$ in good solvent conditions). For large enough P, the correction due to tube renewal becomes negligible and our data $D_{selfsat}$ represent a pure reptation limit. The difference $D_{self} - D_{sat}$ should scale like P^{-3}. More experiments are under way to test both concentration and molecular weight dependences predicted by eq. (1) .

Conclusion

Using the FRS technique, and varying the polymerization index of the labeled chain (N) with respect to the unlabeled one (P), we have put into evidence

1) a limit P/N > 5, where pure reptative motion is observed ;
2) a regime P/N < 5, where tube renewal gives an important contribution to the motion of the chains.

Our results seem qualitatively in agreement with a transposition to entangled polymer solutions of the evaluation of tube renewal in melts of Daoud and de Gennes. Further experiments testing this point in more detail, as well as the derivation of Eq. (1), will be presented in a forthcoming paper .

References

1. Hervet, H., Léger, L., Rondelez, F. : Phys. Rev. Lett. 42, 1681 (1979).
2. Léger, L., Hervet, H., Rondelez, F. : Macromolecules 14, 1732 (1981).
3. Coutadin, J., Sillescu, M., Voelkel, R. : Makromol. Chem., Rapid Commun. 3, 649 (1982).
4. Wesson, J., Takezoe, H., Yu, H., Chen, S.J. : J. App. Phys. 53, 6513 (1982).
5. de Gennes, P.G. : J. Chem. Phys. 55, 572 (1971).
6. Daoud, M., de Gennes, P.G. : J. Polymer Sci. Polymer Phys. Ed. 17, 1971 (1979).
7. Klein, J. : Macromolecules 11, 852 (1978).
8. de Gennes, P.G. : Scaling Concepts in Polymer Physics, Cornell University Press, Ithaca, N.Y. (1979).
9. Millet, M.F., Léger, L. : to be published.

DYNAMICS OF POLYELECTROLYTES IN AQUEOUS SOLUTIONS PROBED BY ELECTRO-OPTICAL RELAXATION

S. Wijmenga, F. van der Touw and M. Mandel*
Dept. of Physical and Macromolecular Chemistry, Gorlaeus Laboratories, University of Leiden, P.O.Box 9502, 2300 RA Leiden, the Netherlands.

Introduction

This is a preliminary report on an electric birefringence study of aqueous solutions of Na-poly(styrene sulfonate) (Na-PSS) in the presence of 0.01 M NaCl. The full details of this investigation, which includes the experimental determination of the Kerr constant and the birefringence decay or relaxation time as function of macromolecular concentration, molar mass and electric field frequency, will be published elsewhere (1). Here we shall present only some of the results concerning the molar mass and macromolecular concentration dependence of the birefringence decay observed after application of electric field pulses of rectangular shape. These results are, in a certain sense, complementary to those obtained in a study of the quasi-elastic light scattering by analogous systems in this laboratory (2). The latter have been shown to be in reasonable agreement with recent predictions, partially based on a scaling approach, of polyelectrolyte theory (3). In particular it was

Physical Optics of Dynamic Phenomena and Processes in Macromolecular Systems

demonstrated that the difference in polyelectrolyte behaviour between dilute and semi-dilute regime can be observed experimentally, and that for flexible chains of sufficient length in a solution with excess low molar mass electrolyte, the polyelectrolyte will be characterized by a coillike behaviour in the dilute regime in contrast to the semi-dilute regime where the entanglements between the chains play a dominant role. It is the purpose of this report to show that the results of the electric birefringence decay give additional confirmation to these findings.

Experimental

The Na-PSS samples investigated were of commercial origin (Pressure Chemical Co) and had been prepared by sulfonation of nearly homodisperse poly(styrene). According to the manufacturer's specification M_w/M_n is smaller than 1.1 and the degree of sulfonation close to unity. The molar mass (M_w) of the five different samples used extended from 1.8×10^5 to 10.3×10^5 g mol^{-1}. The materials as purchased were exhaustively dialyzed against deionized water and then freeze-dried. All solutions were prepared with deionized water and with NaCl of analytical grade (Merck). The salt concentration of all solutions was 0.01 M. The polyelectrolyte concentration of every solution investigated was determined spectrophotometrically at 261 nm. The values of the specific extinction coefficients for the different samples of Na-PSS have been determined separately by calibration with Beer's law using corrections for the water content of NaPSS. The birefringence measurements equipment is of standard composition (5). The birefringence Δn, defined as the difference between the refractive index in a direction parallel and that in a direction perpendicular to the direction of the applied field, is determined from the relative change of the intensity of a polarized laser beam emerging from the solution investigated (and having passed through a quarter-wave plate and an analyser in a fixed orientation) in the presence of the electric field with respect to the intensity in the absence of that field (when the solution is isotropic and $\Delta n=0$). The electric field pulses were obtained by applying voltage pulses produced by a pulse generator and amplifier (Cober Electronics) to the parallel platinum electrodes of a Kerr cell with accurately known spacing (of the order of 1.7 mm). The direction of the laser beam (from a 5mW He-Ne laser from Spectra Physics) which probes the birefringence of the solution, is parallel to the electrodes and perpendicular to the direction of the electric field produced. The Kerr cell is kept at 21.0°C with the help of water from a thermostat circulating in the cell jacket. The electric field was determined from the voltage measured at the cell electrodes and their known spacing, assuming a linear voltage drop across the solution and applying no corrections for electrode effects. The digitized values of both the birefringence and the applied field were stored into the memory of a Biomation transient recorder and fed into an on-line computer (PDP 11) for storage of the results on disc and analysis of the signals. The change of the birefringence of the solution with time, $\Delta n(t)$, can be followed during the duration of the field pulse and after the latter has been cut off (Fig.1.) The duration of the pulse for each solution was chosen in such a way that the birefringence reached a constant, stationary value Δn_s. The field strengths applied could be varied. For polyelectrolyte concentrations $C < 0.1$ g dm^{-3} they never exceeded 2.6 kV cm^{-1}. For higher concentrations the conductance of the solutions increased considerably and the maximum field strength applied had to be reduced to prevent

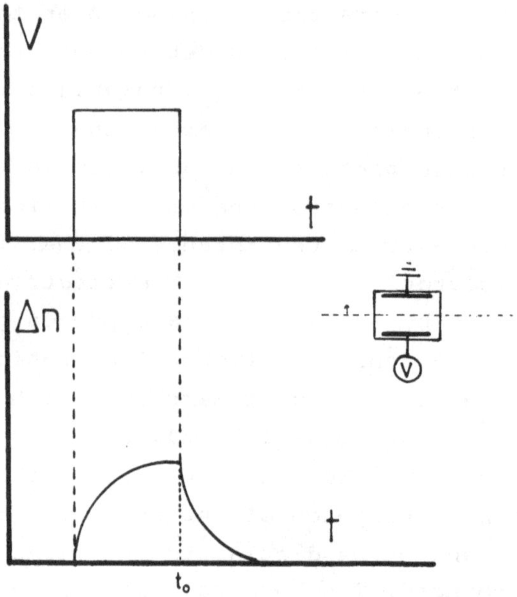

Fig.1. Change of the birefringence during and after the application of electric field or potential pulse.

heating of the solutions. At the highest concentrations ($C < 10$ g dm^{-3}) this maximum came close to 0.1 kV cm^{-1}. None of the solutions investigated did show any dependence of the decay curves on the electric field strength within the range explored.

We shall present here and discuss the decay of the birefringence measured after disappearance of the electric field. The moment at which the field is cut down is arbitrarily put equal to t=0. The results are presented as normalized birefringence curves, $\Delta n_n(t)=\Delta n(t)/\Delta n_s(t>0)$. These normalized decay curves, after correction if necessary for the contribution of water birefringence and the finite response time of the photodetection system, have for all systems investigated been found to be definitely not simple exponentials. As an example an typical normalized decay curve is represented in Fig.2 in a semi-logarithmic plot. All solutions show decay curves exhibiting the same trend characterized by deviations from a linear semi-logarithmic plot at the lower end of the time scale. Decay curves which cannot be represented by a simple exponential with a single relaxation time are usually approached as a superposition of

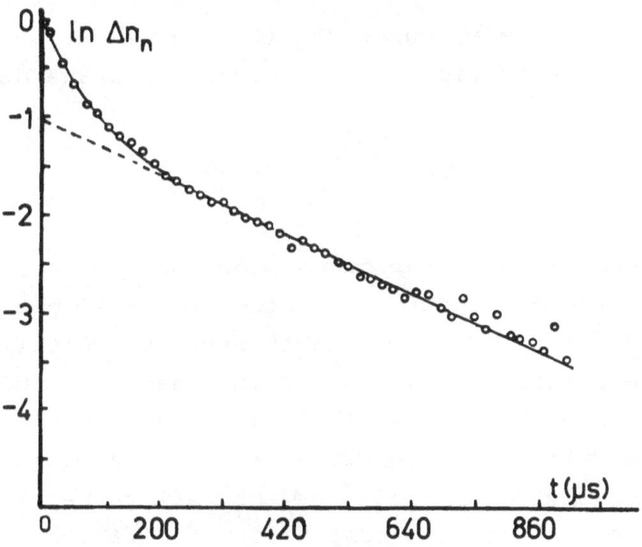

Fig.2. Change of the normal birefringence with time (t>0) in a semi-logarithmic plot (M_w=1,030,000 g/mol, C=0.38 g/l, dilute regime)

several exponential functions

$$\Delta n_n(t) = \sum_{(k)} a_k \exp(-t/\tau_k) \tag{1}$$

where a_k is the relative contribution of the exponential with relaxation time τ_k. Because of normalization we have

$$\sum_{(k)} a_k = 1 \tag{2}$$

provided there is no parasitic baseline. In principle it is possible to fit the experimental normalized birefringence decay curve to an equation such as 1, using a non-linear least squares method. In practice it becomes quite unrealistic to try a fit with more than three superposed exponentials (6 parameters, or 7 if a floating baseline is used).

Another method often used to characterize non single exponential decay or relaxation curves, is to determine the integral under the decay curve Δn_n (t). In as far as eqn.1 is valid, this integration would yield the average decay time $\bar{\tau}$ defined by

$$\bar{\tau} = \sum_{(k)} a_k \tau_k \quad (3)$$

In general there is no unambiguous proof of the validity of eqn.1 in case the decay curve deviates from a simple exponential function. Sometimes a theoretical justification for it may be available, such as is the case if a mode analysis for the motion of a flexible Gaussian chain is assumed to be valid (6). Therefore the physical meaning of the experimentally determined $\bar{\tau}$ may not always be simple. Nevertheless a quantitative comparison of different decay curves through their respective $\bar{\tau}$ values remains always possible.

Results and discussion

It has been found that for all systems investigated, the decay curves could be fitted with a superposition of three exponentials. The relative contribution of the longest relaxation time τ_1 was in all case approximately the same and of the order of 40%, independent of molar mass or concentration. It was found that under the experimental conditions used the ratio $\tau_1/\bar{\tau}$ is practically constant and of the order of 1.7 (Fig.3.).

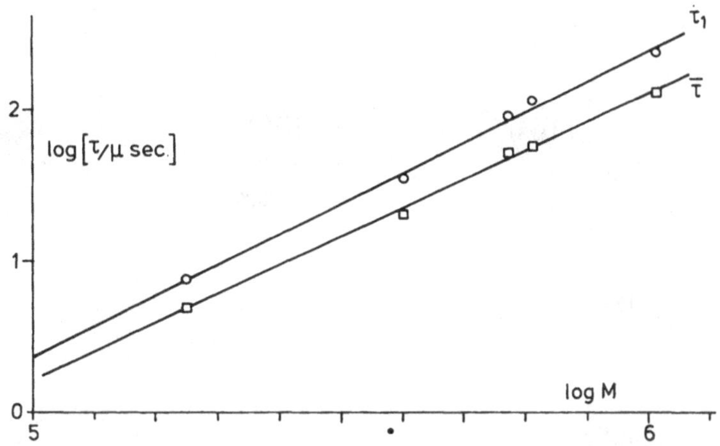

Fig.3. Comparison of the molar mass dependence for τ_1 and $\bar{\tau}$ at the lowest concentration investigated.

Therefore the concentration and molar mass dependence of τ_1 and $\bar{\tau}$ should be the same.
In Fig. 4 the values of $\bar{\tau}$ are compared for a large number of different systems. It can be seen that for solutions containing the same Na-PSS sample, the change of $\bar{\tau}$ with concentration is roughly spoken different at low and at high concentrations. At the low concentration end the value of $\bar{\tau}$ is nearly constant, for higher concentrations $\bar{\tau}$ definitely increases with concentration, the increase being the more pronounced the larger M_w. There is a striking resemblance between the observed concentration dependence of $\bar{\tau}$ and the concentration dependence which has been found for the effective translational diffusion coefficient D_{eff} as determined with the help of quasi-elastic light scattering using analogous systems (2). This is clearly demonstrated in Figs. 5 and 6 where the results for $\bar{\tau}$ and D_{eff} have been compared for two different samples of Na-PSS

in 0.01 M NaCl. Both parameters can be seen to be nearly concentration independent or to increase considerably with concentration in approximately the same concentration ranges.

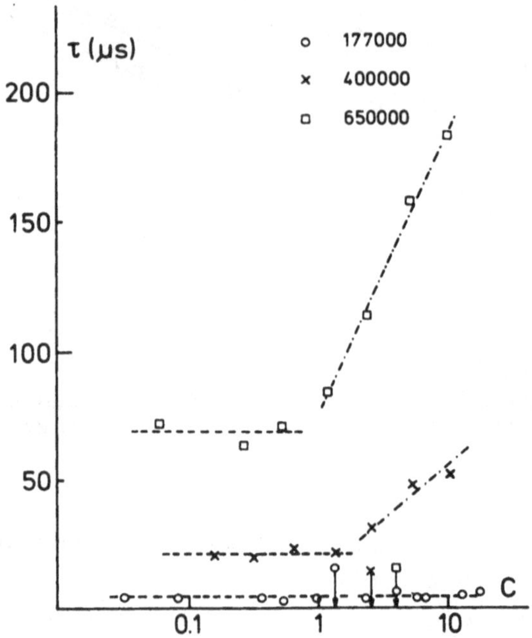

Fig.4. Change of average decay time $\bar{\tau}$ with concentration for three samples of Na-PSS of different molar mass; concentration expressed in g/l on a logarithmic scale.

For D_{eff} it has been shown (2) that the results are consistent with the theoretical analysis in which a distinction is made between the dilute concentration and the semi-dilute concentration regime, the former corresponding to $C \ll C^*$, the latter to $C \gg C^*$. The critical concentration C^* (expressed in mass per unit volume) is, according to the scaling approach (4), given by

$$C^* = M_m (\ell/A)\, R_F^{-3} \qquad (4)$$

Here M_m is the mass of a monomeric unit, ℓ the contour length of the macromolecular chain, A the length of a monomeric unit and R_F the Flory radius, representing the root mean square radius of gyration of a non-ideal chain with the limiting expression ($\ell \to \infty$) for the value of the expansion factor α_s. In the wormlike chain approach to charged macromolecule (7,8), based on the assumption that interactions between charges fixed on a polymer are characterized by a potential of mean force corresponding to the Debye-Hückel screened Coulomb potential, the Flory radius is approximately given by the following expression (3)

$$R_F \cong \ell^{3/5} (L_t/\kappa)^{1/5} \tag{5}$$

where κ is the Debye screening parameter and L_t the total persistence length of the polyelectrolyte. The former is defined by

$$\kappa^2 = 8\pi \, Q \, I \tag{6}$$

with $Q \equiv q^2/4\pi\varepsilon_o \varepsilon kT$ representing the Bjerrum length, q being the value of the elementary charge, ε and ε_o the relative permittivity of the solvent and the absolute permittivity of free space respectively, k the Boltzmann constant, T the temperature, and I the ionic strength of the solution. For the persistence length it has been assumed that it consists of two independent contributions (7,8): the bare or intrinsic persistence length L_p of the chain in the absence of charge interactions and the contribution L_e arising from electrostatic interactions. In the presence of an excess of low molar mass electrolyte, L_e can be expressed as follows if it is assumed that each monomeric unit carries an identical elementary charge (as is the case for PSS):

$$L_e = (Q/4\kappa^2A^2)f^2 \qquad (7)$$

Here f is a factor which is introduced to take into account the possibility that some polyelectrolyte-specific screening may occur reducing the true linear charge density of the chain (proportional to A^{-1}) to an effective value through "association" or "condensation" of counterions to the surface of the macromolecular chain (9). It is accepted that f will satisfy the following condition, $A/Q \leqslant f \leqslant 1$, where the lower limit A/Q corresponds to the predictions of the simple condensation theory in the case of monovalent counterions (10,11) and the upper limit to the case where no such specific screening is expected (as will be the case for $Q/A < 1$). It follows from eqn.7 that L_e, and thus also the total persistence length L_t, will increase with increasing charge density (possibly up to a certain limiting level), and with decreasing ionic strength.
In principle it is thus possible to estimate C^* as a function of molar mass and ionic strength, provided A is known (for PSS it may be put equal to 0.25 nm which is smaller than the value of $Q \cong 7$ nm at room temperature) and some choice is made for the value of f. In Figs.5 and 6 the theoretical values of C^*, calculated assuming $f=A/Q$, and $L_p=10$nm, have been indicated by arrows on the concentration scales. They seem to fall into the concentration range in which a transition in concentration dependence of either D_{eff} or $\bar{\tau}$ is observed.
From the similarity between the behaviour of both quantities it may be concluded that the experimentally determined electric birefringence decay time is also consistent with the distinction between behaviour in the dilute and behaviour in the semi-dilute concentration regime. From this it may be inferred that in the former regime the decay times

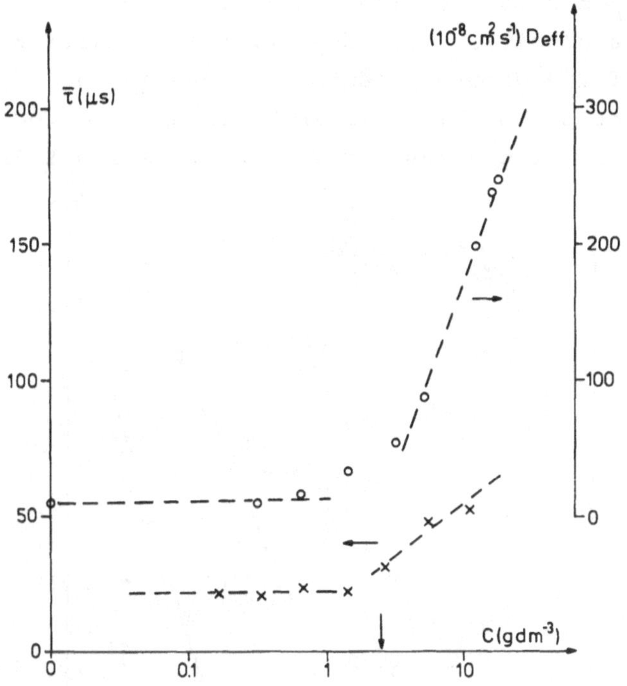

Fig.5. Comparison between the concentration dependence of the average decay time $\bar{\tau}$ (x) and of the effective translation diffusion coefficient D (0) determined from quasi-elastic light scattering (2) for Na-PSS (M_w=400,000 g/mol): vertical arrow gives position of the theoretical value of C^* on the logarithmic concentration scale.

observed should be related to the behaviour of individual macromolecules. Without going into the details of the mechanism which is responsible for the appearance of the birefringence in the case of polyelectrolyte solutions - a mechanism which remains difficult to be fully understood (1) - it is reasonable to assume that in the dilute regime the decay of this birefringence is related to the motions of the macromolecular chain after some orientational constraints imposed by the electric field have disappeared. A prerequisite condition is, however, that the lifetime of at least one of the dipole moments involved in the establishment of the birefringence, is much longer than the characteristic time necessary for the rotational diffusion

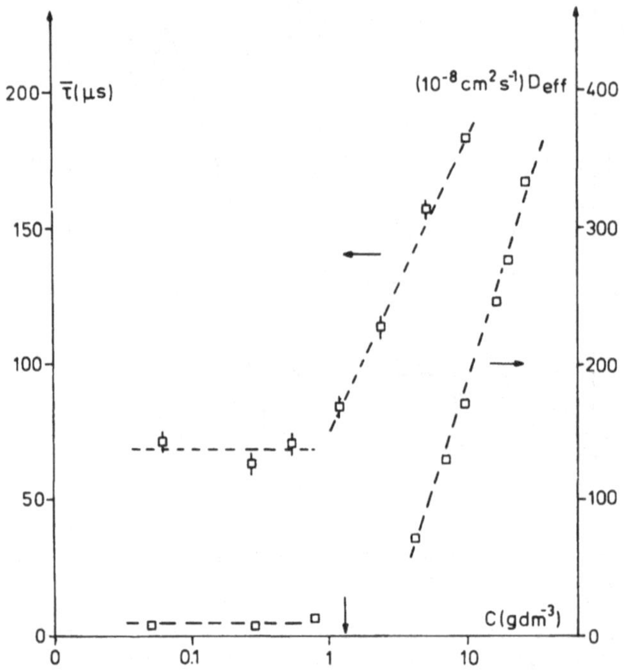

Fig.6. Comparison between the concentration dependence of $\bar{\tau}$ (⌶) and D (□) from quasi-elastic light scattering (2) for Na-PSS (M_w=650,000 g/mol); vertical arrow as in Fig.5.

of the macromolecule, a condition which seems to be satisfied in the case of the polyelectrolytes investigated (1). Furthermore, for the macromolecule at fixed center of mass and fixed end-to-end vector, this dipole moment should have a component parallel to this vector.

Under such conditions, the longest relaxation time of the birefringence decay τ_1 is expected to be inversely proportional to the rotational diffusion coefficient D_R^0 in the dilute regime and, using the Kirkwood-Riseman approximation (12), therefore also to be proportional to the cube of the r.m.s. radius of gyration (4). Within the framework of the scaling approach R_F may be used instead of the r.m.s. radius of gyration leading to the following relation for the mean decay time in the dilute regime:

$$\bar{\tau} \sim \tau_1 \sim \ell^{9/5} (L_t/\kappa)^{3/5} \qquad C << C^* \qquad (8)$$

From a linear least squares fit for log $\bar{\tau}$ as function of log M_w an exponent 1.98 ± 0.05 is found for ℓ , which is in reasonable agreement with the theoretical value 1.8 according to eqn.8. Small deviations between experimentally found exponents and theoretical predictions are not uncommon, also for uncharged macromolecules. They do not invalidate qualitatively the basic assumptions of the theoretical model used. Furthermore, in the present case, eqn.8 was tested using the average value of $\bar{\tau}$ obtained at the lowest concentrations investigated for each PSS sample which still may contain a small concentration dependent contribution. For the semi-dilute regime the interpretation of the mean birefringence decay is less simple. It should be noted that there is an essential difference in the behaviour observed by quasi-elastic light scattering and by electric birefringence decay in this regime. For the former D_{eff} has been found to be molar mass independent (2) in agreement with the theoretical predictions. They state (4) that in the semi-dilute regime a cooperative diffusion coefficient is probed by the light scattering experiment which is determined by the molar mass independent average correlation length ξ of the dynamic network arising through the chain entanglements. In contrast the birefringence decay time in the same regime definitely shows a considerable molar mass dependence (see Fig.2). This is related to the fact that in electric birefringence decay measurements not translational but rather more complicated chain motions are probably probed. In the semi-dilute regime the relaxation of the chain from its perturbed conformational state, determined by the application of an electric field, to unperturbed conditions is possibility related to a reptation mechanism as introduced by Edwards (13) for a macromolecular chain in a network or in a concentrated solution. In the reptation model a characteristic time is the reptation time τ_r, which is the average time required for a complete renewal of the chain configuration (it being assumed that after one time interval τ_r all memory of the original conformation has been lost). A plausible assumption is therefore to consider $\bar{\tau}$ to be of the order τ_r.

For uncharged macromolecules de Gennes (4) has derived a scaling relation for the reptation time

$$\tau_r \cong (N/g)^3 \; \xi^3 (6\pi\eta_s/kT) \qquad C \gg C^* \qquad (9)$$

Here N is the number of statistical elements in the chain, g the average number of statistical elements in a "blob" of size ξ and η_s the viscosity of the solvent. Using eqn.9 together with the expressions for ξ and g as derived by Odijk (3) for wormlike polyelectrolytes in the semi-dilute regime, the following relation can be found between the reptation time, molar mass or contour length and concentration, in the case of a polyelectrolyte solution in the presence of an excess low molar mass electrolyte,

$$\tau_r \sim \ell^3 \; A^{3/2} (L_t/\kappa)^{3/2} C^{3/2} \qquad C \gg C^* \qquad (10)$$

where the constant numerical factor has been omitted. If this theoretical expression is valid and the condition $\bar{\tau} \cong \tau_r$ is satisfied, one expects the average birefringence decay time in the semi-dilute regime to increase with

concentration and molar mass as power laws with exponents 1.5 and 3 respectively.
In comparing this theoretical prediction to experimental results, it should be kept in mind however that in the semi-dilute regime the ionic strength of the solution may be affected by the counterions arising from the dissociation of the polyelectrolyte molecules themselves (2). For a mono-valent-monovalent salt of concentration c_s (in molecules per unit volume), the ionic strength of the polyelectrolyte solution is given by

$$I = (c_s + cf/2) \qquad C >> C^* \qquad (11)$$

c being the concentration of the polyelectrolyte in monomeric units per unit volume and f the factor introduced previously to take eventually into account "association". It follows that I is now to some extent polyelectrolyte concentration dependent, so that κ and L_t will no longer be constants in the semi-dilute regime, even if c_s is kept constant. Therefore in order to check the concentration dependence of $\bar{\tau}$ according to the theoretical eqn.10 it is easier to correct for this ionic strength variation by introducing the auxiliarly quantity B_b defined as follows

$$B_b \equiv \bar{\tau}\,(L_t/\kappa)^{-3/2} \sim \ell^3\, C^{3/2} \qquad C >> C^* \qquad (12)$$

In Fig.7 plots of $B_b{}^{1/3}$ vs $C^{1/2}$ for different molar masses are represented, L_e being calculated assuming f=A/Q. A linear relationship in agreement with (12) is found. However the slope of the different lines do not depend linearly on ℓ as expected from eqn.12. In fact $B_b{}^{1/3}$ depends on a power of M which is smaller than unity and comes close to 0.7. Therefore it may be concluded that the agreement between the reptation model and the experiments is only semi-quantitative, the molar mass dependence, though still rather strong, being less than theoretically expected. It should be noted, however, that eqn.9 for reptation cannot be expected to hold under all circumstances in semi-dilute solutions. It has been shown by Klein (14) that it will be valid only if the number of "blobs" of size ξ per chain in the given network is large with respect to some critical

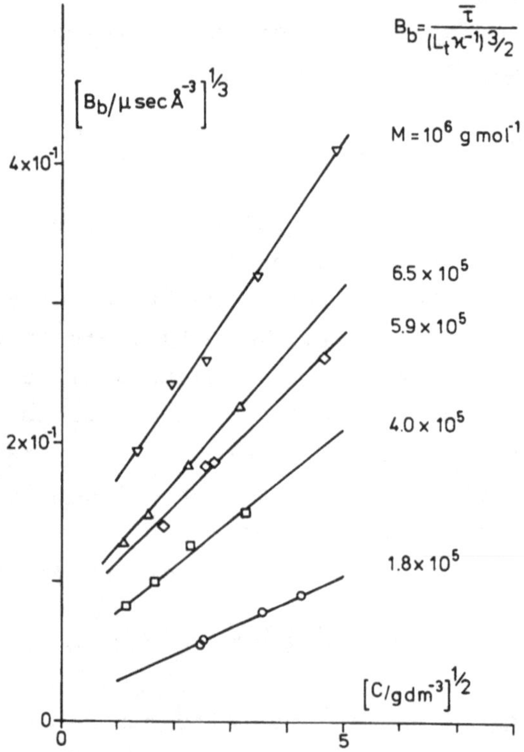

Fig.7. Change of $B_p^{1/3}$ as function of $C^{1/2}$ for different samples of Na-PSS (see text).

value depending on the nature of the chain and the concentration. With decreasing number of "blobs" per chain, the power of (N/g) in the expression for τ_r may decrease from 3 to 2. In the case of the systems investigated, where

(N/g) is relatively small (< 20), a weaker molar mass dependence of τ_r than predicted by eqns.10 or 12 can therefore not be excluded. Furthermore by using $\bar{\tau}$ as the quantity for which the C- and M-dependence is studied, we have to keep in mind that other influences than the reptationlike motion of the complete chain may have to be taken into account.
Finally the same kind of assumptions about the dipole moments responsible for the occurrence of birefringence, has to be made in the case of the semi-dilute as for the dilute solutions.

Conclusions

It can be stated that the birefringence decays of Na-PSS solutions in aqueous 0.01M NaCl are consistent with the hypothesis that increasing concentration leads to a crossing from a dilute into a semi-dilute regime. In the former the decay can be explained by the rotational diffusion of isolated chains with a molar mass dependence of the mean relaxation time which comes close to theoretically predictions. In the semi-dilute regime the concentration dependence is consistent with predictions based on a reptation model for the polyelectrolyte chain as being responsible for the disappearance of the birefringence after the cut-off of the electric field. The assumption has to be made that the average decay time is proportional to the reptation time of the entangled chains forming a dynamic network. The molar mass dependence of the mean relaxation time is definitely too weak with respect to the equation resulting for that reptation model, which assumes much longer chains and higher concentrations than used in the present investigation.

Acknowledgements

This investigation has been carried out under the auspices of The Netherlands Foundation for Chemical Research (SON) and with financial aid from the Netherlands Organization for the Advancement of Pure Research (ZWO).

References

1. Wijmenga, S., Van der Touw, F., Mandel, M., to be published.
2. Koene, R.S., Mandel, M.: Macromolecules, 16, 220-227 (1983).
3. Odijk, T.: Macromolecules, 12, 688-693 (1979).
4. de Gennes, P.G.: Scaling Concepts in Polymer Physics, Cornell University Press, Ithaca 1979.
5. Fredericq, E., Houssier, C.: Electric Dichroism and Electric Birefringence, Clarendon Press, Oxford 1973.
6. See e.g. Jernigan, R.L., Thompson, D.S.: in Molecular Electro-Optics, C.T. O'Konski editor, M. Dekker, New York 1976, Part I, Chapter 5.
7. Skolnick, J., Fixman, M.: Macromolecules, 10, 944-948 (1977).
8. Odijk, T.: J.Polym.Sci., Polym.Phys.Ed., 15, 477-483 (1977).
9. See e.g. M. Mandel in: Macromolecular Electro-Optics, S. Krause editor, Plenum Press, New York, London 1981, p.285-307
10. Oosawa, F.: Polyelectrolytes, M. Dekker, New York 1971.
11. Manning, G.S.: J.Chem.Phys., 51, 924-930, 1969.

12. a. Kirkwood, J.G., Riseman, J.,: J.Chem.Phys., 16. 563-573 (1949)
b. Riseman, J., Kirkwood, J.G.: J.Chem.Phys., 17, 442-446 (1948).
13. Edwards, S.F., Grant, J.W.: J.Phys., A 6, 1169-1186 (1973).
14. Klein, J.: Macromolecules, 11, 852-858 (1978).

DYNAMIC LIGHT SCATTERING FROM A ROD-LIKE POLYMER IN SEMI-DILUTE SOLUTION

K. H. Langley and P.S. Russo*+

Departments of Physics and Astronomy and +Polymer Science and Engineering, University of Massachusetts, Amherst, Massachusetts 01003, USA

Introduction

In a solution of rod-like molecules, one would expect that as the concentration is increased, the rods begin to interfere with the free diffusional motion of one another, reducing the translational and rotational diffusion coefficients to values less than those at infinite dilution. Doi and Edwards[1] first considered these rod-rod interactions in the semidilute regime $L^{-3} << N << (dL^2)^{-1}$ (L and d are the rod length and diameter, respectively and N is the number concentration). On the basis of a model in which the rod moves freely along its length but is completely restricted laterally, the translational diffusion coefficient was predicted to be independent of concentration and equal to approximately one-half that at infinite dilution, $D \approx D^o/2$. Similarly, the restriction of angular motion led to a prediction of the rotational diffusion coefficient $\Theta = \beta\Theta^0/(NL^3)^2$ where Θ^0 is the free (dilute solution) coefficient and $\beta^{1/2} = (NL^3)^* \simeq 10$ represents the onset of solution crowding (in rods per box of volume L^3).

We have explored this regime by dynamic light scattering analysis of isotropic solutions of poly-γ-benzyl-L-glutamate of molecular weight 150,000 and 300,000 (PBLG-150 and PBLG-300 respectively) in the good solvent N,N-dimethylformamide (DMF) at 30°C

* Present address: Department of Chemistry, Louisiana State University, Baton Rouge, Louisiana 70803., USA

Physical Optics of Dynamic Phenomena and Processes in Macromolecular Systems

over the wide concentration range $4 < NL^3 < 1500$ $(0.04 < NdL^2 < 15)$. A number of other authors have also recently examined PBLG in dichloroethane[2] and DMF[3,4]. In our experiments, for $NL^3 < 100$ a second cumulants fit adequately represents the dynamic light scattering correlation function and D may be determined either by the initial correlation function decay rate or by the initial slope of the average decay rate vs q^2. In apparent contrast to the Doi-Edwards theory the measured value of D is found to increase with polymer concentration. This increase is discussed in terms of a repulsive thermodynamic contribution to diffusion parallel to the rod axis. For $NL^3 \simeq 100$ the initial decay rate continues its increase with concentration, and a slow decay enters. At higher concentration, $NL^3 \gtrsim 300$, the correlation function is more complicated. In terms of a three-exponential analysis, there is a fast decay which vanishes at low scattering angle while the intermediate and slow decays show substantial strength at all angles. While it is still not clear why diffusion remains so rapid in semidilute solution, we suggest that bending of the imperfectly rigid molecule on a time scale short compared to diffusion may provide an appropriate mechanism.

Results

Total intensity light scattering from the PBLG samples resulted in normal Zimm plots which yielded molecular weight, radius of gyration, and second virial coefficient accurate to about 5%. For example, in the sample labelled $M = 150{,}000$ by the manufacturer we obtained $M = 179{,}000$, $R_G = 353$ Å and $B_2 = 3.1 \times 10^{-4}$ $cm^3 mol\ g^{-2}$. B_2 is within 3% of the excluded volume limit calculated[5] for a rod of 16 Å geometric diameter.

The concentration dependence of D obtained from a second order cumulants fit to data collected at $\theta = 90°$ $(qL = 4)$ in PBLG-300 is shown in Fig. 1. The large error bars indicated at the highest concentrations arise from baseline uncertainty due to

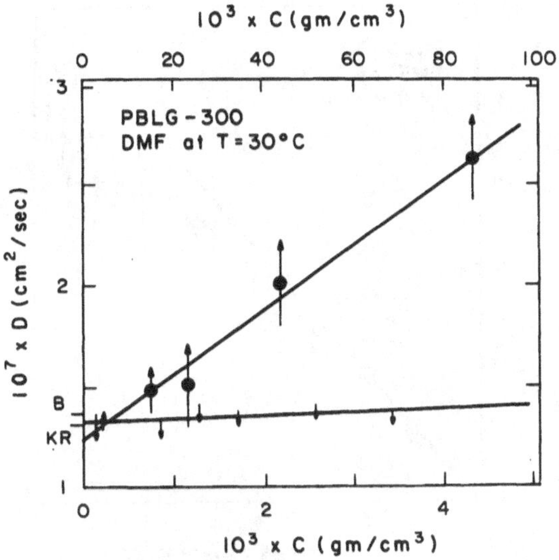

Fig. 1. Concentration dependence of D. Arrows point to the applicable abscissa scale; length is the uncertainty. B and KR are the Broersma and Kirkwood Riseman predictions.

the presence of a slow decay mode. Extrapolation to infinite dilution is within 10% of the value predicted by the Kirkwood-Riseman[6] or Broersma[7] equations. It should be emphasized that second cumulants is an excellent fit to the correlation function at concentrations up to at least 6 mg/cm^3 ($NL^3 = 110$), firmly establishing the trend of D increasing with concentration.

In Figs. 2(a)-(c) the average correlation function decay rate, $\overline{\Gamma}$, is plotted against q^2 at three concentrations: (a) $NL^3 = 4.4$; (b) $NL^3 = 260$; and (c) $NL^3 = 404$. At low concentration, Fig. 2(a), there is an upward curvature above $\theta = 90^\circ$ ($qL = 4$) probably due to intramolecular interference. With increasing concentration, Fig. 2(b) and 2(c), the initial slope of $\overline{\Gamma}$ vs q^2 is larger but there is increasing downward curvature (due to the slow mode) which means that the data in Fig. 1 probably underestimate the increase of D with concentration, At $NL^3 \lesssim 300$ these problems are negligible; either $\overline{\Gamma}/q^2$ at $\theta = 90^\circ$ or the initial slope of $\overline{\Gamma}$ vs q^2 approximates D, and the main conclusions of our experiment are well established.

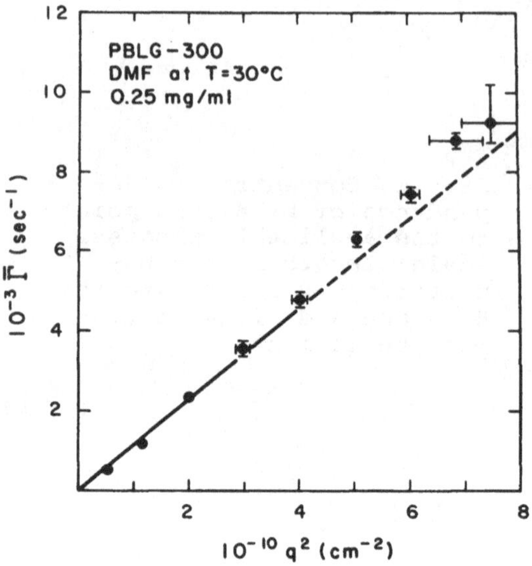

(a)

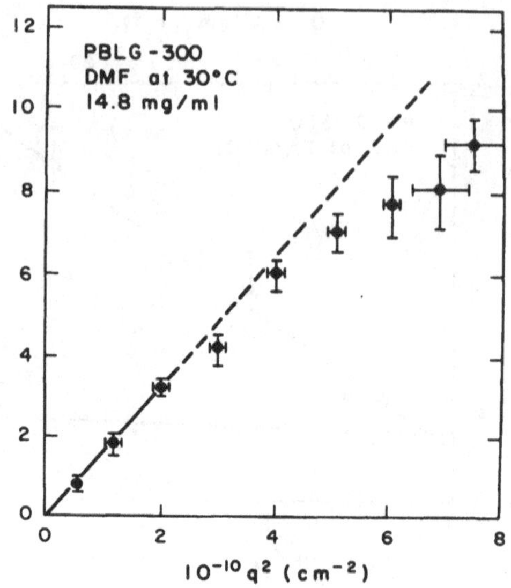

(b)

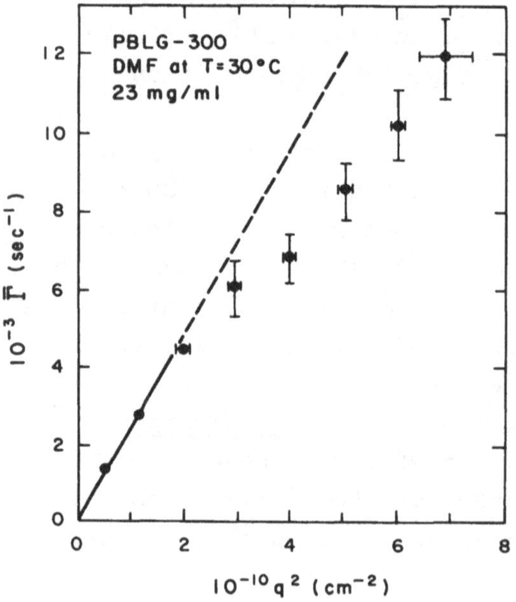

(c)

Fig. 2. $\overline{\Gamma}$ vs q^2 (second order cumulant fit) at three concentrations.

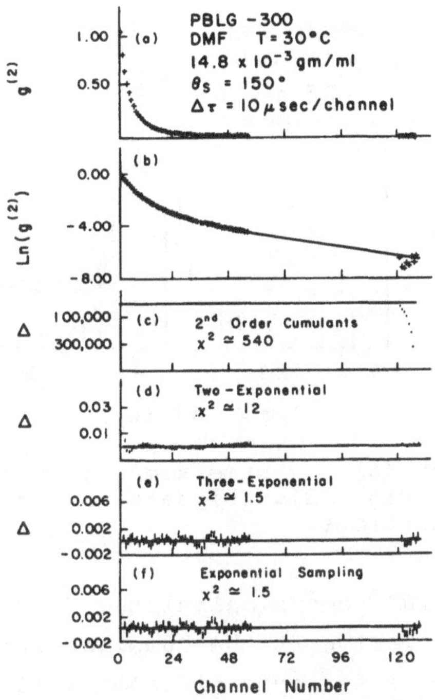

Fig. 3. (a) Typical normalized nonexponential correlation function at $NL^3 = 260$, (b) semilogarithmic plot and exponential sampling fit, (c)-(f) error of the cumulants, two-exponential, three-exponential, and exponential sampling fits.

Nonexponential correlation functions, clearly evident at high concentration and large qL, were analyzed in several different ways. Figure 3 shows a typical example at $NL^3 = 260$, $qL = 5.5$; the curvature in the logarithmic plot (Fig. 3b) displays the complex nature of the decay. The error plots reveal the inadequacy of the second order cumulant and two-exponential fits (c and d). Plots (e) and (f) emphasize the improvement of the three-exponential over the two-exponential fit and show that the data can be fitted equally well by the sum of three exponentials (e) or by the exponential sampling technique of Ostrowsky and Pike[8] (see also Fig. 4a). The only difference in solution conditions between Figs. 4(a) and 4(b) is that 4(b) is from dilute solution where a single peak (or 2 exponentials) is an excellent fit to the data. Two different analyses of the decay rate at concentration $NL^3 = 260$ vs q are shown in Figs. 5 and 6. Utilizing the three-exponential fit of Fig. 5, we identify the intermediate decay rate with translational diffusion (in spite

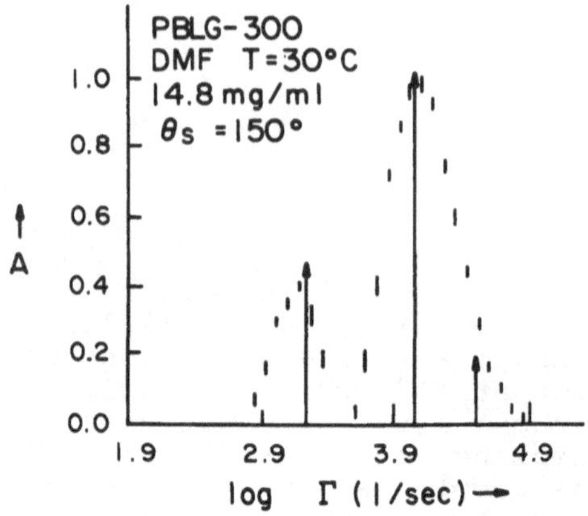

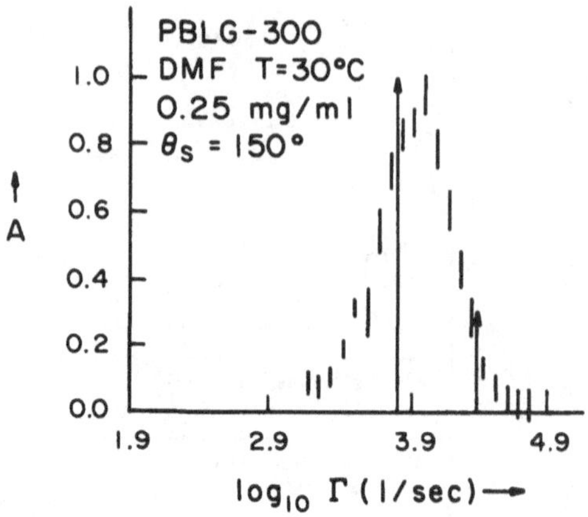

Fig. 4. Exponential sampling fits to (a) the same semidilute solution correlation function as in Fig. (3), (b) data from a dilute solution under identical conditions.

of the slope of 1.64) as $D = 1.57 \times 10^{-7}$ cm^2/s calculated from the initial slope of $\overline{\Gamma}$ vs q^2 agrees well with the cumulant fit value at $\theta = 90^{\circ}$. We note that the fastest mode disappears at smaller angles. The two average decay rates from the exponential sampling analysis (see Fig. 4a) shown in Fig. 6, show similar clear trends. The large error bars approximate the full width (at half maximum) of the decay rate distributions which are inherently rather wide.

Discussion and Conclusions

No comprehensive calculation simultaneously accounts for all the complexities of this system: thermodynamic polymer-solvent interaction, polydispersity, rod flexibility and coupled rotation-translational diffusion. We therefore restrict our discussion here principally to the more dilute regime where correlation functions are characterized by a single decay which presumably represents simple mutual diffusion, complicated only by thermodynamic effects.

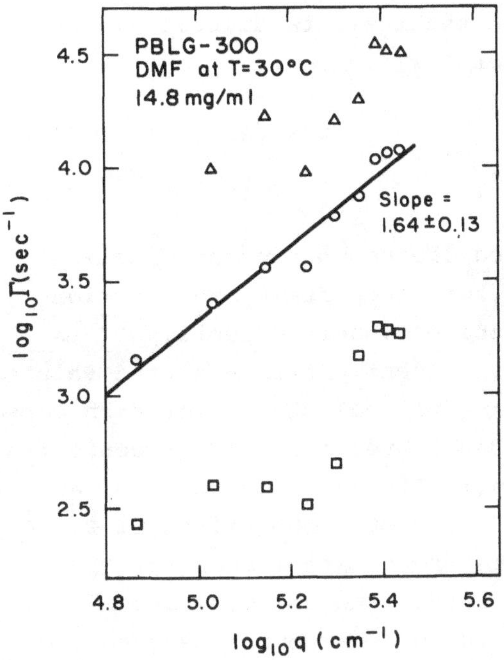

Fig. 5. Angular dependence of the fast (triangles), intermediate (circles) and slow (squares) decay rates from a three-exponential fit to data in the semi-dilute regime (NL^3= 260).

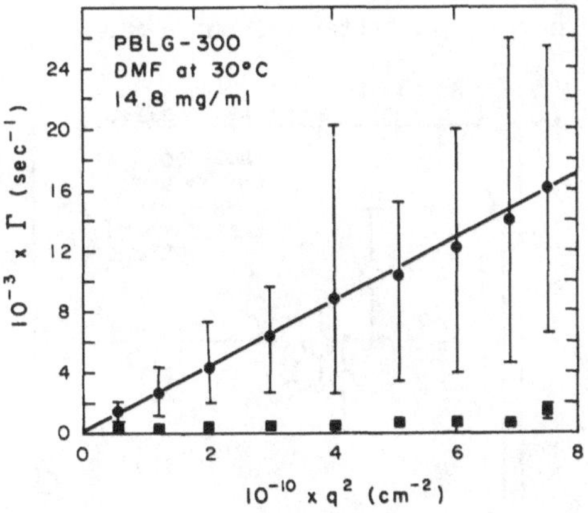

Fig. 6. Angular dependence of the fast (circles) and slow (squares) decay rates from an exponential sampling fit to data in the semi-dilute regime (NL^3= 260).

The measured D can be related to the infinite dilution value (to first order in c in the numerator) by

$$D = D^{o}\{\frac{1-\bar{v}_2 c}{1+k_f c} + \frac{2MB_2 c}{1+k_f c}\} = D^{I} + D^{II} \qquad (1)$$

where D^{II} represents the combined thermodynamic and friction effects while D^{I} represents the reference frame and friction contribution in the absence of thermodynamic effects. D^{I} is what one would expect in a theta solvent. From measured values of D^{o}, v_2, M and B_2, a value of k_f was calculated for each concentration at which D was measured. Over most of our concentration range $k_f \approx 170\ cm^3/g$ as shown in Fig. 7. Computing D^{I} as $D^{o}\{(1-\bar{v}_2 c)/(1+k_f(c)c)\}$ and D^{II} as $D-D^{I}$, the values of D, D^{I}, and D^{II} are shown in Fig. 8. The trend established for $NL^3 <$ 300 is clear. In DMF, a good solvent, the thermodynamically driven repulsive increase in diffusion overcomes the enhanced friction and the measured D increases. If the thermodynamic effects (D^{II}) are accounted for, the simulated theta solvent diffusion coefficient D^{I} does decrease, perhaps more strongly than the Doi-Edwards theory, i.e. to a value even less than $D^{o}/2$.

In common with other authors we find the onset of slowed diffu-

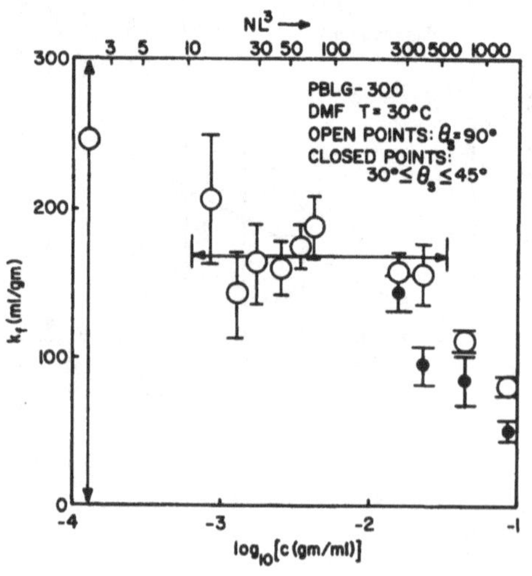

Fig. 7. Friction coefficient k_f calculated from measured D^{o} v_2, M and B_2 and Eq. (1).

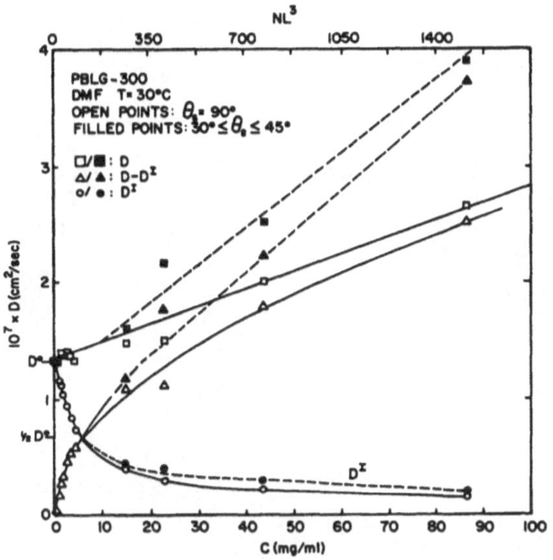

Fig. 8. Measured diffusion coefficient D (squares), calculated combined thermodynamic and friction contributions D^{II} (triangles), and calculated theta solvent behavior D^{I} (circles) vs concentration.

sion in a rod-like molecule associated with the semidilute region is delayed to much higher concentration than one would expect (β seems greater than several hundred). A possible explanation as regards rotational diffusion is suggested by the mechanism illustrated in Fig. 9. One can calculate using the Fujime-Maruyama theory[9] that the imperfectly rigid PBLG may execute tens to hundreds of bending cycles during the time longitudinal diffusion carries a molecule a distance on the order of its length. The strict conditions of the Doi-Edwards theory (that a rod must diffuse longitudinally until it no longer hinders the rotation of its neighbor) may be ameliorated by an escapement process due to bending because of flexibility. Odijk[10] has recently formulated a quantitative description.

Nonexponential correlation functions in the semidilute region are not completely understood. The presence of a slow relaxation is particularly intriguing. Translation-rotation coupling almost certainly plays a role, and it is possible we may be observing effects due to self diffusion in this somewhat polydisperse, strongly interacting system. We have recently presented a fuller discussion of these details[11].

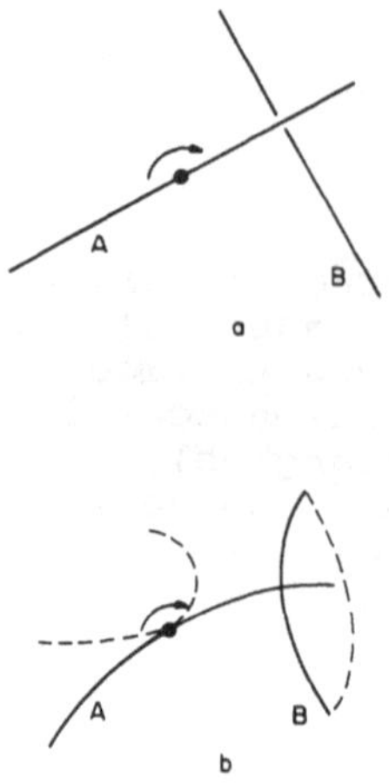

Fig. 9. Rotational enmeshment of rigid rods (a), and flexible rods (b). In (b) rotation of rod A is enabled by bending of A or B or both.

References

1. Doi, M., Edwards, S.F.: J. Chem. Soc. Faraday Trans. 2 74, 560 and 918 (1978).
2. Zero, K., Pecora, R.: Macromolecules 15, 87 (1982).
3. Kubota, K., Chu, B.: Biopolymers 22, 1461 (1983).
4. Schmidt, M.: Macromolecules 17, 553 (1984).
5. Yamakawa, H.: Modern Theory of Polymer Solutions (Harper and Row, New York, 1971).
6. Riseman, J., Kirkwood, J.G.: J. Chem. Phys. 18, 512 (1950).
7. Broersma, S.: J. Chem. Phys. 32, 1626 (1960).
8. Ostrowsky, M., Sornette, D., Parker, P., Pike, E.: Opt. Acta 28, 1059 (1981).
9. Fujime, S., Maruyama, M.: Macromolecules 6, 237 (1973).
10. Odijk, T.: Macromolecules 16, 1340 (1983).
11. Russo, P.S., Karasz, F.E., Langley, K.H.: J. Chem. Phys. 80, 5312 (1984).

LASER LIGHT SCATTERING OF DILUTE AND SEMI-DILUTE HYDROSOLUBLE BIOPOLYMER AND SYNTHETIC POLYELECTROLYTE SOLUTIONS

Guy Muller

UA 500 du CNRS "Polymères,biopolymères,membranes"
Faculté des Sciences de l'Université de Rouen Haute-Normandie
76130 Mont-Saint-Aignan,France

Introduction

Hydrosoluble polymers are often polyelectrolytes carrying a large number of charge bearing groups distributed along the contour length L_c of the polymer chain. They can be of natural origin (polysaccharides, nucleic acids) or obtained by synthesis (polyacrylic acid,polyacrylamides) and their aqueous solutions are known to behave in a very specific way as compared to solutions of uncharged macromolecules due to charge interactions.
We are concerning here with the solution behaviour,particularly the static and dynamic light scattering behaviour, of *extracellular polysaccharide xanthan* and flexible synthetic *uncharged and charged polyacrylamides*. Polysaccharides (as xanthan) are characterized by stiff chains of rather large persistence length q_t whereas vinylic polyelectrolytes such as polyacrylamides are flexible polymers with a persistence length q_t small compared with the contour length of the chain. However it is well known that rodlike polymers of rather large contour length are not strictly rigid but have some flexibility and behave more as worm-like chains. Also whereas flexible polyelectrolytes have more or less coiled conformation at low charge densities or at high salinity, they behave as worm-like chains when the ionic strength is decreased and in the absence of salt they are assumed to nearly possess a rod-like structure. Therefore it is not unrealistic to assume that polyelectrolytes can be regarded as worm-like chains of total persistence length $q_t=q_p+q_e$ where q_p is the intrinsic persistence length (in the absence of electrostatic interactions i.e. at high salinity) and q_e is the electrostatic persistence length arising from the screened charge interactions along the chain which increase its stiffness (1,2). For stiff polymers therefore q_p is generally larger than q_e (except at

Physical Optics of Dynamic Phenomena and Processes in Macromolecular Systems

very low salinity) whereas for flexible ionic polymers $L_c >> q_t$ and $q_e > q_t$(at high salinity $q_t = q_p$).
Two different types of problems will be discussed in the present paper. *In the first part,* information about the exact state in solution and stiffness of xanthan molecule (still a debated question) will be deduced from the combination of viscosity,static and quasi-elastic light scattering data. *The second part* is devoted to the effect of molecular weight,concentration,charge density and ionic strength on the translational diffusion behaviour (as probed by quasielastic light scattering measurements) of dilute and semi-dilute polyacrylamide solutions.

Experimental

Polymer samples.
Two xanthan microgel-free samples were studied (index of polydispersity:1.4) together with xanthan fractions obtained by the surface-exclusion chromatographic method previously described (3). The solutions obtained by diluting fermentation broths were clarified and filtered in order to remove both bacteria cells and any possibly remaining microgels according to a method previously described (4).
Low molecular weight (LMW) and high molecular weight (HMW) *polyacrylamide* samples of different charge density were respectively investigated in aqueous solutions containing various extents of added NaCl.
Measuring methods.
Viscosity was measured in the low shear rate limit by using a Contraves low shear viscometer.
The average molecular weights $\bar{M}_w$,the virial coefficients A_2 and the radii of gyration R_g were determined from *static light scattering measurements* using a Chromatix KMX6 photometer (scattering angle 5°) and a Fica 42000 photometer (scattering angles from 30 to 135°).
Quasielastic light scattering measurements were performed with a Chromatix KMX6 low angle photometer equipped with a real time,multibit 64 Channel autocorrelator Malvern K7025. Scattered light was collected at low angle (5°) and then focused through an aperture (10-50 μm in diameter) on to the photomultiplier. The z-average translational diffusion coefficient D_t and

normalized second moment of the correlation function were obtained by conventional procedures (5,6).

Results and discussion

I. Extracellular polysaccharide xanthan

The extracelular polysaccharide xanthan shows unusual solution properties of interest in a wide range of applications thus explaining extensive research investigations during the last decade. The primary structure is well established and structural studies indicate that the backbone consists of a linear β-1,4 linked glucan substituted on every glucose residue by a charged trisaccharide side chain (7). The terminal mannose of the side-chain may be substituted with a pyruvate group, the actual pyruvate substitution being variable and dependent mainly on the fermentation condition (8, 9). Also it is well known that xanthan undergoes a cooperative conformational transition between an ordered form and a disordered one which is temperature, ionic strength & pH dependent (10-15). However some discrepancies exist in the literature concerning the exact polymer chain conformation and whether xanthan under ordered form exists as a single or multistranded helix (16-22). The difficulty to know with accuracy the exact solution state of xanthan molecules largely resides in the fact that xanthan solutions often contain aggregates or microgels of high molecular weight which are very difficult to remove by usual filtration techniques (23). The poor quality of solutions due to the presence of such large particles is partly responsible for high and variable molecular weights reported in the literature (24,25) and therefore for debates concerning both the real mass per unit length M_L and the persistence length of xanthan molecules.

Molecularly dispersed xanthan under ordered conformation (5g/1 NaCl)

At room temperature the xanthan ordered conformation is independent on the ionic strength over 10^{-2}M NaCl. X-Ray fiber diffraction data (26) indicate that xanthan adopts a fivefold helical conformation with a projection on the helical axis of 0.94 nm per repeat unit. The charge density parameter $\lambda=7.14\ b^{-1}$ (b, in Å, is the spacing between the projections of the fixed charges of xanthan on its contour length) is dependent on the degree of pyruvate substitution DS_{pyr} and on the conformational state (for a dimeric ordered structure λ is twice as high than for a single chain). For xanthan samples investigated here (Table 1), λ≃1.5 (sample A, DS_{pyr}≃1) and λ≃1.1 (sample B, DS_{pyr}≃0.5). As the molecular weight of the repeat unit is close to 1000 daltons (under Na form), the linear mass density $M_L = M_w/L_c$ for a hypothetical single-stranded conformation would be close to 1000 daltons/nm whereas $M_L \simeq 2000$ daltons/nm for a hypothetical double-stranded helical chain. The exact knowledge of $\bar{M}_w$ is therefore of primary importance in so far as an overestimation of the molecular weight due to the poor quality of solutions (microgels, aggregates) would lead to suspect a multistranded structure. In this regard we have observed that the process of lyophylization under disordered conformation (water) alters the solubility of xanthane and that solutions prepared from lyophylized material are not true solutions as evidenced by high and non reproducible measured molecular weights probably due to the presence of superstructures generated during lyophylization (27). Therefore all data reported here have been obtained on *molecularly dispersed xanthan solutions* prepared according to the procedure described in the experimental section.

Table 1. Experimental and theoretical parameters of xanthan
5 g/1 NaCl

Parameters	Xanthan A	Xanthan B
$\bar{M}_w$ (g/mol)	1.8×10^6	3×10^6
$\|\eta\|$ (ml/g)	4300	6900
R_g (nm)	170	-
γL_c [1]	20	30
D_t (c→0) (cm^2/s) (measured)	$2.4_5 x10^{-8}$	$1.8_5 x10^{-8}$
D_t (Y-F) [2] (cm^2/s)	$2.6_5 x10^{-8}$	$1.9_3 x10^{-8}$
D_t (rod) [3] (cm^2/s)	$1.6_3 x10^{-8}$	$1.1_3 x10^{-8}$

[1] calculated with a persistence length q=50 nm and M_L=1000 nm^{-1}

[2] calculated according the Yamakawa-Fujii theory (28) with d=1.9 nm (monohelix)

[3] calculated using the expression of Newman et al. (29) for a rod of diameter d=1.9 nm.

Figure 1 shows the concentration dependence of the scattered light and of the translational diffusion coefficient D_t for dilute xanthan solutions at a very low forward scattering angle (5°) from which extrapolation at infinite dilution leads to $\bar{M}_w$, A_2 and $D_t(c\to 0)$ (or the hydrodynamic radius R_h) whereas the radius of gyration R_g is obtained from the slope of the angular dependence of the seattered light.

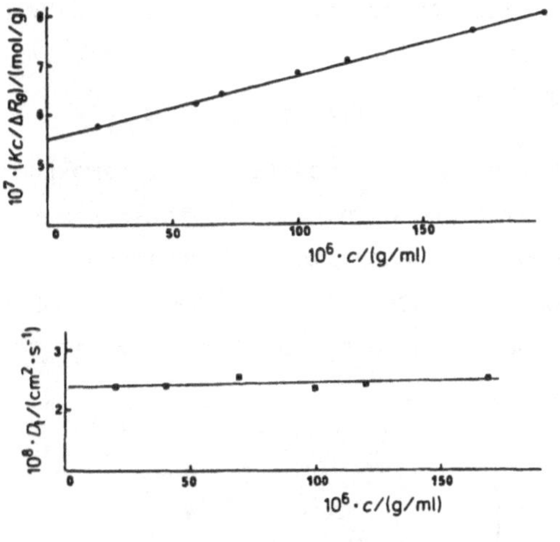

Figure 1. Concentration dependence of $Kc/\Delta R_\theta$ for the xanthan A sample in the presence of 5g/1 NaCl and plot of the translational diffusion coefficient $D_t(q,c)$ against the concentration (q,the scattering vector,is 1.12×10^4 cm^{-1}).

For rigid chain polymers modelled by a worm-like chain the radius of gyration R_g is related to the contour length L_C and to the persistence length q_t (or the Kuhn's length $\gamma^{-1}=2q_t$). The rigidity of the chain is therefore directly related to the flexibility parameter γL_c (reduced contour length). From the Benoit-Doty expression (30) for R_g of the Kratky-Porod worm-like chain (31) the ratio R_g/L_c is a function of γL_c which decreases monotonically from $1/\sqrt{12}$ to zero as $L_c/2q$ increases from zero to infinity. Moreover R_g/L_c is $1/2\sqrt{12}$ at $L_c/2q=6.394$ regardless of q and M_L. Therefore we have

$$H \equiv \lim R_g/M_w \ (L/2q\to 0) = 1/\sqrt{12}M_L$$

and

$$M^* = 12.79qM_L$$ where M^* is the $\bar{M}_w$ for which R_g/M_w is equal to H/2

As shown for *schizophyllan* (a neutral polysaccharide with a triple helical conformation in aqueous solution) it is therefore possible to evaluate q and M_L if H and M^* can be determined by experiment (32). Figure 2 illustrates the plot of $R_g/\bar{M}_w$ vs log $\bar{M}_w$ for a series of xanthan fractions of $\bar{M}_w$ in the range 8×10^4-19.5×10^5 (22) and for the unfractionated xanthan A investigated here. From the value of $H=2.8_8\times10^{-4}$nm (determined by extrapolation) and $M^*=6.3\times10^5$ it is found $M_L\cong1000$ nm^{-1} and q≈50 nm. Recently we reported the $|\eta|$ vs $\bar{M}_w$ dependence for xanthan fractions obtained by hydrodynamic chromatography and the following relation for $10^6<\bar{M}_w<5\times10^6$ was established in the presence of 5g/1 NaCl : $|\eta|\,(ml/g)\simeq6.3\times10^{-3}\bar{M}_w^{\,0.93}$

The comparison of the measured viscosities with Yamakawa's theoretical values computed for different q values (33) has shown that the best agreement between theory and experiment is observed for q≈50 nm, $M_L\simeq1000$ nm^{-1} and d=1.9 nm (molecular diameter).In the case of a double stranded structure no theoretical curve has been found to adequately fit the experimental data (34).

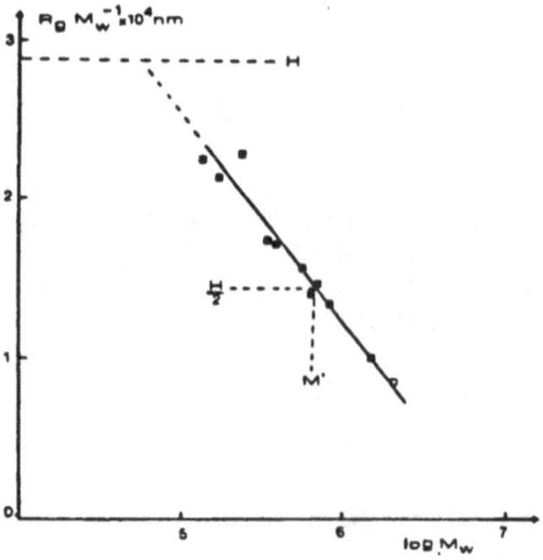

Figure 2. Plot of $R_g/\bar{M}_w$ vs log $\bar{M}_w$ for xanthan in 5g/1 NaCl
■ series of fractions of xanthan samples (from reference 22)
□ unfractionated xanthan A (polydispersity index 1.4)

As shown in Figure 1 the translational diffusion coefficient of xanthan (ordered conformation) is found only slightly concentration dependent in the range investigated here contrary to some literature data (21,35). Such a linear concentration dependence suggests that the xanthan solutions are in the dilute solution region though the concentration range is far beyond

the overlap concentration for rod polymer solutions ($C^{*} \approx L^{-3}$) according to the Doi-Edwards theory (36) but such a theory tends to underestimate the beginning of the semi-dilute domain (37) and as shown here xanthan has some flexibility in such a way that the critical concentration should be much higher. In Table 1 are reported the limiting values of the measured $D_t(c \to 0)$ for the two xanthan samples; in the same Table we have reported the theoretical D_t computed according the Broesma treatment for a rod (38) and the Yamakawa-Fujii theory for a worm-like chain (28).As shown the agreement is rather good between the experimental D_t values and those calculated according to the Yamakawa's theory (D_t{Y-F}) for a single chain.
Figure 3 shows a log-log plot of D_t vs molecular weight for xanthan in 5g/1 NaCl.The hollow triangle is measured D_t using quasielastic light scattering whereas the black squares denote values of D_t computed from s_o and $\bar{M}_w$ data reported by Holzwarth (17).On the same plot is also reported D_t for a heat-treated 4M urea sample (39). As it appears the experimental D_t lie on the theoretical curve based on the Yamakawa theory (solid line) and it is observed that $D_t \sim M_w^{-0.7}$ whereas for an infinitely long rigid rod the exponent should be unity (broken line).

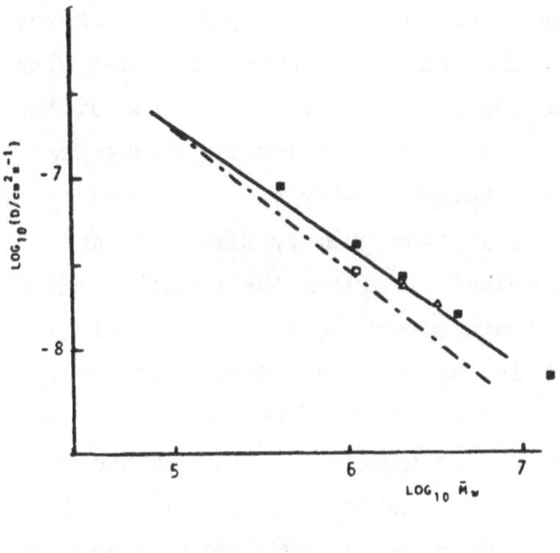

Figure 3.Log-log plot of D_t vs molecular weight for xanthan in 5 g/1 NaCl
solid line : theoretical curve based on the Yamakawa theory using M_L=1000 nm^{-1},q=50 nm and d=1.9 nm
broken line : rod limit

II. Dilute and semi-dilute uncharged and charged polyacrylamide solutions: diffusion behaviour

Development of quasielastic light scattering has offered an accurate method for determining the diffusion of macromolecules in solution (40). In the case of neutral polymers theoretical predictions show that according to the concentration domain considered different diffusion coefficients should be measured corresponding to the motion of the entire isolated chains (dilute domain) or of segments of chains (semi-dilute domain) (41). A noticeable departure of the concentration dependence of the diffusion coefficient is observed at a critical concentration C^* where the polymer coils start to overlap. Above C^* (interpenetration) the characteristic length is no longer the radius of gyration R_g but the *correlation length* ξ which should be independent of the molecular weight but concentration dependent according to power laws with different exponents depending on the solvent conditions (42,43). Whereas the essential features of the scaling concept have been confirmed (at least qualitatively) in the semi-dilute domain for neutral polymers at both good and θ solvent conditions (44-48) data concerning static and dynamic phenomena in semi-dilute polyelectrolyte solutions were rather limited until recently (49-55). Results in agreement with the predictions of scaling theory have been reported concerning the diffusion behaviour of highly charged polystyrene sulfonates with and without added salt (51). Here we report some features of the diffusion behaviour of unhydrolyzed and partly hydrolyzed polyacrylamides:such polyelectrolytes are interesting as their charge density may be varied (hydrolysis and/or copolymerization) while keeping the primary structure and the molecular weight constant. For such polyelectrolytes the solution behaviour (and therefore the diffusion behaviour) are governing by the dimensionless charge density parameter $\lambda=7.14\,b^{-1}$ (b is the spacing between COO^- groups along the contour length) related to the extent of hydrolysis (τ) and to the degree of dissociation of carboxylic groups (α) through the relation $\lambda=2.82\alpha\tau$ (56). The two different polyacrylamide samples (Na^+ form) investigated have $\bar{M}_w \simeq 1.6\times10^5$ (LMW) and $\bar{M}_w \simeq 7.5\times10^6$ (HMW) and charge densities in the range $0<\lambda<1.41$ ($0<\tau<50\%$). For each sample several concentrations were prepared below and above the critical concentration C^* for overlapping. For coiling macromolecules (neutral polymers or polyelectrolytes in the presence of a large extent of added salt) C^* is of the order of $C_1^* \simeq M_w R_g^{-3} N_a^{-1}$ or according to Odijk (57) $C_2^* \simeq M_w R_F^{-3} N_a^{-1}$ where R_F is the Flory limit of the radius of gyration expressed as : $R_F \simeq L_c^{3/5} (q_t/\kappa)^{1/5}$

q_t, the total persistence length, is $q_t = q_p + q_e$ where κ is the reciprocal of the Debye screening length and q_e, the electrostatic persistence length, is $q_e = Q/4\kappa^2 b^2$ where Q is the Bjerrum length (Q≃7Å) and b as defined above. For polyacrylamide q_p (in the presence of a large extent of added salt) reasonnably approximates $q_p \simeq 10$Å (58). For fully elongated chains (rod polymer or polyelectrolyte in water) it is clear that $C^* \sim C_3^* \sim M_w L_c^{-3} N_a^{-1}$.

Figure 4 illustrates the concentration dependence of the translational diffusion coefficient D_t (obtained from the average decay time $\bar{\Gamma}$ of the autocorrelation function of the scattered light at low q value from the relation $\bar{\Gamma} = 2D_t q^2$ valid for homodyne detection) for two polyacrylamide samples, differing by their M_w and charge density λ, in the presence of 20g/1 NaCl. The extrapolation of D_t at infinite dilution by assuming validity of the Stokes-Einstein relation leads to the hydrodynamic radius R_h which should be related to R_g as determined from the angular dependence of the scattered light. It has been shown that the ratio R_g/R_h is of interest concerning the shape of molecules (59) ; also of interest is the comparison between the coefficient K_D of the concentration dependence of D_t (in the dilute domain) and the second virial coefficient A_2 obtained from static light scattering measurements.

The two polyacrylamide samples partly hydrolyzed at an extent of 30% ($\lambda \simeq 0.86$) have the following characteristics:

<u>HMW polyacrylamide in 20g/1 NaCl</u> : $\bar{M}_w \simeq 7.5 \times 10^6$ $|\eta| = 3800$ ml/g (at low shear rate) $A_2 = 6 \times 10^{-4} cm^3 g^{-2}$ $(R_g)_z \simeq 200$ nm (Flory-Fox) 240 nm (Zimm plot) $(R_h)_z = 150$ nm (from the limiting value of D_t) $R_F \simeq 380$ nm $C_2^* \simeq 2.4 \times 10^{-4}$ g/ml $K_D \simeq 1370$ ml/g

<u>LMW polyacrylamide in 20 g/1 NaCl</u> : $\bar{M}_w \simeq 1.6 \times 10^5$ $|\eta| = 130$ ml/g $A_2 = 1.6 \times 10^{-3} cm^3 g^{-2}$ $(R_g)_z \simeq 18$ nm (Flory-Fox) $(R_h)_z \simeq 13$ nm $R_F \simeq 37$ nm $C_2^* \simeq 5 \times 10^{-3}$ g/ml $K_D \simeq 97$ ml/g (<u>in 2g/1 NaCl</u> : $|\eta| = 242$ ml/g $A_2 = 6 \times 10^{-3} cm^3 g^{-2}$ $(R_g)_z \simeq 24$ nm $K_D \simeq 177$ ml/g).

As indicated $R_h < R_F$ in all cases. However the ratio R_g(measured)$/R_h$ is in rather good agreement with the expected coil shape of partly hydrolyzed polyacrylamide in the presence of rather high amount of salt (20g/1 NaCl).

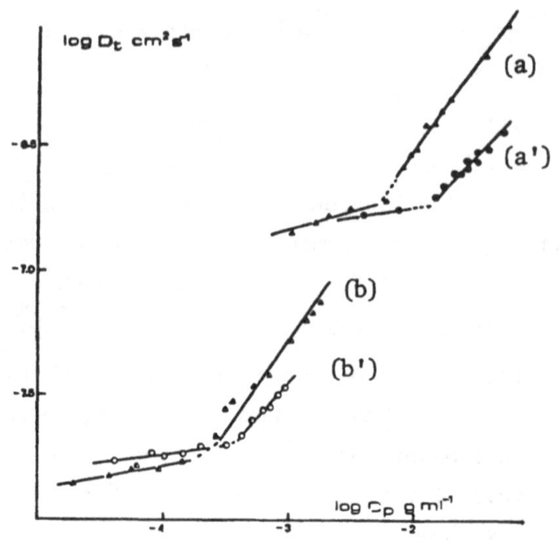

Figure 4. Concentration dependence of translational diffusion coefficient D_t for LMW polyacrylamide (λ=0.86{a} and λ=0{a'}) and HMW polyacrylamide (λ=0.86{b} and λ=0{b'}) in the presence of 20g/1 NaCl.

As clearly evidenced on Figure 4,the concentration dependence of D_t shows two distinct behaviours depending on the concentration range thus confirming that different interactions dominate in dilute solution (intramolecular interactions) and at higher concentration (intermolecular interactions). *In the low concentration range* $(C<C_1^*,C_2^*)$, D_t is only slightly changing with the polymer concentration C_p in agreement with the theoretical predictions for dilute solutions.D_t is observed to change with C_p according to $D_t(C)=D_0(1+K_DC_p)$ where D_0 is the limiting value of D_t as $C_p \to 0$ and K_D,the second virial coefficient,is related to both thermodynamic and hydrodynamic interactions of the polymer in solution.The larger value of K_D at lower salinity (see LMW above) indicates that the thermodynamic term in K_D dominates (A_2 is large at low salinity). In this concentration range D_t is dependent on the $\bar{M}_w$ of polymer samples and the hydrodynamic radii R_h are observed of the order of magnitude of the mean radii of gyration of individual chains.

A noticeable departure from the initial concentration dependence is observed at higher concentration in a transition region roughly located at $C^* \approx C_2^* \sim R_F^{-3}$

where the polymer coils start to overlap and that is experimentally found $\bar{M}_w$,salinity and charge density dependent, i.e. depending on the conformational state of polyions.
Above C^* where intermolecular interactions predominate D_t is observed to increase with C_p; in this domain D_t is nearly independent of the molecular weight but dependent on the concentration according to a power law with an exponent close to 0.7 for the two partly hydrolyzed polyacrylamides of different $\bar{M}_w$ but of identical charge density (λ=0.86) whereas D_t is found to scale with concentration with a lower exponent (near 0.5) for both uncharged polyacrylamides. The value of the exponent for the two charged polyacrylamide samples is compatible with scaling predictions for flexible polymers at good solvent conditions that is in accordance with the values of the virial coefficients A_2 and with the value of the Mark-Houwink exponent (0.8) determined for partly hydrolyzed polyacrylamides in the presence of 20g/1 NaCl . The experimentally found smaller exponent (0.5) for uncharged polyacrylamides deviates from the predictions for good solvent conditions and may be tentatively explained by the fact that marginal solvent conditions prevail as it has been reported that aqueous solution is a poor solvent for uncharged polyacrylamide ; but more experimental data are needed and work is in progress.

On Figure 5 is reported the dependence of D_t on polymer concentration for LMW polyacrylamide of different charge densities in the presence of 2g/1 NaCl. As can be seen the concentration dependence of D_t shows the same profile as observed and described above. At salinity (2g/1) and molecular weight constant, C^* is observed to be shifted to lower concentration as the charge density is increased as expected by considering the effect of charges on the expansion of polyion. The same holds for the extrapolated diffusion coefficients and also at a lesser extent for K_D both of them depending on the charge density parameter in a normal way. In the semi-dilute regime it is again observed that D_t scales with the concentration as $D_t \approx C^{0.7}$ for charged polyacrylamides and as $D_t \approx C^{0.5}$ for the uncharged polyacrylamide. It is interesting to remark that in the dilute domain the diffusion coefficients decrease with increasing charge density whereas the reverse is observed in the semi-dilute domain . Recently a "critical" polyelectrolyte concentration has been reported for polyacrylamides at which the diffusion coefficients were independent of both the polyelectrolyte charge density and the extent of added salt (55) ; such a behaviour is not observed here.

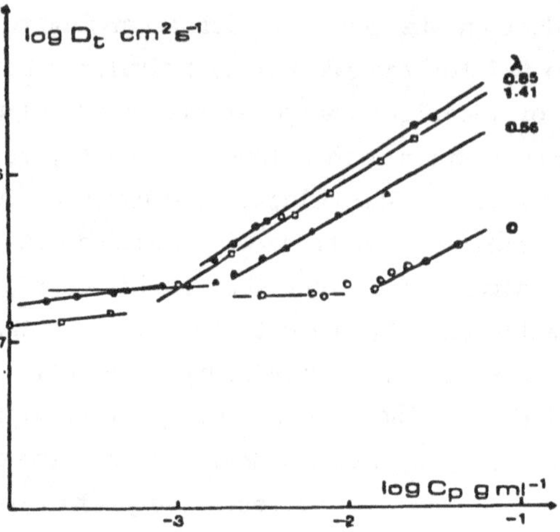

Figure 5. Dependence of D_t on polymer concentration for LMW polyacrylamide of various charge densities λ in the presence of 2g/1 NaCl.

Conclusion

Useful information concerning the state in solution and stiffness of molecularly dispersed xanthan (xanthan microgel-free) are obtained by combining various techniques including low-shear rate viscosity,static and quasi-elastic low angle light scattering indicating that xanthan under ordered conformation behaves as a semi-flexible chain of persistence length close to 50 nm.

On the other hand from low angle quasi-elastic light scattering experiments on polyacrylamide-NaCl systems (differing by their molecular weight and their charge density parameter) different diffusion behaviours are evidenced below and above the critical concentration for overlapping of polymeric coils as expected by the theory. In the semi-dilute domain the experimental data are found qualitatively in agreement with the predictions of the scaling theory.

References

1. Skolnick,J., Fixman,M.: Macromolecules 10, 944 (1977).
2. Odijk,T.: J.Polym.Sci. 15, 477 (1977).
3. Lecourtier,J., Chauveteau,G.: Macromolecules (to appear 1984).
4. Chauveteau,G., Kohler,N.: Paper SPE 9245, 55th Ann.Techn.Conf.and Exh., Dallas, (1980).
5. Koppel,D.E.; J.Chem.Phys. 57, 4814 (1972).
6. Brown,J.C., Pusey,P.N., Dietz,R.: J.Chem.Phys. 62, 1136 (1975).
7. Jansson,P.E., Kenne,L., Lindberg,B.: Carbohydr.Res. 45, 275 (1975).
8. Cadmus,M.C., Knutson,C.A., Lagoda,A.A., Pittsley,J.E., Burton,K.A.: Biotechnol.Bioeng. 20, 1003 (1978).
9. Cadmus,M.C., Rogovin,S.P., Burton,K.A.; Pittsley,J.E., Knutson,C.A., Jeanes,A.: Can J.Microbiol. 22, 942 (1976).
10. Holzwarth,G.: Biochemistry 15, 4333 (1976).
11. Morris,E.R., Rees,D.A., Young,G., Walkinshaw,M.D., Darke,A.: J.Mol.Biol. 110, 1 (1977).
12. Milas,M., Rinaudo,M: Carbohydr.Res. 76, 189 (1979).
13. Norton,I.T., Goodall,M.M., Morris,E.R., Rees,D.A.: J.Chem Soc.,Chem. Commun. 545, (1980).
14. Southwick,J.G., Jamieson,A.M., Blackwell,J: in Brant,D.A. Ed. "Solution Properties of polysaccharides";American Chemical Society: Washington, 1981; ACS Symp.Ser.,N° 150.
15. Rees,D.A.: Pure Appl.Chem. 53, 1 (1981).
16. Holzwarth,G., Prestridge,E.B.: Science 197, 757 (1977).
17. Holzwarth,G.: Carbohydr.Res. 66, 173 (1978).
18. Rinaudo,M., Milas,M.: Biopolymers 17, 2663 (1978).
19. Maret,G., Milas,M., Rinaudo,M.: Polym.Bull. 4, 291 (1981).
20. Southwick,J.G., Lee,H., Jamieson,A.M.,Blackwell,J.: Carbohydr.Res. 84, 287 (1980).
21. Southwick,J.G., McDonnell,M.E., Jamieson,A.M., Blackwell,J.: Macromolecules 12, 305 (1979).
22. Paradossi,G., Brant,D.A.: Macromolecules 15, 874 (1982).
23. Kohler,N., Chauveteau,G.: 53th Annual SPE Fall Meeting, Paper SPE 7425 (1978).
24. Dintzis,F.R., Babcock,G.E., Tobin,R.: Carbohydr.Res. 13, 257 (1970).
25. Holzwarth,G.M.: in D.A.Brant (Ed) " Solution properties of polysaccharides" ACS Symp.Ser. 150, 15 (1981).
26. Moorhouse,R., Walkinshaw,M.D., Winter,W.T., Arnott,S.: in J.C.Arthur (Ed),Cellulose Chemistry and Technology,ACS Symp.Ser. 48, 133 (1977).

27. Muller,G: XIIth International Carbohydrate Symposium,Utrecht,July 1-7, 1984.
28. Yamakawa,H., Fujii,M.: Macromolecules 6, 407 (1973).
29. Newman,J., Swinney, H.L., Day,L.A.: J.Mol.Biol 116, 593 (1977).
30. Benoit,H., Doty,P.: J.Phys.Chem. 57, 958 (1953).
31. Kratky,O., Porod,G.: Rec.Trav.Chim.Pays Bas 68, 1106 (1948).
32. Kashiwagi,Y., Norisuye,T., Fujita,H.: Macromolecules 14, 1220 (1981).
33. Yamakawa,H., Yoshizaki,T.: Macromolecules 13, 633 (1980).
34. Muller,G., Lecourtier,J., Chauveteau,G., Allain,C.: Makromol.Chem., Rapid Commun., 5, 203 (1984).
35. Southwick,J.G., Jamieson,A.M., Blackwell,J.: Carbohydr.Res. 99, 117 (1982).
36. Doi,M., Edwards,S.F.: J.Chem.Soc.Faraday Trans. 2,74, 560 (1978).
37. Kubota,K., Chu,B.: Biopolymer 22, 1461 (1983).
38. Broesma,S.J.: J.Chem.Phys. 32, 1626 (1973).
39. Morris,V.J., Franklin,D., Fanson,K.: Carbohydr.Res. 121, 13 (1983).
40. Berne,B.J., Pecora,R.: "Dynamic light scattering" Wiley,New York (1976).
41. de Gennes,P.G.: Macromolecules 9, 587 (1976).
42. de Gennes,P.G.: "Scaling Concepts in Polymer Physics" Cornell University Press, Ithaca N.Y. (1979).
43. Daoud,M., Janninck,G.: J.Phys.(Paris),Lett. 36, 4 (1975); 39,L-299 (1978).
44. Brochard,F., de Gennes,P.G.: Macromolecules 10, 1157 (1977).
45. Adam,M., Delsanti,M., Janninck,G.: J.Phys.Lett. 40, 435 (1979).
46. Adam,M., Delsanti,M.: Macromolecules 10, 1299 (1977).
47. Nyström,B., Roots,J.: Progress Polym.Sci. 8, 333 (1982).
48. Yu,T.L., Reihanian,H., Jamieson,A.M.: Macromolecules 13, 1590 (1980).
49. Roots,J., Nyström,B.: Polymer 22, 573 (1981).
50. Koene,R.S., Smit,H.W.J., Mandel,M.: Chem.Phys.Letters 74, 176 (1980).
51. Koene,R.S., Mandel,M.: Macromolecules 16, 220 (1983).
52. Koene,R.S., Nicolai,T., Mandel,M.: Macromolecules 16, 227 (1983).
53. Hara,M., Nakajima,H.: Polym.J. 12, 711 (1980).
54. Muller,G.: Polym.Bull. 11, 391 (1984).
55. Kowblansky,M., Zema,P.: Macromolecules 15, 788 (1982).
56. Fenyo,J.C., Lainé,J.P., Muller,G.: J.Polym.Sci.Polym.Chem.Ed. 17,193 (1979).
57. Odijk,T.: Macromolecules 12, 688 (1979).
58. Fisher,L.W., Sochor,A.R., Tan,J.S.: Macromolecules 10, 955 (1977).
59. Bantle,S., Schmidt,M., Burchard,W.: Macromolecules 15, 1604 (1982).

STATISTICAL THEORY OF STREAMING-BIREFRINGENCE IN TEMPORARY POLYMER NETWORKS

E. Kröner, R. Takserman-Krozer

Institut für Theoretische und Angewandte Physik,
Universität Stuttgart und Max-Planck-Institut für
Metallforschung, Stuttgart, D 7000 Stuttgart

Abstract

A review on our statistical theory of temporary polymeric networks [1,2] is presented. Of particular interest is the mobility gained by the network when its junctions are allowed to decay and form. This mobility causes specific viscoelastic effects and leads to birefringence. The transition probabilities and their inverses, the relaxation times, of the kinetic processes are fundamental for the flow behaviour. They depend on the stresses and are calculated together with these by a simultaneous equation system. One of these equations is a diffusion equation within a generalized spring-bead model; generalized in order to allow for the decay and formation of the beads (representing the junctions). We discuss in particular the various simplifications introduced to make the problem manageable. A computer program which will allow us to test the theory with the help of experiments is almost finished.

1. Introduction

The term "temporary" is here understood in the sense of former works of Green and Tobolsky, Scott and Stein, Lodge, Yamamoto

Physical Optics of Dynamic Phenomena and Processes in Macromolecular Systems

and others. The essential point is that the junctions of temporary networks are not permanently present but they may form or decay in the time interval of observation. The corresponding kinetic processes occur with certain probabilities per unit time, called transition probabilities. In condensed polymer systems such as bulk polymers, melts, concentrated polymer solutions and gels there is a high probability of inter- and intramolecular contacts. Some of these have the character of junctions in the sense of the temporary network. This is the case when the contact is rather stable so that one may speak of a life time which is large compared to typical times ($\sim 10^{-12}$s) on the atomic scale. In particular we postulate that junctions do not glide (with or without friction) along the polymer chains. If there are at all gliding contacts, then the pertaining friction effect will be described by effective friction constants. These are not of special interest here because we wish to explore the effect of the mentioned kinetic processes on the mechanical and optical properties of such networks.

It is clear that the junctions represent a kind of bound state. Very often the binding forces will be of van der Waals type. Chemical binding also occurs. But even rather different situations are imaginable, for instance when the junctions are tiny crystallites within an otherwise amorphous network. Such crystallites can form and dissolve in typical time intervals.

Streaming birefringence is a useful method also for studying temporary networks. To think of something definite assumes that our physical system is a temporary network in solution. As long as the fluid is at rest, no birefringence is observed. In flow, however, the network becomes anisotropic and birefringence arises. It is determined by the polarisability (a material-characteristic 2nd rank tensor) along the polymer chains. Therefore, to calculate the birefringence from the polarizability one needs information about the distribution of the chain segments in the network. As has been shown by the previous workers in this field, not the whole statistical information is required. It rather suffices to know the 2nd moments of the chain distribution. This is the same information that is also needed to calculate the stress tensor. Once the 2nd moments are known, the calculation of both stress and birefringence is just routine.

In order to find the 2nd moments, given for instance the velocity gradient (tensor), $\underline{\underline{q}}$, temperature Θ, and mass concentration c as macroscopic (external) variables plus some molecular parameters, we apply a generalized spring-bead model, in which the beads represent junctions and the springs are polymer chains between junctions. This model is generalized insofar as the beads are not conserved: they form and decay. In this way we are led to a generalized diffusion equation, described in section 2. For the case of a fluid at rest, i.e. in thermal equilibrium, exact results can be obtained. A few of these are quoted in section 3. To solve the diffusion equation, the relevant transition probabilities, or the inverse quantities, the relaxation times, will be needed. The calculation of these from a model to be explained is described in section 4. In analogy to Boltzmann's kinetic equation the diffusion equation is treated by the relaxation time approach. The integro-differential equation for the diffusion now reduces to a differential equation which is solved simultaneously with the equation for the transition probabilities in section 5. Birefringence and stresses are now easily calculated. In the concluding section 6 we list all variables and parameters which determine the numerical values of the birefringence and viscoelastic functions (input for computer program).

2. The generalized diffusion equation

We start from the following balance equation:

$$\frac{\partial W}{\partial t} + \sum_i^N \nabla_i \cdot (\dot{\underline{r}}_i W) = \iint (W' \vec{p} - W \overleftarrow{p}) \, d\underline{R}' \, dZ' \qquad (1)$$

Eq.(1) is an integro-differential equation for the distribution function W. The integration is taken over the whole configuration space and $W' \equiv W(\underline{R}', Z')$. The symbols $\underline{R}$ and Z denote sets $\{\underline{r}_i\}$ and $\{z_{ij}\}$, where $\underline{r}_i$ is the position vector of the i-th bead and z_{ij} is the number of elementary units (of length l_o) of polymer, e.g. the number of monomers in the chain connecting beads i and j (i,j = 1,2,3....N). Thus, $z_{ij}l_o$ is the contour length of the chain (ij). Furthermore

$$\vec{p}\,dt \equiv p(\underline{R}', Z' \rightarrow \underline{R}, Z)dt, \quad \overleftarrow{p}\,dt \equiv p(\underline{R}, Z \rightarrow \underline{R}', Z')dt \qquad (2)$$

are transition probability densities (densities with regard to the unprimed ($\underline{R}$)- and primed (R')-variables respectively for the indicated transitions in the configuration space. The first integral in (1) gives us the rate at which configurations $\underline{R}$, Z are formed whereas the second integral yields their decay rate. Thus, eq.(1) is almost self-evident.

Eq.(1) does not suffice to calculate the distribution function W, from which birefringence and stresses follow, even if the transition probabilities were given. In fact, assume for the moment that these are zero so that the network is permanent. Eq.(1) then reduces to a pure continuity equation and, therefore, has to be complemented by an equation that specifies the physics of the problem. The required additional equation is known as Langevin's equation. Since, as is common in the spring-bead model, inertia forces are neglected, the Langevin equation states that the resulting force acting on any bead vanishes. In this paper we consider the following forces: friction and Brownian (stochastic) forces acting from the fluid onto the beads and elastic forces transmitted through the springs. For later we plan to include also the forces of the hydrodynamic interaction and the excluded volume effect.

With the just indicated restriction the Langevin equation obtains the form

$$\zeta_i(\dot{\underline{r}}_i - \underline{r}_i \cdot \underline{\underline{q}}) + kT \ln W + \sum \kappa_{ij} \underline{r}_j = 0 \tag{3}$$

where ζ_i is the (effective) friction constant of the i-th bead and κ_{ij} the elasticity matrix occurring in the potential energy

$$E_{pot} = \frac{1}{2} \sum_{i,j} \kappa_{ij} \underline{r}_i \cdot \underline{r}_j = \frac{1}{2} \sum_{(ij)}^{2N} k_{ij} h_{ij}^2 , \quad \underline{h}_{ij} = \underline{r}_j - \underline{r}_i \tag{4}$$

The three mentioned forces can easily be read off from eq.(4).

In our actual calculations we have used the same (effective) friction constant, say ζ, for all beads and taken the spring constants k_{ij} independent of the $\underline{r}$'s (see e.g. [3] and [4]):

$$k_{ij} = \frac{\alpha}{z_{ij}} , \quad \alpha = \frac{3kT}{l\, l_0} \times \frac{1}{\sigma} \times \frac{1-\cos\Psi}{1+\cos\Psi} \tag{5}$$

where l is the bound length, l_0 the length of the elementary unit, Ψ the bound angle, and σ a factor depending on the structure of the polymer molecule. The approximation (5) is good as long as the network chains are not "almost stretched".

If eq.(4) is substituted into (1), then we obtain the (generalized) diffusion equation of the temporary network. Since, however, the transition probabilities are not given a priori, they must be found from another consideration. It is obvious that $\vec{p}$ and $\overleftarrow{p}$ depend on the stress in the network. In fact, the more the chains adjacent to a junction are strained, the easier can the decay occur. This implies that equations for $\vec{p}$ and $\overleftarrow{p}$ must contain the stress or strain, thus also the distribution function. In other words, the mathematical problem to be solved is one of coupled equations for W, $\vec{p}$ and $\overleftarrow{p}$.

3. Thermal equilibrium: the fluid at rest

The transition probabilities are the topic of section 4. The only problem where we do not need know them is that of thermal equilibrium. In fact, the transition probabilities determine the rate of approach to equilibrium but not the equilibrium itself. In equilibrium, the two sides of eq.(1) vanish separately and so does $\partial W/\partial t$. Assuming as a fundamental statistical postulate that all microstates with equal energy, number of junctions and number of elementary units possess equal probability we derived the equilibrium statistics of the temporary network in ref. [5]. In particular, we obtained the distribution function W_{eq} and the partition function, called C, from which the equations of state follow. It was shown that the distribution W_{eq} of the contour lengths z of the network chains in equilibrium obeys a Wien-type formula, namely

$$W_{eq}(z) = \frac{\lambda}{6kT} z^3 \exp\left\{-\frac{\lambda z}{kT}\right\} \tag{6}$$

where λ is related to the chemical potential of the elementary unit l_o. Recall that zl_o is the contour length of the considered chain.

4. The transition probabilities

The exact determination of the transition probability densities defined by eqs.(2) is not possible. In fact, if we compare the junctions with particles, as is done in the spring-bead model, then $\vec{p}$ and $\overleftarrow{p}$ are many-particle quantities which require a treatment within the frame of a many-particle theory. A similar statement is valid for the Boltzmann's transport problem. Also there transition probabilities, or, equivalently, scattering cross sections, are to be calculated. That this is not possible in the case of many-particle collisions limits the applicability of the theory to dilute systems. These, however, are treated well by the Boltzmann's equation, although only 2-particle collisions are considered.

In the same spirit we now restrict ourselves to "dilute" networks. By this term we understand networks with a low density of junctions so that a process at one junction does not influence the processes at the other junctions. This means that we now consider only one-junction processes: thus we have a drastic reduction of the difficulties. Even so the problem is not simple. We therefore introduce a further simplification by demanding that any junction is reformed immediately after decay. In this approximation, decay and reformation are treated as a single process with the particular consequence that the number N of junctions is now conserved. We plan to give up this simplification later. Here we try to keep the damage small by choosing for N the equilibrium number of junctions which we have calculated in ref. [5]. The approximation seems to be reasonable also insofar, as the chains forming a decaying junction shall have an increased mobility and therefore also an increased probability to form a new junction. We expect that in average the formation occurs in such a manner that the network around the junction gives way to the stress in an optimum fashion. That means that the four chains forming the junction will have different contour lengths before and after the process. Thus by each process the network suffers a deformation which clearly is inelastic, in fact viscoelastic.

Using these ideas and applying activation enthalpy arguments we have calculated in ref. [6] one-process transition probabilities p_i depending essentially on the chemical potential of the junction, on the local elastic potential of the junction and on the local elastic energy, i.e. on the spring constants k_{ij} and the end-to-end vectors $\underline{h}_{ij}$ of the involved four chains (ij). The transition probability occurring in eq.(2) can now be written as

$$p(\underline{R}, Z \rightarrow \underline{R}', Z') = \sum_{i}^{N} p_i \delta_{[i]} \tag{7}$$

where $\delta_{[i]}$ is the delta function with respect to all chains except those forming the junction i. Eq.(7) implies that a change in the configuration occurs always in such a way that only one junction is concerned. Since eq.(7) is derived for any configurations $\underline{R}$,Z and $\underline{R}'$,Z', the result (7) means that, in the one-process approximation, we know both $\vec{p}$ and $\overleftarrow{p}$. Therefore we now have, in some approximation, the equation which is to be combined with the diffusion equation to obtain the complete equation system for the solution of our problem.

5. Relaxation time approach

Although the equation system (1), with (3), and (7) can be solved in principle by some iterative procedure, the actual calculation is extremely complex - inspite of the simplifications introduced so far. Again we take up an idea common in the treatment of Boltzmann's equation. There a method named the relaxation time approach was extremely successful. The idea is that a system not in equilibrium will, when left alone, tend to equilibrium according to an exponential law defining a relaxation time τ. In the better version of the theory, this time may depend on the particle velocities or, in our case, on the (nonequilibrium) configuration considered.

In our theory, the relaxation time approach consists in the omission of the first integral in (1). A qualitative argument of justification goes as follows: When left alone, the processes would, in average, go from the nonequilibrium configurations towards the equilibrium configurations whereas much less processes would result in additional nonequilibrium configurations, so that the first integral in (1) should be small compared to the second. Speaking of equilibrium and nonequilibrium configuration is, of course, not very precise, because not single configurations, but rather ensembles of configurations form an equilibrium and nonequilibrium state. We think however, that the assumed diction is good enough for our qualitative argument.

With the relaxation time approach the diffusion equation reduces to

$$\frac{\partial W}{\partial t} + \sum_{i}^{N} \nabla_i \cdot (\dot{\underline{r}} W) = -\bar{p}\,(W - W_{eq}) \tag{8}$$

with

$$\bar{p} = \bar{p}(\underline{R}, Z) = \iint p(\underline{R}, Z \rightarrow \underline{R}', Z')\, dR'\, dZ' \tag{9}$$

The physical meaning of $p(\underline{R}, Z)$ is the probability per unit time that the configuration $\underline{R}, Z$ will decay into any other configuration. Correspondingly, $\tau = 1/\bar{p}$ is the relaxation time for these processes. Obviously, there is a whole spectrum of relaxation times.

It was shown in ref. [2] that the relaxation time approach implies that the distribution of contour lengths w(z) in the state of flow is the same as that in the equilibrium state. It follows that also the number of junctions is the same in the two cases. The latter statement has been taken as an approximation in the calculation of the transition probability. Obviously this assumption is consistent with the relaxation time approach. Any better approximation has to deal with the integro-differential equation (1).

For the calculation of birefringence and stresses not the whole distribution W but only its 2nd moments are needed. Equations for these are derived if the diffusion equation is multiplied by the dyadic $\underline{r}_k \underline{r}_l$ and then integrated over the whole $\underline{R}$-space. Because p depends on $\underline{R}$, the equation for the

2nd moments depends also on higher moments so that we obtain a hierarchy of equations rather than a closed equation. This unpleasant situation is avoided in a farther-going approximation, when namely $\underline{R}$ in $\bar{p}(\underline{R},Z)$ is replaced by $\langle \underline{R} \rangle$, the ensemble average of $\underline{R}$. It is difficult to judge the goodness of this assumption. The final result for $\bar{p}$ is now

$$\bar{p} = \nu^* \left(\frac{kTll_o}{\alpha}\right)^3 \sum_i^N (z_{ia} + z_{ib})^{\frac{5}{2}} (z_{ic} + z_{id})^{\frac{5}{2}} \exp\left\{\frac{1}{2kT} \sum_\beta^4 k_{i\beta} h_{i\beta}^2\right\} \tag{10}$$

where ν^* is some frequency which depends on molecular parameters only and the subscript β refers to the chains forming the i-th junction. z_{ia}, z_{ib}, z_{ic} and z_{id} are the contour lengths of these chains. The numbering must be chosen such that z_{ia} and z_{ib} belong to the one molucule and z_{ic} and z_{id} to the other of the two molecules forming the junction.

We have not yet written down the equation for the 2nd moments. When doing so, we find that in this equations also mixed moments of the form $\underline{r}_k\underline{r}_l$ appear. We can avoid such a situation by passing over to normal coordinates, say $\underline{s}_{i'}$ (i'=1,2,3...N). This means that we go into the mode-representation. The wanted equation then reads, with $\underline{\underline{S}}_{i} \equiv \underline{s}_{i'}\underline{s}_{i'}$

$$\frac{\partial \underline{\underline{S}}_{i'}}{\partial t} - \underline{\underline{b}}_{i'} \cdot \underline{\underline{S}}_{i'} - \underline{\underline{S}}_{i'} \cdot \tilde{\underline{\underline{b}}} = kT \left(\frac{1}{\zeta} + 2\bar{p}\frac{kT}{\alpha\lambda}\right) \underline{\underline{I}} \tag{11}$$

where $\underline{\underline{I}}_{i'}$ is the unit tensor in the space of the i'-th mode,

$$\underline{\underline{b}}_{i'} = \tilde{\underline{\underline{q}}} - \left(\frac{1}{2}\bar{p} + \frac{\lambda_{i'}}{\zeta}\right), \quad \tilde{q}_{ij} = q_{ji}, \quad \tilde{b}_{ij} = b_{ji} \tag{12}$$

and $\lambda_{i'}$ is the i'-th eigenvalue of the matrix κ_{ij}. Applying the same transformation to $h_{i\beta}^2$ yields

$$h_{i\beta}^2 = \sum_{i'}^N (q_i^{i'} - q_{i\beta}^{i'})^2 \, \mathrm{Spur}\, \underline{\underline{S}}_{i'} \tag{13}$$

with the q's as the eigen vectors belonging to the eigen values. Substituting eq.(13) in eq.(10) we obtain the equation which together with eq.(11) permits us to calculate simultaneously the 2nd moments, represented by S_i, and the transition probability p as a function of the $\underline{s}_i$, and the z_{ij}.

6. Conclusion

This paper is a review of our recent work on temporary polymer networks [1,2]. It has its own value because we were able to improve several arguments, in particular those referring to the quality of the approximations involved. The main ideas of our theory are represented. Certain details, both in physical arguments and mathematical performance have been omitted - as we believe, without impairing the clarity. A computer program is almost finished which permits us to obtain numerical values for the 2nd moments and transition probabilitites. With these we are able to compute both birefringence and stresses. As our equations show, this program has a certain complexity. We hope that it can be used soon to compare the results of our theory with available experiments, both on streaming birefringence and on the viscoelastic flow. The program is developed for stationary simple shear and elongational flow. Since eq.(11) is time-dependent, also creep-type behaviour of temporary networks can be treated.

Corresponding to the complexity of the problem, the theory contains in its flow part a number of simplifications. We plan to remove at least part of these in later work.

In conclusion we list up all those quantitites which are needed in the computer program to calculate the numerical values of birefringence and viscoelastic functions.

μ	chemical potential of one junction = energy of formation of one junction (a negative quantity)
ε	energy of trans state relative to gauche state
ℓ	elementary bond length
ℓ_o	length of elementary unit (here monomer)
Ψ	bond angle
σ	structure factor
v	volume of solvent molecule
M	molecular weight of polymer
M_{mon}	molecular weight of monomer
T	temperature
c	mass concentration (= mass of polymer portion/V)
$q^{11}, q^{22}, q^{33}, q^{13}$	components of velocity gradient tensor
η_o, ξ_o	limit ($\underline{\underline{q}}\to 0$) of material functions for shear flow
η_s	(Newtonian) viscosity of solution when polymer bonds are removed (no more chains, but monomers still present)

Acknowledgement

Our work on temporary polymer networks has been supported by the Deutsche Forschungsgemeinschaft. We are obliged to Dipl. Phys. D. Chassapis for discussions on birefringence and on the computer program.

References

[1] Takserman-Krozer, R., Kröner, E.: Rheol. Acta 23, 1 (1984).

[2] Kröner, E., Takserman-Krozer, R.: Rheol. Acta 23, 139 (1984).

[3] Volkenstein, M.W.: Statistical Mechanics of Chain Molecules, Interscience, New York 1963.

[4] Flory, P.J.: Statistical Mechanics of Chain Molecules, Interscience, New York 1969.

[5] Takserman-Krozer, R., Krozer, S., Kröner, E.: Colloid & Polymer Sci. 257, 1033 (1979).

[6] Kröner, E., Takserman-Krozer, R.: Proc. 4th Int. Conf. Continuum Models of Discrete Systems, Stockholm, O. Brulin, R.K.T. Hsieh (eds.), North-Holland, Amsterdam 1981.

QUASI-ELASTIC LIGHT SCATTERING OF AN IRREVERSIBLE SOL-GEL TRANSITION

S.J.Candau, M.Ankrim, J.P.Munch

Laboratoire de Spectrométrie et d'Imagerie Ultrasonores (formerly Laboratoire d'Acoustique Moléculaire), E.R.A. au C.N.R.S., Université Louis Pasteur, 4, rue Blaise Pascal, 67070 Strasbourg Cedex, France.

P.Rempp, G.Hild, R.Okasha

Centre de Recherches sur les Macromolécules, C.N.R.S., 6,rue Boussingault, 67083 Strasbourg Cedex, France.

Introduction

The recent application of percolation (1,2) and kinetic (3-5) theories to the description of the sol-gel transition has renewed the interest for the study of gelation mechanisms (6). These theories lead to several predictions about the properties of branched polymers formed in the pregel state of an irreversible gelation, which differ drastically from those of the earlier Flory-Stockmayer model (7,8) or its extension, the cascade model (9,10). Depending upon the characteristics of the gelling system, either the classical or the percolation behavior should be observed.

Recently, kinetic investigations of the free radical crosslinking copolymerization of styrene and different biunsaturated comonomers present in small proportions (with respect to styrene concentration) have been reported (11-13). In these studies, the comonomers used were divinylbenzene (meta-DVB and para-DVB), ethylene dimethacrylate (EDM) and diisopropenyl benzene (meta-DIB and para-DIB).

The physico-chemical properties of the copolymers formed in the pregel state were investigated as a function of the monomer conversion, at different stages before the appearance of the gel point (14).

However, the analysis of the data does not provide an unambiguous test of validity for any given gelation model, as it requires some assumptions concerning the relationship between monomer conversion and crosslinking

Physical Optics of Dynamic Phenomena and Processes in Macromolecular Systems

probability. On the other hand, Daoud et al.(15) have recently pointed out that the study of the molecular weight dependence of either the radius of gyration or the diffusion coefficient of the branched species, diluted in a good solvent, should allow to distinguish between a percolation and a Flory-Stockmayer distribution.
The analysis presented in this paper, which uses our previous results and new ones, leads to a variation law of the z-average diffusion coefficient $\langle D\rangle_z$ versus the weight average molecular weight M_w which is in agreement with the predictions of Daoud et al.

Experimental section

1) Sample preparation. The free radical crosslinking copolymerization of styrene and a biunsaturated comonomer was performed in benzene solution at T = 60°C, using 2-2'-azobis-isobutyronitrile (AIBN) as the initiator. The details of the copolymerization procedure are given in previous papers (11-13). The initial monomer-comonomer solution was divided into a series of sealed tubes. The soluble copolymers obtained after given times were separated, weighed, dried at room temperature to constant weight, and redissolved at low concentration in benzene. The conversion degree p_m, representing the ratio of the consumed monomers (styrene + comonomer) to the total initial amount of monomers was measured as a function of time by vapor phase chromatography. Table I summarizes the principal characteristics of the investigated systems. The gel point t_c was determined by visual observation.

2) Light scattering measurements. The weight average molecular weight M_w of the soluble copolymers formed was determined in benzene solution using a photogoniodiffusometer FICA equiped with a laser (λ : 632.8 nm). The time autocorrelation function of the scattered light intensity was measured by means of a 96-channel clipped digital auto-correlator. The scattering angle was varied from 20° to 90°. The experiments were performed at a temperature of 25°C, which was held constant within ± 0.01°C.

Table I

Characteristics of the investigated systems

Sample	Comonomer	Initial Composition (10^{-2}g cm^{-3})			Gel point t_c (h)
		Styrene	Comonomer	AIBN	
I	p-DIB	41.6	1.248	1.266	17.5
II	m-DIB	39.86	1.248	1.266	26.5
III	EDM	41.6	1.97	1.266	5.5
IV	m-DVB	40	1.04	1.266	7.5
V	m-DVB	41.86	1.045	1.266	8.5
VI	m-DVB	40.84	1.168	1.266	6.75
VII	m-DVB	18.62	1.026	0.201	17.33

Intensity correlation data were routinely processed using the method of cumulants (16,17) to provide the average decay rate $\bar{\Gamma}$. The diffusion coefficient was determined from the average decay rate $\bar{\Gamma}$ according to

$$\bar{\Gamma} = 2 K^2 D$$

The magnitude of the scattering vector K is given by :

$$K = 4 \pi n \sin(\theta/2) / \lambda$$

where θ is the scattering angle, λ is the wavelength of the incident light in vacuo, and n is the index of refraction of the scattering medium.

Results

The present investigation is focused on the behavior of measurable quantities as a function of the reduced parameter $(\Delta p_m/p_{mc}) = (p_{mc} - p_m)/p_{mc}$

where p_{mc} is the conversion degree at the gel point t_c. The value of p_{mc} was obtained by extrapolation at t_c of the curve $p_m(t)$. This extrapolation was made easier by using a semi-log plot of p_m versus t, as the kinetics of conversion obey first order laws, for all systems (except the styrene - p.DIB in which unexpected deviations were observed (13)). For the series VI and VII, the actual values of p_{mc} were not measured but determined from $\Delta p_m/p_m = f(\Delta t/t_c)$ plots obtained for similar systems (14).

Figures 1 and 2 show log-log plots of M_w versus $\Delta p_m/p_{mc}$ for the different systems investigated. The variation of $M_w = f(\Delta p_m/p_{mc})$ can be represented by straight lines, except for the early stages of the process, which correspond to weight average molecular weights less than about 40.000. A least square fit procedure leads to the values listed in table II for the exponent γ_{exp} of the power law $M_w \sim (\Delta p_m/p_{mc})^{-\gamma_{exp}}$.

Table II

Values of the γ exponent for the investigated systems.

Systems	γ_{exp}	
Styrene p-DIB (I)	1.68	
Styrene m-DIB (II)	1.55	
Styrene -EDM (III)	1.57	$M_w \sim \left(\frac{\Delta p_m}{p_{mc}}\right)^{-\gamma_{exp}}$
Styrene -mDVB (IV)	1.75	
Styrene -mDVB (V)	1.68	

The z-average diffusion constant $\langle D\rangle_z$ was determined from the double extrapolation to zero concentration, and to zero scattering angle of the average decay rate $\overline{\Gamma}$. The latter extrapolation was required because, in the vicinity of the gel point, $\overline{\Gamma}/K^2$ becomes K-dependent, due to a significant contribution of the internal deformation modes of the macromolecules (14).

Fig.3 shows log-log plots of $\langle D\rangle_z$ versus $\Delta p_m/p_{mc}$ for three styrene-m-DVB systems.The variations of $\langle D\rangle_z$ with $\Delta p_m/p_{mc}$ can be satisfactorily represen-

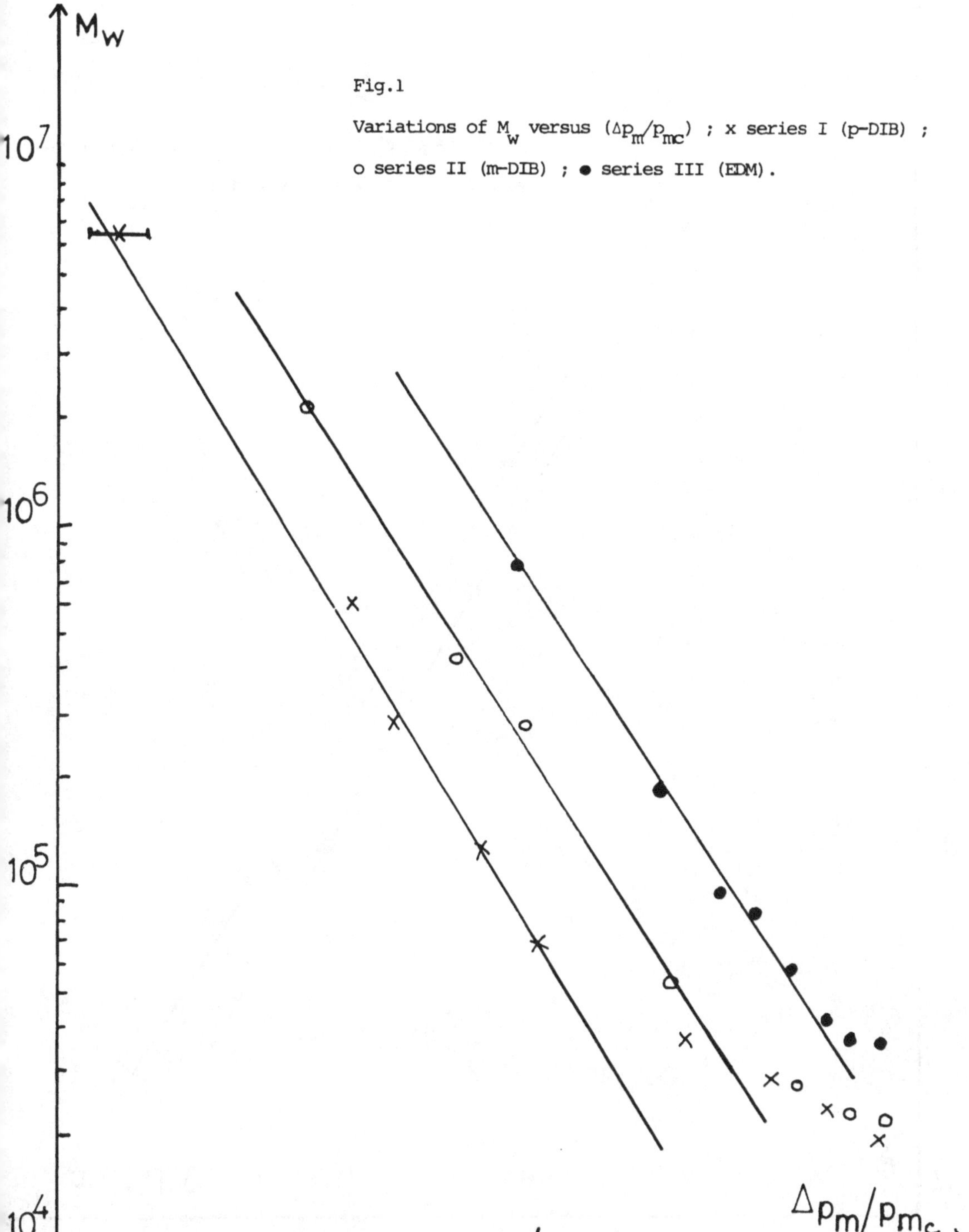

Fig.1

Variations of M_w versus $(\Delta p_m/p_{mc})$; x series I (p-DIB) ; o series II (m-DIB) ; ● series III (EDM).

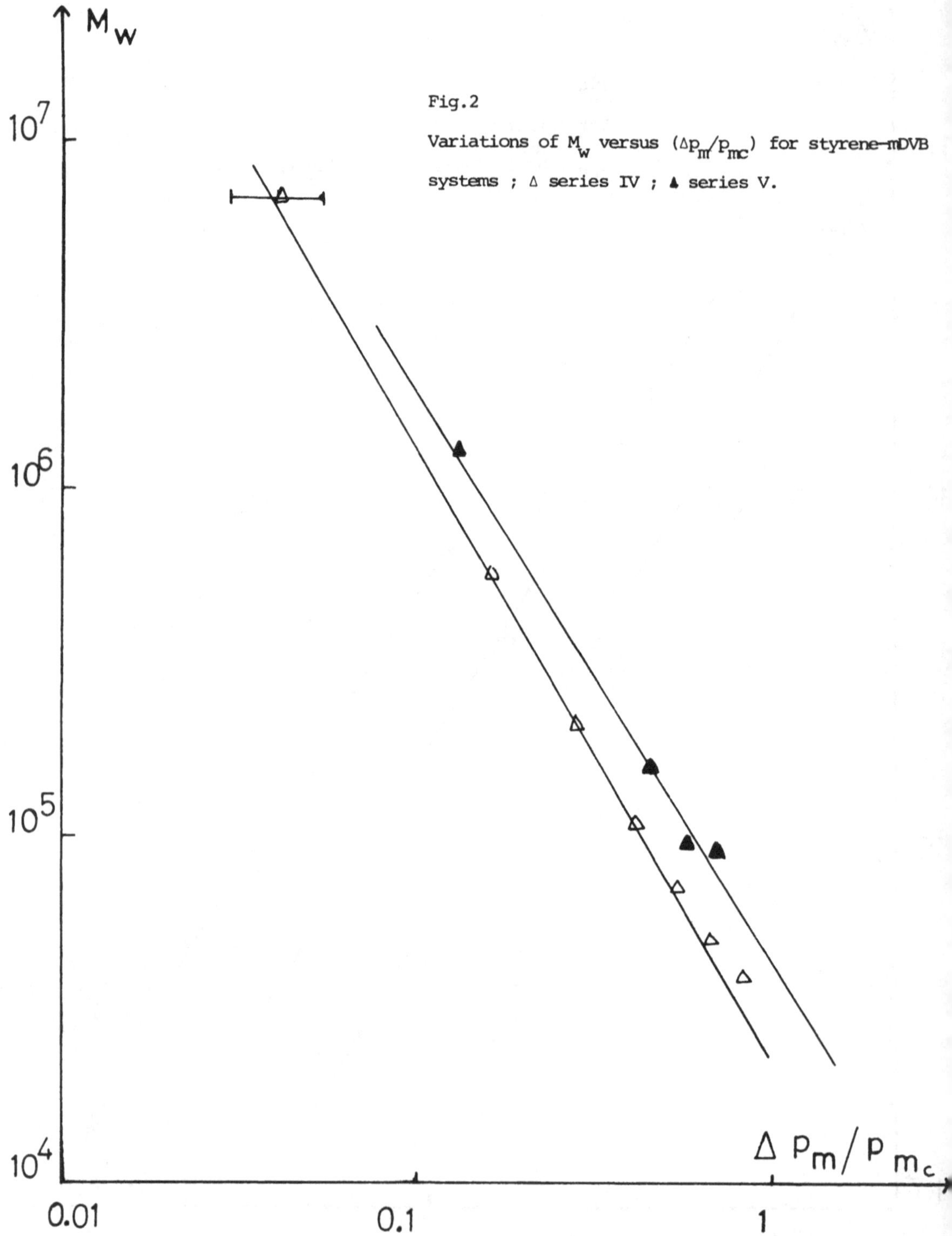

Fig.2

Variations of M_w versus ($\Delta p_m/p_{mc}$) for styrene-mDVB systems ; △ series IV ; ▲ series V.

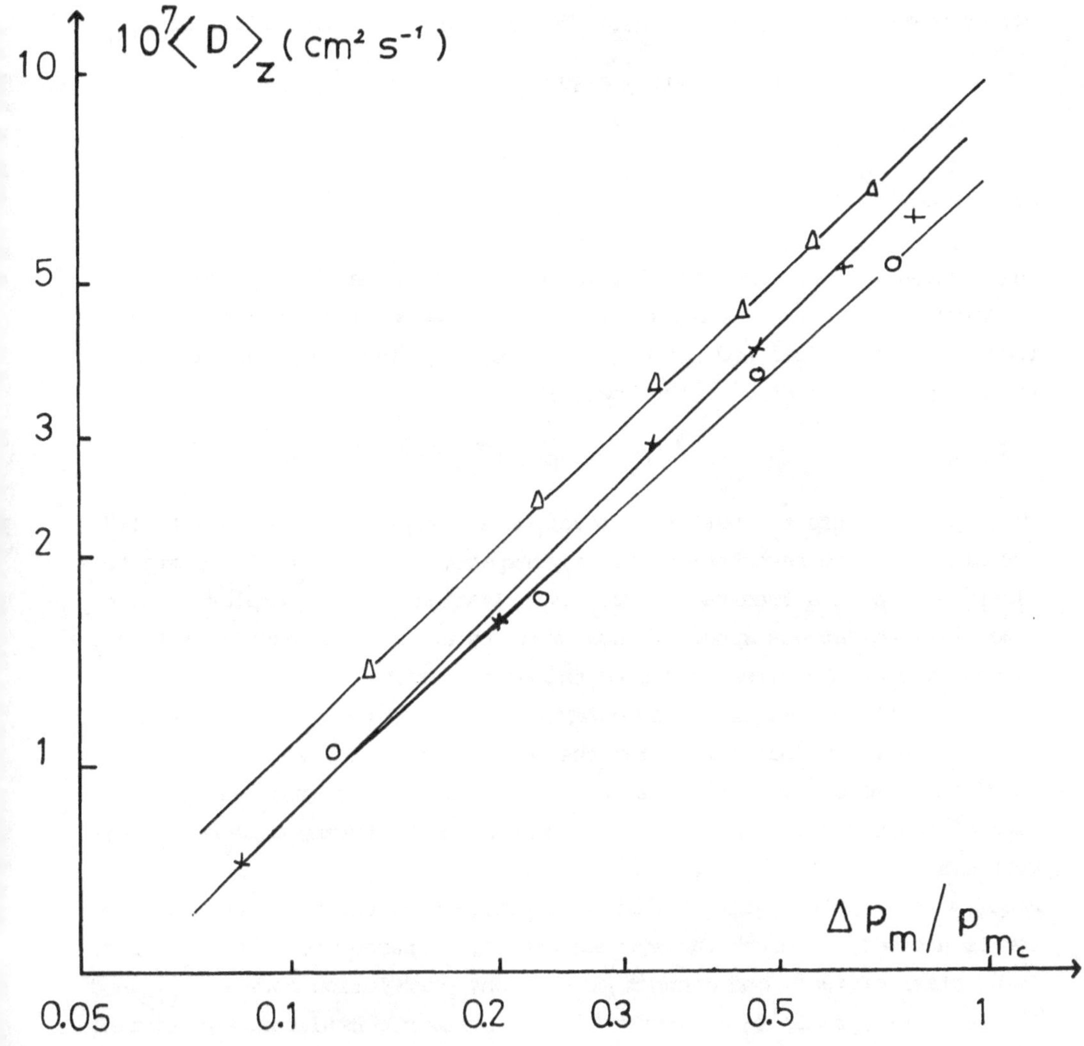

Fig.3 Variations of $\langle D\rangle_z$ versus $(\Delta p_m/p_{mc})$ for styrene-mDVB systems ; Δ series V ; + series VI ; o series VII.

ted by power laws $\langle D\rangle_z \sim (\Delta p_m/p_{mc})^{\nu_{exp}}$ with the following values of ν_{exp} :

0.92 (VII) ; 0.97 (V) ; 1.02 (VI).

Discussion

Until recently, the theories of gelation have been mainly applied to the description of f-functional random polycondensation (10). For such reactions, both classical and percolation models predict a divergence of M_w at the gel point according to the power law :

$$M_w \sim (\Delta p/p_c)^{-\gamma}$$

where p represents the extent of reaction and γ is equal to 1 and to 1.7 for classical and percolation models, respectively (10, 6). The characterization of p in a free radical copolymerization is more complex because generally the various kinds of unsaturations are not isoreactive (11-13). In the present study, we have used the overall conversion degree p_m as characteristic parameter of the extent of the reaction. In spite of different reaction probabilities for the double bonds of styrene and for those of the comonomer, p_m represents also to a good approximation the fraction of reacted double bonds, since the comonomer/styrene mole-ratio is very small.

A quite remarkable result is that γ_{exp} varies only slightly from one system to another, although the systems exhibit different reactivity ratios. As a matter of fact, one doesn't observe any correlation between γ_{exp} and the reactivity ratio r_s of styrene in the systems considered. For instance, r_s is approximately the same for the styrene-m DVB and styrene-EDM systems (13) whereas the values of γ_{exp} differ significantly (1.57 for series III, $\sim$ 1.7 for series IV and V). Therefore, it seems reasonable to assume that the fluctuations in the value of γ_{exp} result from experimental uncertainty (mainly in the determination of p_{mc}). Moreover, the value of γ_{exp} is, as a first approximation, independent of r_s. The average value $\gamma_{exp} \simeq 1.65$ is quite close to the prediction of the percolation model.

The molecular weight dependence of $\langle D\rangle_z$ provides a better test for the

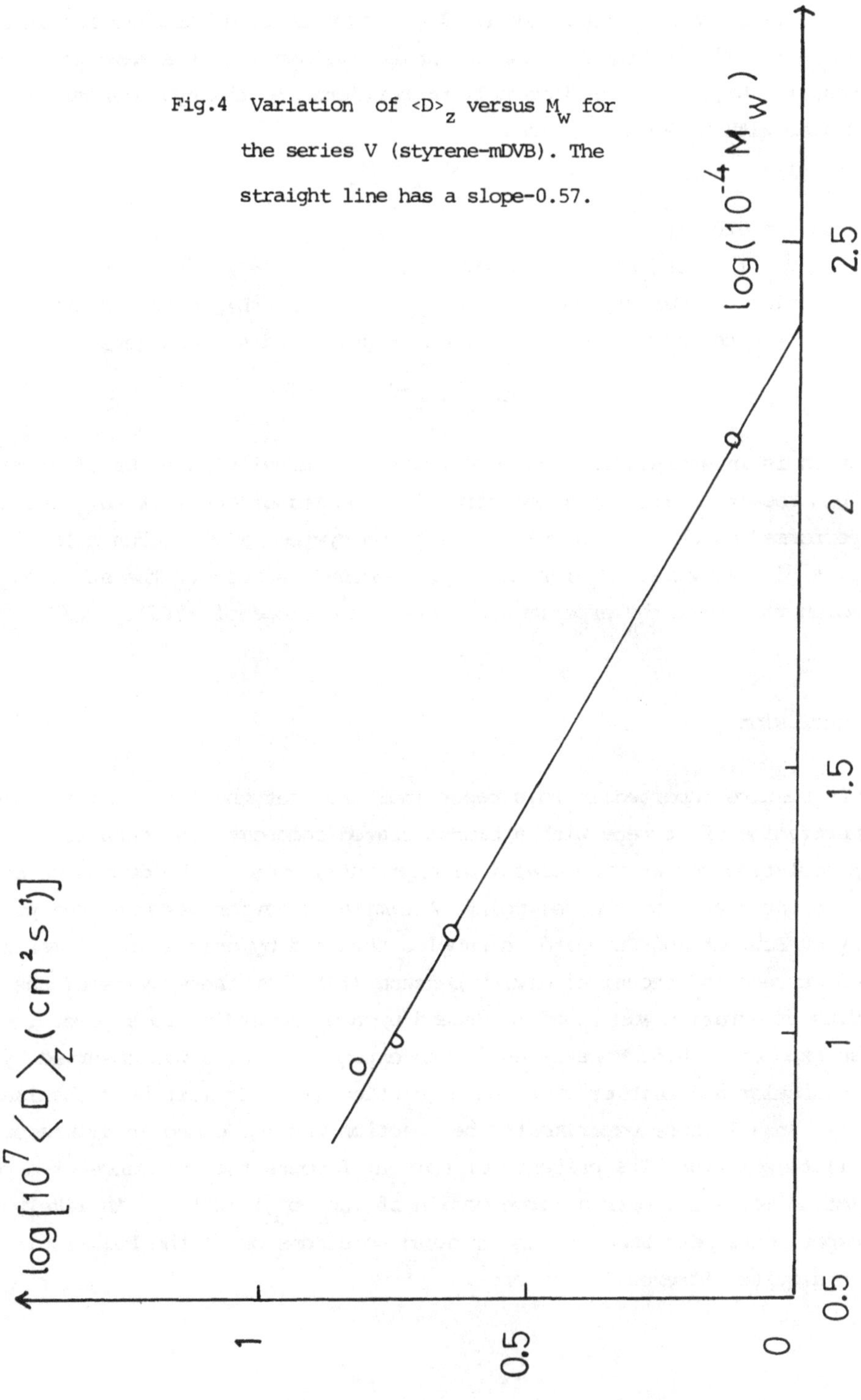

Fig.4 Variation of <D>$_z$ versus M$_w$ for the series V (styrene-mDVB). The straight line has a slope-0.57.

theoretical models. Daoud et al (15) assume that, after dilution in a good solvent, all the branched polymer molecules swell in the same way. As a result, $\langle D\rangle_z$, which is inversely proportional to the hydrodynamic radius, scales with M_w according to :

$$\langle D\rangle_z \sim M_w^{-\nu}$$

with ν = 0.63 in the percolation theory.

ν = 1 in the Flory-Stockmayer model.

Combining the variations with $\Delta p_m/p_{mc}$ of M_w and $\langle D\rangle_z$ obtained experimentally for the styrene-m.DVB systems (Figs. 3 and 4), one gets :

$$\langle D\rangle_z \sim M_w^{-0.57 \pm 0.05}$$

which is in satisfactory agreement with the prediction of Daoud et al. This result is also obtained directly from measurements of $\langle D\rangle_z$ and M_w performed on the same samples of a given system. Fig.4 shows a log-log plot of $\langle D\rangle_z$ versus M_w for the styrene-m.DVB, series V. The straight line which fits best the experimental data has a slope of -0.57.

Conclusion

The results reported in this paper indicate that the free radical copolymerization of styrene with a biunsaturated comonomer is characterized by percolation rather than classical exponents, even in the domain of conversion far away from the gel point. A similar behavior was observed previously by Schmidt and Burchard on samples obtained by anionic copolymerization of styrene and technical divinylbenzene (18). For these systems, the radius of gyration was found to depend upon M_w according to a power law with an exponent $\sim$ 0.5. More recently, an exponent of 0.58 was observed by Schosseler and Leibler in a chain gelation (19). It must be emphazised that in all these experiments the reaction was performed in dilute or semidilute solution. The presence of solvent favours the occurence of a percolation mechanism over a large domain of $(\Delta p_m/p_{mc})$ (20). It is likely that experiments performed in concentrated solutions or in the bulk would lead to a quite different behavior.

References

1. De Gennes, P.G.: J. Phys. Lett. 37, L1 (1976).
2. Stauffer, D.: J. Chem. Soc., Faraday Trans. 2, 72, 1354 (1976).
3. Ziff, R.M., Hendriks, E.M., Ernst, M.H.: Phys. Rev. Lett. 43, 593 (1982).
4. Manneville, P., de Seze, L.: in "Numerical Methods in the Study of Critical Phenomena", Ed. Della Dora, J., Demongeot, J., and Lacolle,B., (Springer Verlag, Berlin, 1981).
5. Herrmann, H.J., Landau, D.P., Stauffer, D.: Phys. Rev. Lett. 49, 412 (1983).
6. See for instance the review article : Stauffer, D., Coniglio, A., Adam, M.: in Adv. Polymer Sci. 44, 103 (1982).
7. Flory, P.J.: "Principles of Polymer Chemistry", Chap.IX (Cornell University Press, Ithaca, N.Y. 1953).
8. Stockmayer, W.H.: J. Chem. Phys. 12, 125 (1944).
9. Gordon, M.: Proc. Roy. Soc. A268, 240 (1962).
10. Burchard, W.: Adv. Polym. Sci. 48, 1 (1983).
11. Okasha, R.: Thesis, Strasbourg, (1983).
12. Hild, G., Rempp, P.: Pure Appl. Chem. 53, 1541 (1981).
13. Hild, G., Okasha, R.: Makromol. Chem., submitted to publication.
14. Munch, J.P., Ankrim, M., Hild, G., Okasha, R., Candau, S.: Macromolecules 17, 110 (1984).
15. Daoud, M., Family, F., Jannink, G.: J. Physique Lett. 45, 193 (1984).
16. Koppel, D.E.: J. Chem. Phys. 57, 4814 (1972).
17. Brown, J.C., Pusey, P.N., Dietz, R.: J. Chem. Phys. 65, 1101 (1975).
18. Schmidt, M., Burchard, W.: Macromolecules 14, 370 (1981).
19. Schosseler, F., Leibler, L.: J. Physique Lett., to be published.
20. Daoud, M.: J. Physique Lett. 40, L201 (1979).

MICROPHASE SEPARATION IN POLY(ACRYLAMIDE-BISACRYLAMIDE) COPOLYMERIZED GELS

Erik Geissler and Anne-Marie Hecht

Laboratoire de Spectrométrie Physique*,
Université de Grenoble I, B.P. 68,
38402 St Martin d'Hères Cedex, France.

Abstract

Radically copolymerized gels are in general inhomogeneous. A model of the phase separation has been proposed by Weiss, Van Vliet and Silberberg for the system poly(acrylamide-bisacrylamide). Their predictions are, however, in disagreement with light and neutron scattering measurements. Here, we describe an alternative mode of phase separation in this system that compares favourably with the scattering observations.

Introduction

Gels copolymerized from aqueous solutions of acrylamide and bisacrylamide undergo phase separation. Weiss, Van Vliet and Silberberg (1) and Janas, Rodriguez and Cohen (2) have investigated the permeability and shear modulus of these gels; they find, on the basis of certain assumptions, that a form of microsyneresis (3) takes place, whereby there are regions in which the polymer content is drastically depleted. The variations that they deduce for the concentration and characteristic dimensions of the constituent phases are inconsistent with neutron scattering experiments (4) as well as with observations of turbidity. Here, we examine the possible existence of other modes of phase separation.

Theoretical Section

Bansil (5) reports evidence from Raman spectroscopy of grouping among bisacrylamide molecules in this system. In line with this picture, we assume that the gel consists of a weakly cross-linked phase which can swell (phase 1), plus a heavily cross-linked phase which cannot (phase 2). We make the following further assumptions:

1) Simple additivity of the shear moduli of each phase,

$$G = \beta G_1 + (1 - \beta) G_2 \qquad (1)$$

* CNRS associate laboratory.

Physical Optics of Dynamic Phenomena and Processes in Macromolecular Systems

where β is the volume fraction of phase 1. G is assumed to limit the swelling of the whole sample in the free solvent. The consequences of the somewhat drastic assumption contained in Equation 1 are discussed briefly in the last section of this paper.

2) Within each phase, associations of bisacrylamide molecules are random.

3) Direct links between bisacrylamide molecules are discounted from the total number of active chains contributing to the shear modulus (e.g. a bisacrylamide dimer is simply counted as a six-functional cross-link). Straightforward binomial combination theory then gives for the shear modulus of each phase

$$G_j = \frac{RT}{m} \frac{a_j b_j}{a_j + b_j/2} \qquad (2)$$

where m is the molecular weight of the acrylamide molecule and R is the gas constant. The factor ½ multiplying b_j in the denominator of Equation 2 takes account of the molecular weight of bisacrylamide being approximately 2m.

Conservation equations govern the total concentrations a and b of acrylamide and bisacrylamide respectively:

$$a = \beta a_1 + (1 - \beta)a_2 \qquad (3)$$

$$b = \beta b_1 + (1 - \beta)b_2 . \qquad (4)$$

On swelling from the prepared state to that of swelling equilibrium in excess solvent, the total swelling ratio is V, and the concentration of acrylamide in phase 1 decreases from a_1 to a*, where

$$a^* = \frac{a_1}{V - 1 + \beta} , \qquad (5)$$

since phase 2 is assumed not to swell. a* can be estimated from the Flory equation for the swelling pressure of a gel

$$0 = - RT\{\log(1 - \phi) + \phi - \chi\phi^2\} - G/V^{1/3} \qquad (6)$$

or an equivalent expression. Here $\phi = a^*/\rho$, the volume fraction of polymer within phase 1, and χ the Flory-Huggins parameter (ca. 0.45 to 0.48). The results do not depend qualitatively on the value of χ chosen. We thus have four measurable quantities, a, b, G and V, giving four independent equations. There are, however, five unknowns, namely a_j, b_j and β, so that the equations are, as usual, underdetermined. Equations 1 to 5 can be combined to give

$$A\beta^2 + B\beta + C = 0 \tag{7}$$

where the coefficients can be expressed in the form

$$\begin{aligned} A &= 2a^*a - 2a^{*2}b - a^{*2}a - ax^2 + f(a^* + x)^2 \\ B &= 2ax^2 + 2a^*ax(V - 2) - 4a^{*2}b(V - 1) - 2a^{*2}a(V - 1) + 2a^2(a^* - x) \\ &\quad + 3aa^*b - abx - 2f(a^* + x)^2 + f(a^* + x)(2a^*V + b) \\ C &= 3a^*ab(V - 1) - 2a^{*2}b(V - 1)^2 - 2a^*ax(V - 1) + 2a^*a^2(V - 1) \\ &\quad + abx - a^2b + 2a^2x - a^3 - ax^2 - a^{*2}a(V - 1)^2 + f(a^*V - a^* - x)^2 \\ &\quad + fb(a^*V - a^* - x) + fa(a + b). \end{aligned} \tag{8}$$

The total concentration of the dense phase 2 is

$$x = a_2 + b_2, \tag{9}$$

and the parameter f is defined as

$$f = mG/RT$$

Solutions of Equation 7 with real positive values of a_j, b_j and β are confined to an extremely narrow region in the vicinity of

$$x, \beta \simeq 1, \tag{10}$$

i.e., the dense phase really is dense, as well as being a minority phase. Within the narrow range (10), a large number of solutions are unsuitable due to their instability to osmotic swelling ($a_1/b_1 < a_2/b_2$). Such unwanted solutions are avoided by requiring that the two solutions of Equation 7 be equal. This simplifying procedure is equivalent to assuming that there is thermodynamic equilibrium in the distribution of matter between the two phases, an assumption which may only have limited justification.

Experimental Section

A series of polyacrylamide-water gels of constant total concentration $a + b = 0.08$ g cm^{-3} was prepared, having different values of bisacrylamide concentration ($b = 1, 1.5, 2, 2.5, 3, 4$ and 5×10^{-3} g cm^{-3}). 0.70 mg/ml of ammonium peroxodisulfate and 0.28 µl/ml of TEMED were used as catalysers. The gels were left for two weeks to allow for completion of the reaction, then removed from their mould, weighed, and their shear modulus measured.

This was done by observing through a microscope the deformation of the gel under a known pressure. The samples were then placed in water and allowed to swell to equilibrium at room temperature for three weeks, whereupon they were weighed.
The light scattering measurements were performed with a standard FICA apparatus, and the turbidity measurements with a BECKMAN spectrophotometer.

Results and Discussion

The results, shown in Table 1 and Figure 1 are in agreement with the neutron scattering observations[4] which suggest that the majority phase is increasingly depleted in polymer as cross-linking is increased.
The model described above makes no predictions as to the spatial distribution of the microphase separation: this remains a free parameter, to be determined experimentally. The scattering function, obtained from the static light scattering experiments, rules out the possibility of lamellar or rod-like structures. At the lowest cross-linking densities, two distinct lengths can be discerned, one of the order of a few microns in length, and the other of the order of tens of nanometers. As the cross-linking density increases, intermediate lengths appear, and at the highest values of b, the Zimm plot becomes a straight line again. Table 1 contains the shortest apparent length visible in these spectra, R_G.

Table 1. Summary of results

b (g/l)	G (dyn/cm^2) $\times 10^5$	V	a_1 g cm^{-3}	a_2 g cm^{-3}	b_1 g cm^{-3} $\times 10^{-3}$	b_2 g cm^{-3}	β	Turbidity at 632.8 nm(cm^{-1})	R_G apparent(nm)
1	0.73	1.71	0.076	0.968	0.902	0.032	0.997	----	22.1
1.5	0.96	1.52	0.077	0.832	1.27	0.167	0.999	----	19.7
2	1.18	1.31	0.075	0.898	1.59	0.102	0.996	----	30.8
2.5	1.24	1.21	0.071	0.902	1.80	0.098	0.993	0.09	62.7
3	1.47	1.15	0.073	0.846	2.19	0.153	0.995	0.22	86.8
4	1.11	1.04	0.059	0.895	2.31	0.088	0.980	0.46	125.7
5	1.09	1.005	0.057	0.721	2.74	0.088	0.973	2.62	134.5

N.B. Value of χ taken to be 0.47 in Equation 6

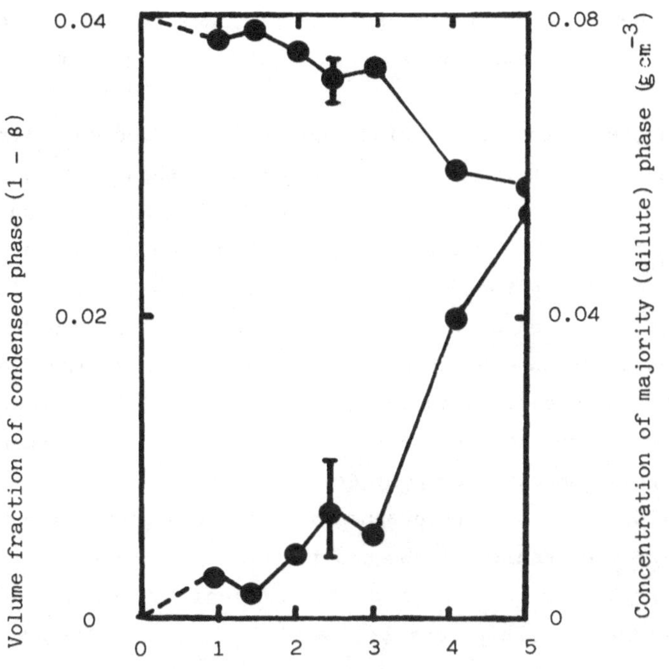

Figure 1. Upper curve: calculated acrylamide concentration of majority phase 1 (right hand scale) as a function of total bisacrylamide content. Lower curve: calculated variation of the volume fraction of the condensed phase 2, (1 - β). The error bars shown correspond to the estimated ±10% error in the measurement of the shear modulus G. Total acrylamide + bis-acrylamide content, 0.08 g cm^{-3}.

In Table 1 are also listed the values of the turbidity where it was large enough to be measured. It can be seen that this parameter increases with increasing bisacrylamide: this behaviour contrasts with the results of reference 1 from which the opposite variation can be calculated for samples similar to those investigated here.

The turbidity is caused by the difference in refractive index n between the two phases, and since the concentration of each phase is known, so is n. One may imagine a model of the gel in which the condensed phase is confined inside isolated spheres of radius R_G. On inserting the numerical values of the turbidity, one finds that the volume fraction of the condensed phase inside these spheres is the same, within experimental error, as that calculated for the whole sample, namely (1 - β). We conclude that the condensed phase permeates the whole sample, and may better be described by a range of characteristic lengths rather than a single correlation length.

Recent small angle neutron scattering experiments performed at the D 11 facility of the Institut Laue Langevin in Grenoble show that gels similar to those of Table 1 display an intense forward scattering, in addition to that caused by the density fluctuations in a homogeneous polymer solution (6). Variation of the proton/deuterium ratio in the water contained in the gels produces a variation in contrast for the objects causing this excess scattering. The observed behaviour and intensity of this signal are not compatible with the model (1) in which the inhomogeneities are composed of pockets of solvent, but, on the contrary, indicate that the minority phase consists of highly concentrated polymer.

Although the model outlined above appears to be consistent with the principal features of the results of the scattering experiments, some caution is called for regarding certain of the assumptions. In particular, one of the more criticizable assumptions is the additivity of the shear moduli (Equation 1): compared to standard models of composite elastomers containing solid particulate fillers (7), Equation 1 overestimates G. This has the effect of shifting β to slightly too high values. In view of the complex structure that we find for the dense phase, however, it is not certain that the classical models for filled elastomers give a better description than Equation 1. The present analysis is not, in fact, very sensitive to the precise form of the coefficients of G_1 and G_2 in Equation 1, except for the most strongly cross-linked samples. If, instead of Equation 1, Kerner's equation (8) is used for the total shear modulus, the resulting decomposition also consists of a highly cross-linked phase together with a phase depleted in polymer, with parameters very similar to those obtained above. However, the mathematical solutions are no longer confined to a single value of $x = a_2 + b_2$, but admit of a broad range, characteristic of multiphase decomposition.

Our general conclusion is that although microphase separation of the type described in reference 1 may exist in certain systems, the experimental evidence for the samples investigated here is in better agreement with a mode in which there is a highly concentrated and densely cross-linked phase coexisting with a majority phase that is slightly depleted in polymer.

References

1. Weiss, N., Van Vliet, T., Silberberg, A.: J. Polymer Sci. Polymer Phys. Ed. 17, 2229-2240 (1979).

2. Janas, V.F., Rodriguez, F., Cohen, C.: Macromolecules 13, 977-983 (1980).

3. Dusek, K., Prins, W.: Adv. Polymer Sci. 6, 1-102 (1969).

4. Geissler, E., Hecht, A.-M., Duplessix, R.: J. Polymer Sci. Polymer Phys. Ed. 20, 225-233 (1982).

5. Bansil, R., Gupta, M.K.: Ferroelectrics 30 63-71 (1980).

6. Duplessix, R., Hecht, A.-M., Geissler, E.: in preparation.

7. Dickie, R.A.: J. Polymer Sci. Polymer Phys. Ed. 14, 2073-2082 (1976).

8. Kerner, E.H.: Proc. Phys. Soc. 69 B, 808-813 (1956).

RAYLEIGH AND BRILLOUIN SCATTERING FROM LINEAR POLYMER DURING GELATION

Takashi Igarashi and Shingo Kondo

Research Institute of Composite Materials, Gunma University
1-5-1, Tenjin-cho, Kiryu 376 Japan

Introduction

Linear polymers having initially fluidity become rubbery solids as a result of crosslinking. At a gel point elastic modulus increases remarkably accompanying a maximum in energy dissipation (Figure 1). This behaviour is usually observed by "torsional braid analysis" or by coaxial type torsional vibration method and can be explained on the basis of theories of rubber elasticity and of relaxation in polymer chains. Since Brillouin spectra reflect the molecular motion on the time scale of order of 10^{-10} s, it may be interesting to investigate the effects of gelation on the propagation velocity and attenuation of hypersonic sounds (phonons with long

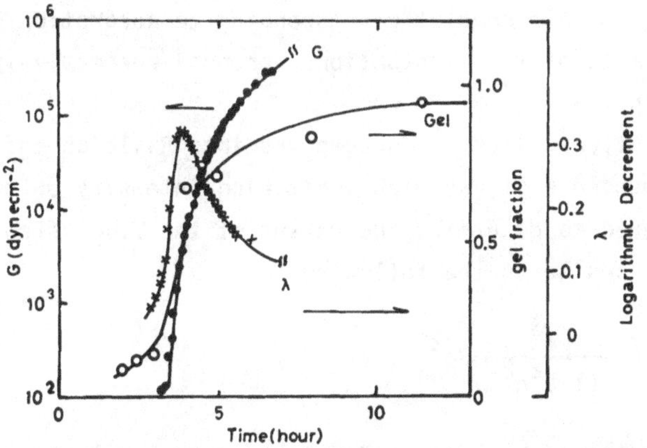

Figure 1. Change in elastic properties measured in low frequency (coaxial type torsional vibration method) during gelation.

Physical Optics of Dynamic Phenomena and Processes in Macromolecular Systems

wavelengtn) which are the Fourier components of the thermal motion of molecules. This is also a problem of whether the relaxation in polymer chains have any meaning on this time scale. On the other hand, high resolving power for the spectral lines and light amplification technique adopted in Brillouin spectroscopy enable us to detect "slight" density fluctuations associated with the gelation of linear polymers.

Rayleigh and Brillouin Scattering

By a spectroscopy system having a sensitivity of order of 10^8 Hz (0.1 GHz) one can split the scattered light into triplet and can measure Rayleigh and Brillouin spectral intensities I_C and I_B, Brillouin shift, ν_B, and Brillouin spectral line width, $\Delta\nu_B$.

The light generally is scattered by fluctuations in dielectric constant of the medium, which are accompanied with fluctuations in the density in this experiment. Frequency of the light scattered from static fluctuations is not affected by them. This is elastic scattering. Contrary, frequencies of the scattered light from time dependent fluctuations are shifted up or shifted down into doublet corresponding to absorption or emission of quantized wave mode of the fluctuations (phonon) respectively. This is inelastic scattering.

After the scattered light is decomposed into Rayleigh and Brillouin components the dependence of Rayleigh scattering intensity on the scattering angle is measured to determine the extent of the fluctuated cloud, "a", and density, ρ, according to the following formula

$$I_r(q) = A \frac{a^3}{(1+a^2q^2)} \delta\rho^2 \qquad (1)$$

This formula was proposed by Debye and Bueche[1] assuming that the correlation function for the static fluctuation have the form of $\langle\delta\rho(r),\delta\rho(0)\rangle \sim e^{-|r|/a}$, where $I_r(q)$ is the intensity of the scattered light, "a" is the correlation length of the fluctuation, q is the magnitude of the scattering vector, A is a constant determining the intensity of the scattered light and $\delta\rho(r)$ is the density deviation at r. The symbol $\langle\cdots\cdots\rangle$ means average.

In the case of the inelastic scattering of light by phonon, i.e. Brillouin scattering, conditions for conservation of energy and of momentum lead to relations:

$$\omega_1 = \omega_0 \pm \omega_s \tag{2}$$

and

$$k_1 = k_0 \pm k_s \tag{3}$$

where ω_1 and k_1 are the angular frequency and wave number of the scattered light respectively, ω_0 and k_0 denote those of the incident light and ω_s and k_s those of phonon. Further in the case if phonon damping is taking place during the propagation in the medium, broadening of the Brillouin lines is observed. When the phonon damping obeys the form of $e^{-\Gamma t}$, the scattering law, $S(q,\omega)$, is finally given as

$$S(q,\omega) \sim \frac{\Gamma(q)}{(\omega - (\omega_0 \pm \omega_s))^2 + \Gamma(q)^2} \tag{4}$$

where q is the magnitude of scattering vector and Γ is the damping coefficient of phonon per unit time. The damping coefficient per unit distance, α, is given by

$$\alpha = \Gamma/v_s \tag{5}$$

where v_s is the propagation velocity of phonon. The damping coefficient per one wavelength of the phonon, $\alpha\lambda_s$, is Γ/ν_s, where λ_s is the wavelength of the phonon and ν_s is its frequency. Since Brillouin shift, $\nu_B = |\omega_0 - \omega_1|/2\pi$, should agree with the frequency of the interacting phonon: $\nu_s = \nu_B$, then $\alpha\lambda_s$ becomes Γ/ν_B. On the other hand, it is known that v_s is related to ν_B as

$$v_s = \lambda_0 \nu_B / 2n_r \sin(\theta/2) \tag{6}$$

where λ_0 is the wavelength of the incident light, n_r is the refractive index of the specimen and θ is the scattering angle. From the Loretzian profile of Brillouin spectrum represented by equation (4), one can easily estimate half width at half height, $\Delta\nu_B$, as

$$\Delta\nu_B = \Gamma/2\pi = \alpha v_s/2\pi$$

Materials

α,ω-divinylpolydimethylsiloxane (100 parts) and polydimethylsiloxane (5.2 parts) are reacted at 25°C. The mixture with 3000 cstokes becomes rubber with shear modulus of 3×10^6 dyn/cm^2 in about 10 h.

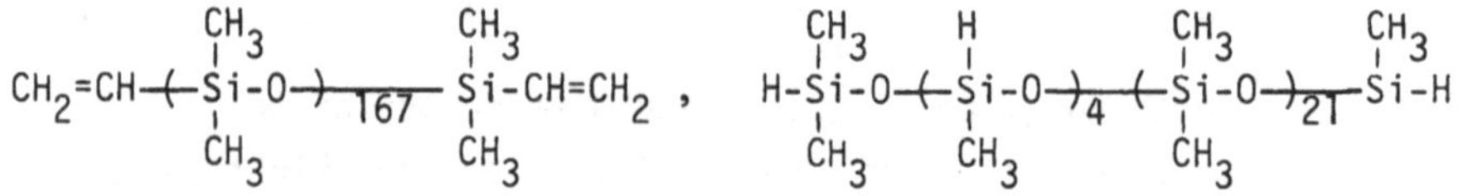

Results

Gel components in the reacting system are detected after 3-4 h, increase quickly and level off. Keeping the pace with the gel fraction, quick increase in elastic modulus is observed. This time may be regarded as the "gel point", when the energy dissipation represented by logarithmic decrement, λ, is observed (Figure 1).

Figure 2 shows that Rayleigh scattering intensity increases prior to the formation of gel and the first maximum appears at the gel point (a), while change in Brillouin scattering intensities cannot be observed (a). Change in Brillouin shift and in phonon damping, $\alpha\lambda_s$, also cannot be observed (b). These results mean that light (6328 Å) is scattered by static density fluctuation emerging when infinite network are formed, while phonons with the comparable wavelength (3000 Å) cannot be scattered by them. Prior to detection of gel, brilliant "speckle pattern" is observed in the specimen, which is quantitatively represented by standard deviation, σ, of its intensities. The change in σ during the gelation seems to be parallel to Rayleigh intensity (Figure 3). The density and the refractive index increase monotonically during the reaction, and any bend or jump in these curves cannot be observed.

The values of $I_C/2I_B$ are found to depend on the scattering angle, which bring knowledge of the distribution of the density fluctuation. As mentioned above, assuming that the density correlation is represented by $e^{-|r|/a}$, one can estimate local density fluctuation, $\delta\rho^2$, and their spatial

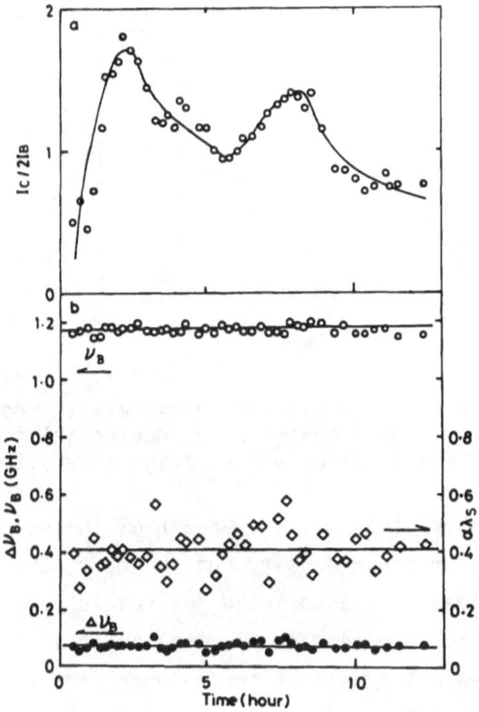

Figure 2. Properties obtained from Rayleigh and Brillouin spectra during gelation. I_C:Rayleigh spectral intensity. I_B:Brillouin spectral intensity. ν_B:Brillouin shift. $\Delta\nu_B$:Brillouin spectral line width. α:phonon damping coefficient per unit length. λ_s:wavelength of phonon.

extent according equation (1).

Discussion

Considering that the gel fraction reflects the rate of infinite network production and the above obtained results, it is suggested that local density fluctuations due to partial network structures emerge prior to de-

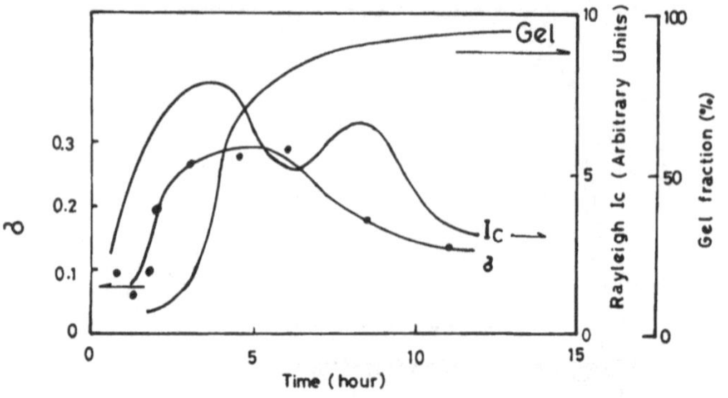

Figure 3. Change in speckle pattern intensity represented by the standard deviation, σ, during gelation compared with gel fraction and Rayleigh intensity.

tection of the gel by extraction method. Growth of the network parts, accompanying the increase in their density and in their number, may lead to associate among themselves. The associated "microgels" may give the minimum correlation length and maximum density fluctuation shown in Figure 5. As the result of the subsequent growth of the network structure, gel parts may become "majority", and the "minority" of "open space" may scatter the light instead of gels. This is duality theorem or Babinet's theorem. The "open space" parts which scatter light finally become narrow and rare parts (Figure 6).

Next, it should be noted that the life of the phonon with long wavelength is not affected by the above mentioned density fluctuation. The density fluctuation in our "microgels" are of the size of the wavelength of the scattered light, and the phonons interacting with the light also have comparable wavelength to the size of "microgels". If the phonons are scattered by them inelastically, the life of phonons will be affected by emerging of the "microgels" as the energy of torsional oscillation is remarkably dissipated at the gel point. The values of $\alpha\lambda_s$ are observed to be constant during the gelation. Change in Brillouin shift during the gelation also cannot be observed. These mean that the elastic modulus determining the propagation velocity of phonons is not affected by the gelation, and the elastic modulus is thought to be independent of rubber elasticity.

Studies on the propagation velocity of phonons were made on the basis

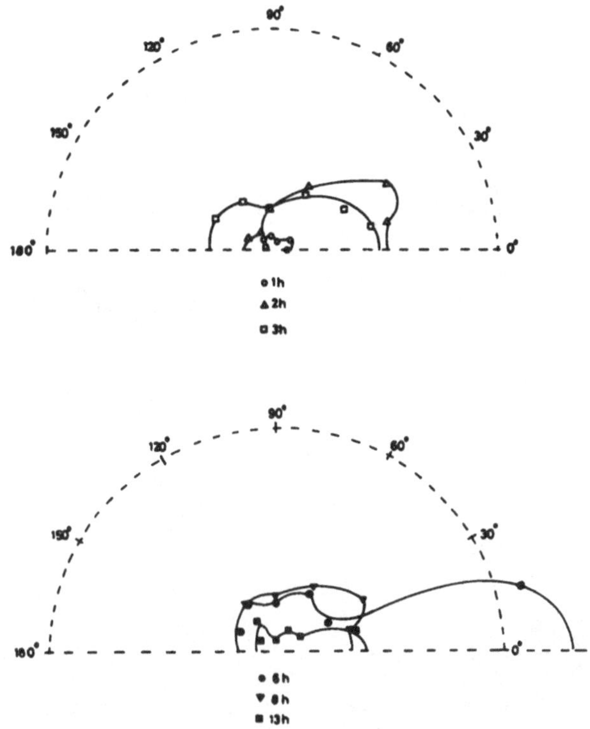

Figure 4. Angular distribution of $I_C/2I_B$ in the horizontal plane Electric field of the incident light is in the vertical direction. The appearance of directivity is associated with formation of density fluctuations with the size of the wavelength of the light.

of intermolecular potential (i.e. energetic elasticity).[2] On the other hand, hydrodynamics tells us that damping coefficient of sound is given by

$$\Gamma = \frac{1}{2}\frac{1}{\rho}\left(\frac{4}{3}\eta + \zeta\right)|k|^2 \tag{8}$$

where η is the shear viscosity and ζ is the volume viscosity. The value of $\alpha\lambda_s(=\Gamma/\nu_B)=0.4$ observed for phonon (Figure 2) gives the value of $4/3\cdot\eta+\zeta=0.6$ poise according to equation (8), if the value of λ_s is regarded as 3000Å, i.e. $k=2\times10^5\text{cm}^{-1}$. This value of 0.6 poise cannot be the value deduced from motion of long chain molecules in rubbery state. Phonon damping also may be led from anharmonicity of intermolecular potential.

ally perfect dielectric crystals under the condition $\omega_s\tau<<1$,

$$\alpha\lambda_s = \frac{2\pi C_v\gamma^2 T\tau\omega_s}{3\rho v_s^2} \tag{10}$$

where C_v is the isochoric specific heat, T is the absolute temperature and ρ is the density. In gas-kinetic theory one finds the following expression for the thermal conductivity, κ, of solids:

$$\kappa = \frac{1}{3} C_v v_s^2 \tau \quad , \tag{11}$$

$\alpha\lambda_s$, therefore, becomes

$$\alpha\lambda_s = \frac{2\pi\kappa\gamma^2 T\omega_s}{\rho v_s^4} \tag{12}$$

They discussed attenuation of sound of order of 10^9 - 10^{10} Hz in crystalline quartz at low temperature and suggested that equation (10) well predicts the experimental results, particularly dependence of $\alpha\lambda_s$ on temperature.

In the case of glassy polymers at room temperature, however, Jackson et al. [5] pointed out that values of $\alpha\lambda_s$ given by equation (10) are far too small compared with observed values. They expected that the discrepancy is due to the scattering of phonons by static inhomogeneities or by static imperfections. They[6], further, made experiment with gelatin solution and found out that the increase in line width is proportional to the gelatin concentration in the range from 0 to 0.2 g/cm^3.

There are some questions concerning the explanation that the static inhomogeneities or the static strain field round static imperfections may scatter the phonons. Figure 7 shows that $\alpha\lambda_s$ obtained from the line width of Brillouin spectra decrease monotonically from 0.5 at the beginning to 0.2 during the hardening of epoxy resin, while the $I_C/2I_B$ reflecting the static inhomogeneities increase. The values of $\alpha\lambda_s$ which are at low temperature usually of order of 10^{-5} for perfect crystals may become of order of 10^{-1} at room temperature when their perfection is lost as in fused quartz.[7]

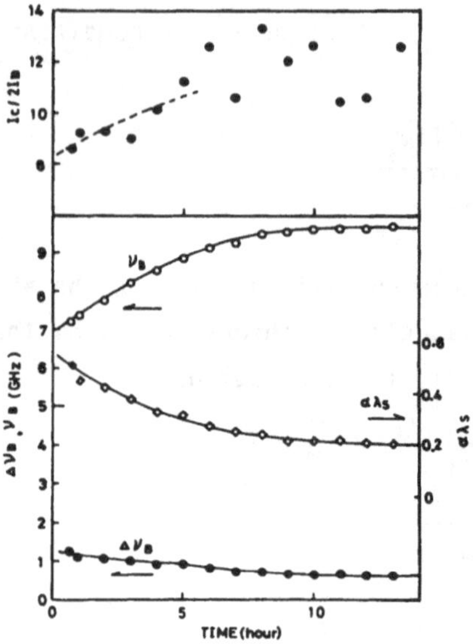

Figure 7. Properties of epoxy resin during isothermal (25°C) hardening process. Resin is hardened with 1,3-bisaminomethylcyclohexane and allylglycidylether.

It is known that water has at room temperature $\alpha\lambda_s$ of 0.2. No strain field exists in water. However, epoxy resins with remarkable residual strain field[8] has also $\alpha\lambda_s$ of 0.2. Woodruff's equation (12) gives from observed values of κ (for water 6×10^4 and for epoxy resin 4×10^5 erg/cm s K): $\alpha\lambda_s$=0.02 for water and 5×10^{-4} for epoxy resin. Thus estimated value of $\alpha\lambda_s$ for water is of the order of 10^{-1} of the experimental value. It is known that thermal conduction itself gives at high temperature phonon damping having the comparable order of magnitude. This effect, however, may not give the way out of the discrepancy.

Recently, other treatments on the phonon damping are made by applying phonon Green's function method[9] or generalized Langevin equation method[10], and both lead to the same conclusion:

$$\alpha \sim (\phi(r)''')^2 \omega_s T \qquad (13)$$

where ϕ is intermolecular potential and ϕ''' represents the effect of anharmonic term (=6g). Of course, these studies are made to treat crystalline solids. For disordered state such as liquids or amorphous solids, plane waves do not represent exact eigen state. In such a system, additive term, $Q_{kk'}$, proportional to structure factor, S(k-k'), appears in Hamiltonian in plane wave representation, which act as a perturbation on phonon to give rise to transitions between phonon mode of k and k'.[11,12] In other words, disorders absorb phonon k and emit phonon k'. Though this process takes place in inelastic manner, and gels in our system may be regarded as such disorders, Brillouin spectra are scarcely aware of these kinds of disorders. Anharmonic terms, therefore, may directly determine the life of phonons in our systems. The problem of parallelism between thermal expansion coefficient, which is directly related to anharmonicities of the intermolecular potential, and phonon damping is very interesting to us.

References

1. Debye,P., Bueche,A.M.: J.Appl.Phys. 20, 518 (1949).
2. Kondo,S., Igarashi,T.: J.Appl.Phys. 51, 1514 (1980).
3. Akhieser,A.: J.of Phys. 1, 277 (1939).
4. Woodruff,T.O., Ehrenreich,H.: Phys.Rev. 123, 1553 (1961).
5. Jackson,D.A., Stevens,J.R.: Mol.Phys. 30, 911 (1975).
6. Bedborough,D.S., Jackson,D.A.: Polymer 17, 573 (1976).
7. Kittel,C.: Introduction to Solid State Physics, Chapter 6, 2nd Ed. Wiley, New York 1956.
8. Igarashi,T., Kondo,S., Kurokawa,M.: Polymer 20, 301 (1979).
9. Mradudin,A.A., Fein,A.E.: Phys.Rev. 128, 2589 (1962).
10. Lovesey,S.W.: Condensed Matter Physics, Dynamic Correlations, The Benjamin/Cummings Publ. London 1980.
11. Ziman,J.M.: Models of Disorder, Cambridge University Press 1979.
12. Morgan,G.J.: J.Phys. C 1, 348 (1968).

QUASI-ELASTIC LIGHT SCATTERING STUDIES OF THE DIFFUSION OF COMPACT MACROMOLECULES THROUGH GELS

D. B. Sellen
Astbury Department of Biophysics, University of Leeds, Leeds LS2 9JT, U.K.

Introduction

Quasi-elastic light scattering measurements have been made both upon gels of cross-linked flexible polymers (1), and gels of stiff polysaccharide chains (2). In the former case the width of spectral broadening varies as $\sin^2(\theta/2)$, and a diffusion coefficient may be calculated which is the ratio of the longitudinal elastic modulus to the force per unit volume required to maintain unit relative velocity between the polymer network and solvent. For most gels this is of the same order as that of a diffusing globular protein (5×10^{-7} cm^2s^{-1}). Gels formed from stiff polysaccharide chains however, are thought to have a randomly branched fibrous structure which is almost stationary, each fibre consisting of a side-by-side association of a number of polysaccharide chains. The degree of spectral broadening is very small (one percent or less - the scattered light consisting of a stationary speckle pattern which accounts for most of the coherence), and relaxation times are of the order of tens of milliseconds and vary little with angle of scatter. They are thought to be associated with some sort of structural relaxation.

The small degree of spectral broadening and the long relaxation times associated with light scattered from polysaccharide gels, makes possible the measurement of diffusion coefficients of compact marker macromolecules within the gels by investigating heterodyne beat spectra. In this way the interstitial spaces within the gel may be investigated, and estimates made of the mass per unit length of the gel fibres. For gels of this type, partition coefficients may also be investigated by quasi-elastic light scattering. Experiments with agarose and calcium alginate gels will be

Physical Optics of Dynamic Phenomena and Processes in Macromolecular Systems

described briefly, full details having been published elsewhere (2). Experiments of this type with gels of cross-linked flexible polymers are more complicated, due to the fact that the diffusion coefficients associated with the gels are of the same order as those of the diffusing marker macromolecules. Nevertheless the author has carried out some preliminary experiments with polyacrylamide gels using bovine plasma albumin and dextrans as marker macromolecules, and these will be described more fully.

Experimental Methods

For the following reasons it is advantageous to make measurements with each gel cast as a cylinder, supported with its axis vertical and surrounded by a solution of the diffusing macromolecule.

(i) The gel is observed under conditions of equilibrium swelling.

(ii) The gel may be reduced in diameter sufficient to avoid secondary scatter without introducing troublesome reflections of the incident laser light from its surface.

(iii) Measurements may be made on gels into which the marker macromolecules have been allowed to diffuse after the gel has formed.

(iv) In some cases partition coefficients of the diffusing marker macromolecules between the gel and surrounding solution may be determined from the zero time value of the autocorrelation function.

Cylinders of gel may be formed either in glass tubing or in dialysis tubing as appropriate. In view of the time taken to carry out experiments on gels, it is desirable to add a small quantity of bacteriocide (0.01M sodium azide). Solutions should be cleaned by filtering through millipore filters before gelation.

Standard equipment may be used to obtain autocorrelation functions. However, the spectral broadening of light scattered from a gel - whether or not a diffusing marker macromolecule is present - is never complete, as the fluctuations in gelling polymer segment density invariably contain a stationary component, giving rise to a corresponding component in the speckle pattern of the scattered light. This may be easily demonstrated by slowly scanning angles of scatter whilst measuring the time integrated scattered intensity (2). The relative contribution of the heterodyne and homodyne terms to the correlation function, and also the zero delay time value relative to the baseline, depends upon which part of the stationary component of the speckle pattern is being observed. In the case of a solution of freely diffusing macromolecules, the measured correlation function is given by;

$$G^{(2)}(\tau) = N_s <n>^2 \left[1 + \beta g^{(1)}(\tau)^2\right] \quad (1)$$

if a multibit full correlator is used. N_s is the number of samples, $<n>$ is the mean counts per sample time, τ is the correlation time and $g^{(1)}(\tau)$ is the first order normalized correlation function of the scattered electric field. β is the spatial coherence factor which has a maximum value of unity, and depends upon the area of speckle pattern (which in this case is entirely mobile) observed by the scattered light receiver system. In the case of a gel exhibiting partial spectral broadening, equation 1 becomes:

$$G^{(2)}(\tau) = N_s <n>^2 \left[1 + \beta\{2r(1-r)g^{(1)}(\tau) + r^2(g^{(1)}(\tau))^2\}\right] \quad (2)$$

$$\text{where } r = <n_b>/<n> \quad (3)$$

$<n_b>$ is the counts per sample time associated with the broadened component and $<n>$ the total number of counts. $<n>$ however varies spatially such that:

$$\left[<<n>^2>_{sp} - (<<n>>_{sp})^2\right]/(<<n>>_{sp})^2 = \beta(1-R)^2 \quad (4)$$

$$\text{where } R = <n_b>/<<n>>_{sp} \quad (5)$$

The subscript sp indicates a spatial average and R is the degree of spectral broadening of the scattered light. Figure 1 shows correlation functions obtained from light scattered at two angles very near 90° from a polyacrylamide gel using a multibit correlator (greater precision could of course be obtained with longer integration times). It can be seen that the baseline changes by a factor roughly equal to the square of the relative change in the difference between zero time value and baseline. This is to be expected for small degrees of spectral broadening where the heterodyne term predominates. From Figure 1 there is no way of determining β. Therefore r cannot be determined and $g^{(1)}(\tau)$ can be determined only on the assumption that the optical beating is entirely heterodyne or entirely homodyne. It is possible to determine β by making measurements upon a macromolecular solution using as near as possible the same geometrical optics. For a freely diffusing solution the ratio of the zero time value of the correlation function to the baseline is $1+\beta$. For a gel it is $1+\beta(2r-r^2)$, so that r may be determined. If it is appropriate to describe $g^{(1)}(\tau)$ entirely

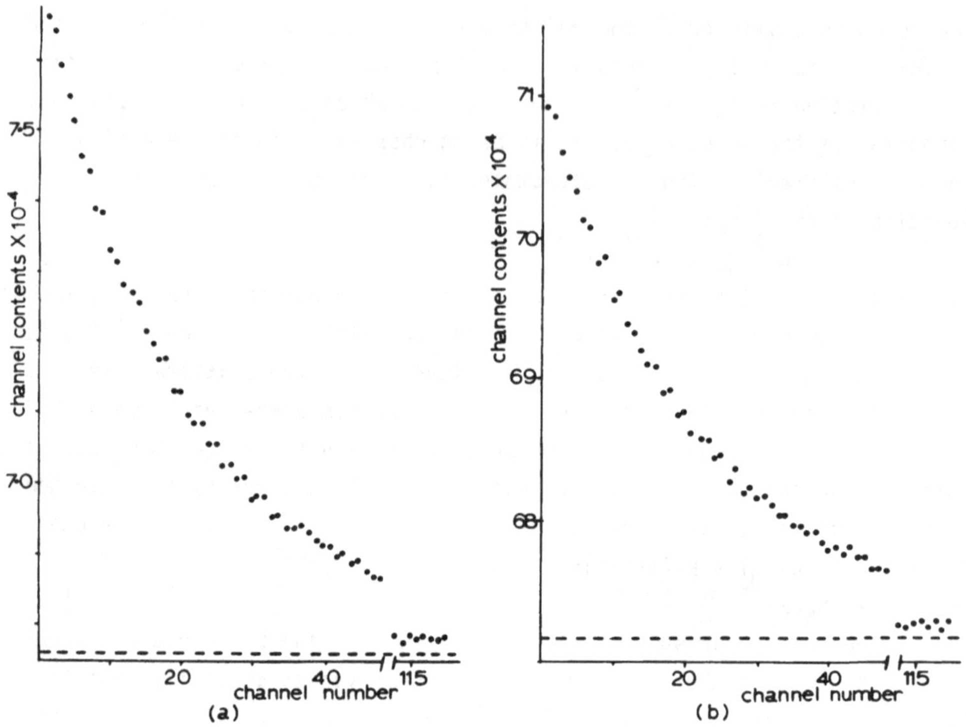

Figure 1. Correlation functions obtained from light scattered at a) 90.5^0 b) 89^0, from a 7.5% polyacrylamide gel (bis/acrylamide ratio 1%) using a multibit correlator. Sample time 3μs, integration time 160s, incident light 50mw, 633nm. The exact angles of scatter have no significance (see text). The broken lines are the computer baselines.

in terms of diffusion coefficients as would be the case for a gel of a cross-linked flexible polymer, then the normalized measured correlation function ϕ, may be expressed in terms of the various averages of diffusion coefficient by the series:

$$\ln \phi = \frac{1}{2-r}\left[-2K^2 D_z \tau + K^4 \tau^2 \{(D^2)_z - \frac{r^2-2r+2}{2-r}(D_z)^2\} \ldots\ldots\right] \qquad (6)$$

It is also possible to determine β by slowly scanning angles of scatter whilst running the photon correlator. In this case;

$$G^{(2)}(\tau) = N_s(\langle\langle n\rangle\rangle_{sp})^2 \Big[1 + \beta(1-R)^2(g_{sp}^{(1)}(\tau d\theta/dt))^2 + \beta\{2R(1-R)g^{(1)}(\tau) + R^2(g^{(1)}(\tau))^2\}\Big] \qquad (7)$$

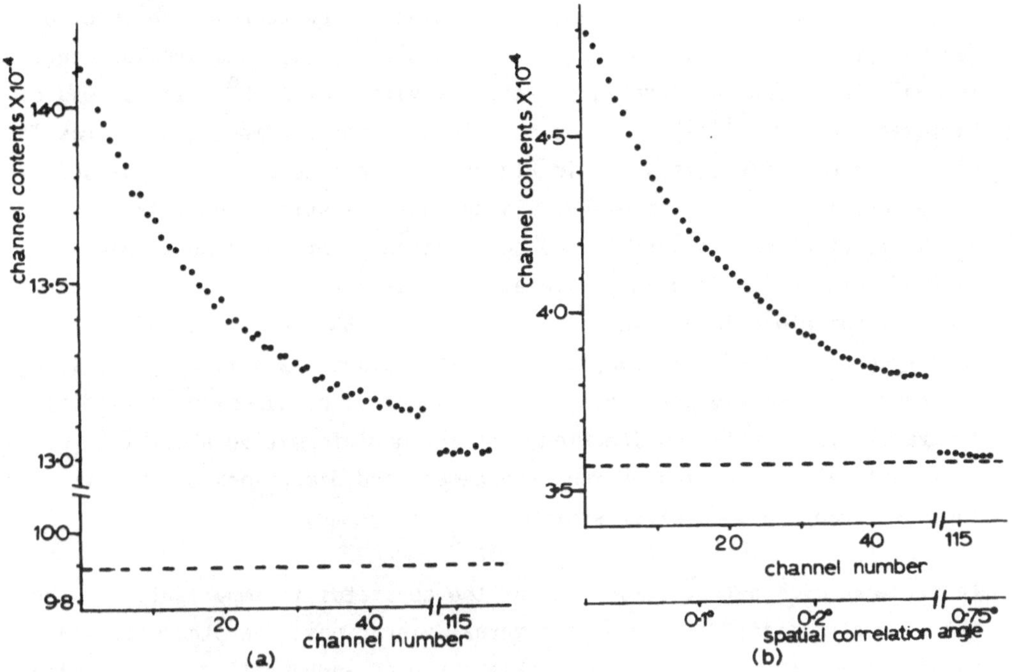

Figure 2. Correlation functions obtained from light scattered from a 7.5% polyacrylamide gel (bis/acrylamide ratio 1%) using a multibit correlator whilst scanning angles of scatter at 16s per degree. Incident light 50mw, 633nm. The broken lines are the computer baselines.
a) scan from 85^0 to 95^0, sample time 3μs.
b) scan from 80^0 to 115^0 six times with gel in slightly different positions (see text), sample time 0.1s, prescaling factor 128, light attenuation 170.

where $d\theta/dt$ is the scanning rate and $g_{sp}^{(1)}(\theta)$ is the first order normalized spatial correlation function. There is an apparent short term baseline at $N_s(\langle\langle n\rangle\rangle_{sp})^2\left[1 + \beta(1-R)^2\right]$ whilst the zero time value is $N_s(\langle\langle n\rangle\rangle_{sp})^2(1+\beta)$ and the computer baseline is $N_s(\langle\langle n\rangle\rangle_{sp})^2$. Thus $g^{(1)}(\tau)$ and both the spatially integrated total count rate and the count rate associated with the broadened component may be found from one experiment. In equation 6, r is replaced by R. Figure 2a shows a correlation function obtained with a polyacrylamide gel by scanning from 85^0 to 95^0. From the zero time value relative to the computer baseline β=0.43. It has been found convenient with gels to work with this relatively low coherence factor in order to avoid critical alignment of the system. Distortion of the correlation function due to systematic variation in $\langle\langle n \rangle\rangle_{sp}$ and $g^{(1)}(\tau)$ over the angular range is negligible, and the data is effectively valid for $\theta=90^{o}$. It is possible to investigate $g_{sp}^{(1)}(\theta)$ by using much longer sample times, and the result of such an experiment is shown in Figure 2b where the equivalent values of θ are indicated. The problem here is to obtain a sufficiently large sample of the fluctuations in $\langle n \rangle$ for the correlation function to be properly defined, without introducing systematic variations in $\langle\langle n \rangle\rangle_{sp}$. The correlation was therefore taken over six scans between 80° and 115°, and between each scan the cylindrical sample was twisted about its axis slightly so as to change the speckle pattern. On the assumption of Gaussian statistics however, this correlation function contains no information about the gel. It is entirely a function of the arrangement and dimensions of the stops in the scattered light receiver system.

In all work with gels the mounting of the apparatus is important. If the gel is set in oscillation due to external vibrations, the otherwise stationary component of the random speckle changes, and oscillations, usually at the frequency of oscillation of the gel, appear in the correlation function. This phenomenon has been used to investigate the mechanical properties of gels (3,4), but in the type of work being discussed here it is important to avoid these oscillations. This may be achieved most effectively by mounting the apparatus on a heavy base which is then placed on an air bed.

Polysaccharide Gels

Agarose and calcium alginate gels have been investigated (2,5,6). The Rayleigh ratio of light scattered from both types of gel is some two orders of magnitude greater than that from a "clear" gel such as polyacrylamide. Nevertheless provided that the dimensions of the specimen are reduced sufficiently to avoid secondary scatter, the Debye theory is obeyed, i.e. if $\lambda^{4}R_{\theta}$ is plotted against $\sin^{2}(\theta/2)/\lambda^{2}$ for different wavelengths, using incoherent light, a single plot is obtained, indicating that the scatter does not arise from macroscopic fluctuations in gel density.

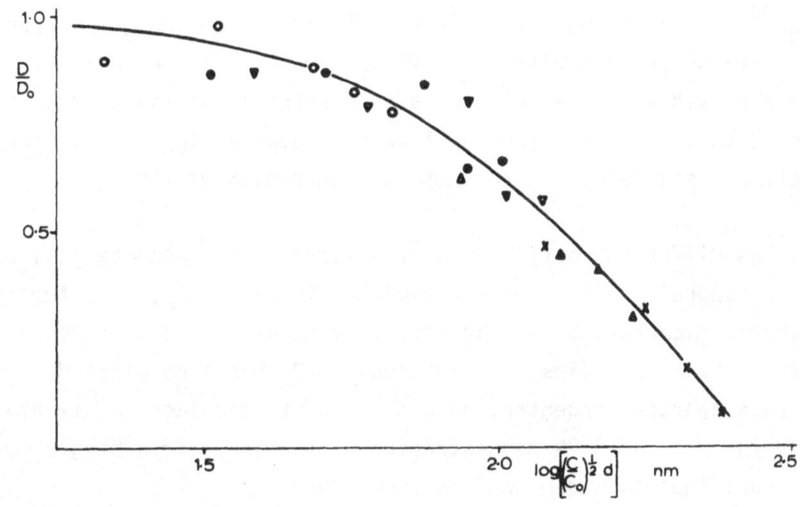

Figure 3. Diffusion coefficient, D, relative to that in dilute solution, D_0, of dextran fractions within agarose gels of various concentration, C, as a function of dextran hydrodynamic diameter, d. The plots at the various concentrations have been superimposed by displacing horizontally (C_0=0.7%). O 0.3% ● 0.7% ▽ 1% △ 2% X 4%

Degrees of spectral broadening have been found to be very small,one percent or less, falling as low as 0.01% for a 4% agarose gel. Correlation functions are non-exponential and relaxation times are of the order of tens of milliseconds and do not vary as $[\sin^2(\theta/2)]^{-1}$, much less variation being observed. It is not appropriate therefore to describe $g^{(1)}(\tau)$ for the light scattered from these gels in terms of diffusion coefficients. However when dextran fractions of suitable concentration are allowed to diffuse within these gels, separate relaxation times which vary as $[\sin^2(\theta/2)]^{-1}$ can be easily identified and measured, and the diffusion coefficients of the dextrans within the gels determined. Because of the high level of unbroadened scatter from these gels it is safe to assume that the optical beating is entirely heterodyne, so that it is not necessary in this case to know β in order to determine diffusion coefficients. However β must be known in order to determine partition coefficients. The zero time value of that part of the correlation function associated with the diffusing marker macromolecule (after subtracting the baseline and contributions from the gel) is

$2\beta N_s<n><n_D>$. The partition coefficient is found by comparing the counts per sample time associated with the diffusing macromolecule $<n_D>$, with the count rate observed when the cell contains a solution of the diffusing macromolecule only. A correction must be made however for attenuation of the incident and scattered light due to the turbidity of the gel.

Figure 3 shows diffusion coefficients of dextran within agarose gel, D, relative to corresponding diffusion coefficients in water, D_o, as a function of hydrodynamic diameter, D, and agarose concentration. d is calculated from D_o using the Stokes-Einstein equation. D/D_o has been plotted against log d for each agarose concentration and the plots displaced horizontally relative to the data at 0.7% concentration, (C_o), to yield a single "master" plot. The fact that this can be done shows that:

$$D/D_o = f[g(C)d] \quad (8)$$

g(C) is a function which can be found from the displacements and f is the function generating the master curve. A double logarithmic plot of the displacement against gel concentration shows that:

$$g(C) = a\ C^b \quad (9)$$

with b = 0.5. Since by dimensional analysis D/D_o must be a function of $\rho_1{}^{1/2}d$ where ρ_1 is the length of gel fibre per unit volume, and $C=\rho_1 m$ where m is the mean mass per unit length of fibre, b=0.5 is just the value expected if m is independent of concentration. A plot of very similar shape was obtained with calcium alginate gel except that values of d for given values of D/D_o were seven times smaller, indicating that the mass per unit length of gel fibre is some fifty times smaller. The only available theory for the function f is due to Ogston et al (7) who give the relation:

$$D/D_o = \mathrm{Exp}\left[-\pi^{1/2}\rho_1{}^{1/2}d/2\right] \quad (10)$$

If plotted as a function of log d, an exponential is similar in shape to the curve shown in Figure 3, but the transition from the freely diffusing to the completely immobilized state takes place much more slowly. The validity of equation 10, and possible modifications to it, have already been discussed extensively(2). It is thought that it is valid for conditions near complete immobilization, and on this basis molecular weights per unit length of gel fibre of 25,000 and 590g.$mol^{-1}nm^{-1}$ have been calculated for agarose and calcium alginate gels respectively. The latter figure is in excellent agreement with the model proposed for calcium alginate gel by

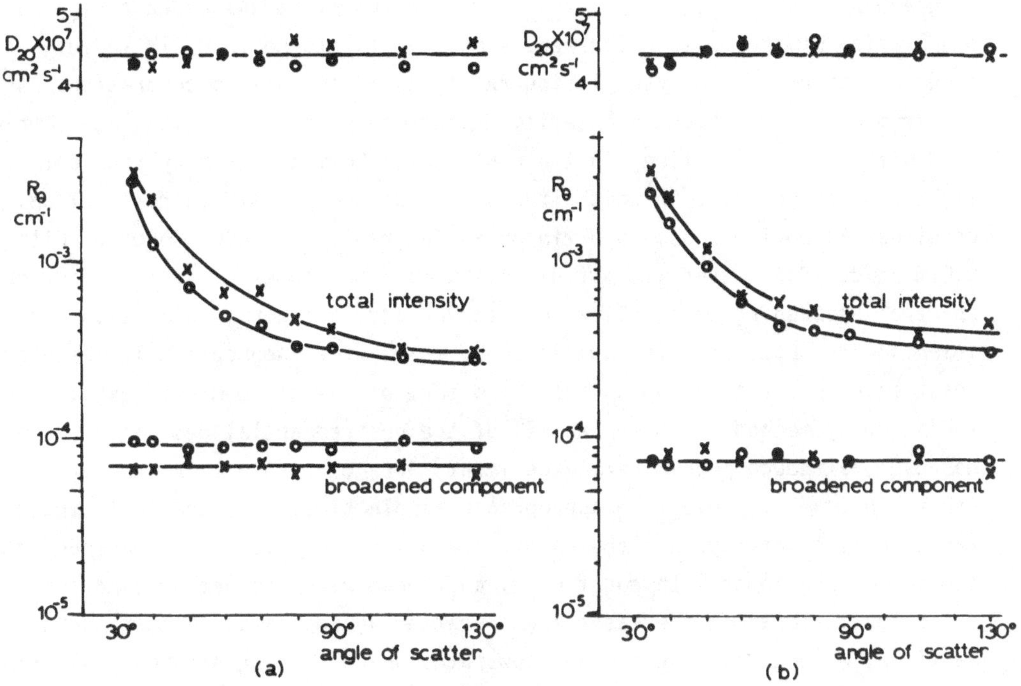

Figure 4. Diffusion coefficients and intensities of light scattered from a 7.5% polyacrylamide gel (bis/acrylamide ratio=1%). X indicates the gel alone, O indicates the gel surrounded by, a) a 4% solution of bovine plasma albumin, b) a 4% solution of a dextran fraction of weight average molecular weight 70,000 $g.mol^{-1}$. a) and b) represent different gel samples.

Morris et al(8). The figure for agarose suggests that there are of the order of fifty agarose chains associated together in each gel fibre. Partition coefficients determined for agarose are consistent with the diffusion coefficients on the basis of the theory of Ogston et al for gel concentrations above 1%. At lower concentrations smaller partition coefficients are found, possibly due to dextrans becoming temporarily trapped by mobile agarose chains (2).

Polyacrylamide Gels

Polyacrylamide gels were made up as cylinders of 5mm in diameter using the standard procedure (9), except that 0.01M sodium azide was added during the preparation of the gels, which were subsequently stored in 0.01M sodium azide solution. The bis/acrylamide ratio was 1% and gel concentrations ranged from 4% to 13%. Light scattering measurements were made with each sample held with its axis vertical in the centre of a 20mm diameter cylindrical light scattering cell, in which it was surrounded by 0.01M azide solution, or either 4% bovine plasma albumin or 4% dextran (M_w=70,000) solution with 0.01M azide added. Even without diffusing macromolecules present degrees of spectral broadening up to 30% were observed and, in general, diffusion coefficients were obtained with equation 6 by means of an appropriately weighted least squares fit going up to the third term of the polynomial. Usually the scanning method (from -5° to $+5^{o}$ of the desired angle) described above was used, although preliminary experiments were carried out in the normal manner in order to select an appropriate sample time. The spatially integrated total count rate and the count rate associated with the broadened component were found with equation 7, and these were converted to Rayleigh ratios by calibrating the laser light scattering apparatus against a conventional Aminco light scattering apparatus which in turn had been calibrated with Ludox. The usual scattering volume correction was applied. Small corrections for systematic variations in the various parameters with angle of scatter within the scanning range were made for angles below 50°. A Malvern Instruments 64 channel multibit correlator was used for all measurements with polyacrylamide gels.

Figure 4 shows data obtained at various angles of scatter for a 7.5% gel and Figure 5 shows data for various gel concentrations. Diffusion coefficients were converted to 20°C assuming a variation proportional to the ratio of absolute temperature to the viscosity of water. Considering first polyacrylamide gels in the absence of diffusing macromolecules; calculated diffusion coefficients were found to be independent of angle of scatter, and the deviation of the correlation function from a single exponential was not usually more than that expected from the fact that the optical beating is neither completely heterodyne nor completely homodyne. The maximum value of $\left[(D^2)_z\right]^{1/2}/D_z$ found was 1.02. Whereas the total Rayleigh ratio increased markedly at low angles, the Rayleigh ratio of the spectrally broadened component was found to be independent of angle of scatter. Measurements of

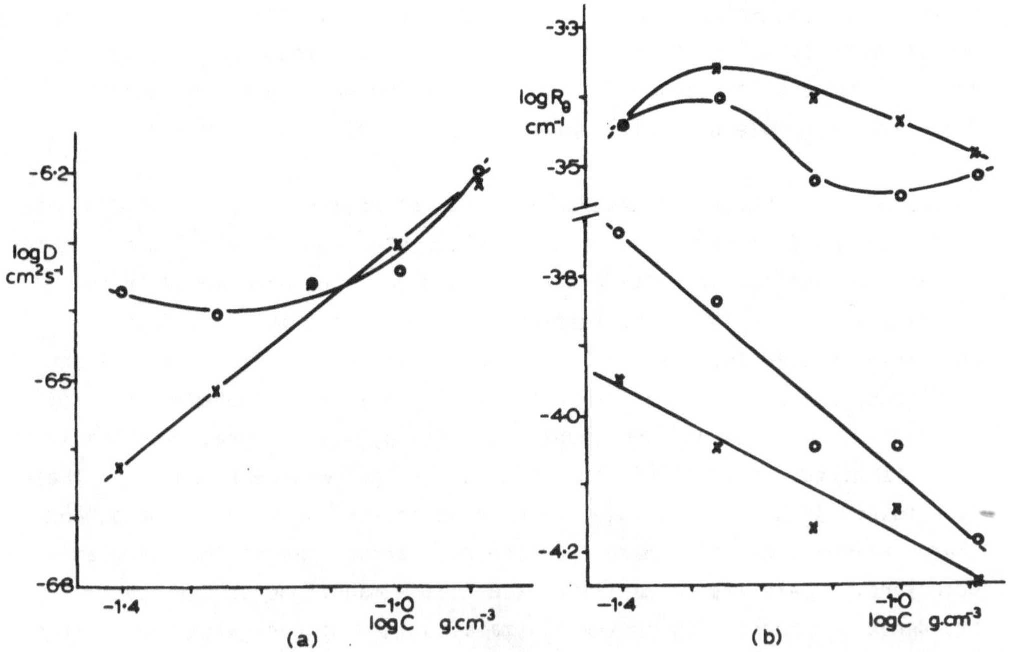

Figure 5. a) Diffusion coefficients and, b) intensities of light scattered from polyacrylamide gels of various concentration. X gel alone, O gel surrounded by 4% bovine plasma albumin solution. In b) the upper plots are the total Rayleigh ratios at 90° angle of scatter, whilst the lower plots are the Rayleigh ratios of the spectrally broadened component which were found to be independent of angle of scatter.

the total Rayleigh ratio were also made with incoherent light in an Aminco apparatus using the wavelengths 546nm and 436nm. Plots of $\lambda^4 R_\theta$ against $\sin^2(\theta/2)/\lambda^2$ yielded a single plot for the two wavelengths showing that the stationary component in the fluctuations in gel density does not in the main arise from macroscopic causes and that there are genuine static fluctuations in polyacrylamide density at a molecular level. A double logarithmic plot of diffusion coefficient against gel concentration yielded a straight line of slope 0.8. However the concentrations indicated in Figure 5 have not been corrected for swelling and as this is likely to increase with decreasing concentration it is likely that 0.8 is too high. Similarly the slope of the double logarithmic plot of the Rayleigh ratio of the spectrally broadened component against concentration 0.55, is also probably too high.

Laser light scattering from polyacrylamide gels has already been extensively investigated by Geissler and Hecht (10,11) and the results discussed in terms of current theories. The results for the gels in the absence of diffusing macromolecules will not therefore be discussed further here.

Bovine plasma albumin and dextran fraction of molecular weight 70,000 have diffusion coefficients in water of $5.8 \times 10^{-7} cm^2 s^{-1}$ and $3.4 \times 10^{-7} cm^2 s^{-1}$ respectively (no buffers or salts other than sodium azide were added in these experiments). Thus they have respectively higher and lower diffusion coefficients than that of a 7.5% polyacrylamide gel ($4.4 \times 10^{-7} cm^2 s^{-1}$) and 4% solutions have Rayleigh ratios respectively five times and twice that of the spectrally broadened component. Surprisingly therefore, essentially the same diffusion coefficient as that of the gel was measured after these macromolecules had been allowed to diffuse in for twenty four hours, and there was no evidence of more than one relaxation time in the correlation functions. There was a small increase in the magnitude of the spectrally broadened component with bovine plasma albumin, but none with the dextran fraction. The most marked effect in the case of a 7.5% gel was a fall of some 30% in the total scattered intensity. The fall in Rayleigh ratio is of the same order as the Rayleigh ratio of the solutions surrounding the gel, and it was checked by making measurements with incoherent light at two wavelengths as described above.

The effect of a 4% solution of bovine plasma albumin on gels of various concentration is shown in Figure 5. The measured diffusion coefficient falls with increasing gel concentration until it equals the diffusion cofficient of the gel, after which it rises as the diffusion coefficient of the gel. Even when the measured diffusion coefficient is markedly different from that of the gel at the lower gel concentrations, there was no evidence of more than one relaxation time in the correlation functions. The proportionate increase in the spectrally broadened component of the scattered light increases with decreasing gel concentration, whilst the fall in the total scattered intensity seems to be greatest at intermediate concentrations. At gel concentrations where no change in diffusion coefficient was measured, evidence that the protein had actually permeated the gel is provided by the fact that there was a marked effect on the total scattered intensity. This was however confirmed by dyeing the protein within the gel with aniline blue.

Discussion and Conclusion

It is obvious that polysaccharide gels and polyacrylamide gels are quite different as regards both the dynamics of their structure, and the mechanisms by which compact macromolecules diffuse within them. In the case of polysaccharide gels the picture we have is one of an almost stationary fibrous structure, in which compact macromolecules can diffuse subject to being impeded by the geometry of the structure, but with well defined diffusion coefficients which can be clearly identified and measured by quasi-elastic light scattering. In the case of polyacrylamide gels we have a highly mobile structure, which itself has a diffusion coefficient corresponding to the diffusion of fluctuations in the density of the gelling material. It appears that whereas diffusing compact macromolecules are impeded by the gel structure their diffusion coefficient does not fall below that of the gel itself and that at higher gel concentrations the diffusing macromolecules are carried along by the diffusion fluctuations in the density of the gelling material, presumably following the migration of the less dense regions, provided of course that they are small enough to enter the gel in the first instance. Thus the diffusion processes are not independent and it is not possible to assign a well defined diffusion coefficient to the diffusing macromolecules. Thus partition coefficients cannot be determined by quasi-elastic light scattering. The fact that the diffusing macromolecules actually give rise to a marked fall in the overall scattered intensity, but not the broadened component, indicates that they show a preference for the regions of the gel which are permanently less dense, so rendering the system more homogeneous. Further studies of diffusion within polyacrylamide gels are in progress.

Acknowledgements

The author wishes to thank Mr S. Hunter of the Astbury Department of

Biophysics at Leeds, who prepared the polyacryalmide gels and carried out the protein dyeing experiments. The photon correlation equipment was obtained with a grant from the Science and Engineering Research Council.

References

1. Tanaka, T., Hocker, O., Benedeck, G.B.: J.Chem.Phys. 59, 5151-5159 (1973), and many subsequent publications by these and other authors.
2. Key, P.Y., Sellen, D.B.: J.Polym.Sci.Polym.Phys.Ed. 20, 659-679 (1982).
3. Brenner, S.L., Gelman, R.A., Nossal, R.: Macromolecules 11, 202-212 (1978).
4. Gelman, R.A., Nossal, R.: Macromolecules 12, 311-316 (1979).
5. Mackie, W., Sellen, D.B., Sutcliffe, J.: Polymer 19, 9-16 (1978).
6. Sellen, D.B.; Polymer 19, 1110 (1978).
7. Ogston, A.G., Preston, B.N., Wells, J.D.; Proc.R.Soc. London A333, 297-316 (1973).
8. Morris, E.R., Rees, D.A., Thom, D., Boyd, J.: Carbohydrate Research 66, 145-154 (1978).
9. Shapiro, A.L., Vinuela, E., Maizel, J.V.: Biochem.Biophys.Res.Comm. 28, 815-820 (1967).
10. Hecht, A.M., Geissler, E.: Journal de Physique 39, 631-638 (1978).
11. Geissler, E., Hecht, A.M.: Macromolecules 14, 185-188 (1981).

SLOW DIFFUSION OF LABELED MACROMOLECULES STUDIED BY A HOLOGRAPHIC GRATING TECHNIQUE

Markus Antonietti, Jochen Coutandin, Dietmar Ehlich, Hans Sillescu
Institut für Physikalische Chemie der Universität Mainz
Jakob-Welder-Weg 15, D-6500 Mainz

Introduction

Experimental methods for investigating slow translational diffusion are subject to limits given by the Einstein relation $D = \langle r(t)^2 \rangle /6\,t$ where $\langle r(t)^2 \rangle$ is the mean square average displacement of the diffusant in the time interval t. Most techniques imply an interface between two parts of the sample one containing the diffusant at the beginning of the experiment. Examples are the radioactive tracer (1), the IR-scanning (2), and the Rutherford backscattering (3) techniques. Although it has been possible to prepare thin polymer films of thickness down to a few μm the difficulties of obtaining a well defined interface are considerable and have so far limited most applications to diffusion coefficients $D \gtrsim 10^{-10}$ cm^2s^{-1}. The NMR pulsed gradient techniques (4) are applicable to homogeneous samples, however, are limited by the requirement that the diffusion time t must be shorter than the longest spin relaxation time which again yields a limit of $D \gtrsim 10^{-10}$ cm^2s^{-1}. This limitation can be overcome by the holographic grating technique in which, by interference of two coherent beams, a grating is induced in a sample where the diffusant molecules are labeled with a photosensitive dye. The diffusive decay of this grating is monitored by measuring the intensity of forced Rayleigh scattering (FRS) from the sample. The first application of this technique to polymer diffusion was an investigation of labeled polystyrene (PS) in solution (5). We have used a similar experimental setup for measuring self diffusion coefficients in molten PS (6, 7, 8). In order to extend the range of accessible D values we are using photolabels having a lifetime of some days with respect to thermal decay. Furthermore, we have increased the angular range of the two

Physical Optics of Dynamic Phenomena and Processes in Macromolecular Systems

intersecting beams to $\theta \sim 90^0$ thus reducing the grid spacing $d = \lambda/\ 2\sin(\theta/2)$ to ~ 300 nm for $\lambda = 458$ nm. Since the FRS intensity is roughly (cf.ref.7) proportional to $\exp(-\ 8\pi^2 Dt/d^2)$ it decays to 1% of its initial value in $t \lesssim 1$ h for $D \gtrsim 10^{-14}\ cm^2 s^{-1}$. Thus, diffusion of macromolecules in polymer melts can easily be investigated within acceptable diffusion times. The present contribution provides experimental data for testing de Gennes' reptation model (10) of polymer chain diffusion.

Experimental

The fluorescein label used in our previous applications (6-8) gives rise to secondary reactions that may influence the FRS decay curves (7,11). Furthermore, it exists as a dimer in PS, and affects the diffusion of labeled PS in the region of low molecular weights (7,8). Thus, we have prepared a stilbene label by the following procedure (12,13). p-Tolunitrile is treated with a 3:1 mixture (by volume) of H_2SO_4 (98%) and HNO_3 (65%) at $T < 30^0C$. Condensation of the resultant o-nitro-p-tolunitrile with an equimolar quantity of p-dimethylamino-benzaldehyde occurs within 2 h at $120\text{-}130^0C$ in a mixture containing $\sim 5\%$ piperidine as catalyst. One obtains a black solid, 2-nitro-4-cyano-4'-dimethylamino-stilbene, which is recrystallized from acetic acid (m.p. 178^0C). The nitrile is hydrolysed by refluxing over night in a water-ethanol-KOH mixture. The Cs salt of the product is reacted with the chloromethyl end group of PS as described in ref. (7). The main label reaction

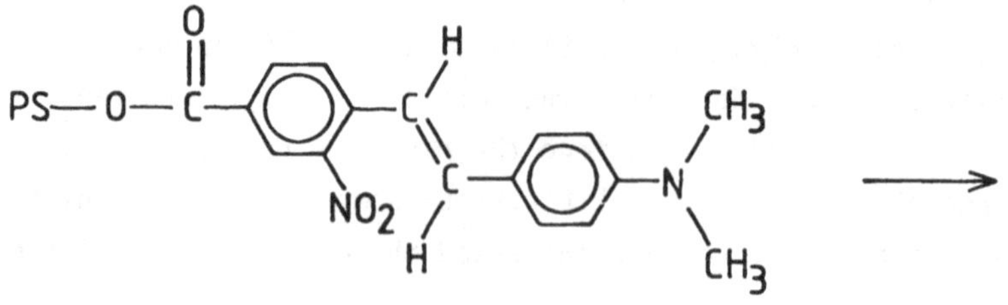

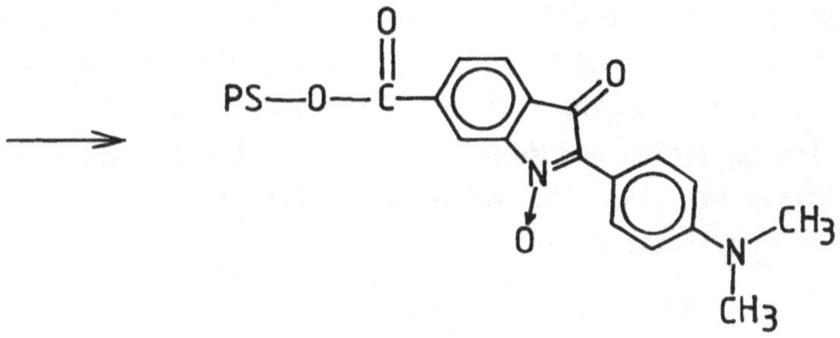

occurs in two steps where the first photochemical step leads to a colourless intermediate which reacts thermally to the blue final product. The thermal step is very rapid above ~ 150°C, and very slow below ~50°C. In between, it occurs on the same time scale as the diffusion process which can, however, easily be separated since the FRS decay by diffusion depends upon the grid spacing d.

Crosslinked PS was obtained by a Friedel-Crafts reaction with p-dichloroxylylene in C_2H_4Cl solution (14):

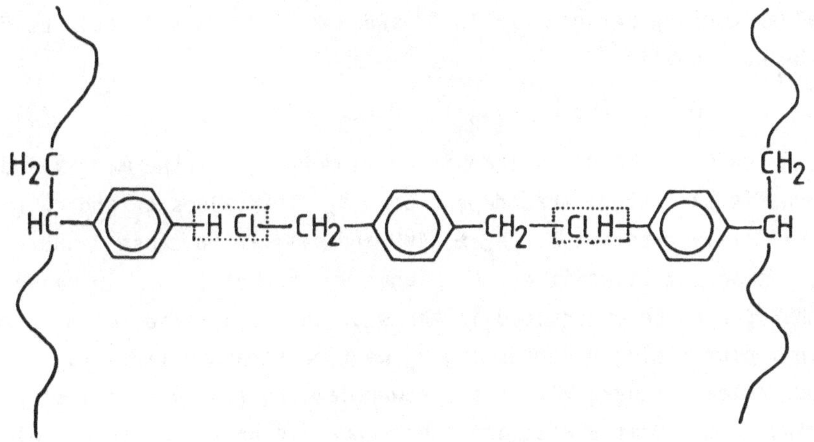

The soft gel obtained from a 20% solution was passed through a fine sieve, treated with THF/HCl, precipitated from CH_3OH, and dried at 175°C in order to obtain particles of crosslinked PS having a size of ~ 100 μm (14).

These particles were mechanically mixed with 2-5% of a fine powder of labeled linear PS, pressed to a 0.3 mm thick pellet, and annealed in vacuo for 4 days at 180°C. During this annealing period the labeled chains diffuse into the particles and form a solution in the crosslinked PS matrix. For details see ref.(14). The diffusion coefficients could be determined as described for linear PS in ref.(7).

Results and Discussion

The molecular weight dependence of polymer chain diffusion is usually discussed in terms of a power law

$$D = D_0(T)M^{-\alpha} \qquad (1)$$

where $\alpha = 2$ is predicted by the reptation model. In order to test this prediction, we have measured D values of 9 samples of linear PS labeled with fluorescein having a number average degree of polymerization P_n between 177 and 1 550 ($18\,000 \leq M_n \leq 161\,000$; $M_w/M_n \lesssim 1.02$). Except for the highest P_n values of 1 100 and 1 550, we have dissolved the labeled PS in a PS matrix having $M_w = 110\,000$. At temperatures between 135 and 208°C, the D values ranging between 2×10^{-14} and 3×10^{-10} $cm^2 s^{-1}$ could be fitted by the WLF equation

$$D = D_\infty \exp\left[- c_1 c_2 \ln 10/(c_2 + T - T_g)\right] \qquad (2)$$

where $T_g = 104°C$ is the glass transition temperature of the matrix and D_∞, c_1, c_2 are fit parameters that depend upon M_n. The values c_1 and c_2 (8) are approximately constant for $P_n \lesssim 400$ and similar to those found for the zero shear gradient viscosities. At higher P_n, our c_2 values increase in a manner that cannot be understood in terms of the usual free volume assumptions. In a plot of log D versus log P_n we find straight lines for $P_n \gtrsim 300$. A least square fit of the power law, Eq.(1), yields a slope α that varies from 2.0 at 208°C to values above 2.4 at temperatures below 167°C. We have checked whether this behavior originates from a label influence by doing some experiments using the stilbene label described above. For short chains, $P_n = 177$, we find D values increased by 20-50% in comparison with fluorescein labeled PS. However, for $P_n \gtrsim 300$ the differences become smaller, and the label influence upon D can safely be neglected. At present, we have no explanation for the unexpected temperature dependence of α. The fact that α approaches the reptation value of 2 at the highest temperature confirms the reptation model for diffusion of sufficiently flexible chains in an entangled polymer melt.

Polymer chain diffusion can be forced to curvilinear motion along the chain contour by chemical crosslinking of the surrounding matrix. In particular, the diameter of the "tube" defined in the reptation model (10) can be varied by changing the average number N_c of monomer units between crosslinks. If one assumes about 200 monomer units between entanglements in molten linear polymers (10,15) a choice of $N_c < 200$ provides a test of whether the reptative motion depends critically upon a tube diameter defined by the root mean square distance between entanglements. Since entanglements are largely eliminated in dilute solutions one can prepare networks without physical entanglements by performing the crosslinking reaction in dilute solution as described above. In our experiments (14) we have varied N_c between 16 and 400, and we have measured D values of PS chains having chain lengths P_n between 177 and 572 at temperatures of 177, 185 and 194^oC. The D values were found reduced with respect to diffusion in uncrosslinked PS by factors less than 4. The reduction is smaller in the microgel systems prepared from dilute solution. Apparently, the fixation of physical entanglements by crosslinking in concentrated entangled polymer solutions leads to networks where reptative motion is more difficult. The measured D values in a network with $N_c = 50$ were proportional to $P_n^{-1.8}$ at all 3 temperatures. The main result of our study is the remarkably small decrease of D in crosslinked as compared with linear molten PS which provides strong support of the reptation model.

References

1. Bueche, F.: J. Chem. Phys. 48, 1410 (1968) and references cited therein.
2. Klein, J.: Nature 271, 143 (1978).
3. Kramer, E.J., Green, P., Palmstrom, Ch.J.: Polymer 25, 473 (1984).
4. von Meerwall, E.D.: Adv. Polym. Sci. 54, 1 (1983).
5. Hervet, H., Léger, L., Rondelez, F.: Phys. Rev. Lett. 42, 1681 (1979).

6. Coutandin, J., Sillescu, H., Voelkel, R.: Makromol. Chem. Rapid Commun. 3, 649 (1982).

7. Antonietti, M., Coutandin, J., Sillescu, H.: Macromolecules 17, 798 (1984).

8. Antonietti, M., Coutandin, J., Sillescu, H.: Makromol. Chem. Rapid Commun., in press.

9. Ehlich, D.: Diplomarbeit, Mainz 1984.

10. de Gennes, P.G.: "Scaling Concepts in Polymer Physics", Cornell University Press, Ithaca 1979.

11. Coutandin, J.: PhD-thesis, Mainz 1984.

12. Pfeiffer, P.: Ber. 48, 1777, 1808 (1915).

13. Splitter, J.S., Calvin, M.: J. org. Chem. 20, 1086 (1955).

14. Antonietti, M., Sillescu, H.: submitted for publication.

15. Ferry, J.D.: Viscoelastic Properties of Polymers, 3rd Ed., Wiley, New York 1981.

MECHANICAL INSTABILITY OF GELS UNDERGOING LARGE SWELLING

Yoshitsugu Hirokawa*, John Kucera, Shao-Tang Sun§
and Toyoichi Tanaka

Massachusetts Institute of Technology, Cambridge, MA 02139, USA

*Permanent Address: Research and Development Center, Nippon Zeon Company, Kawasaki, Japan

§Hercules Company, Wilmington, Delaware, USA

Abstract

When an ionic polymer gel undergoes extensive swelling, a regular buckling pattern temporarily appears on its surface due to a mechanical instability. The characteristic length scale of the pattern increases with time until it becomes comparable to the overall size of the gel and then it disappears leaving the gel exactly similar to its original shape. In this paper we demonstrate a method for creating the buckling pattern in equilibrium and study its dependence on gel size and osmotic pressure.

I. Introduction

The kinetics of swelling and shrinking gels is an important scientific and technological problem which is not yet fully understood. It was demonstrated that, for small swelling, gel expansion can be described as a collective diffusion process of a polymer network (1). The concept of collective diffusion of a polymer network is totally different from the previous, but incorrect, idea that the gel swelling and shrinking is caused by the diffusion of solvent molecules. The new point of view on the kinetics of gel swelling was supported by the quantitative agreement between the collective diffusion coefficient of a polymer network determined by laser light scattering spectroscopy and macroscopic swelling experiments (1,2). The values obtained by independent experiments were $3(\pm0.5)\times10^{-7}cm^2/s$ and $3.2\times10^{-7}cm^2/s$ which were two orders of magnitude smaller than that of water molecules.

For large swelling compared to the original volume the linear kinetic theory no longer applies. Interesting, non-linear behavior emerges in the form of mechanical buckling on the surface. First seen as a transitory phenomenon in spherical gels, this buckling has now been produced in equilibrium in thin slabs affixed to a base by chemically treated film. We describe here how the instability depends on thickness and osmotic pressure.

II. Gel Preparation

Samples of gels were prepared in free radical copolymerization at room temperature. 5.00 grams of acrylamide (main constituent molecule), 0.133 grams of N,N'-methylenebisacrylamide (crosslink molecule) and 0-0.64 grams of sodium acrylate (ionizable molecule) were dissolved in 100 ml of pure water. 240 µl of N,N,N',N'-tetramethylethylenediamine (accelerator) and 40mg of ammonium persulfate (initiator) were added to initiate the gelation. The solution was then immediately transferred to a glass mold and allowed to set for at least one hour in a nitrogen atmosphere. Spherical gels were cast in glass bulbs 4cm in diameter. Slabs of gel from 0.2 to 2mm thick were made by pouring pre-gel solution onto 7.5cm square glass plates with spacers and covering with commercially available treated film (GelBond® PAG) glued onto square acrylic backings. The film chemically bonds to the gel during gelation, preventing that side from swelling when immersed in solvent. After curing, the gels were immersed in pure water and kept for 24 hours to 3 days to reach equilibrium.

III. Kinetic Experiments

Figure 1 shows the time evolution of a spherical gel immersed in pure water. Initially the gel surface was smooth and glossy. Immediately after immersion there appeared a fine, regular pattern similar to that of frosted glass. The size of the pattern increased and became deeper until after one day a single pattern element became comparable in size to the original radius of the gel. Eventually, the pattern engrossed the entire gel forming again a glossy sphere. Growth continued until the gel became about four times its original diameter with about 20% to 30% of the growth occurring after the buckling had disappeared.

The gel swells to such a large extent because of the osmotic pressure of Na^{+} ions introduced by the sodium acrylate. Immediately after immersion, its surface is released from osmotic stress and "diffuses" out into the solvent. As it diffuses, it expands, while deeper layers remain fixed creating a large shear stress. This shear causes buckling which forms a regular pattern with a length scale limited by the thickness of the expanded layer. The osmotically released layer diffusively thickens as the square root of time decreasing the expansion gradient until finally the gel becomes homogeneous again and the buckling disappears.

IV. Pattern Found in Equilibrium

Slab gels of various thicknesses but same sodium acrylate concentration were formed with one face mechanically anchored to an acrylic square as previously described. Figure 2 shows some of these gels after they were allowed to swell to equilibrium in pure water. With one side fixed the pattern remains. Shape similarity is not maintained between initial

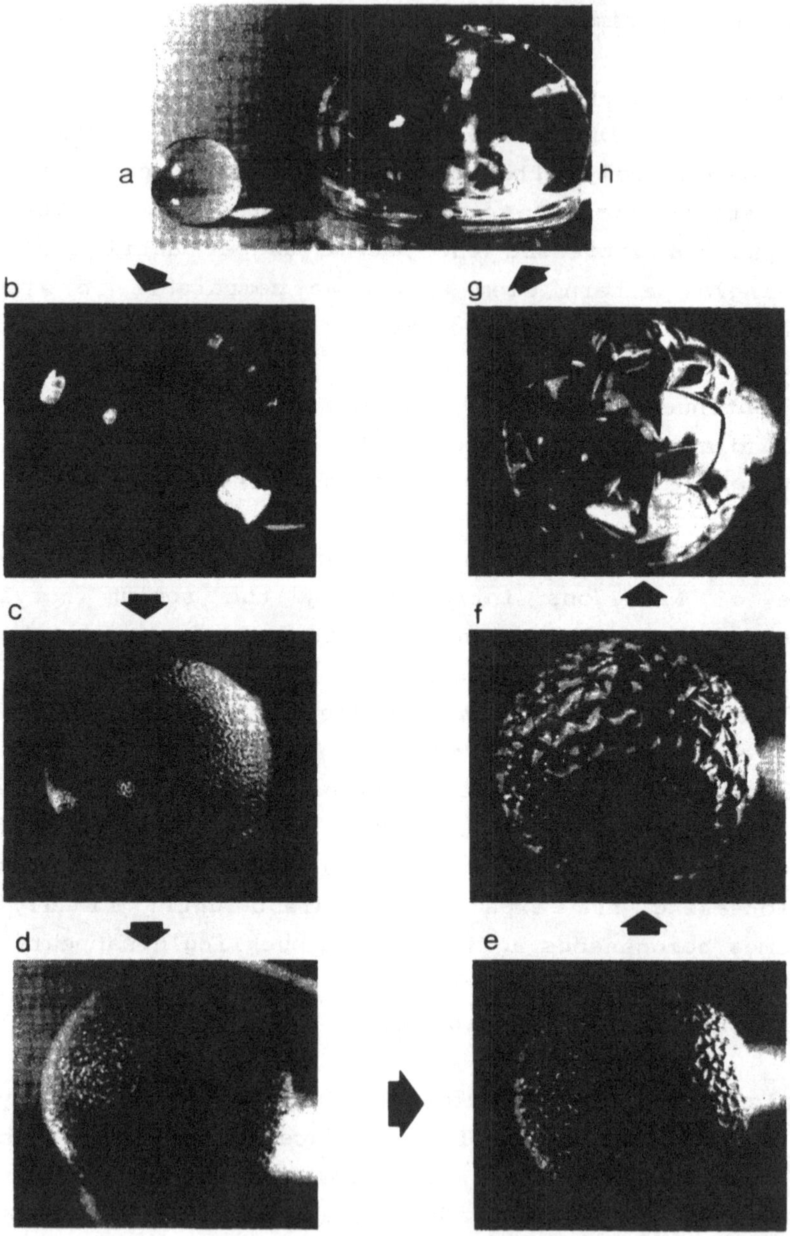

Figure 1: Swelling of a spherical gel. Swelling time (a) 0 min, (b) 0 min, (c) 3 min, (d) 11 min, (e) 30 min, (f) 90 min, (g) 26 hours, (h) 3 days.

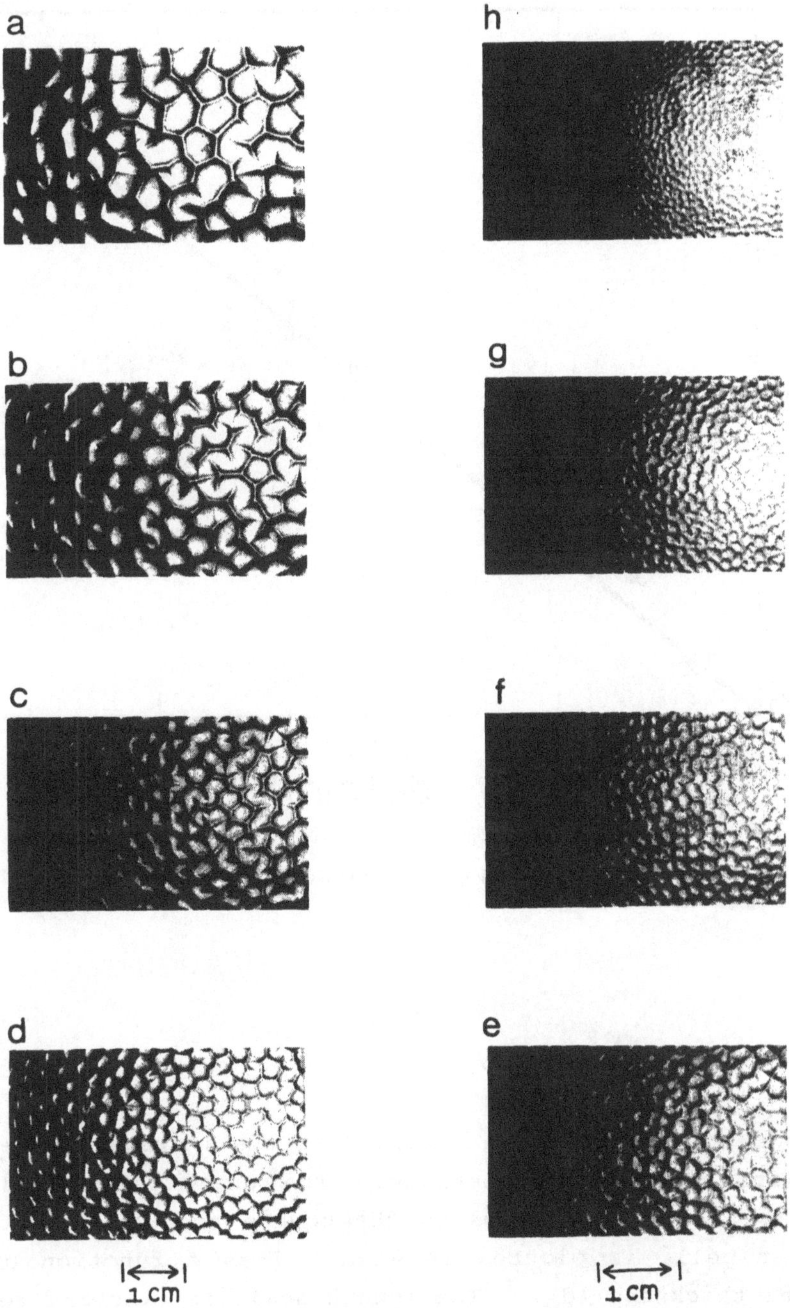

Figure 2: Swelling of slab gels with different thicknesses and same concentration of sodium acrylate (0.15g/100ml water). Thickness of gels: (a) 2.00mm, (b) 1.65mm, (c) 1.33mm, (d) 1.01mm, (e) 0.83mm, (f) 0.66mm, (g) 0.50mm, (h) 0.32mm. Note scale difference for (e), (f), (g) and (h).

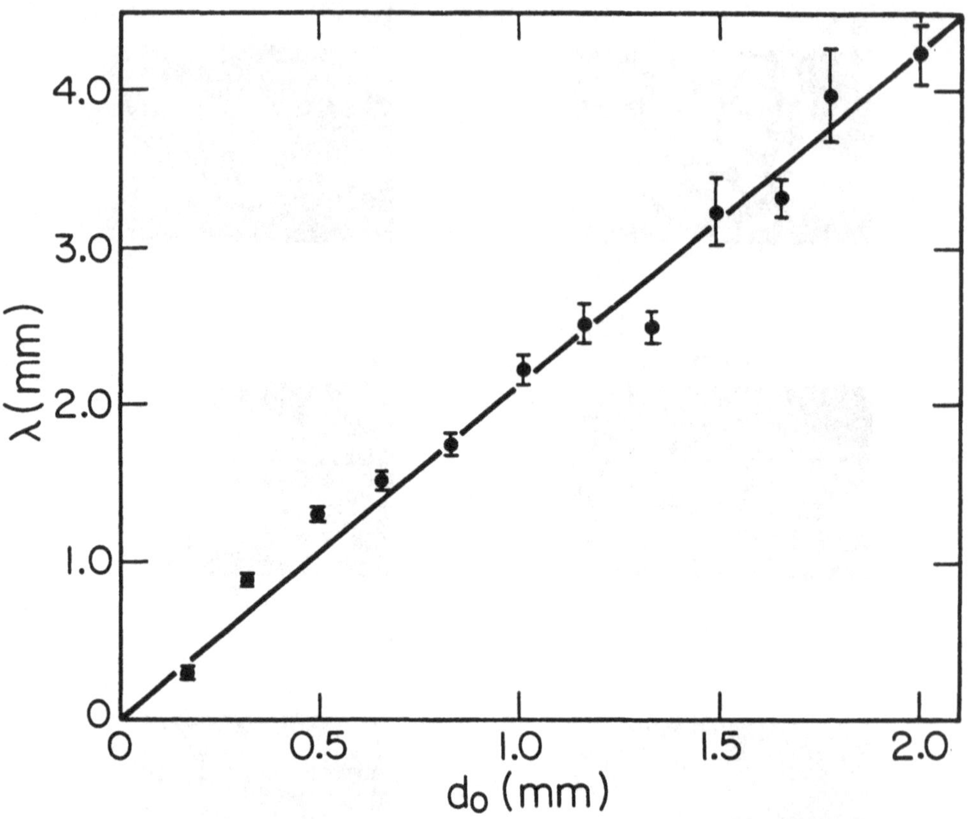

Figure 3: Relation between characteristic wavelength and initial thickness of gel.

and final states. One can clearly see the increase in pattern size with increasing thickness. The characteristic length scale (λ), obtained by taking the square root of the average area per cell, is plotted in Figure 3 as a function of the original thickness (d_o). The length scale is linearly related to original thickness in this instance with slope 1.95 (±0.04) and intercept 0.19 (±0.03).

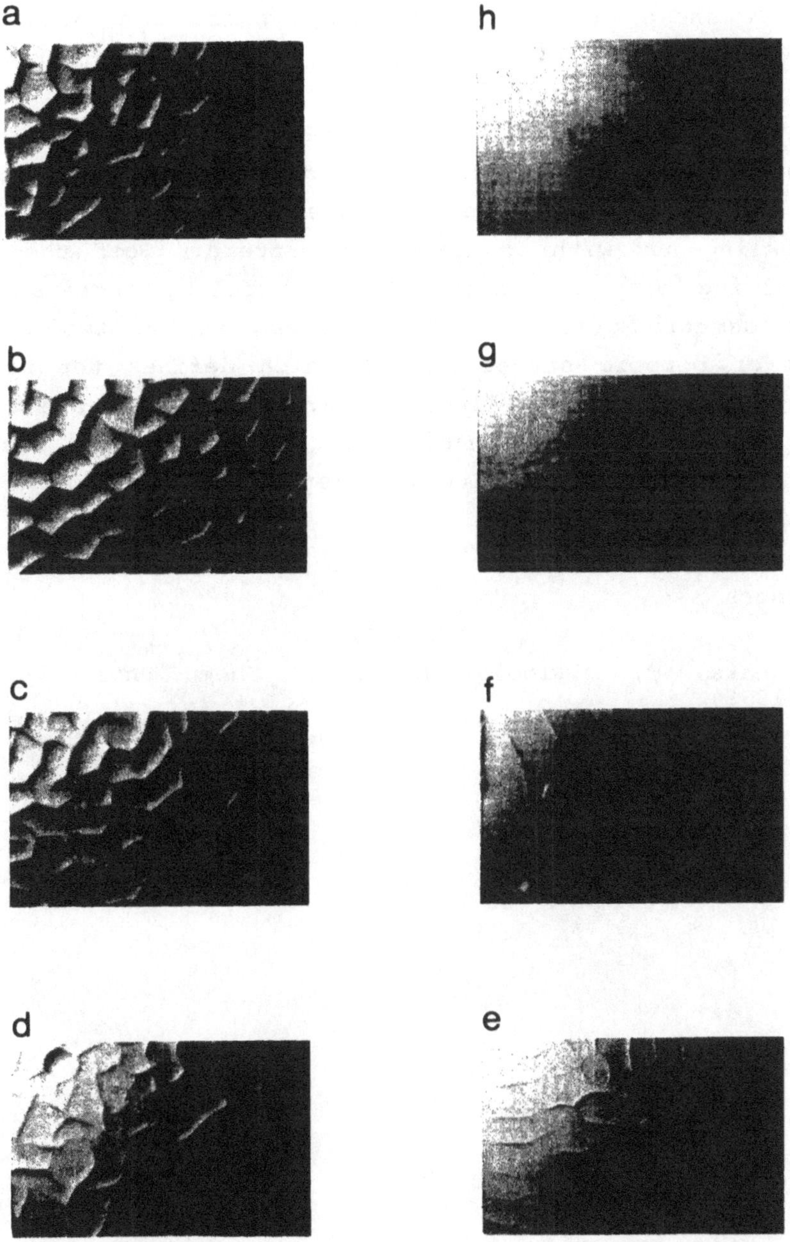

Figure 4: Swelling of slab gels with same thickness (2.00mm) and different concentrations of sodium acrylate. Concentrations (g/100ml water): (a) 0.12, (b) 0.10, (c) 0.09, (d) 0.08, (e) 0.07, (f) 0.06, (g) 0.05, (h) 0.04.

Figure 4 shows the effect of varying the concentration of sodium acrylate. The gels having high concentrations of sodium acrylate (high osmotic pressure) show a regular buckling and large swelling whereas the gels having low concentrations (low osmotic pressure) show no buckling and less swelling. Those intermediate show the early stages of buckling but with an irregular pattern. Both the amount of swelling and the depth of the buckling increases with higher osmotic pressure. There seems to be two critical values of osmotic stress one of which defines the boundary between linear and non-linear behavior, and the other which determines if the pattern is regular. Further study is necessary to understand this phenomenon.

References

1. Tanaka, T, Fillmore, D.J.: J. Chem. Phys. 70, 1214 (1979).

2. Tanaka, T., Hocker, L., Benedek, G.B.: J. Chem. Phys. 59, 5151 (1973)

PHOTON CORRELATION SPECTRA OF AMORPHOUS LOW MOLECULAR WEIGHT POLYETHYL METHACRYLATE ABOVE THE GLASS TRANSITION TEMPERATURE

George Fytas*

Max-Planck-Institut für Polymerforschung Mainz
c/o Institut für Physikalische Chemie der Universität Mainz
Postfach 3980, D-6500 Mainz, FRG

Abstract

The photon correlation functions of the polarized component of the scattered light from two fractions of poly(ethyl methacrylate) (PEMA) of weight average molecular weight M_w 1.6 × 10^4 and 7.3 × 10^4 were measured in the temperature range from 70 to 107°C and at pressure from 1 to 1000 bar. An analysis of the data using only one relaxation α-process yields an unexpected temperature, pressure and M_w dependent distribution of relaxation times. In contrast to this, a double relaxation process representation of the experimental time correlation functions leads to relaxation parameters which have physically meaningful values for both PEMA fractions. The characteristic time of the two processes approach each other at high temperatures, low pressures and hence the separation is larger for the higher M_w PEMA fraction. The present results are in agreement with the findings of dielectric and mechanical experimental methods.

---

*Permanent address: Department of Chemistry, University of Crete, Iraklion, Greece

Physical Optics of Dynamic Phenomena and Processes in Macromolecular Systems

Introduction

During the last few years, photon correlation spectroscopy (PCS) has been extensively applied to study bulk polymer dynamics above the glass transition temperature T_g (1). Several general features emerge from these studies, which indicate universality of the dynamics of the relaxation process probed by PCS. 1. A main glass-rubber relaxation, also known as primary or α-mode from dielectric relaxation measurements (which has a tremendous effect on the mechanical properties), dominates the time correlation function of the polarized scattered intensity arising from slow density fluctuations. 2. The experimental correlation functions have a highly non-exponential shape and can be represented well by the empirical Williams-Watts function (2) $(\exp(-t/\tau_0)^{\dot{\beta}})$ with distribution parameter $\dot{\beta} \sim 0.4$ for all studied polymers except poly(alkyl methacrylates). 3. The mean relaxation time $\bar{\tau}(= \frac{\tau_0}{\dot{\beta}} \Gamma(\dot{\beta}^{-1})$ with Γ being the gamma function) conforms to the VFT or the equivalent WLF temperature equation which describe well mechanical shift factors (3). While the shape of the time correlation functions is insensitive to pressure ($\dot{\beta}$ is constant), the relaxation time $\bar{\tau}$ exhibits a strong pressure dependence and the activation volume corresponds to the volume of few monomer units (4). For poly(propylene glycol), $\bar{\tau}$ and $\dot{\beta}$ are found to be independent of molecular weight and of the probing wavelength (5). These findings suggest that localized segmental motions of the polymer chain grossly affect the density fluctuations in undiluted polymers near T_g.

For poly(alkyl methacrylates), however, where two relaxation processes have been observed in dielectric and mechanical measurements (3,6) the light scattering results do not conform to the previous scheme (8-10). The relatively low value of $\dot{\beta}$ ($\sim$ 0.2), which is found to be temperature and pressure dependent distorts significantly the mean time $\bar{\tau}$. One could fit, of course, a very broad distribution with two less distributed functions; however, the solution should be unique and physically meaningful results should be obtained. Since we do not yet have a theoretical model describing the measured correlation spectra of bulk polymers near T_g, a significant contribution to this problem can be achieved by selecting suitable experimental conditions i.e. temperature, pressure and molecular weight, and by comparison with other methods. In poly(butyl methacrylate) (PBMA) (10) we have shown that the most appropriate analysis of the data near T_g and at ambient pressure involves two distinct relaxation processes.

In this paper we attempt to explore the applicability of PCS on broad distributions in amorphous polymers and report a photon correlation study of the polarized scattered light from two fractions of poly(ethyl methacrylate) (PEMA) of weight average molecular weight (M_w) 1.6×10^4 (T_g=47°C) and 7.3×10^4 (T_g=65°C) in the temperature range from 70°C to 107°C and at a pressure from 1 to 1000 bar. It is known, mainly from dielectric measurements (6), that the separation on the time scale of the two processes varies rapidly with temperature and pressure. Since similar response is also expected by changing T_g, the present pressure,temperature and molecular weight dependent study is further justified.

Theoretical Background

The quantity measured in PCS is the auto correlation function of the scattered electric field $g(q,t)=\langle\vec{E}(q,t)\cdot\vec{E}^*(q,o)\rangle/\langle|\vec{E}(q,o)|^2\rangle$ where q is the scattering wave vector and the angular brackets denote an ensemble average. The function g(q,t) depends on the polarization of the scattered light relative to the incident beam. In the VV-geometry, the scattered intensity includes both isotropic I_{iso} and anisotropic I_{VH} contributions (V and H denote perpendicular and parallel polarization to the scattering phase respectively). For weakly anisotropic scatterers, which is the case for poly(alkyl methacrylates), the VV intensity is almost equal to the isotropic component. This arises mainly from density fluctuations in undiluted polymers and has the correlation function:

$$g(q,t) = \langle\delta\rho(q,t)\delta\rho^*(q,o)\rangle/\langle|\delta\rho(q,o)|^2\rangle \qquad (1)$$

where $\delta\rho(q,t)$ is the q th mode of the density fluctuations. Equation (1) can also be written in the form (5):

$$g(q,t) = \langle\sum_{m,n}\sum_{i,j} \exp(iq\cdot(r_j^{(n)}(t)-r_i^{(m)}(0)))\rangle \qquad (2)$$

where $r_j^{(n)}$ is the position of the center of mass of scattering unit j of polymer n at time t. Only the segment center of mass motion affects the isotropic scattering component in eq.(2). However, due to the intra- and interchain interactions in bulk polymers, there is no rigorous calculation of the general expression entering in eq.(2).

The normalized first order correlation function g(t) is related to the directly measured single clipped intensity correlation function $G_k(t)$ in an homodyne experiment through the equation:

$$G_k(t) = A\left|1+b|g(t)|^2\right| \quad (3)$$

where A is the baseline and b is an unknown parameter used in the fitting procedure which depends on the relaxation strength i.e. the fraction of the scattered intensity associated with the slow ($t>10^{-6}$s) density fluctuations.

Experimental

Photon correlation functions at different temperatures (70-107°C) and pressures (1-1000 bar) were obtained at a scattering angle of $\theta=90^{o}$ with the apparatus described elsewhere (4). The light source was an argon ion laser operating at $\lambda=514.5$ nm and power up to 300 mW. The incident beam was polarized perpendicular to the scattering plane and since there was very little depolarized intensity, no polarizer was used in the scattered light.

The single-clipped photocount auto-correlation function $G_k(t)$ was measured by means of a 96-channel Malvern correlator. The method of matching the sections with different delay time $\Delta\tau$ to construct the composite correlation function over 6 decades in time was described in ref.(5).

The two samples used in this study (kindly provided by Dr.W.Wunderlich, Röhm & Haas GmbH Darmstadt, W.Germany) were poly(ethyl methacrylate) (PEMA) having $M_w=1.6\times10^4$ and 7.3×10^4 respectively. The ratio M_w/M_n for both samples amounts to 2.36. The glass transition temperature determined from DTA data amount to 48 and 65°C for the low and high molecular weight fraction respectively.

We have prepared clear, optically homogeneous, dust free and strain free samples using the procedure described elsewhere (10). A strong confirmation of the suitability of the bulk PEMA samples for photon correlation measurements was provided by the low value of the Landau-Plazcek intensity ratio (11).

Data analysis

The highly nonexponential time correlation functions of PEMA can be represented by the Williams-Watts function (2):

$$g(t) = \exp\,(-t/\tau_0)^{\dot{\beta}} \tag{4}$$

however, the distribution parameter $\dot{\beta}$ is found to decrease with decreasing temperature (T) at constant pressure (P) and with increasing P at constant T. For given P,T,$\dot{\beta}$ is smaller for the higher molecular weight PEMA fraction and varies systematically in the range 0.17-0.4. For the other bulk polymers investigated so far, $\dot{\beta}$ is found to be $\sim$ 0.4 insensitive to T and P provided that a single relaxation mechanism dominates g(t). The present results mirror those obtained for poly(methyl methacrylate) (PMMA) (9) and indicate the presence of at least two relaxation processes, as is the case of PBMA (10). Moreover, we know from dielectric and mechanical studies of poly(alkyl methacrylate) that the separation on the time scale of the two processes is larger in PEMA than in PBMA at given P,T.
For these reasons we performed a double-relaxation process fit of the equation:

$$((G_k(t)-A)/A)^{1/2} = b_1 \exp(-t/\tau_1)^{\dot{\beta}_1} + b_2 \exp(-t/\tau_2)^{\dot{\beta}_2} \tag{5}$$

to the experimental time correlation functions as described elsewhere (10). The numerical subscripts are used to denote the relaxation parameters of the fast (1) and slow (2) component respectively. A further support for the present type of data analysis is provided by the internal consistency of the obtained relaxation parameters for the two PEMA fractions.

Results and Discussion

1. Molecular weight dependence

First we consider the time correlation functions of the two PEMA fractions as a function of temperature at atmospheric pressure. Fig.1 shows two normalized correlation functions g(t) at 85.5°C for PEMA. It is apparent, that g(t) of the higher M_w-fraction (hereafter referred as II)

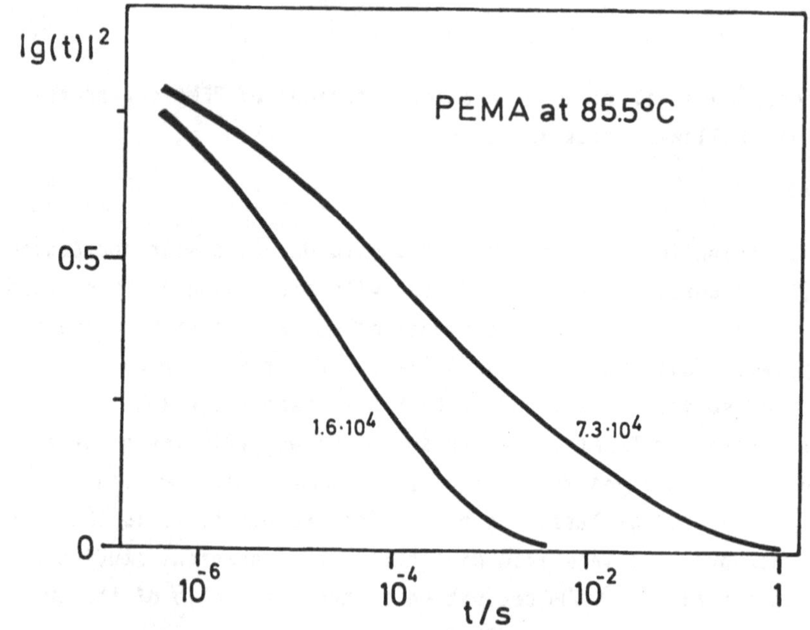

Fig.1: Normalized composite correlation functions for the two fractions of poly(ethyl methacrylate) at 85.5°C. The thick solid lines represent fits of eq.(4) and eq.(5) to the experimental functions for the low and high molecular weight fractions respectively. The experimental points (340 for the low and 430 for the high molecular weight) are within the thickness of the lines.

is shifted to long delay times and shows a broader shape as compared to PEMA-I (low M_w fraction). Whilst the former effect should be expected because of the higher T_g value of PEMA-II, the change of the relaxation time distribution can be ascribed to the double relaxation distribution feature of g(t). For polystyrene (12), where one relaxation process dominates the slow density fluctuations, the shape of g(t) is indeed insensitive to the molecular weight and the correlation functions are therefore superposable.

The mean relaxation times $\bar{\tau}_1$ and $\bar{\tau}_2$ obtained from the fit of eq.(5) to the experimental correlation functions at atmospheric pressure, are plotted vs 1/T in Fig.(2). Up to about 80°C for sample I and 100°C for sample II the deviation plot of the fit to eq.(4) shows a systematic error and the data have consequently been represented by eq.(5). In this temperature range, the distribution parameters $\dot{\beta}_1$ and $\dot{\beta}_2$ are equal to 0.36 ± 0.2 and 0.34 ± 0.2 for the fast and slow process respectively and the ratio b_1/b_2 amounts to 0.43 ± 0.04. At higher temperatures the relaxation time $\bar{\tau}_1$ is close to $\bar{\tau}_2$ (Fig.(2)), so that the single Williams-Watts equation (4) represents the experimental functions yielding the slow relaxation time $\bar{\tau}_2$ (10).

The slow time $\bar{\tau}_2$ for the two PEMA fractions exhibit a large temperature dependence typical for α-dielectric times and mechanical shift factors of amorphous polymers near and above T_g. Hence in these systems $\bar{\tau}_2$ conforms to the WLF equation (3):

$$\log \frac{\bar{\tau}_2}{\bar{\tau}_g} = - \frac{C_1(T-T_g)}{C_2+T-T_g} \quad (6)$$

where C_1 and C_2 are the WLF constants and $\bar{\tau}_g$ is the mean relaxation time at T_g. For PEMA the superposition of dynamic shear compliance data yields $C_1=18$ and $C_2=65K$ for $T_g=335K$ (3). Using the measured values for T_g (321 and 338K for PEMA I and II respectively) and molecular weight independent $C_2(= 65K)$ (12,13) we obtain $C_1=21.4\pm0.4$ from the fit of eq.(6) to the times $\bar{\tau}_2$ of Fig.2. This finding suggests that the difference between $\bar{\tau}_2$ for the two PEMA fractions is mainly due to the difference in the fractional free volume. Concerning the temperature dependence of $\bar{\tau}_2$, there is a fair agreement between the results derived from light scattering and mechanical relaxation data. As for the absolute value of $\bar{\tau}_2$, one has to compare the photon correlation functions g(t) with compressional compliance measurements (14).

A new pertinent aspect of the relaxation times in Fig.2 is the invariance, within experimental uncertainty, of the fast relaxation time $\bar{\tau}_1$, on the molecular weight and hence on T_g. We feel therefore justified in fitting an Arrhenius temperature dependence to the times $\bar{\tau}_1$. The obtained activation energy amounts to 22±5kcal/mol, which lies in the reported range of the activation energy for the β-dielectric process. On the grounds of this agreement we ascribe the fast light scattering mode to the β-relaxation. For poly(alkyl methacrylates),the β-dielectric mode is assigned to the hindered rotation of the -COOR side group around the bond linking it to the main chain (6,15). The slow and fast relaxation processes in Fig.2 seem to merge at high temperatures where the shape of the experimental correla-

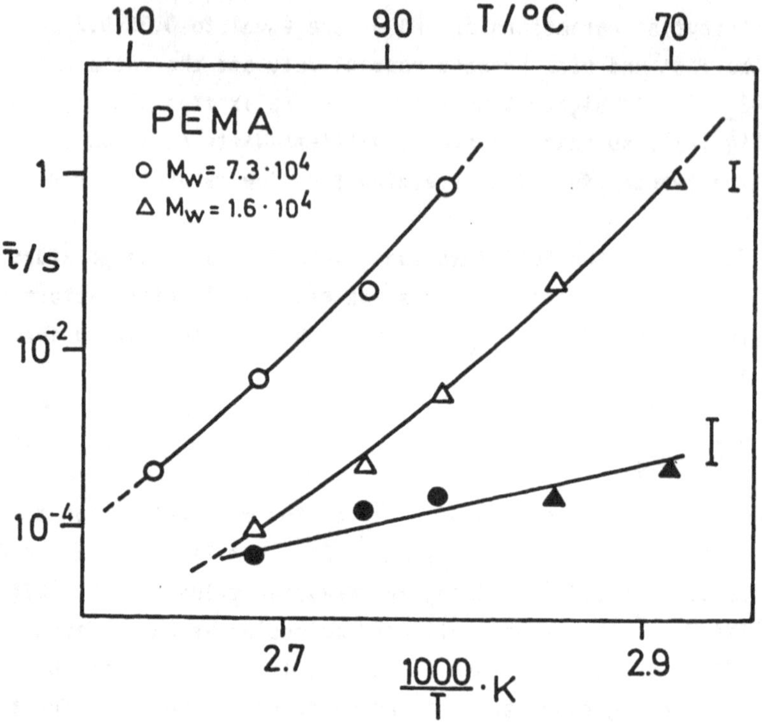

Fig.2: Temperature dependence of the mean relaxation time of the slow process (o,Δ) and the fast process (solid symbols) in PEMA.

tion function become narrower. Finally, the difference between $\dot{\beta}$ for the two PEMA functions and the strong temperature dependence of $\dot{\beta}$ as well is strong evidence of the additional fast mode.

2. Pressure dependence

While the hydrostatic pressure has a pronounced effect on the mean α-relaxation time for poly(ethyl acrylate) (PEA) (4), PS (16), poly(methyl acrylate) (PMA) (17) and poly(phenylmethylsiloxane) (18) the shape of the time correlation functions is essentially insensitive to pressure ($\dot{\beta}$ is constant). For PEMA however, the shape of g(t) is changed with pressure in a similar manner as with decreasing temperature. Fig. 3 illustrates this situation for PEMA-II at 100°C, 1 bar and 800 bar,

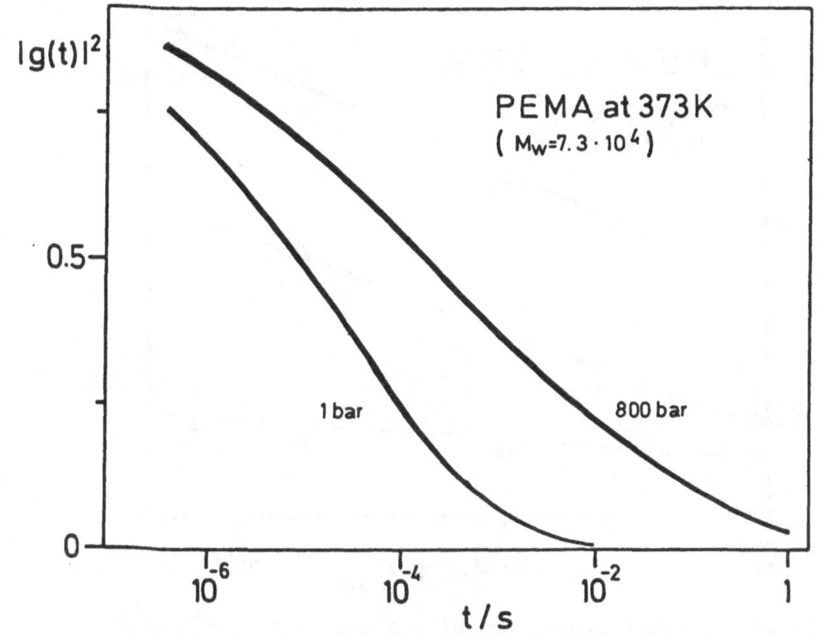

Fig.3: Normalized composite correlation functions for PEMA at 373K at two different pressures. The thick lines represent fits of eq.(5) to the experimental functions. The 340 and 430 experimental points at 1 and 800 bar respectively are within the thickness of the lines.

which is a further support for the double relaxation distribution in the experimental correlation functions for PEMA. We have consequently fitted eq.(5) to the experimental data and the obtained values of times $\bar{\tau}_1$ and $\bar{\tau}_2$ at 373K are plotted vs pressure in Fig.4.

If we consider the relaxation process as an activated process, the slope of the lines in Fig.4 is proportional to the activation volume $\Delta V^{\#}$ ($=2.303\ RT(\partial \log\bar{\tau}/\partial P)_T$), which represents the difference between the molar volume of the transition state and the molar volume of the initial state. We see that $\Delta V_1^{\#}$ for the fast relaxation process is much smaller then $\Delta V_2^{\#}$ for the slow process in both PEMA fractions. The relaxation time $\bar{\tau}_2$ exhibits a strong pressure dependence, which resembles strongly the results obtained for the glass-rubber (α) relaxation in bulk polymers above T_g. $\Delta V_2^{\#}$ at 373K amounts to 204 and 234 cm^3/mol for PEMA-I and II respectively. This small difference is due to the decrease of the cooperativity in the

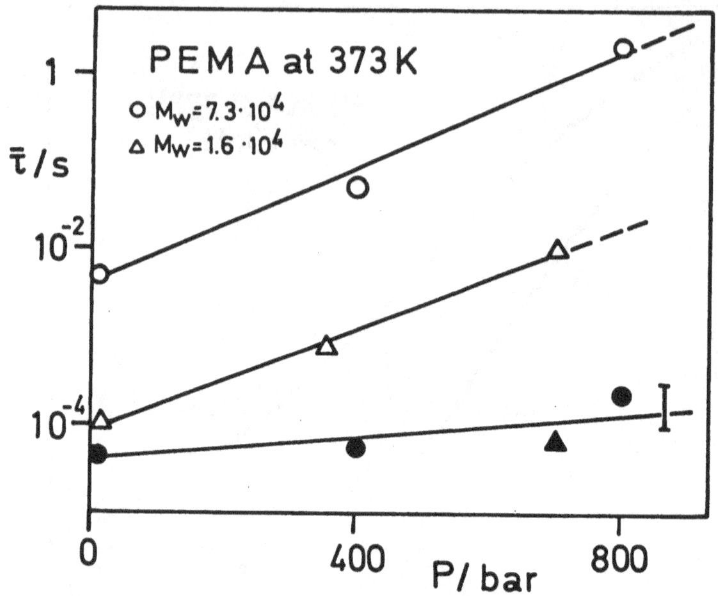

Fig.4: Pressure dependence of the mean relaxation time of the slow process (o,Δ) and the fast process (solid symbols) in PEMA.

high molecular weight fraction grossly represented by $T/(C_2+T-T_g)$ (4). The value of $\Delta V_2^{\#}$ corresponds to the molar volume of a few monomer repeat units (the monomer volume is 103 cm^3/mol at 20°C). Thus, the slow (α)mode can be understood on a molecular basis as a process involving isomerization of the main chain restricted to only a small number of monomer repeat units (5,19). Alternatively, the weak pressure dependence of the fast relaxation time $\bar{\tau}_1$ in Fig.4 is in accord with the physical picture of the hindered side group rotation. Finally, the analysis outlined above has yielded unique values for the relaxation parameters of the two processes and hence led to an adequate description of the slow density fluctuations in amorphous PEMA above T_g.

Acknowlegement

This work is based on the experimental data obtained in the Department of Chemistry at the University of Bielefeld, West Germany. The financial support from the Max-Planck-Gesellschaft is gratefully acknowledged. Thanks are due to Dr.I.G.Voigt-Martin for reading the manuscript and to Mrs. I. Schiller for typing it.

References

1. Patterson, G.D.: Adv.Polym.Sci.: 48, 125 (1983).
2. Williams, G., Watts, D.C.: Trans.Faraday Soc. 66, 80 (1970).
3. Ferry, J.D.: Viscoelastic Properties of Polymers, John Wiley & Sons, New York 1980.
4. Fytas, G., Patkowski, A., Meier, G., Dorfmüller, Th.: Macromolecules 15, 870 (1982).
5. Wang, C.H., Fytas, G., Lilge, D., Dorfmüller, Th.: Macromolecules 14, 1363 (1981).
6. Mc Crum, N.G., Read, B.E., Williams, G.: Anelastic Properties and Dielectric Effects in Polymeric Solids, John Wiley & Sons, New York 1967.
8. Patterson, G.D., Stevens, J.R.,Lindsey, C.P.: J.Macromol.Sci., Phys. B18, 641 (1980).
9. Patterson, G.D., Carrol, P.J., Stevens, J.R.: J.Polym.Sci., Polym.Phys. Ed. 21, 613 (1983).
10. Meier,G., Fytas,G., Dorfmüller, Th.: Macromolecules 17, 957 (1984).
11. Fytas, G., Li, B.Y., Wang, C.H.: to be published.
12. Mittag, U. Diplomarbeit, Universität Bielefeld (1983).
13. Plazek, D.J., O'Rourke, V.M.: J.Polym.Sci. A-2, 9, 209 (1971).
14. Fytas, G., Wang, C.H., Fischer, E.W., Meier, G., to be published
15. Heijboer, J.: Intern.J.Polymeric Mat. 6, 11 (1977).
16. Patterson, G.D., Caroll, P.J., Stevens, J.R.: J.Polym.Sci., Polym.Phys. Ed. 21, 605 (1983).
17. Fytas, G., Patkowski, A., Meier, G., Dorfmüller, Th.: J.Chem.Phys. 80, 2214 (1984).
18. Fytas, G., Dorfmüller, Th., Chu, B.: J.Polym.Sci., Polym.Phys.Ed. 22, 000 (1984).
19. Hall, C.K., Helfand, E.: J.Chem.Phys. 77, 3275 (1982).

RAYLEIGH-BRILLOUIN SCATTERING OF AMORPHOUS POLYMERS

C.H.Wang

Department of Chemistry, University of Utah, Salt Lake City, Utah 84112 / USA

B. Stuhn*

Institut für Physikalische Chemie der Universität Mainz, 6500 Mainz, FRG

* present address: Chemische Werke Hüls, Marl, FRG

Abstract

The technique of determining the longitudinal stress modulus over a wide dynamic range using Rayleigh-Brillouin scattering has been studied. A correct relation between the Rayleigh-Brillouin spectrum owing to density fluctuations and the mechanical relaxation spectrum of linear viscoelastic polymer is provided. The method of computing the longitudinal modulus and the compliance spectra from the time correlation of density fluctuations of PPG, as determined by the photon correlation spectroscopic technique, is illustrated.

Physical Optics of Dynamic Phenomena and Processes in Macromolecular Systems

Introduction

Rayleigh-Brillouin scattering in bulk polymers occurs as a result of fluctuations of local density and orientation of chain segments. The dynamics of the long wavelength fluctuations in the frequency range of 0.1 - 10^{11} Hz can be investigated by the dynamic light scattering technique using Fabry-Perot interferometry and optical mixing spectroscopy.

At the 90° scattering angle, the scattered light observed in the VV scattering configuration is a superposition of the density fluctuation and orientational fluctuation components, whereas the scattered light observed in the VH scattering configurations is due to orientational fluctuations. Thus, by measuring both the VV and VH spectra, it is possible to obtain the component owing to density fluctuations. To study the Rayleigh-Brillouin spectrum owing to density fluctuations is equivalent to studying the Fourier transform of $C(\tilde{q},t)$ given by

$$C(\tilde{q},t) = \langle \delta\rho(\tilde{q},t)\delta\rho^*(q) \rangle \qquad (1)$$

where $\delta\rho(\tilde{q},t)$ is the spatial Fourier transform of the fluctuation of the number density from the equilibrium value within scattering volume V. The quantity $\tilde{q}$ is the scattering vector and its amplitude is related to the scattering angle θ by Bragg relation: $q = 4\pi n \sin(\theta/2)/\lambda$, where λ is the wavelength of the incident light in vacuum, and n is the refractive index of the scattering medium. In the amorphous polymer system the light scattering spectrum depends only on the amplitude of the scattering vector and not on its direction. In the light scattering experiment using visible light, q is of the order of 10^5 cm^{-1}; thus, in an amorphous medium the fluctuation wavelength of density fluctuations given by $\lambda_f = 2\pi/q$ is long compared with the intermolecular distance. This permits using the method of continuum mechanics to calculate $C(\tilde{q},t)$, the time correlation function of density fluctuactions. In our previous work[1-4] we were mainly interested in the interpretation of the Rayleigh-Brillouin spectrum in terms of the longitudinal modulus. In this paper we consider the reverse problem and we hope to extract information about the longitudinal modulus spectrum in the 0.1 to 10^6 Hz frequency range from the time correlation function of density fluctuations, a quantity which can be measured by the photon correlation technique. Although our theoretical result is valid above and below the glass transition temperature, we argue that above T_g the longitudinal modulus as determined by the photon correlation technique mainly reflects the bulk modulus, due to the fact that above T_g the shear modulus is smaller than the bulk modulus by one to two orders of magnitude. Since the bulk modulus is difficult to measure by the usual mechanical technique, light scattering provides an effective method for determining the bulk modulus. In this paper we shall first provide the relationship between the imaginary part of the longitudinal modulus M'' and the Rayleigh-Brillouin spectrum using the method of continuum mechanics. Computation of M'' from the dynamic light scattering spectrum is illustrated by the photon correlation function data of poly(propylene glycol) above T_g published previously[5].

Longitudinal Modulus and the Dynamic Light Scattering Spectrum

The most general constitutive equation describing the stress-strain relation of an amorphous linear viscoelastic system is given by[6]

$$\sigma_{ij}(t)=-P\delta_{ij}+\int_{-\infty}^{t}dt'\{K(t-t')\dot{\Delta}(t')+2G(t-t')[\dot{\gamma}_{ij}(t')-\tfrac{1}{3}\dot{\Delta}(t')\delta_{ij}]\} \qquad (2)$$

where P is the hydrostatic pressure and is a function of space and time owing to density and temperature fluctuations; $\dot{\gamma}_{ij}(t)$ is the strain rate tensor and can be written as the velocity gradient tensor by $\dot{\gamma}_{ij}(t)=[(\partial v_i/\partial x_j)+(\partial v_j/\partial x_i)]/2$. Δ is the volume dilatation given by $\Delta = \delta V/V = \gamma_{xx}+\gamma_{yy}+\gamma_{zz}$.

The quantity K(t) is the bulk modulus and G(t) is the shear modulus. In Rayleigh-Brillouin scattering the bulk and the shear modulus appear as a combination of $K(t)+\frac{4}{3}G(t) = M(t)$ known as the longitudinal modulus. The mechanical property of a linear viscoelelastic system is completely characterized if K(t) and G(t) are determined.

In order to relate the bulk and shear moduli to $C(\underset{\sim}{q},t)$, we need the equation of continuity expressing the conservation of mass

$$\frac{\partial}{\partial t}\rho_m = -\frac{\partial}{\partial x_j}(\rho_m v_j) \tag{3}$$

and Newton's equation for the conservation of momentum

$$\rho_m \frac{d}{dt}v_i = \rho_m\left(\frac{\partial}{\partial t}v_i + v_i\frac{\partial}{\partial x_j}v_j\right) = \frac{\delta}{\delta x_j}\sigma_{ij} \tag{4}$$

In Eqs.(3) and (4) the Einstein summation convention for repeated indices is implied.

To calculate $C(\underset{\sim}{q},t)$ from the above equations we need to linearize Eqs. (3) and (4) to be consistent with the linear viscoelastic theory. This corresponds to dropping the non-linear terms in Eqs.(3) and (4).

Substituting Eq.(2) into Eq.(4) and neglecting the non-linear term we obtain

$$\rho_m^o \frac{\partial v_i}{\partial t} = -\frac{\partial P}{\partial x_i} + \int_{-\infty}^{t} dt'\{[K(t-t') + \tfrac{1}{3}G(t-t')]\frac{\partial}{\partial x_i}\nabla\cdot\underset{\sim}{v}(t') + G(t-t')\,\nabla^2 v_i(t')\} \tag{5}$$

The linearized version of Eq.(3) is

$$\frac{\partial}{\partial t}\,\delta\rho_m = -\,\rho_m^o\,\frac{\partial v_i}{\partial x_i} = -\,\rho_m^o\,\nabla\cdot\underset{\sim}{v} \tag{6}$$

Equations (5) and (6) form a set of coupled equations which can be solved to obtain $C(\underset{\sim}{q},t)$, provided that the density dependence of the hydrostatic pressure P is known. Although P is generally a function of density ρ and temperature T, for simplicity we shall neglect the temperature dependence in the present paper. This corresponds to neglecting contribution of the thermal diffusion to the density fluctuations. The effect of thermal diffusion has been considered elsewhere, but it does not affect the final form of the equation [7]. Thus, we write

$$\frac{\partial}{\partial x_i}P = \left(\frac{\partial P}{\partial \rho}\right)_T \left(\frac{\partial \rho}{\partial x_i}\right) = \frac{1}{\rho^o \chi_T}\left(\frac{\partial \rho}{\partial x_i}\right) \tag{7}$$

where ρ^o is the equilibrium number density.
To obtain $C(\underset{\sim}{q},t)$ from Eqs.(5) and (6) and Eq.(7), we can use the usual technique of taking the Fourier transform with respect to the spatial coordinates and the Laplace transform with respect to time on both sides of Eqs.(5) and (6). This procedure will convert Eqs. (5) and (6) into an initial value problem. The results are given by

$$i\,s\,\hat{A}_1 + i\,q\,\hat{A}_2 = A_1(\underset{\sim}{q}) \tag{8a}$$

$$\frac{iq}{\rho_m^o \chi_T}\,\hat{A}_1 + \left(is + \frac{q^2 \hat{M}}{\rho_m^o}\right)\hat{A}_2 = A_2(\underset{\sim}{q}) \tag{8b}$$

where $A_1(\underset{\sim}{q})$ and $A_2(\underset{\sim}{q})$ are the initial values of $\delta\rho(\underset{\sim}{q},t)$ and $v_z(\underset{\sim}{q},t)$, respectively. $\hat{A}_1$ and $\hat{A}_2$ and given by

$$\hat{A}_1 = \delta\hat{\rho}(\underset{\sim}{q},t) = \int d^3r\, e^{i\underset{\sim}{q}\cdot\underset{\sim}{r}} \int_0^\infty dt\, e^{-st}\, \delta\rho(\underset{\sim}{r},t) \tag{9a}$$

$$\hat{A}_2 = \hat{v}_z(\underset{\sim}{q},s) = \int d^3r\, e^{i\underset{\sim}{q}\cdot\underset{\sim}{r}} \int_0^\infty dt\, e^{-st}\, v_z(\underset{\sim}{r},t) \tag{9b}$$

In Eq. (9b) we set q to be along the z axis. The quantity $\hat{M}$ is the Laplace transform of the longitudinal modulus given by

$$\hat{M} = \int_0^\infty dt\, e^{-st}\, [K(t) + \tfrac{4}{3} G(t)] \tag{10}$$

We have not written out the explicit s-dependence in A_1, A_2 and M that appear in Eqs.(8) -(10).
Equations (8a) and (8b) can be solved to yield $\delta\hat{\rho}(\underset{\sim}{q},t)$ in terms of $\delta\rho(\underset{\sim}{q})$ and $v_z(\underset{\sim}{q})$. The Laplace transform of the time correlating function of $C(\underset{\sim}{q},t)$ can be contructed by multiplying $\delta\hat{\rho}(\underset{\sim}{q},t)$ with $\delta\rho^*(\underset{\sim}{q})$ and then taking the ensemble average. The result is

$$\hat{C}(\underset{\sim}{q},s) = [(s + q^2\hat{M}/\rho_m^o)/\Delta(s)]\langle|\delta\rho(\underset{\sim}{q})|^2\rangle \tag{11}$$

where $\Delta(s)=s(s+q^2\hat{M}/\rho_m^o)+\omega_T^2$; $\omega_T=q(\rho_m^o\chi_T)^{-1/2}$ is the isothermal sound frequency.
From Eq. (11) we can solve $\hat{M}$ in terms of $\hat{C}$. This is accomplished by setting next $s=i\omega$ and then identifying the real and imaginary parts on both sides of the equation, we obtain the imaginary part of the dynamic modulus as

$$M''(\omega) = (1/\chi_T)\, \omega\hat{I}/[(1-\omega\hat{H})^2+(\omega\hat{I})^2] \tag{12}$$

where

$$M''(\omega) = \omega\int_0^\infty dt\ M(t)\cos\omega t \tag{13}$$

$$\hat{I}(q,\omega) = \int_0^\infty dt\ \cos\omega t\ C(q,t)/\langle|\delta\rho(q)|^2\rangle \tag{14}$$

$$\hat{H}(q,\omega) = \int_0^\infty dt\ \sin\omega t\ C(q,t)/\langle|\delta\rho(q)|^2\rangle \tag{15}$$

Equation (12) is the basic result of this paper. Note that Eq. (17) differs from that given by Patterson[8]. In ref. 8, Patterson stated that $M''(\omega)$ is proportional to $\omega\hat{I}$, rather than given by Eq.(12). Equation (12) will reduce to Patterson's result only when both $\omega\hat{I}$ and $\omega\hat{H}$ are small compared to 1. However, as to be shown below by an explicit numerical calculation, neither $\omega\hat{H}$ nor $\omega\hat{I}$ is small compared to 1. Clearly, to assume that $M''(\omega)$ is proportinal to $\omega\hat{I}$ is an error.

It should be pointed out that although both $\hat{I}$ and $\hat{H}$ functions may depend on q, $M''(\omega)$ is independent of q. This is possible only if either I and H are q independent or if the q dependence is determined by the ω dependence. This is indeed the case because at low frequency ($< 10^6$ Hz) $C(q,t)$ in the one-component amorphous polymer is found to be q independent, on the other hand at high frequency the q-dependence is reflected by the ω-dependence because it is related to the propagation of the acoustic wave.

<u>Competition of $M''(\omega)$ from the Dynamic Light Scattering Data.</u>

To illustrate the above theoretical result, we shall compute M'' and $\omega\hat{I}$ using the photon correlation spectra of poly(propylene glycol) (PPG) previously published[5]. The complete Rayleigh-Brillouin spectra of PPG at three temperatures are shown in Fig. 1.

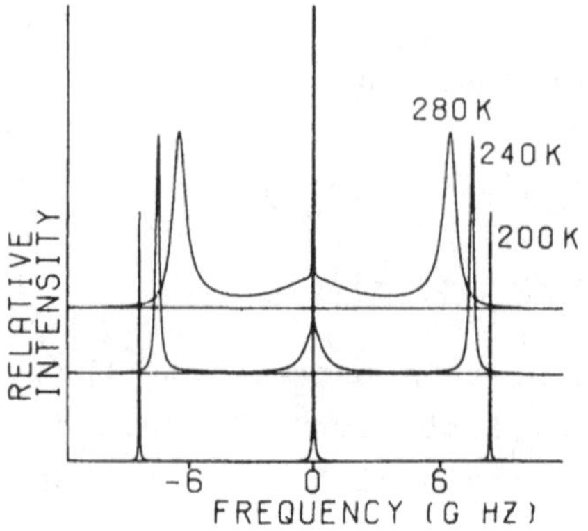

Fig. 1: Rayleigh-Brillouin spectra of PPG at three temperatures. Note the extra intensity distribution between the central and Brillouin components at the 240 and 280 K spectra.

Note that in addition to the shifted side bands, there is also present a considerable distribution of the scattering intensity on both sides of the unshifted central peak. The intensity distribution, known as the Mountain component is closely associated with the relaxation behavior of the longitudinal stress modulus. Near T_g the Mountain component is too narrow to be resolved by the Fabry-Perot interferometer because of its limited resolution; however, this narrow component can be studied using the photon correlation technique[5].

In the homodyne photon correlation experiment, we measure the intensity time correlation function $G_2(t)$, which in the Gaussian approximation is related to the normalized time correlation function of the scattered field g(t):

$$G_2(t) = A\ (1 + b|g(t)|^2) \tag{16}$$

where A is the baseline; b is a parameter related to the contrast. b can be determined in the fitting procedure. The contrast is proportional to the fraction of the scattering intensity that is detected by the correlator. Shown at the top of Fig. 2 is g(t) of PPG at T=-63°C, deduced from the measured $G_2(t)$ using Eq.(16).

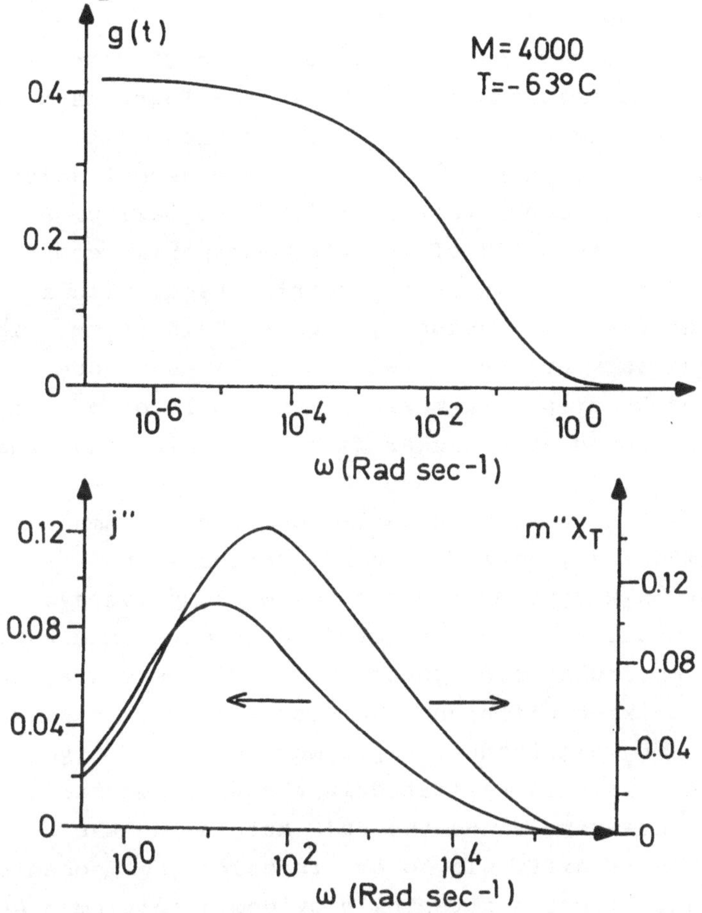

Fig. 2: The correlation function of density fluctuations and imaginary part of the longitudinal relaxation modulus and imaginary part of the longitudinal compliance for PPG at T = -63°C. The simple j" is equal to $\omega\hat{I}$, which is proportional to D".

In PPG the field correlation function g(t) is equal to $\alpha_r C(\underset{\sim}{q},t)/C(\underset{\sim}{q})$, where α_r is the relaxation strength, representing the portion of the total time correlation function of density fluctuations that is detected by the digital correlator operated at the shortest sampling interval. The quality α_r for PPG at various temperatures has been determined and reported in ref. 4.

Using g(t) for C(q,t), we have computed $\omega\hat{I}$ and M" using Eqs. (12) and (14). The results are shown in the lower part of Fig. 2. Since the imaginary part of the dynamics compliance D" is equal to $\chi_T\omega\hat{I}$ [7], where χ_T, is isothermal compressibility. The result shows that in PPG at the maximum D" (at f ≈ 32 Hz) is less than 10% of χ_T, the isothermal compressibility. Both D" and M" have an asymmetric shape, with a traiting tail at the high frequency region. This asymmetric shape reflects a distribution of relaxation times[9]. Note that both M" and D" display a maximum. The position of the maximum in M" is found at a higher frequency value than that of D".

For PPG above T_g the shear modulus is considerably smaller than the bulk modulus[6]; thus M" (or D") reflects the behavior of the bulk modulus (or the volume compliance). Since the bulk modulus is not expected to involve changes in the long-range molecular configuration or contour shape, we expect little striking difference between the polymer liquid of small molecular weight and the polymer solid of high molecular weight. This is reflected in the experimental result of PPG[5]. However, since the bulk modulus or the volume compliance is difficult to obtain using the mechanical technique, dynamic light scattering provides a complementary method for determing the bulk compliance in amorphous polymers.

Comparison of the value of the volume compliance determined by light scattering with that by the mechanical relaxation technique will be published elsewhere[10].

Acknowledgement: C.H. Wang acknowledges the financial support of NSF (DMR 82-16221) and the Alexander von Humboldt foundation for the American Senior Scientist award. We also thank Professor E.W.Fischer for useful discussion.

References

1. Y.Y.Huang and C.H.Wang: J.Chem.Phys. 62, 120 (1974).
2. Y.-H. Lin and C.H.Wang: J.Chem.Phys. 69, 1546 (1978).
3. W.Lampert and C.H.Wang: Macromolecules, 15, 557 (1982).
4. C.H.Wang, G.Fytas and J. Zhang: Rayleigh-Brillouin Scattering Studies os Segmental Fluctuations in Siloxane Polymers, J.Chem.Phys. (submitted).
5. C.H.Wang, G. Fytas, D. Lilge and Th. Dorfmüller: Macromolecules 14, 1363 (1981).
6. J.D. Ferry: Viscoelastic Properties of Polymers, 3rd ed. 1980, John-Wiley and Sons, Inc. New York, Chapt. i.
7. C.H.Wang and E.W. Fischer: J.Chem. Phys.(in press).
8. G.D. Patterson: Advances in Polymer Sciences, 48, 125 (1983) and references cited herein.
9. T. Tetsutani, M. Kakizati and T.Hideshima: Polym.J. 14, 305 (1982).
10. C.H. Wang, G.Fytas and E.W. Fischer (to be published).

BRILLOUIN SCATTERING STUDY OF SEGMENTAL MOTIONS OF POLYALKYL METHACRYLATES (PMMA,PEMA AND PNBMA) AT LOW TEMPERATURE

Eiji Kato and Yoshiro Saji
Kobe University of Mercantile Marine, Fukae-Minami, Higashinada, Kobe 658, Japan

Introduction

Up to now, there have been several Brillouin scattering experiments in polymer solids above and below the glass-rubber phase transition temperature (T_g).[1-4] However the experiment near the glass to glass phase transition temperatures which are seen below T_g has been scarcely done.[5] These transitions are caused by freezing of local motions such as rotations or stretching vibrations of side chains in solid polymers. Therefore Brillouin scattering near these transition temperatures gives information about the motion of segments in terms of side chain motions. The purpose of the present experiments is to examine anharmonic nature of the segment motions in solid polymers in terms of the side chain motions and the side chain length. For this purpose we measured the frequency shift and the half width of the Brillouin components of three polyalkyl methacrylates (PMMA,PEMA and PnBMA) with different side chains from room temperature to about 10 K.

Results and Discussion

We see several changes in slope for the Brillouin shift versus temperature curves at characteristic temperatures as shown in Fig.1. These temperatures are compared with the

Physical Optics of Dynamic Phenomena and Processes in Macromolecular Systems

transition temperatures T_1,T_2 and T_3 obtained by Haldon and Simha[6] with thermal expansion method. They reported that the transition temperatures were associated with the freezing of motions such as rotations or stretching vibrations of side chains. Table 1 shows that the temperatures obtained in the present experiments qualitatively agree with these transition temperatures.

The lattice Grüneisen parameter γ_L is an important parameter which shows anharmonicity of the segmental motions.[7,8] γ_L is given by $\gamma_L = \ln\nu/\ln V = d\nu/\beta\nu dT$. Here ν is the Brillouin shift, V the volume and β the volume expansion coefficient. Using the results shown in Fig.2 and data given by Haldon and Simha, we calculate γ_L . Fig.2 shows the results of γ_L as a function of n for an alkyl side group C_nH_{2n+1}. We see that the values of γ_L of the present samples are very large and increase in proportion to n. Namely the segmental motions become more anharmonic with increasing the length of the side chains. Below the temperature at which all motions of the side chains are frozen, γ_L decreases by about 30 % as shown in Fig.2. This means that the motions of side chains are one of the causes of the anharmonicity of the segmental motions.

Fig.2 also shows that the Brillouin shift at room temperature decreases with increasing the length of the side chains and, on the contrary, near 0 K it increases with increasing of the side chains. Generally it has been believed that the side chains make the intermolecular forces weak. This agrees with the result at room temperature. On the contrary, the result near 0 K means that the stiffened side chains make the intermolecular forces strong.

The hypersonic absorption coefficient α is obtained by $\alpha = \pi\Delta\nu/v$. Here $\Delta\nu$ is the half width of the Brillouin components and v the hypersonic velocity given by $v = \nu\lambda_0/2n \sin\theta/2$. λ_0 is the wavelength of the incident light in a vacuum, n the refractive index and θ the scattering angle. The results obtained are shown in Fig.3. They are similar to the polybutadiene measured by Huang and Wang.[3] But the present results do not show such an explicit relation between α and the glass transition temperature Tg as they reported.

Woodruff and Ehrenreich[9] gave a formula of α due to a mechanism of phonon-phonon interaction : $\alpha = \pi \gamma^2 C_V T \omega / 4 \rho v^3$. Here $\omega = 2\pi\nu$, γ the Grüneisen parameter, C_V the specific heat and ρ the density. Using the values of ν , v and γ_L obtained in the present experiments, we calculate α . The results are shown with solid curves in Fig.3. The curves are in good agreement with the experimental data. Here we used 1.2×10^5 erg/g.K and 1.7×10^5 erg/g.K as the values of C_V for PnBMA and PEMA. These values are much lower than the value of the specific heat of an ordinary polymer solid such as PMMA. This means that the specific heat related to the absorption due to the anharmonicity of the segmental motions is an only part of the specific heat due to the vibrations of segments, not due to all lattice vibrations in the polymer solid.[7]

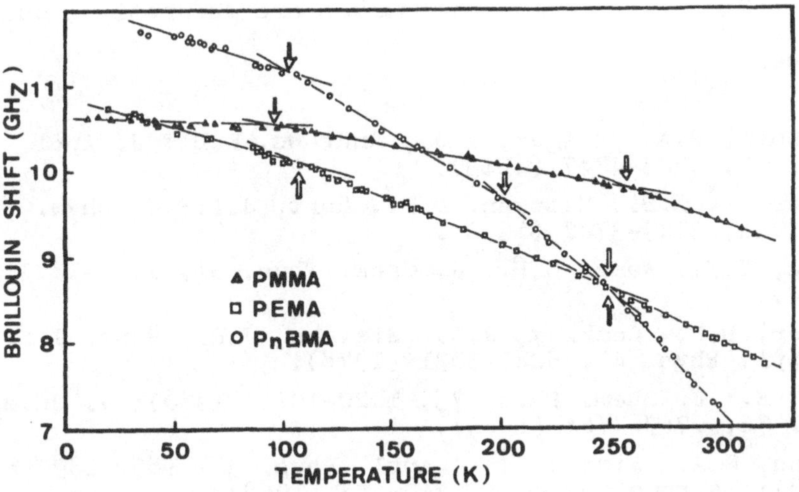

Figure 1. Brillouin shift vs. temperature for the scattering angle 90°. The arrows show the changes in slope.

Table 1. Transition temperatures.
a : Present experiments, b : Reference 6.

Sample	T_1(K)		T_2(K)		T_3(K)	
	a	b	a	b	a	b
PMMA	262	288			98	83
PEMA		298	250	258	112	110
PnBMA	250	248	205	222	105	97

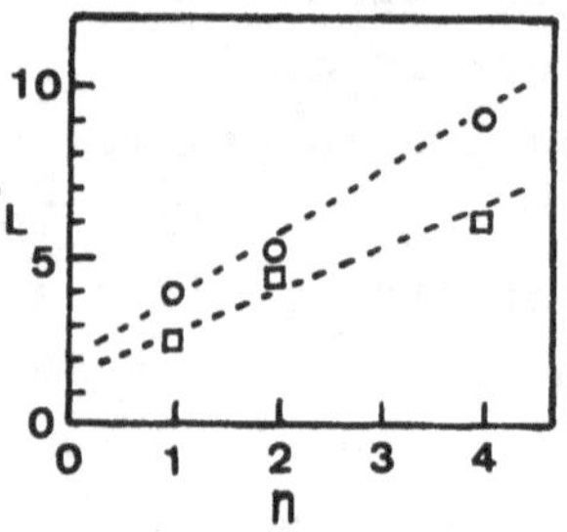

Figure 2. γ_L vs. n for an alkyl C_nH_{2n+1}. n = 1:PMMA n=2: PEMA and n=4:PnBMA.
○: at room temperature
□: below T_3

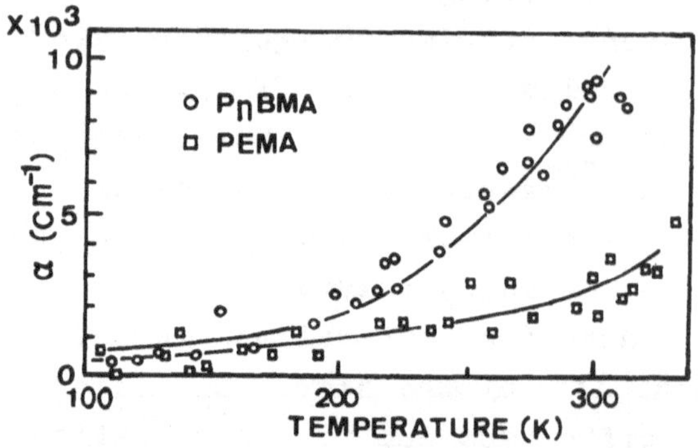

Figure 3. The hypersonic absorption α vs. temperature. The solid curves are theoretical ones.

References

1. Friedman, E.A., Ritger, A.J., Andrews, R.D.: J. Appl. Phys. 40, 4243-4247 (1969).
2. Romberger, A.B., Eastman, D.P., Hunt, J.L.: J. Chem. Phys. 51, 3723-3727 (1969).
3. Huang, Y.Y., Wang, C.H.: J. Chem. Phys. 61, 1868-1873 (1974).
4. Mahler, D.S., Coakley, R.W., Stevens, J.R., Hunt, J.L.: J. Appl. Phys. 49, 5029-5031 (1978).
5. Kato, E.: J. Chem. Phys. 73, 1020-1025 (1980); J. Chem. Phys. 76, 2763-2764 (1982).
6. Haldon, R.A., Simha, R.: J.Appl. Phys. 39, 1890-1899 (1968); Macromolecules 1, 340-343 (1968).
7. Wada, Y., Itani, A., Nishi, T., Nagai, S.: J. Polym. Sci. 7, 201-208 (1969).
8. Mopsik, F.I., Broadhurst, M.G.: J. Appl. Phys. 44, 4261-4264 (1973).
9. Woodruff, T.O., Ehrenreich, H.: Phys. Rev. 123, 1553-1559 (1961).

TIME RESOLVED SMALL-ANGLE X-RAY SCATTERING STUDIES ON KINETICS AND MOLECULAR DYNAMICS OF ORDER-DISORDER TRANSITION OF BLOCK POLYMERS

Takeji Hashimoto
Department of Polymer Chemistry, Kyoto University, Kyoto 606, Japan

Introduction

Block polymers (A-B, A-B-A, etc.) comprising amorphous polymers A and B form microdomain structure in the segregation limit where Flory-Huggins interaction parameter χ between the constituent block chains A and B satisfy

$$\chi N > (\chi N)_c \tag{1}$$

N being total degree of polymerization of the entire block polymer and $(\chi N)_c$ being the critical value of χN which depends on a fraction f of one type of block chain (e.g. A). This microdomain structure is formed as a consequence of microphase separation of polymers A and B and has a unique structure in which microphases comprising one type of polymers (e.g. A) are spatially arranged in the matrix of other (e.g. B) with a long ranged order called as "superlattice". The spacing of this periodic structure originates from connectivity of polymers A and B and is of the order of molecular dimension of the block polymer.

When temperature is raised, χ decreases and the condition of $\chi N < (\chi N)_c$ is satisfied to result in dissolution of the microphase-separated domains into disordered molecular mixture (*order-disorder transition*) (1-6). In this paper we shall concern with equilibrium and kinetic aspects of the order-disorder transition, the relevant order parameter here being spatial concentration variations of segments A and B ($\rho_K(\underset{\sim}{r})$, K = A or B). Such studies would be of importance from an academic viewpoint , especially on effects of *molecular connectivity* on phase-equilibrium and molecular dynamics involved in the transition as well as from an industrial view-

Physical Optics of Dynamic Phenomena and Processes in Macromolecular Systems

point, especially for molecular design of processability of the thermoplastic elastomers.

Experimental Methods

The systems to be studied here are (i) SB/tetradecane and (ii) SBS/dipentene. SB is a diblock polymer comprising of polystyrene (PS) and polybutadiene (PB) having number average molecular weight $\overline{M}_n = 5.2 \times 10^4$ g/mol and weight fraction of PS $W_{ps} = 0.30$. SBS is a triblock polymer comprising of PS-PB-PS having $\overline{M}_n = 5.8 \times 10^4$ g/mol and $W_{ps} = 0.48$. The solvents were added to lower the order-disorder transition temperature (T_r). n-tetradecane (C14) is a solvent poor for PS but good for PB. Dipentene is a solvent neutrally good for both PS and PB. The concentration range covered is from 15 to 40 wt % of polymer for SB-solution. The experimental technique involves *time-resolved* analyses of small-angle X-ray scattering (SAXS) during the order-disorder transition (microphase dissolution) and disorder-order transition (microphase separation) followed by temperature-jump (T-jp) and static SAXS analyses below and above T_r.

Theoretical Background on Time-Resolved SAXS

We shall present here the change of SAXS profiles with time during the order-disorder transition. The change of the profiles during the reverse transition, i.e., disorder-order transition may be described similarly to the transition in polymer blends of A and B, but there are no rigorous theories presented so far which can account for the effect of molecular connectivity on the kinetics of the disorder-order transition.

When the thermodynamic driving force for the microphase separation is suddenly removed by T-jp above T_r for the block polymers having upper critical solution temperature (UCST), the system undergoes the order-disorder transition as schematically shown in Figure 1(a) according to a molecular mechanism as may be envisioned by *reptation* process, a curviliniar diffusion of polymer molecules along its own contour in bulk or in concentrated solution (7, 8). We assume that, according to the reptation, each molecule undergoes Brownian motion independently of others,

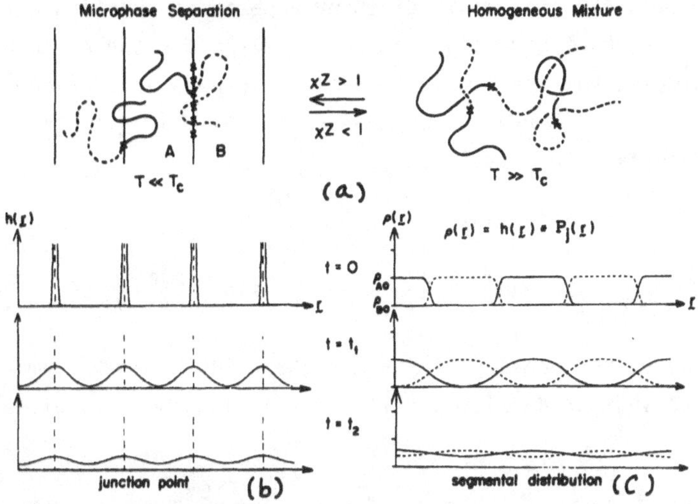

Figure 1. Molecular mechanism of order-disorder transition.

and the spatial distribution of the chemical junction points $h(\underset{\sim}{r})$ becomes broader with time as shown in Figure 1(b). This process may be described by a random stochastic process. As the reptation takes place the segments loose their memories with respect to their position and orientation, resulting in mixing of the unlike segments, increased interfacial thickness and eventually in a essentially uniform segmental density distribution characteristic for the disordered mixture as shown in Figure 1(c). We further simplify our treatment by assuming that the temperature T is raised sufficiently high above T_r, so that the thermodynamic interaction between A and B polymers does not significantly affect the diffusion of polymers.

Under these conditions, the change of the segmental density profile with time $\rho_K(\underset{\sim}{r}, t)$ should be simply given by Fickian (9)

$$\partial \rho_K(\underset{\sim}{r}, t)/\partial t = D_c \nabla^2 \rho_K(\underset{\sim}{r}, t), \quad K = A \text{ or } B \tag{2}$$

where D_c is the self-diffusivity for the block polymer as a whole which should be a function of the self-diffusivity of the constituent polymers $D_{c,K}$. In case when T is not sufficiently high above T_r, eq. 2 should be modified to take into account an effect of finite value of $\chi(T)$. If the initial segment density profile is given by a periodic function with a repeat distance 2ℓ, having domain size 2a and 2ℓ-2a for the A and B domains, respectively, and the domain boundary thickness as characterized by the *characteristic interfacial thickness* (1, 10) $\sqrt{2\pi}\,\sigma_o$ as shown in Figure 1(c) (lamellar microdomain), i.e.,

$$\tilde{\rho}_A(x,t=0) = \rho_A(x,t=0)/\rho_{Ao}$$

$$= \frac{a}{\ell} + \frac{2}{\pi} \sum_{m=1}^{\infty} \exp\left(- \frac{m^2\pi^2}{\ell^2} \frac{\sigma_o^2}{2}\right) \frac{1}{m} \sin\left(\frac{m\pi a}{\ell}\right) \cos\left(\frac{m\pi x}{\ell}\right) \quad (3)$$

where ρ_{Ao} is the number density of the A segment in the homopolymer A, then the reduced segment density profile $\tilde{\rho}_A(x,t)$ at time t is given by (from eqs. 2 and 3 (9))

$$\tilde{\rho}_A(x,t) = \frac{a}{\ell} + \frac{2}{\pi} \sum_{m=1}^{\infty} \exp\left(- \frac{m^2\pi^2}{\ell^2} \frac{\sigma_o^2 + 2D_c t}{2}\right)$$

$$\frac{1}{m} \sin\left(\frac{m\pi a}{\ell}\right) \cos\left(\frac{m\pi x}{\ell}\right) \quad (4)$$

$$= \tilde{\rho}_A(x,t=0,\sigma_o=0) * h(x,t) \quad (5)$$

where h(x, t) represents the spatial distribution of the chemical junctions at time t and is a solution of eq.2 with the initial condition of

$$h(x,\ t = 0) = (2\pi\sigma_o^2)^{-1/2} \exp(-x^2/2\sigma_o^2) \quad (6)$$

for the spatial distribution of junctions (Gaussian approximation). h(x, t) is given by (9)

$$h(x,t) = [2\pi(\sigma_o^2 + \sigma_1^2)]^{-1/2} \exp[-x^2/2(\sigma_o^2 + \sigma_1^2)] \quad (7)$$

where

$$\sigma_1^2 = 2D_c t \quad (8)$$

Similar equation for $\tilde{\rho}_B(x,\ t)$ is obtained by replacing x by x-2a. Now knowing the segmental density profile $\tilde{\rho}_K(x,\ t)$, one can calculate the scattering contrast for X-ray and therefore the change of the scattering intensity with time I(q, t),

$$I(q, t) = I(q, t = 0) \exp[-2R(q)t] \qquad (9)$$

where

$$R(q) = q^2 D_c \qquad (10)$$

$$q = (4\pi/\lambda) \sin \theta \qquad (11)$$

where λ and 2θ are the wavelength of X-ray and the scattering angle, respectively. Based on eqs. 9 to 11, one can study kinetics of the order-disorder transition by the time-resolved SAXS technique.

Equilibrium Aspects of Order-Disorder Transitions

Effects of the order-disorder transition on the SAXS profiles were investigated by Roe et al (5) and Hashimoto et al. (6). Strong scattering maxima in the ordered state arises from a long-range spatial order of the micro-domains and the spacing calculated from the maximum generally decreases with increasing T and decreasing ϕ_p, polymer volume fraction in neutrally good solvents, due to decreasing segregation power (6),

$$D \sim (\phi_p/T)^{1/3} \quad \text{(in ordered state)} \qquad (12)$$

A weak scattering maximum is observed even for block polymers in bulk and concentrated solution in disordered state (5, 6) and is interpreted as the maximum arising from *correlation hole effect* as predicted by de Gennes (11) and Leibler (3) based on the random phase approximation (RPA) and recently by Benoit and Benmouna (12).

The scattering formula presented by Leibler (3) for the bulk disordered block polymers may be rewritten as follows (6)

$$I^{-1}(q) \sim \{F(x)/N - 2A\} - 2B/T \qquad (13)$$

if one can write χ as A + (B/T). We applied eq. 13 to the block polymer solutions with neutral solvents by replacing χ for bulk by $\chi\phi_p$ and found that the linearity between I^{-1} and T^{-1} holds in the disordered state and

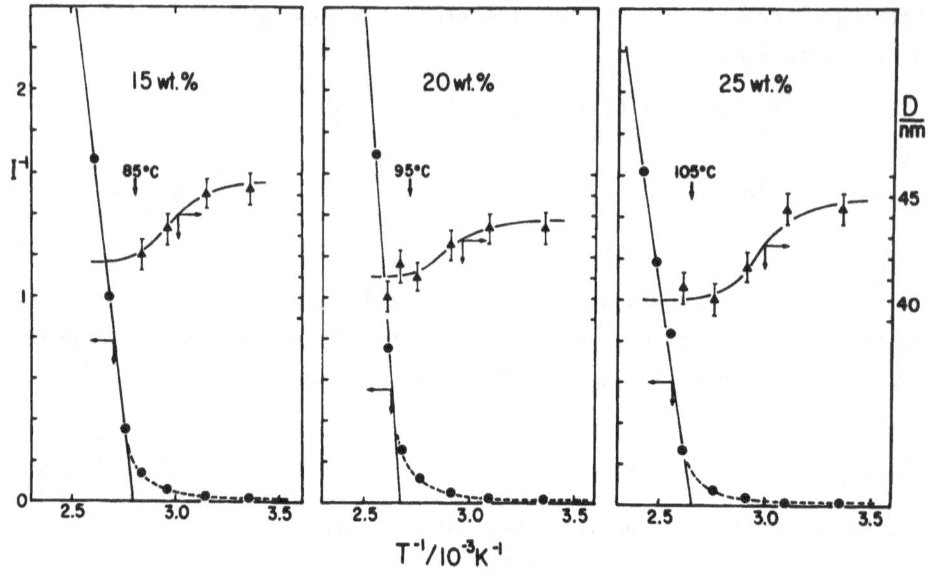

Figure 2. Temperature dependence of scattered intensity I at a given q q_{max} and Bragg spacing D for 15,20,and 30 wt% SB in C14.

the spacing as estimated from the maximum is independent of T and ϕ_p (6),

$$D \sim (\phi_p/T)^0 \qquad \text{(disordered state)} \tag{14}$$

The linear relationship between I^{-1} and T^{-1} and eq. 14 as well were also observed for C14 solutions as shown in Figure 2 at sufficiently high temperature. However upon lowering temperature the deviations from the linear relationship between I^{-1} and T^{-1} and from the constancy of D with T (e.g. eq. 14) were observed, which are believed to be a consequence of onset of the disorder-order transition. The transition temperatures T_r were estimated as a function of concentration from the temperature at which the deviations start to occur. The results indicated that T_r linearly increases with ϕ_p, being 100, 110, 120, 130, 140°C for 20, 25, 30, 35 and 40 wt% of SB in C14, respectively.

Kinetics of Order-Disorder Transition

Figure 3 shows an isometric display demonstrating typical change of the scattering profiles with time after T-jp up to 160°C for 35 wt% SB solution in C14 with T_r = 130°C. Each curve was obtained with 2 s exposure to X-ray. We employed here a repetitive T-jp method with repetition of ten times in order to count for a sufficient number of photons. The scattered intensity is observed to decay with time, indicating that the microdomains are dissolved at a time scale of the order of a few 10 s in this particular system.

Figure 4 shows the decay of the scattered intensity at q = 0.149 nm^{-1} (close to q_{max} giving rise to maximum intensity) with time. The intensity decay with time is nonexponential in the beginning but becomes exponential in the later stage as predicted from eq. 9. The initial nonexponential decay turns out to be an effect of finite time required for T-jp(13). The slopes of the straight lines give rise to D_c from eqs. 9 to 11 as a function of temperature at a given concentration (20 wt%).

Figure 5(a) shows the diffusivity of the SB block polymers in C14 as a function of 1/T at various concentrations (D_{eff} in the figure corresponds to D_c in the text). The activation energies ΔH_a thus measured are shown in the table inserted at upper right corner of the figure. The small

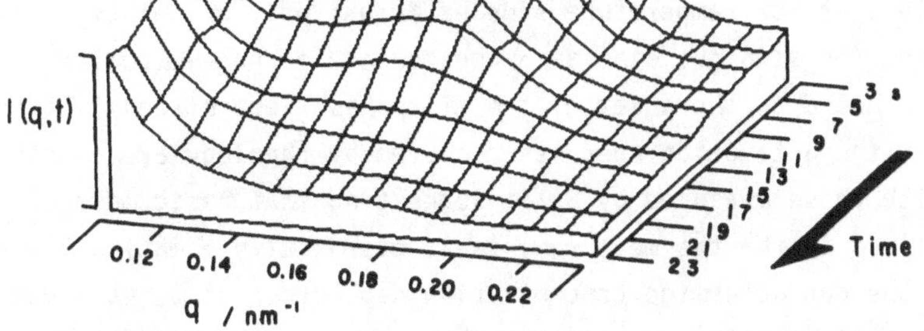

Figure 3. Isometric display of time-resolved SAXS profiles during order-disorder transition.

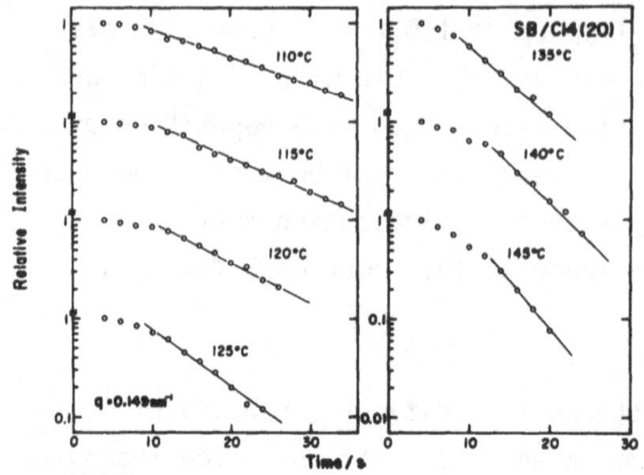

Figure 4. Logarithmus of scattered intensity at a given q with time at various temperatures.

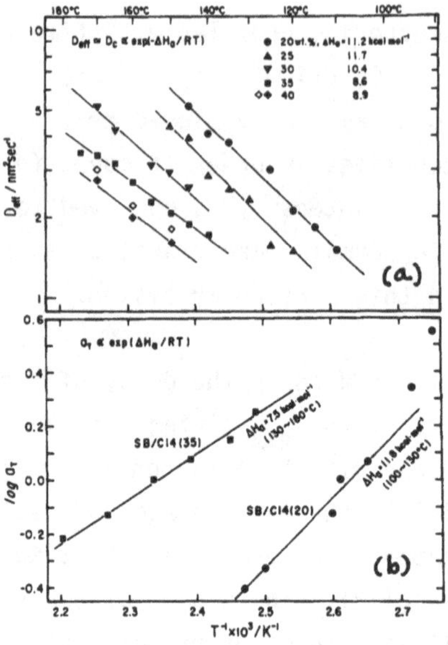

Figure 5. Diffusivities(a) and shift factors(b) as a function of 1/T at various concentrations of SB/C14.

diffusivity of the order of 10^{-14} to 10^{-13} cm^2/s was estimated with this technique. The measured diffusivity is a self-diffusivity of the block polymer as a whole in the concentrated solutions, which may be decribed in terms of the diffusivities of A and B block chains and thermodynamic interaction parameters between A and B, A and solvent, and B and solvent.

Rheological measurements of the solutions in the same temperature and concentration ranges indicated the flow behavior is undoubtedly Newtonian. Figure 5(b) shows the temperature dependence of the shift factor a_T in constructing the master curves of real and imaginary parts of shear moduli (the two points at low temperature side of 20 wt% polymer solution showing deviation from the straight line as a consequence of non-Newtonian flow behavior attributed to the onset of the microphase separation). It should be noted that the activation energies measured by rheology are in good agreement with those measured by SAXS, suggesting that basic molecular process involved in the two measurements is essentially identical. From Figure 5(a) one can determine concentration dependence of D_c at given temperatures. It is rather surprising that the experimental results closely fit with the scaling prediction for semidilute solution, $D_c \sim C^{-1.75}$ even at such high concentration regime covered in this studies. This point deserves further theoretical and experimental studies, especially on reduction of D_c due to thermodynamic interaction between A and B. Residual weak repulsive interaction between A and B will impede mixing of the unlike segments and hence reduce D_c.

We shall consider now kinetics of the microphase separation, i.e., the disorder-order transition. Figure 6 shows typical evolution of scattering profiles after temperature drop (T-drop) from 190 to 30°C for 55 wt% SBS solution in dipentene with T_r = 160°C, the system exhibiting UCST behavior. The scattering maximum appears shortly after T-drop, maximum intensity increases and the peak position q_{max} slightly shift toward smaller angles with time. The intensity increase tends to level off at about 80 s after T-drop for this particular system. Figure 7 shows logarithmus of scattered intensity at various q plotted as a function of time t during

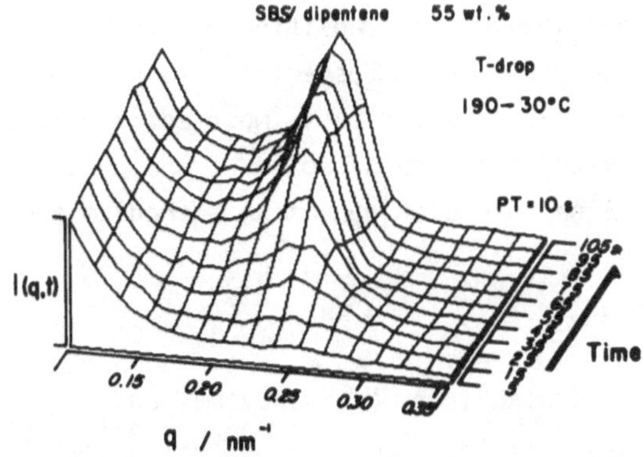

Figure 6. Isometric display of time-resolved SAXS profiles during disorder-order transition of 55 wt% solution of SBS in dipentene.

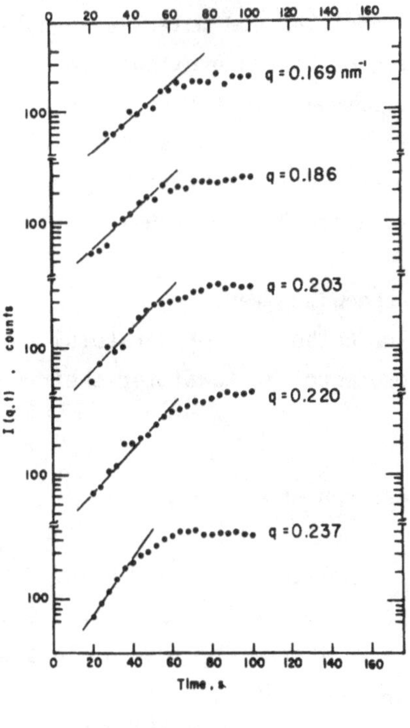

Figure 7. Logarithmus of scattered intensity with time at various q during disorder-order transition of the solution shown in Figure 6.

the isothermal phase transition. For this particular system, intensity initially increases exponentially with time. However in the later stage the intensity increase deviates from the exponential behavior. These phenomena should reflect very fundamental aspects of growth mechanism of spatial segmental density fluctuations in block polymers; (i) in the initial stage, the sinusoidally varying concentration fluctuations of styrene and butadiene monomers will be built up (see Figure 1(c) at $t = t_2$, and consider for example the case of m = 1 in eq. 4), through reptation of block polymer molecules, and (ii) in the later stage, the sinusoidal waveform for the concentration fluctuations may be transformed into the square waveform as may be depicted in Figure 1(c) with elapse of time from t_2 to t_1, and t_2 to t=0, through local rearrangements of molecules caused by segregation power between A and B. The segregation will cause stretching of polymers A and B perpendicular to the interfaces and hence increase the spacing D, which in turn will cause shifting of the q_{max} toward smaller angles. The mechanism and kinetics of this disorder-order transition and molecular dynamics involved in the transition may be analogous, to some extent, to those of polymer blends (especially to *spinodal decomposition* for this particular system). However most important physical factor to be understood in block polymers seems to be roles of *molecular connectivity* between polymers A and B on molecular dynamics, mechanism and kinetics of the transition.

Acknowledgement
The author is very grateful to Mr. K. Kowsaka, Dr. M. Shibayama, and professor H. Kawai for their collaborations.

References

1. Helfand, E., Wasserman, Z.R.: Macromolecules, 9, 879 (1976).
2. Meier, D.J.: J. Polym. Sci., Part C, 26, 81 (1969).
3. Leibler, L.: Macromolecules, 13, 1602 (1977).
4. Noolandi, J., Hong, K.M.: Ferroelectrics, 30, 117 (1980).
5. Roe, R.J., Fishkis, M., Chang, J.C.: Macromolecules, 14, 1091 (1981).

6. Hashimoto, T., Shibayama, M., Kawai, H.: ACS Polym. Prepr., 23(1), 21 (1981): Macromolecules, 16, 1093 (1983).
7. de Gennes, P.G.: J. Chem. Phys., 55, 572 (1971).
8. Doi, M., Edwards, S.F.: J. Chem. Soc., Faraday Trans. II, 74, 1789, 1802, 1818 (1978).
9. Hashimoto, T., Tsukahara, Y., Kawai, H.: J. Polym. Sci., Polym. Lett. Ed., 18, 585 (1980): Macromolecules, 14, 708 (1981).
10. Hashimoto, T., Shibayama, M., Kawai, H.: Macromolecules, 13, 1237 (1980).
11. de Gennes, P.G.: "Scaling Concepts in Polymer Physics", Cornell Univ. Press., N.Y., 1979, Chap.IX.
12. Benoit, H.: invited lecture at 1984 Prague Meeting on Macromolecules.
13. Hashimoto, T., Kowsaka, K., Shibayama, M., Kawai, H.: ACS Polym. Prepr., 24(2), 224 (1983).

THERMOSTIMULATED LUMINESCENCE IN THE RELAXATION REGIONS OF POLYMERS

Josef Pospíšil, Antonín Havránek
Department of Polymer Physics, Faculty of Mathematics and Physics, Charles University
180 00 Praha, Czechoslovakia

Ivars A. Tale
Institute of Solid State Physics, Latvian State University
226 063 Riga, Latvia, USSR

Introduction

Anomalous behaviour of the thermostimulated luminescence has been observed (1, 2, 3) in the temperature regions where the motions of polymer chains or their parts occur. Such regions are called relaxation regions and the most common methods of their detection are viscoelastic and dielectric measurements. The anomaly of the thermostimulated luminescence will be explained by the influence of the macromolecular chain motion upon the luminescence behaviour of polymers. A quantitative model describing this influence will be given.

Experimental and Results

The thermostimulated luminescence (TSL) method consists in the monitoring of the luminescence intensity I versus temperature T curve of the sample which has been excited at a low temperature, usually by X-ray irradiation, and then heat-

Physical Optics of Dynamic Phenomena and Processes in Macromolecular Systems

ed. The trapped carriers are released during the heating of the sample, they recombine and emit light which is detected as the luminescence intensity I (the glow curve) (4, 5, 6). In addition to the simple linear heating, the step-wise heating (Gobrecht's method (7, 8)) has been used in our experiments. This method enables the determination of the mean (over one step) activation energy $\langle E \rangle$ and the mean frequency factor $\langle s \rangle$ of the trapped charge carrier release.
The anomalous large values of $\langle E \rangle$ and $\langle s \rangle$ were obtained if the most common relation

$$n = s \exp(-E/kT) \tag{1}$$

(n ... the frequency of the charge carrier release, k ... the Boltzmann's constant, T ... the temperature) had been used to the evaluation of the experimentally obtained glow curves in the relaxation regions. Namely the values of $\langle E \rangle$ in the range 0.7 to 0.9 eV and of $\langle s \rangle$ in the frequency range 10^{20} to 10^{21} s^{-1} were obtained for the measured butadiene-styrene copolymers (1, 9, 10). These values together with the characteristic temperature T_m of the two measured relaxation regions, the main relaxation region α and the β relaxation region, obtained by various methods for one sample (Krylene 1509) are given in Table 1.

Table 1. Characteristics of relaxation regions (Krylene 1509)

Relaxation region	$\langle E \rangle$ (eV)	$\langle s \rangle$ (Hz)	T_m (K)			
			dilat. + DTA	viscoel.	TSL linear heat.	TSL step heat.
α	0.9	10^{21}	214 +215	219	215	195
β	0.7	10^{20}	+139	135	160	153

Theoretical and Discussion

A plausible explanation of the glow curves avoiding the anomalous large values of $\langle E \rangle$ and $\langle s \rangle$ may be obtained if we use the Tale's (11) model for the interpretation of the glow curves in the relaxation regions. In this model the activation energy E of equation (1) is replaced by the mean value

$$\bar{E} = \frac{w_1 E_1 + w_2 E_2}{w_1 + w_2} \tag{2}$$

as it is assumed that two types of energy barriers for the ionization of traps occur in the relaxation region. The height E_1 of the first type barrier is unaltered by the relaxation process, the height E_2 of the second type barrier diminishes strongly, in some cases even to zero ($E_2 \rightarrow 0$), by the molecular motion responsible for the relaxation process. The first and second type barriers occur with the probabilities w_1 and w_2 . According to the free volume concept (e.g. (12)) it is additionally assumed that

$$w_2 = w_{2o} \exp (-Q/kT) \tag{3}$$

where Q is the activation energy of the appropriate relaxation process, i.e. the value which may be derived e.g. from viscoelastic measurements. If we introduce eq.(3) into eq.(2) we obtain

$$E = E_2 + (E_1 - E_2) \frac{1}{1 + (w_{2o}/ w_1) \exp(-Q/kT)} \tag{4}$$

The approximate equation $\bar{E} \doteq E_1$ is valid at low temperature. The decrease from E_1 to E_2 is most significant in the region where

$$(w_{2o} / w_1) \exp (-Q/kT) \doteq 1 \tag{5}$$

At high temperature $\bar{E} \doteq E_2$ as generally $w_1 \ll w_{2o}$. The intensity glow curve, in the case of monomolecular kinetics and linear time (t) increase in temperature, $T = T_o + \gamma t$, is given by the equation

$$I = -dn/dt = I_o \exp\left(\frac{-\bar{E}(T)}{kT}\right) \exp\left\{-\frac{s}{\gamma}\int_{T_o}^{T} \exp -\frac{\bar{E}(T')}{kT'}\, dT'\right\} \quad (6)$$

where n is the concentration of the ionized luminescent centres. The mean activation energy over one step of the glow curve measured in the step-wise regime is

$$\langle E\rangle = \frac{-d(\ln I)}{d(1/kT)} = (E_1 - E_2)\,\frac{1 + (1 + Q/kT)\, w_{2o}/w_1 \exp(-Q/kT)}{\left\{1 + (w_{2o}/w_1) \exp(-Q/kT)\right\}^2} \quad (7)$$

In the limits $T \rightarrow 0$, $T \rightarrow \infty$, we obtain $\langle E\rangle \doteq \bar{E} \doteq E_1$ and $\langle E\rangle \doteq \bar{E} \doteq E_2$ respectively.

These limit conditions together with the rewritten eq.(5)

$$w_{2o} / w_1 \doteq \exp(Q/kT_m) \quad (8)$$

(T_m ... the temperature of the relaxation maximum-see Tab.1) make possible the determination of all the parameters of eq.(7). Then the experimentally obtained $\langle E\rangle$ vs. T curve may be compared with the theoretical (evaluated by the eq.(7)) one. The result of such a comparison which has been made for the Krylene 1509 sample is given in Fig. 1 . The parameters used in this comparison are listed in Table 2.

Table 2. Parameters of theoretical data used for comparison of experiments

Relaxation region	E_1 (eV)	E_2 (eV)	s (Hz)	w_{2o}/w_1	T_m (K)	Q (eV)	Q_{exp} (eV)
β	0.455	0.395	1.10^{13}	1.10^{14}	153	0.390	0.25
α	0.627	0.561	1.10^{14}	1.10^{14}	195	0.495	0.75

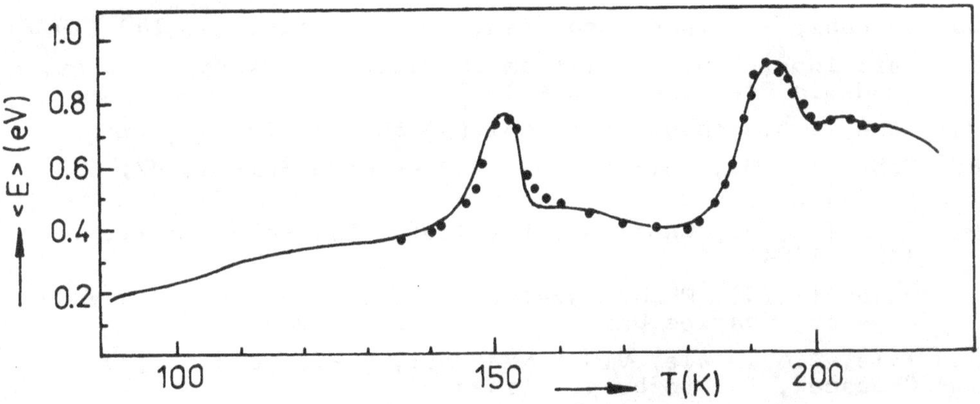

Fig. 1

The dependence of the mean activation energy <E> on temperature (— experimental curve, ••• theoretical curve).

The agreement between the theoretical and experimental curve is quite good, no anomalous (unplausible order of magnitude) values of the parameters need not be used and also the differences between the activation energies Q_{exp} (last column of Table 2) obtained from the viscoelastic experiment and the values Q used for the best fit of the two curves are not too large. The description of the experimentally obtained curves in terms of the given model is satisfactory also for other samples and experimental conditions. Especially, this model can explain the existence, anomalous behaviour, and the observed E and s dependence of the glow curve maxima width in the relaxation regions of polymers.

References

1. Knappe, W., Voigt, G. and Zyball, A.: Colloid Polym.Sci., 252, 673 (1974)
2. Pospíšil, J., Tale, I.A.:Polym.Bull. 9, 40 (1983)
3. Pospíšil, J., Havránek, A.: Proc. 24th microsymposium Copolymers: Structure and Solution Properties, Prague,1983
4. Pender, L.F., Fleming, R.J.: J. Phys. C., 10, 1571 (1977)

5. Linkens, A., Vanderschueren, J.: J.Electr., 3, 149 (1977)

6. Partridge, R.H.: Radiation Chemistry of Macromolecules, Academic Press, New York 1972

7. Tale, I.A.: phys. stat. sol.(a) 66, 65 (1981)

8. Gobrecht, H., Hofmann, D.: J.Phys.Chem.Solids, 27, 509 (1964)

9. Pospíšil, J., Chudáček, I., Donth, E.: Polym. Bull. 12, 149, (1984)

10. Pospíšil, J.: Ph.D. Thesis, Faculty of Mathematics and Physics, Charles University, Prague, 1984

11. Tale, I.A., Tale, V.G., Nagornyi, A.A., Pospíšil, J., Chudáček, I.: to be published

12. Bueche, F.: Physical Properties of Polymers, Wiley, New York 1962

DYNAMIC PHENOMENA, INTERACTIONS, CONFORMATIONAL CHANGES AND CHEMICAL PROCESSES IN MACROMOLECULAR SYSTEMS AS STUDIED BY POLARIZED LUMINESCENCE

Elizaveta Anufrieva and Mark Krakovyak
Institute of Macromolecular Compounds of the Academy of Sciences of the USSR
Leningrad, USSR

Polarized luminescence of macromolecules with covalently bonded luminescent groups

The anisotropy of light of a luminescent group bonded to the polymer chain by a covalent bond i.e. the polarization of luminescence of a solution of a labelled polymer provides information about dynamic phenomena in polymer coils, the changes in intermolecular interactions in the polymer system, the conformational transformations in macromolecules and chemical reactions involving the functional groups of the polymer. The content of luminescent groups in the polymer should not exceed 0.2 mol%. In order to obtain the above information the structure of the luminescent group bonded to the polymer by a covalent bond should meet many requirements considered in detail in refs (1, 2). The structure of the anthracene ring and its optical properties meet all the requirements imposed by the method of polarized luminescence (PL). The rotation of the anthracene ring about the bond attaching its carbon 9 - position to the macromolecule does not affect the polarization of luminescence because the dipole moment of the transition coincides with the rotation axis and the rotation about other following bonds is possible only together with the vibrational motion of the main chain. The time range of relaxation processes studied by polarized lu-

Physical Optics of Dynamic Phenomena and Processes in Macromolecular Systems

minescence is determined by the duration of the lifetime of emission of the luminescent group. Since nanosecond relaxation processes are appreciably affected by the change in intra- and intermolecular interactions (presumably because their times are comparable with the duration of contacts responsible for the change in intermolecular interactions), the lifetime of emission of the luminescent group should be in the nanosecond range. The lifetime of luminescence of anthryl - containing luminescent groups is 5-10 ns (1, 2).

Investigation of dynamic phenomena in macromolecular systems

In order to determine relaxation time τ_w by PL the following equation is used:

$$\tau_w = [(1/P_0' + 1/3)\times 3\ \tau_f] \ / \ (1/P - 1/P_0')$$

where P is the polarization of luminescence being measured for a solution of the labelled polymer, τ_f is the duration of luminescence of the solution and $1/P_0'$ is the value determining the contribution of high-frequency motions of the luminescent group to the relaxation spectrum of the labelled polymer and characterizing the rotation amplitude of the luminescent group i.e. the thermodynamic flexibility of the position of the molecule bearing the luminescent group. The methods of estimation of this value and the factors determining it have been described in ref. (1). The time τ_w characterizes the mobility of the middle parts of the chains if the luminescent group is inserted into the middle of the main chain and that of end parts if it is attached to the chain end. Luminescent groups in long side chains containing more than 10 bonds provide information about the mobility of side chains, whereas those in short side chains provide information about the joint motion of these chains and neighbouring parts of the main chain, i.e. about the intramolecular mobility of the coil. If the molecular weight of the polymer is less than 10 000, then the new nanosecond relaxation process appears; it is related to the rotational motion of the macromolecule as a whole.

The methods of synthesis of polymers and multicomponent systems with covalently bonded luminescent groups developed by us (ref. (2), Ch.3) permit the bonding of luminescent groups not only to a definite part of the polymer chain (middle or end parts or a side group) but also to a definite component of a multicomponent system, for example to a block of a definite chemical structure in three-block copolymers. The components of a multicomponent system can be distinguished with the aid of the luminescent group and thus it is possible to obtain information about each component of this system. This is particularly important in the investigation of block copolymers (3), interpolymer complexes (4-7), exchange or substitution reactions in multicomponent systems, crosslinked polymers with bridge bonds (8) and other multicomponent systems. The method of PL allows the study of the coil dynamics in polymer solutions diluted to 0.003%. Under these conditions the contacts between the coils either virtually do not exist or their appearance is very improbable. This is particularly important in the study of dynamic properties in thermodynamically poor solvents. The PL method makes it possible to investigate the dynamics of polymer coils in solvents of any nature - in polar and non-polar solvents and water at different pH values and ionic strengths. Hence, the range of problems solved by PL is very wide.

Intermolecular interactions and intramolecular mobility

The change in intramolecular interactions in polymer coils in solutions profoundly affects relaxation times in nanosecond relaxation processes related in a certain manner to the motion of the main chain . The change in nanosecond relaxation processes in polymer chains resulting from changes in the thermodynamic strength of the solvent, the formation (breaking) of intramolecular hydrogen bonds, the strengthening (weakening) of hydrophobic interactions in aqueous (aqueous-organic) media and changes in electrostatic interactions has been discussed in detail (1, 2). The next example shows that the nanosecond relaxation processes reflect even smaller differences in intramolecular interactions in the coils caused by the differences in the molecular weights of the polymers investigated (Fig.1).

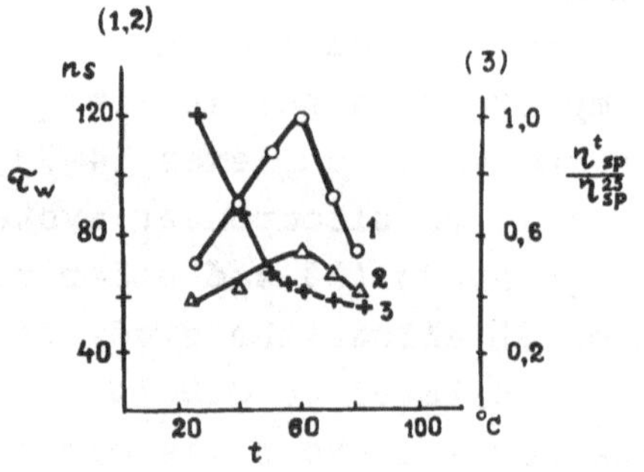

Fig.1. Temperature dependence of 1) relaxation time $_W$ of high molecular weights PMAA in water , M = 330,000 2) relaxation time τ_W of low molecular weight PMAA in water, M = 15000, $C_{polymer}$ = 0.5 mg/ml and 3) relative changes in reduced viscosity of PMAA solution (M = 300000) in water

The data in Fig.1 show that an increase in hydrophobic interactions between non-polar methyl groups of polymethacrylic acid (PMAA) in water upon heating to 60°C leads to an increase in the times τ_W, whereas subsequent heating causing the breaking of intramolecular hydrogen bonds decreases these times. Upon heating in water from 25 to 60°C relaxation times for high molecular weight PMAA (M = 330000) increase by a factor of 1.7 and those for low molecular weight PMAA (M = 15000) increase by a factor of 1.15.

Conformational transformations of macromolecules and relaxation processes in macromolecular systems

The high sensitivity of relaxation times of nanosecond relaxation processes in polymer chains to changes in intramolecular interactions permits the investigation of conformational transformations of macromolecules based on their relaxation characteristics. It was shown that relaxation times of nanosecond relaxation processes are very sensitive to coil-compact structure transition in PMAA molecules in water when the fraction of ionized COO^- groups in PMAA decreases (2). It is known that relaxation times of nanosecond relaxation processes are very sensitive to the coil-α-helix transition in polyglutamic acid molecules (9) or to the coil - α-helix-compact structure transition in the molecules of glutamic acid-leucine copolymer (83:17%) (9). The PL method (relaxation characteristics of nanosecond processes) is widely used for the study of the coil-globule transitions (9-11). The formation of the globule is accompanied by the appearance of a new relaxation process: the rotational mobility of the macromolecule as a whole. The formation of the globule has been observed in the molecules of poly-1,2-dimethoxy ethylene in water upon heating (10), in those of glutamic acid-leucine copolymers (83-17%) in water when the fraction of ionized groups decreased (9) and in PMMA molecules when the fraction of the precipitant (octane) in toluene increased (2). On the basis of the study of relaxation characteristics the relationships of the formation of the globular type structure were established. The molecular weight dependence of the range of globule formation was revealed (the higher the molecular weight, the lower the structure formation factor leading to the formation of the globule). The invariance of the range of globule formation with changing polymer content in solution indicates that the transition is of the intramolecular character.

The approach based on the study of relaxation characteristics by PL was found to be particularly informative in the investigation of conformational transformations in the molecules of liquide-crystalline polymers containing mesogenic groups. By using the PL method it was possible to establish not only the significance of mesogenic groups in structure formation in dilute polymer solutions but also to reveal a new type of structural transition and its relationship to the molecular weight of the liquid-crystalline polymer. This is the coil-isotropic liquid globule - liquid-crystalline globule transition. The samples used were fractions of cholesterol ester of polymethacryloyl-ω-oxyundecane acid (PChM-10) with molecular weights from 0.35×10^6 to 6.6×10^6 (11). Analysis of changes in the mobility of the main and side chains of PChM-10 molecules during structure formation (formation of a compact structure of macromolecules) in heptane upon cooling from 70 to 20°C and the evaluation of changes in the packing density from variations in the amplitude of high-frequency motions (parameter $1/P'_0$) showed that in PChM-10 molecules in the temperature range 55-47°C a compact structure of globular type with hindered main chain ($\tau_{m.ch.}$) and mobile side chains ($\tau_{s.ch.}$) is formed. The packing density of units differs only slightly from that of the coil. The arrangement of mesogenic groups remains isotropic (optical activity of solution remains invariable). These data imply that in the 55-47°C range an isotropic liquid globule is formed. Upon cooling below 47°C the parameters of kinetic ($\tau_{m.ch.}$, $\tau_{s.ch.}$) and thermodynamic ($1/P'_0$) flexibility change cooperatively and approach the values characteristic of the condensed state, and an order in the arrangement of mesogenic groups appears (Fig.2).

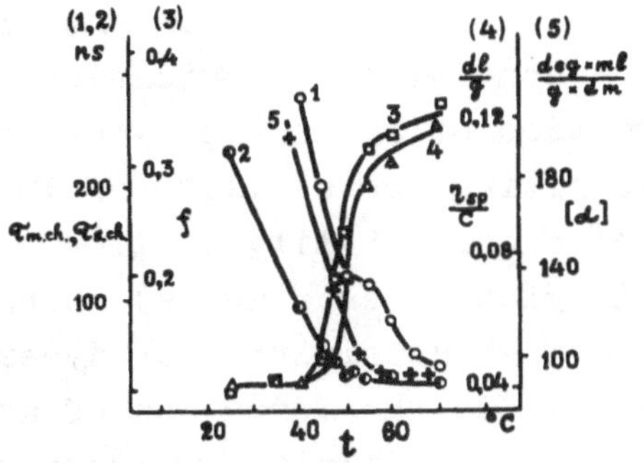

Fig.2. Temperature dependence for PChM-10 in heptane of 1) relaxation time $\tau_{m.ch.}$ characterizing the mobility of the main chain, 2) relaxation times $\tau_{s.ch.}$ characterizing the mobility of side chains, $C_{polymer}$ = 1 mg/ml, 3) contribution of high-frequence motions of the side groups of the polymer, f, 4) intrinsic viscosity [η] and 5) optical activity [α].

These data show that the isotropic liquid globule is transformed into the liquid-crystalline globule.

The PL method has been employed in the study of conformational transitions occurring not only in the molecules of homopolymers or random copolymers but also in those of block copolymers: three-block copolymers consisting of polystyrene (PS) and PMMA blocks in selective solvents (3). The study of relaxation characteristics of PS and PMMA blocks in three-block copolymers has made it possible to observe the formation of intramolecular structures of various types, globular, dumbbell-shaped and ring structures, and to establish the dependence of interblock interactions on the arrangement of PS and PMMA blocks along the chain and on the solvent nature (3).

Dynamics of polymer chains in interpolymer complexes

The PL method provides interesting information on the dynamics of polymer chains composing an interpolymer complex and thus makes it possible to draw conclusions on the structure, composition and the conditions of formation of this complex (4-7). In the formation of an interpolymer complex, relaxation times characterizing the mobility of interacting macromolecules (intramolecular mobility of linear chains or the mobility of protein globule as a whole) change by a factor of 10 or greater. This high sensitivity of dynamic characteristics of macromolecules to the formation of an interpolymer complex (IPC) makes it possible to evaluate with a high precision the contribution of macromolecules contained in IPC and to determine the factors destroying this complex. It has been proposed (4) to use the coincidence of dynamic characteristics of polymer chains in IPC as a criterion for the formation of this IPC. The application of this criterion to the analysis of complexation in an aqueous solution of ionized PMAA and lysozyme (6) has led to the conclusion that the protein globules in the PMMA molecules are distributed irregularly.

The PL method was also employed in the study of stereopolycomplexes formed in DMF by PMMA chains with various stereoregularities: isotactic chains (iso-PMMA) with 63% of isotactic triads and syndiotactic chains (syndio-PMMA) with 78% of syndiotactic triads. In contrast to the complexes of water-soluble polymer chains, e.g. PMMA-PEG (polyethylene glycol), stereopolycomplexes are not formed instantaneously but over a long period of time. Relaxation times in the formation of the complex increase by one order of the magnitude for each polymer chain contained in the stereopolycomplex. The dissociation of the stereopolycomplex is of the cooperative type and occurs upon the variation in solvent nature, e.g. upon the addition of chloroform to dimethylformamide (Fig.3) (7).

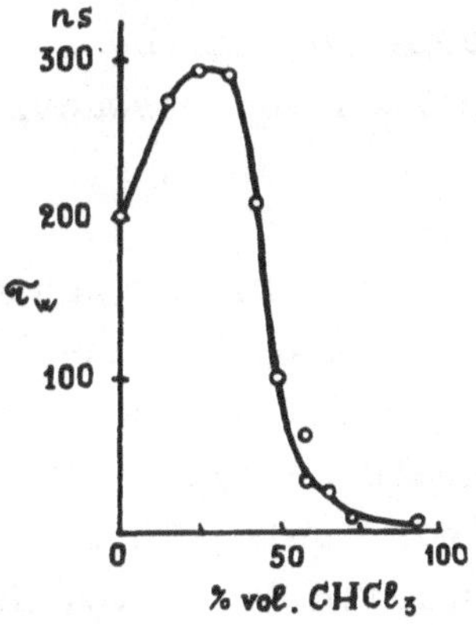

Fig.3. Change in relaxation times τ_w of syndio-78-PMMA when the IPC with iso-63-PMMA in a DMF-chloroform mixture is dissociated.

Chemical processes and dynamic phenomena in macromolecular systems studied by PL

The investigation of a system of model labelled polymers with different chemical structures and varying locations of luminescent markers (in the main chain, at a chain end or in side groups) has provided information about the dynamic properties of various structural fragments of macromolecules distinguished by the attachment of the marker (1, 2). These investigations provided the basis for the application of the PL method to the study of chemical processes involving macromolecules. In fact if low or high molecular weight reagents containing a luminescent (e.g. an anthracene) group are used in the investigation of the formation or chemical transformations of macromolecules, labelled polymer systems are formed. The investigation of dynamic properties of the parts or components of the polymer system containing a luminescent group and their comparison with those of labelled polymers with a definite structure and known location of markers may provide important information in the study of the mechanism of formation of macromolecules or reactions with their participation.

1) Application of PL to the investigation of reactions of macromolecules formation. The investigation of the initiation mechanism for the polymerization of vinyl monomers with aryldiazoalkane type initiators (12) may be taken as an example of using PL for solving this problem. According to the opinion of Imoto, Nakaya and co-workers (13, 14) who have proposed this type of initiators, if, for example, the polymerization of methyl methacrylate is initiated by 9-anthryldiazomethane (ADM), anthracene-containing intermediates of bifunctional character are formed. Their interaction with the monomer molecules should lead to the formation of macromolecules containing anthracene groups (A) either in the side chain (I) or in the main chain (II)

I A II III

To study the mechanism of this iniciation reaction, samples of polymethyl methacrylate (PMMA) with anthracene groups known to be located in side groups (I), in the main chain (II) or at a chain end (III) have been synthesized (2). These polymers, as well as the PMMA obtained with ADM as initiator, have been investigated by PL (12). The investigations have shown that the anthracene groups in PMMA obtained with ADM are located in the most mobile parts of the macromolecule: at chain ends (Fig.4). Hence, the actual mechanism of initiation reaction differs from that proposed above (13, 14).
Another example of using PL for the investigation of the mechanism of formation of macromolecules has been offered by the study of free-radical isomerization copolymerisation with the participation of 9-vinylanthracenes (15).

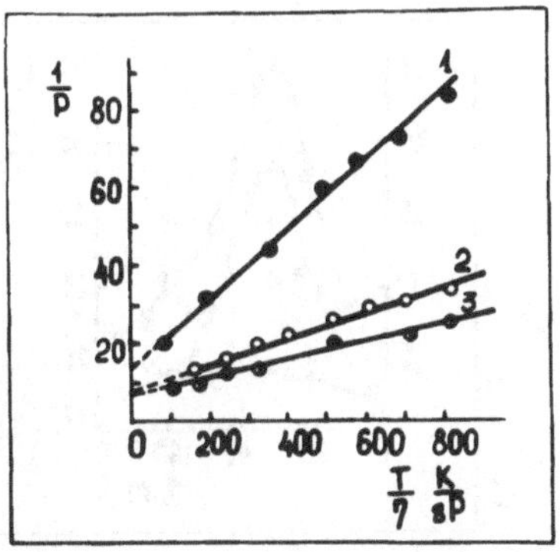

Fig.4. Reciprocal value of polarization of luminescence 1/P vs T/η for poly(methyl methacrylate) with different locations of attached anthracene groups in methylacetate-triacetine mixtures with various compositions (curve 1 - PMMA obtained by the initiation of polymerization with 9-anthryldiazomethane , curve 2 - PMMA with anthracene groups in side chains, curve 3 - PMMA with anthracene groups in the main chain).

2) Application of PL to the investigation of reactions of macromolecules with low molecular weight reagents. The application of PL to the study of reactions between the functional groups of macromolecules and the molecules of anthracene-containing low-molecular-weight reagents makes it possible to obtain information not only on the local reactivity of various parts of polymer chains but also on the dynamic characteristics of those parts which contain the functional groups participating in the reaction. It has been revealed in the study of the Friedel-Krafts reaction between the phenyl groups of styrene and α-methyl styrene copolymers with various compositions that the dependence of the reactivity of copolymers on their composition is described by a bell-shaped curve (Fig.5, curve 1).

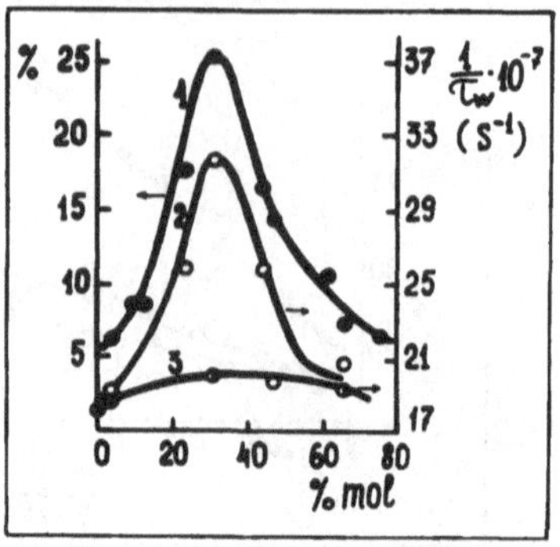

Fig.5. Degree of transformation of 9-chloromethylanthracene in the reaction with styrene- α -methylstyrene copolymers (curve 1) and local dynamic characteristics (1/P) of the copolymer molecules[(curve 2) - when the markers are attached in the process of the reaction investigated and(curve 3)- at a random distribution of the markers] vs content of α - methylstyrene units in the copolymers.

The type of this dependence is in good agreement with that of local dynamic properties of these copolymers on their composition (curve 2). These properties were determined by PL, the anthracene groups attached in the course of the reaction being used as markers. It is interesting to note that the intramolecular mobility of the same copolymers determined by PL with the aid of markers randomly located along the copolymer chains is only to a slight extent dependent on the composition of these copolymers (curve 3). This fact implies that during this reaction the reagent selects for an attack the phenyl groups located in the most mobile parts of polymer chains.

3) Application of PL to the investigation of the formation of branched and crosslinked polymers. The topologic structure and many properties of branched and crosslinked polymers are determined to a considerable extent by the

content of crosslinking bridges and their distribution in the polymer system. The formation of these bridges both in in the copolymerization of mono- and bifunctional monomers and in the crosslinking of linear polymer chains occurs in two stages. First, polymer chains with "pendant" double bonds or functional groups are formed (structure R in Fig.6) and then a certain part of these reactive groups is transformed into crosslinking bridges (structure B).

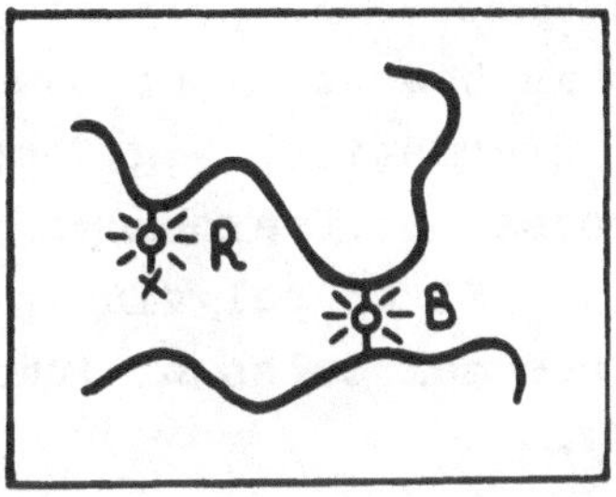

Fig.6

The use of anthracene-containing crosslinking agents (bifunctional monomers or reagents) makes it possible to apply the PL method to the study of two-stage processes of the formation of crosslinking bridges. The sensitivity and informative significance of PL in the investigation of these chemical reactions are due to the fact that the value of the 1/P parameter of luminescence of anthracene groups in type R structures is several times higher than that for the structures of type B.
The application of PL to the study of the formation of branched and crosslinked polymers in combination with other methods of investigation makes it possible to obtain new data on the relationships of the formation of molecular topologic structure of branched and crosslinked polymers (8).

4) Application of PL to the study of intermolecular reactions. Interpolymer complexes (IPC) formed as a result of the interaction between the molecules of polymers with different chemical structure play an important part in biological systems and seem to have a great future in the field of the preparation of synthetic polymers. The course of interpolymer reactions (formation of IPC, exchange and substitution reactions involving macromolecular components of IPC and dissociation of IPC) leads to a ten- and even hundred-fold change in the relaxation times characterizing the nanosecond polymer processes in polymer chains. Hence, in the study of these reactions, the PL method is very informative. Fig.7 shows the data characterizing the processes of exchange reactions in a solution containing an IPC of polyvinylpyrrolidone and polyacrylic acid and free molecules of luminescent polyacrylic acid (PAA*) (5).

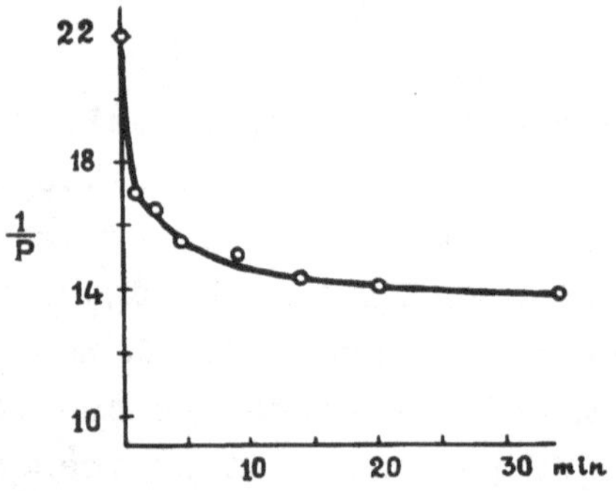

Fig.7. Exchange reaction in the following systems: IPC (polyvinylpyrrolidone-polyacrylic acid)-luminescent polyacrylic acid (PAA*) in water. Time dependence of 1/P for PAA* molecules after their mixing with the IPC solution.

The data show that PL is very sensitive to the processes occurring in intermolecular reactions and that the approach based an the application of labelled and non-labelled macromolecules to the study of these reactions is very promising.

References

1. Anufrieva, E.V.: Pure and Applied Chemistry 54, 533-548 (1982).
2. Anufrieva, E.V., Gotlib, Yu.Ya.: Advances in Polymer Science 40, 1-68 (1981).
3. Anufrieva, E.V., Krakovyak, M.G., Gromova R.A., Milovskaya, E.B., Rudkovskaya, G.D., Zamoiskaya, L.V., Lushchik, V.B., Sheveleva, T.V.: Vysokomol.Soedin., Ser.A23, 2667-2673 (1981).
4. Anufrieva, E.V., Pautov, V.D., Geller, N.M., Krakovyak, M.G., Papisov, I.M.: Dokl.Akad.Nauk SSSR, 220, 353 (1975)
5. Anufrieva, E.V., Pautov, V.D., Papisov, I.M., Kabanov, V.A.: Dokl.Akad.Nauk SSSR, 232, 1096-1100 (1977).
6. Pautov, V.D., Kuznetsova, N.P., Mishaeva, R.N., Anufrieva, E.V.: Vysokomol.Soedin., A25, 1599 (1983).
7. Nekrasova, T.N., Anufrieva, E.V., Krakovyak, M.G., Lushchik, V.B., Korshun, A.M.: Vysokomol.Soedin., ser.A25, 133 (1983).
8. Krakovyak, M., Ananieva, T., Anufrieva, E., Gotlib, Yu., Nekrasova, T., Skorokhodov, S.: Makromol.Chem., 182, 1003 (1981).
9. Anufrieva, E.V., Bychkova, V.E., Krakovyak, M.G., Pautov, V.D., Ptitsyn, O.B.: FEBS Letters 55, 46-49 (1975).
10. Anufrieva, E.V., Volkenshtein, M.V., Gotlib, Yu.Ya., Krakovyak, M.G., Pautov, V.D., Stepanov, V.V., Skorokhodov, S.S.: Dokl.Akad.Nauk SSSR, 207 1379 (1972).
11. Anufrieva, E.V., Pautov, V.D., Freidzon, Ya.S., Shibaev, V.P., Plate, N.A.: Vysokomol.Soedin., Ser.A24, 825-830 (1982).
12. Krakovyak, M.G., Anufrieva, E.V., Lushchik, V.B., Shelekhov, N.S., Skorokhodov, S.S.: J.Macromol.Sci.-Chem., A12, 789 (1978).
13. Nakaya, T., Ohashi, K., Imoto, M.: Makromol.Chem., 114, 201 (1968).
14. Imoto, M., Nakaya, T.: J.Macromol.Sci., C7, 1 (1972).
15. Krakovyak M.G., Anufrieva, E.V., Shelekhov N.S., Skorokhodov, S.S.: European Polymer J., 10, 685 (1974).

INTRAMOLECULAR EXCIMER FORMATION AND POLYMER DYNAMICS: THE COIL-GLOBULE TRANSITION

Carla Cuniberti and Angelo Perico
Istituto di Chimica Industriale dell'Università di Genova,
Centro Studi ChimicoFisici di Macromolecole Sintetiche e
Naturali del C.N.R.,Corso Europa 30,16132 Genova,Italia

Introduction

Linear chain polymers with non-specifically interacting units may collapse from the coil state to a "globular" state as a simple consequence of the nonspecific attractive potential field established in bad solvents (1).Different scaling laws for the chain size and for the chain average density are predicted in the globular state with respect to the ideal coil, the transition between the two states being expected as a cooperative conformational change at temperature below theta conditions.The sharpness of the transition,as well as the transition point,are predicted to increase with increasing chain stiffness and chain length. In the limit of infinite length the transition should occur at $T = \Theta$ with first order behavior in polymers with flexible chain (2,3). Since the globular state is favoured only if the probability of intramolecular contacts overwhelm that of the intermolecular ones,the globular form exists in the region of the temperature-concentration phase diagram squeezed between the zero concentration axis and the left branch of the phase separation curve. The restricted range of concentration where homogeneous solutions may be obtained arises serious difficulties for measuring polymer dimensions below Θ. Only recently in fact,light scattering data on high molecular weight polystyrene (4) gave a clear evidence of the cooperative decrease of the radius of gyration and of the hydrodynamic radius below the Θ temperature.

The collapse of the chain size however emphasizes only one of the aspects of the transition. By definition (2),the globular state -in contrast with the coil state- is a thermodynamically defined spatial structure characterized by density fluctuations

Physical Optics of Dynamic Phenomena and Processes in Macromolecular Systems

less than the density itself. This change of fluctuation regime should show up in dynamical properties measured in extremely dilute solutions. In this paper we shall give evidence of the change in the polymer internal dynamics below theta conditions by means of the intramolecular excimers technique.

Experimental method

The formation of an excited complex or "excimer" between an electronically excited molecule (M^*) and its ground state analog (M) is a well known photophysical event whose kinetics is easily led under diffusion control. In this case, the reaction between two chromophores belonging to remote units of a polymer chain directly probes the encounter frequency of these units which in its turn is determined by the viscoelastic properties of the chain (5). According to the usual kinetic scheme:

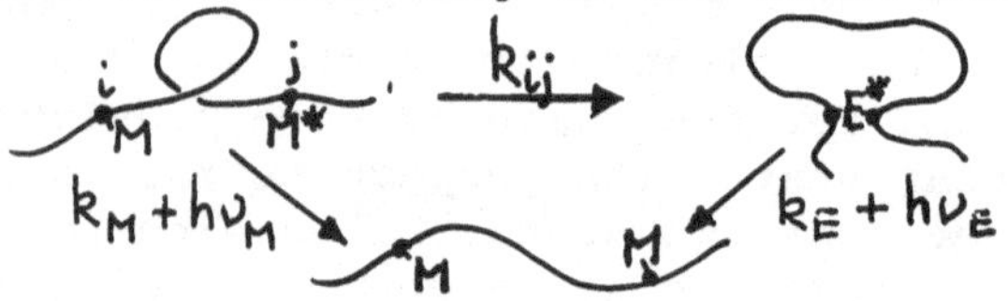

the photostationary ratio F_E/F_M of the excimer to monomer fluorescence emissions gives the relaxation time of the chain segment joining the two units i,j

$$F_E/F_M \sim k_{ij} \sim \tau_{ij}^{-1} \tag{1}$$

In case of several chromophores M present on the same chain, concurrent reactions take place and the derived rate constant, k, is averaged over the distribution of the reactive configurations. If the M units are located at large regular interval of n monomeric units along the chain, the global first order rate constant is easily obtained by scaling of the relaxation frequencies from the end-to-end result and approach asymptotically with increasing molecular weight the value (6)

$$k = (2AK_BT/6\pi\eta_0 \ell_{eff}^3)\, w^{3\epsilon}\, \zeta(3\nu) \tag{2}$$

Here A/ℓ_{eff}^3 is related to the polymer rigidity, ϵ and ν are excluded volume exponents, η_0 is the solvent viscosity, $w = 1/n$ and, $\zeta(x)$ is the Riemann zeta function. We have shown (6) eq.2 to hold also in case of randomly distributed chromophores of long lifetime at low w by studying dilute solutions of pyrenyl-labeled polyvinyl acetate with w=0.01. At fixed molecular weight, the solvent dependence of $\bar{k}$ was given at $T \geq \Theta$ by

$$\bar{k} \sim T/\eta_o \alpha_\eta^3 \qquad (3)$$

To the purpose of studying the behavior below Θ, two features of the pyrenyl excimers method are important: (a) the spectroscopic characteristics of the pyrenyl moieties affording the use of very dilute polymer solutions, (b) the inverse proportionality of $\bar{k}$ to the volume swelling factor providing a sensitive probe of any departure from the coil behavior.

Results and discussion

The results here presented refer to two samples of the same pyrenyl-labeled polyvinyl acetate (PVA-PBA) used in the investigation of the dynamics above Θ ($MW1=1.08x10^5$, $MW2=6.2x10^5$). The relevant properties of PVA-PBA solutions in the three solvents employed are listed in Table 1 (T_p is the light scattering phase separation point of $1x10^{-4}$g/dl solutions, T_E is the point that separates diffusion from equilibrium controlled regime in the F_E/F_M vs. T diagram, T_{max} is the maximum temperature value at which the $\bar{k}$ were calculated in each solvent). Mixtures of MOH-EOH and MOH-POH at 20°C were used as well. The Θ-temperature of solvents and solvents mixtures were obtained by the turbidimetric method (7), the viscosities measured by capillary viscometry and fluorescence spectra obtained with a Perkin-Elmer MPF44 spectrofluorimeter.
The temperature and solvent dependence of the proportionality constant in eq.3 result negligible in respect of the friction and swelling effects. A reduced average relaxation time, $\bar{\tau}_{red} = (\bar{k}_{red})^{-1} = 2.46 (F_E/F_M)(\eta_o/T) 10^{-12}$ seconds, is obtained taking as reference T/η_o for MOH at 6°C (Θ). Figure 1 shows $\bar{\tau}_{red}$ relative to its value at Θ plotted against the reduced temperature. Two regions with very distinct behaviors are evident.

Table 1. Characterization of PVA-PBA solutions in single solvents

Solvent	Θ (°C)	T_p (°C)		T_E (°C)	T_{max} (°C)	η_o^{20} (cP)
		$1.08x10^5$	$6.2x10^5$			
Methanol (MOH)	6[a]	<0	<0	45	20	0.60
Ethanol (EOH)	57	3	25.5	75	45	I.20
Propanol (POH)	81	24	-	90	65	2.37

a. Literature data for PVA

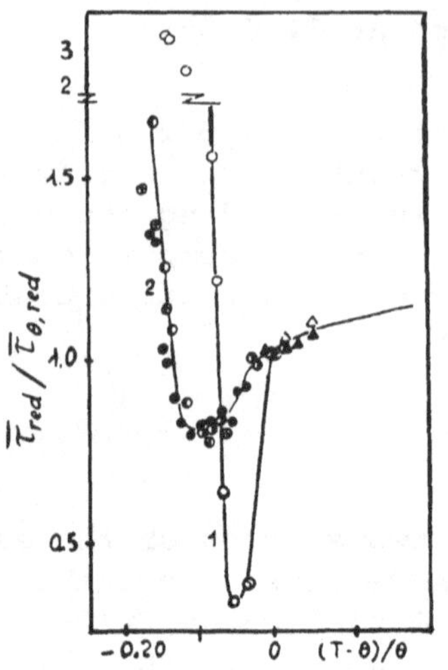

Fig. 1. The reduced relative relaxation time against the reduced temperature for the two PVA-PBA samples;curve 2: MW1, $c=1x10^{-3}$g/dl, curve 1: MW2, $c=4.2x10^{-4}$g/dl. (▲,△)MOH,(●,○)EOH,(⊕)POH (◑)MOH-EOH, (◐)MOH-POH The line at positive reduced temperature describes the behavior found previously(6).

Since the contribution of intermolecular excimers has been observed to reduce the apparent relaxation time steadily on lowering the temperature beyond the cloud point,the sharp increase observed at high dilution describes an intramolecular effect. The presence of the minimum in the two curves indicates a change of dynamical regime within the coil occurring after a substantial contraction.While at $T \geqslant \theta$ the reduced relaxation times are molecular weight independent,large differences show up below θ. Taking the temperatures of the minima as the transition points to the "globular" state ,the ratio (T2-θ)/(T1-θ) is only 1.2 larger than the expected $(MW1/MW2)^{1/2}$(2).
On concluding,the excimers technique appears to afford to gain new information on the single chain behavior in conditions difficultly of access with more usual experimental methods.

References

1. Stockmayer,W.H.:Makromol.Chem.35,54(1960).
2. Lifshitz,I.M.,GrosbergA.Yu.,Khokhlov,A.R.:Rev.Mod.Phys.50,683 (1978).
3. Williams,C.,Brochard,F.,Frisch,H.: Ann.Rev.Phys.Chem.32,433(1981).
4. Sun,S-T.,Nishio,I.,Swislow,G.,Tanaka,T.: J.Chem.Phys.73,5971(1980).
5. Cuniberti,C.,Perico,A.:Prog.Polym.Sci.10,271 (1984).
6. Cuniberti,C.,Perico,A.:Eur.Polym.J.16.887 (1980).
7. Cornet,C.F.,van Ballegooijen,H.:Polymer 7,293 (1966).

EXCIMER FORMATION IN POLYSTYRENE SOLUTIONS: EFFECT OF THE COIL-GLOBULE TRANSITION

Petr Štěpánek, Čestmír Koňák, Blahoslav Sedláček
Institute of Macromolecular Chemistry, Czechoslovak Academy of Sciences, 162 06 Prague 6, Czechoslovakia

Introduction

The formation of excimers in polystyrene solutions can be expected to follow one or several alternative mechanisms present: (a) excimer formation between adjacent chromophores which are separated by a main chain segment of three atoms; this is known as Hirayama's "n=3 rule" (e.g., (1)); (b) excimer formation between two distant chromophores on the same chain close to each other as a result of folding of the polymer chain; (c) excimer formation between chromophores of different chains.

Mechanism (a) manifests itself in the molecular weight dependence of the excimer (I_E) to monomer (I_M) intensity ratio Q, cf. e.g. (1). Intermolecularly formed excimers raise slightly the value of Q in concentrated solutions (2).

The aim of the present contribution is to study the effect of density changes of polymer coils in the coil-globule transition below the Θ temperature on excimer formation.

Experimental

Fluorescence measurements were carried out in cyclohexane solutions of polystyrene standards having molecular weights (M_w) in the range from 2.1×10^3 to 6.77×10^6, see Table 1. All reported values of Q are for air-equilibrated solutions at 40°C.

Physical Optics of Dynamic Phenomena and Processes in Macromolecular Systems

A mercury lamp was used as the exciting source. The exciting light (253.7 nm) was isolated by a monochromator. The Suprasil sample cell had a 1 cm path length. Fluorescence was analyzed by a monochromator positioned at 90° to the exciting source with spectral resolution of 5 nm. The analyzed light was then processed through a photomultiplier (Quantacon-RCA) and a photon counting system (Ortec-Brookdeal, type 5C1). Temperature regulation of the sample cell was better than ± 0.1 K. All reported Q measurements are the ratio of the intensity at 335 nm (I_E) to that at 283 nm (I_M).

Results

The high molecular weight standards (M_w>1.26x10^6) were not suitable for such studies as the polymer contains fluorescently active contaminants (probably stabilizers).
Arrhenius plots of Q for several molecular weights are shown in Figure 1. The full straight lines and the broken lines correspond to the poor solvent region and phase separation region, respectively. The arrows above the curves identify the cloud points. The apparent activation energies E_{app} obtained from the relation $\ln(I_E/I_M) = -E_{app}/RT$, where R is the gas constant and T is the absolute temperature, are collected in Table I. While in a good solvent the activation energy of excimer formation is independent of M_w (1), in a poor solvent E_{app} decreases with increasing M_w. This decrease is probably due to an increasing formation of intramolecular excimers (mechanism b) during the coil collapse which is more pronounced at higher molecular weights (3). Changes in E_{app} with temperature cannot be distinguished. The same mechanism (b) can be used for explanation of the pronounced increase in Q (>10%) at phase separation.

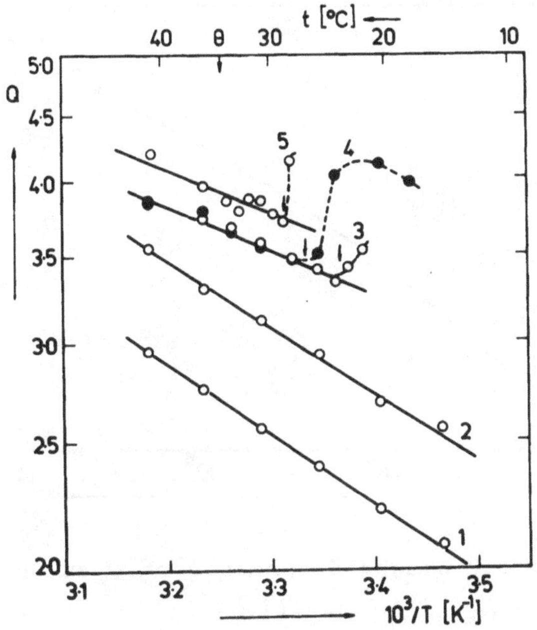

Fig. 1. Arrhenius plot for $Q = I_E/I_M$. M_w: 1 2.1×10^3; 2 2.04×10^4; 3 3.9×10^5; 4 4.2×10^5; 5 1.26×10^6.

Table 1: M_w, Q and E_{app} values for polystyrene samples PC (Pressure Chemicals) and TS (Toyo Soda).

Sample	M_w	c, g/ml	Q at 36°C	E_{app}, kcal/mol
PC	2.1×10^3	1.24×10^{-3}	2.77	2.5
PC	2.04×10^4	1.06×10^{-3}	3.30	2.4
PC	3.9×10^5	2.75×10^{-4}	3.73	1.6
TS	4.2×10^5	3.12×10^{-4}	3.81	1.7
TS	1.26×10^6	2.59×10^{-4}	3.96	1.3
PC	1.8×10^6	contaminated		
PC	2.0×10^6	contaminated		
TS	3.84×10^6	contaminated		
TS	6.77×10^6	contaminated		

Molecular weight dependence of Q at 36°C is shown in Figure 2.

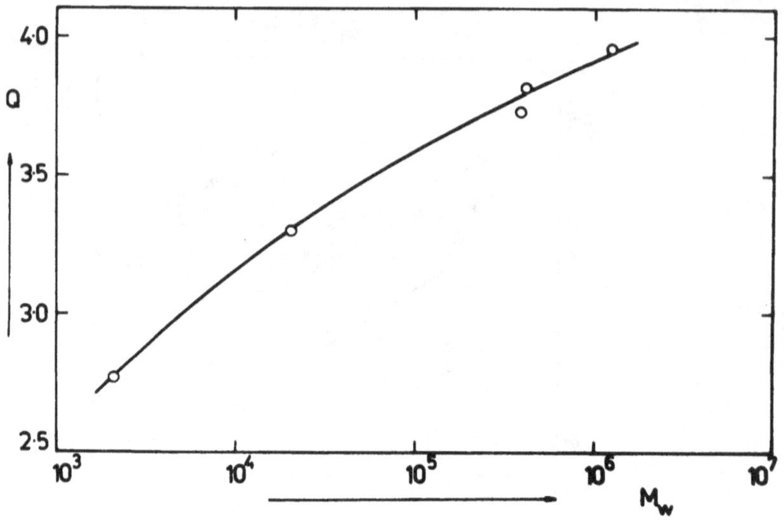

Fig. 2. Molecular weight dependence of Q at 36°C

Q increases with M_w in a qualitative agreement with published data (1). Smaller values of Q in our measurements are probably due to a stronger quenching of excimer emission by oxygen in the samples used.

For polystyrene solutions, the changes in Q are too small to enable a quantitative analysis of the coil-globule transition to be made.

References

1. Ishii, T., Handa, T., Matsunaga S.: Macromolecules 11, 40 (1978).
2. Torkelson, J.M., Lipsky, S., Tirrell, M., Tirrell, D.A.: Macromolecules 16, 326 (1983).
3. Sun, S.T., Nishio, I., Swislow, G., Tanaka, T.: J.Chem.Phys. 73, 5971 (1980).

THE INFLUENCE OF CHROMOPHORIC GROUPS ROTATION ON THE DECAY OF FLUORESCENCE OF POLYSTYRENE IN SOLUTION

Krzysztof Sienicki and Czesław Bojarski
Institute of Physics, Technical University of Gdańsk
80-952 Gdańsk, Poland

Introduction

Studies of intramolecular motion in dilute polymer solutions have been carried out by several physical methods (1). Among other things, fluorescence depolarization (2) and studies of fluorescence decay (3-4) have proven to be quite useful as a probe of polymer structure and its dynamics.

Results

In order to describe the excited-state kinetics of a polymer, we assume that excimers are formed as a result of two processes. The first is the energy migration (EM) with a rate constant in the form (5)

$$k_{EM}(t) = 4\pi\Lambda r_{DD}[1 + r_{DD}(\pi\Lambda t)^{-1/2}] \quad (1)$$

where r_{DD} is the capture radius of excitation by excimers and Λ is the EM coefficient, and the second is the chromophoric groups rotation with a rate constant k_{rot}.

Assuming that excimer dissociation is negligible, which is true for certain polymers, the rates of concentration changes of excited monomers ($N_M(t)$) and excimers ($N_D(t)$) upon pulse excitation of a system at t = 0 can be described by the following equations:

$$dN_M(t)/dt = -\tau_{OM}^{-1} N_M(t) - k_{EM}(t)\, q\, N_M(t) - k_{rot}\, q\, N_M(t) \qquad (2)$$

$$dN_D(t)/dt = -\tau_{OD}^{-1} N_D(t) + k_{EM}(t)\, q\, N_M(t) + k_{rot}\, q\, N_M(t) \qquad (3)$$

where τ_{OM} and τ_{OD} are the monomer and excimer lifetimes in the excited state, respectively, and q is a fraction of excimer sites (FES). With initial conditions $N_M(0) = N_o$ and $N_D(0) = 0$ at $t = 0$, the monomer $i_M(t)$ and excimer $i_D(t)$ fluorescence response functions have the form

$$i_M(t) = N_o \exp\{-t/\tau_M - 2\Gamma_M(t/\tau_M)^{1/2}\} \qquad (4)$$

$$i_D(t) = (4\pi\Lambda r_{DD} + k_{rot})q \exp(-t/\tau_{OD})\Big\{ \frac{1}{b}[1-\exp(-bt-2ct^{1/2})] - (\pi/b)^{1/2}\exp(c^2/b)(a+c/b)\,[\mathrm{erf}(b^{1/2}t^{1/2} + cb^{-1/2}) - \mathrm{erf}(cb^{-1/2})]\Big\} \qquad (5)$$

where $\mathrm{erf}\, x = 2/\pi^{1/2}\int_0^x \exp(-u^2)du$, $\tau_M^{-1} = \tau_{OM}^{-1} + (4\pi\Lambda r_{DD} + k_{rot})q$, $\Gamma_M = 4(\pi\Lambda\tau_M)^{1/2} r_{DD}^2\, q$, $a = r_{DD}(\pi\Lambda)^{-1/2}$, $b = \tau_M^{-1} - \tau_{OD}^{-1}$ and $c = \Gamma_M \tau_M^{-1/2}$.

Soutar et al. (6) obtained an empirical equation for the monomeric fluorescence decay of PS in dichloromethane (DCM) in the form $i_M(t) = 1.15 \exp(-t/0.88\text{ ns}) + 0.004 \exp(-t/14.9\text{ ns})$ (see Fig. 1a, ∘ dots). Fig. 1a also contains a solid line which represents the best fit of Eq. (4) to the experimental values. The fitting was obtained using the method of smallest squares for the following values (5,7) $\tau_{OM} = 13$ ns, $\Lambda = 10.3\times10^{-9}\ m^2s^{-1}$, $r_{DD} = 6.65$ Å and $k_{rot} = 3.54\times10^9\ s^{-1}$ obtained from measurements of ESR (8). The best fit obtained for FES equals to $q = 1.72\times10^{-2}$ and for which Eq. (4) takes the form $1.241 \exp\{-t/0.985\text{ ns} - (t/24.94\text{ ns})^{1/2}\}$. The values of FES obtained from fluorescence decay are close to the values determined from the stationary measurements of PS luminescence ($q = 2.29\times10^{-2}$ (5)) and the Monte

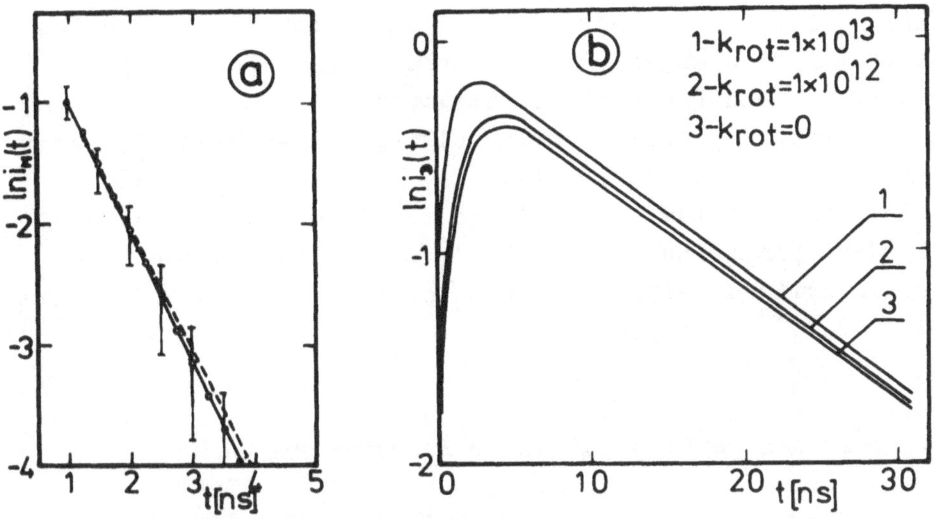

Fig. 1. Monomer $i_M(t)$ and excimer $i_D(t)$ fluorescence response functions of polystyrene in dichloromethane.

Carlo simulation ($q = 2.48 \times 10^{-2}$ (9)) .

In Fig. 1a the dotted line represents the PS monomer fluorescence decay when $k_{rot} = 0$. As one can see, neglecting the rotation of phenyl groups has a negligible influence on the PS fluorescence decay. This is understandable considering the relation $\tau_{rot} = k_{rot}^{-1} \gg \tau_j$ where τ_j is the time of energy localization on the phenyl group. We have shown recently (5) that in case of PS, τ_j is of order of picoseconds. Therefore rotation of the phenyl groups at room temperature has a secondary meaning in the description of PS monomer fluorescence decay and the factor deciding about the character of decay is the EM process leading to the formation of excimers.

Figure 1b shows the time dependence of $\ln(i_D(t))$ computed from Eq. (5) at a fixed energy transport rate and concentration of FES ($q = 1.72 \times 10^{-2}$) . As one can see, in this case too, the rotation of chromophoric groups only negligibly modifies the excimer fluorescence decay. The especially characteristic feature (Fig. 1b) is the shift of maximum of $i_D(t)$ to shorter times with increasing rotation rate. Similar changes of $i_D(t)$ were shown recently by Fredrickson and Frank (3) on the basis of a model utilising the t-matrix approximation in solving the problem of energy transport in a macromolecule.

From the fitting of Eq.(4) to the fluorescence decay of poly(1-vinylnaphthalene) and poly(2-vinylnaphthalene) we have determined values of FES which are equal to 6.26×10^{-2} and 8.02×10^{-3} , respectively.

This work was carried out under Research Project MR.I-5.

References

1. Papers presented at this Conference.
2. Anufrieva, E.V., Gotlib, Y.Y.: Adv. Polym. Sci. 40 , 1 (1981).
3. Fredrickson, G.H., Franck, C.W.: Macromolecules 16 , 572-577 (1983).
4. Itagaki, H., Horie,K., Mita, I.: Macromolecules 16 , 1395-1397 (1983).
5. Sienicki, K., Bojarski, C. in preparation.
6. Soutar, I., Phillips, D., Roberts, A.J., Rumbles, G.: J. Polym. Sci., Polym. Ed. 20 , 1759-1770 (1982) .
7. Sienicki, K., Bojarski, C.: Polym. Photochem., 4 , 000 (1984) .
8. Friedrich, C., Laupretre, F., Noël, C., Monnerie, L.: Macromolecules 14 , 1119-1125 (1981) .
9. Gelles, R., Franck, C.W.: Macromolecules 15 , 741-748 (1982) .

SMALL-ANGLE SCATTERING OF POLYELECTROLYTE SOLUTIONS

Ryuzo Koyama
College of General Education, Kyoto University
Kyoto 606, Japan

Introduction

The small-angle neutron and X-ray scattering curves of polyelectrolyte solutions have various characteristic forms according to the physical conditions(1 ~ 5). In the solutions without added salts the scattering intensity I(q) has a broad peak in the region $10^{-2} \sim 10^{-1}$ Å^{-1}, where q is the absolute value of the scattering vector $\underset{\sim}{q}$

$$q = (4\pi/\lambda)\sin(\theta/2) \tag{1}$$

(λ is the wavelength and θ is the scattering angle).
But when some salts are added to the solution, I(q) increases at small q, and becomes monotonically decreasing with q at sufficiently high concentration of the salts. These scattering curves have been calculated in a previous paper(6), assuming a strong electrostatic repulsion between rodlike segments in a random coil molecule. The following calculation derives these by a more realistic chain polymer model(7).

Calculation of Scattering Intensity

In this calculation we neglect the internal structure of the monomer unit, and regard it as a point scatterer of the radiation. The scattering intensity per unit volume of the polymer solution can be written as the Fourier transform of the radial distribution function of the monomers g(R) ,

Physical Optics of Dynamic Phenomena and Processes in Macromolecular Systems

$$I(q) = C_I n [1 + n \int_{-\infty}^{+\infty} (g(R) - 1) \exp(-i\underset{\sim}{q}.\underset{\sim}{R}) \, d\underset{\sim}{R}] \qquad (2)$$

where C_I is a constant and n is the average number of monomers in unit volume. Since $n g(R)$ gives the probability of finding a monomer at the distance $\underset{\sim}{R}$ from another monomer at the origin, this is the sum of the probability of finding a monomer of the same polymer with that at origin $n g_1(R)$ and that of finding a monomer of different polymer from that at origin $n g_2(R)$, i.e.

$$g(R) = g_1(R) + g_2(R) \qquad (3)$$

When using the normalized intramolecular distribution function $f_{ij}(R)$ of the i-th and j-th monomers in the polymer, $ng_1(R)$ can be written as

$$ng_1(R) = N^{-1} \sum_i^N \sum_j^N f_{ij}(R), \qquad (i \neq j) \qquad (4)$$

Therefore the Fourier transform of $g_1(R)$ can be given by the molecular scattering factor of the polymer P(q)

$$g_1(q) = n^{-1}(N P(q) - 1) \qquad (5)$$

$$P(q) = N^{-2} \sum_i^N \sum_j^N f_{ij}(q) \qquad (6)$$

where $f_{ij}(q)$ is the Fourier transform of $f_{ij}(R)$. Eqs 4 and 5 do not contain the special terms with $i = j$

$$f_{ii}(R) = \delta(\underset{\sim}{R}), \quad f_{ii}(q) = 1 \quad (i = 1,2..,N) \quad (7)$$

but eq 6 includes them. Here $\delta(\underset{\sim}{R})$ is the δ function.

Next, considering the strong electrostatic repulsion between the polyelectrolyte chains, we assume that the probability of finding two monomers from two different polymer molecules within a distance R_o is negligible, but beyond this distance this probability is almost constant. Then if a monomer of a polymer A is somewhere in a spherical region of radius R_o, any monomer from polymers other than A is scarcely found at the center of this spherical region. Therefore, after introducing the origin of a coordinate system at the i-th monomer of polymer A, one can write approximately the probability of not finding the monomer of the other polymers at $\underset{\sim}{R}$ as (cf. Fig.1)

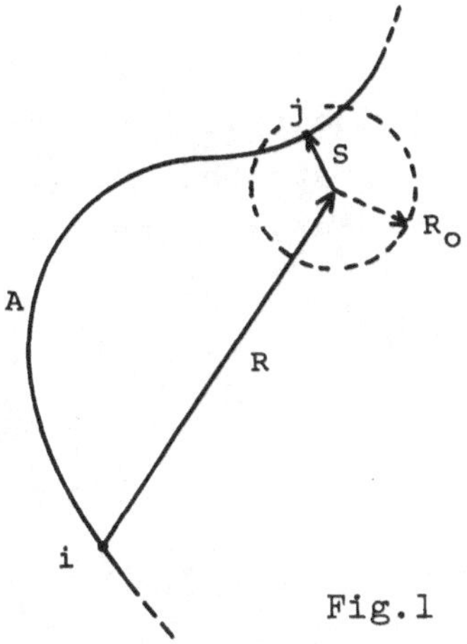

Fig.1

$$p_i(R) = C_o(\ell_o/R_o) \sum_j \int_{(S<R_o)} f_{ij}(\underset{\sim}{R}+\underset{\sim}{S})\, d\underset{\sim}{S} \tag{8}$$

$$C_o < 1 \tag{9}$$

where ℓ_o is the length of the chain per monomer, C_o is a constant, and the integral is carried out in the spherical region of radius R_o $(S < R_o)$. The factor $C_o(\ell_o/R_o)$ is added to eliminate an overestimation that comes from taking the sum over all the monomers of polymer A, because the spherical region can contain a chain segment of mean length R_o consisting of about R_o/ℓ_o monomers (7). The same probability can be defined by introducing the origin at an arbitrary monomer of polymer A. So the average probability that no monomers of the other polymers are at a distance $\underset{\sim}{R}$ from a monomer of polymer A is

$$p(R) = N^{-1} \sum_i^N p_i(R) \tag{10}$$

On the other hand when all the monomers of polymer A are outside this spherical region, the monomer of the other polymer is found at the center of the sphere with constant probability n by the assumption , and this case occurs with the probability $1 - p(R)$. Therefore the average probability of finding a monomer of the other polymers at distance $\underset{\sim}{R}$ from a monomer of polymer A is approximately

$$n\,g_2(R) = n\,(1 - p(R)) \tag{11}$$

To calculate the Fourier transform of $p(R) = g_2(R) - 1$, we approximate the integral of eq 8 by another form physically more natural, introducing a function $\phi(S)$ as follows

$$\int_{(S<R_o)} f_{ij}(\underset{\sim}{R} + \underset{\sim}{S})\, d\underset{\sim}{S} = \int_{-\infty}^{+\infty} f_{ij}(\underset{\sim}{R} + \underset{\sim}{S})\phi(S)\, d\underset{\sim}{S} \tag{12}$$

$$\phi(S) = G_o \exp(-S^2/R_o^{\,2}), \quad G_o \approx 1 \tag{13}$$

Using eqs 6 ~ 13, one can calculate the Fourier transform of $-p(R) = g_2(R) - 1$ as

$$-p(q) = n^{-1} N\, h\, \phi(q)\, P(q) \tag{14}$$

$$h = \pi^{3/2} C_o G_o n\, \ell_o\, R_o^{\,2} \tag{15}$$

$$\phi(q) = \exp(-X^2/4), \quad X = R_o q \tag{16}$$

By eqs 3 and 11, I(q) of eq 2 can be given by $1 + n(g_1(q) - p(q))$.

Therefore using eqs 5 and 14, we obtain the equation for I(q)

$$I(q) = C_I\, n\, N\, P(q)\, [\, 1 - h\, \phi(q)\,] \tag{17}$$

For the particular case $q = 0$ we have $P(0) = \phi(0) = 1$ (from eqs 6, 7 and 16), and eq 17 becomes

$$I(0) = C_I\, n\, N\, [\, 1 - h\,] \tag{18}$$

This I(0) should not be negative, hence

$$h \le 1 \tag{19}$$

For various polymer models, P(q) is generally a decreasing function of q, but $1 - \phi(q)$ is a monotonically increasing function converging to unity at large q. Therefore for very small h ($\ll 1$), I(q) of eq 17 is a decreasing function similar to P(q), but when h is large (≈ 1), this I(q) may have a maximum.

Calculation for Wormlike Chain Polymer

Scattering curve

We assume that the polymer molecule has a wormlike chain conformation, and use the result of a previous calculation of P(q) of this chain (8). Figures 2 and 3 show scattering curves I(q) for L/a = 4, calculated using this P(q) in eq 17, where L is the length of the polymer and a is the persistence length of this chain. In the figures s denotes the ratio

$$s = R_o/a \tag{20}$$

Approximate calculation

The previous calculation shows for sufficiently large L q

$$P(q) \approx \pi/L\,q \qquad (L\,q >> 1) \tag{21}$$

where L can be given by

$$L = N\,\ell_o \tag{22}$$

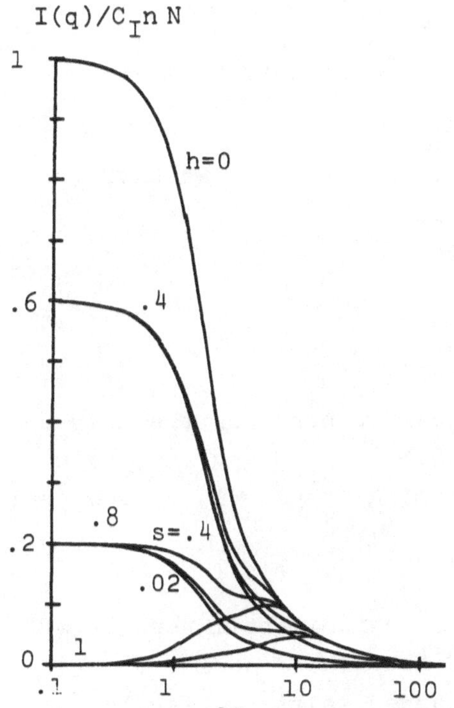

Fig.2

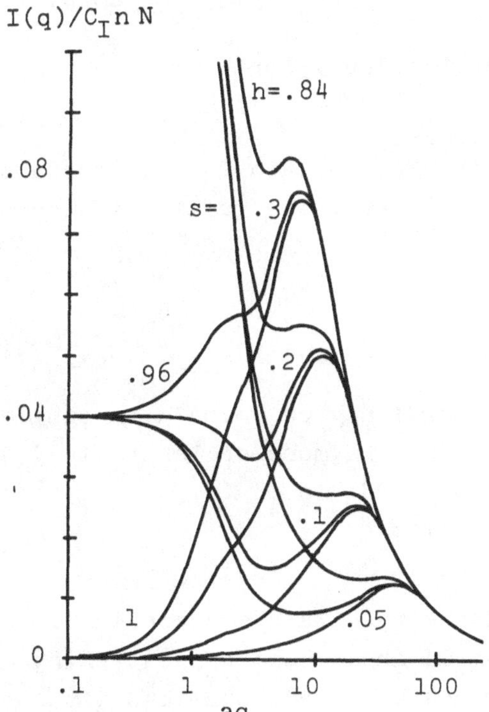

Fig.3

When this P(q) is introduced into eq 17, we obtain an approximate equation valid for large L q

$$I(q) = \pi C_I n \ell_o R_o [1 - h\exp(-X^2/4)]/X \tag{23}$$

The maximum point $X_m(h)$ as a function of h of this I(q) can be determined numerically

$$X_m(h) = R_o q_m \tag{24}$$

where R_o is given by eq 15

$$R_o = \pi^{-3/4}(h/C_o G_o \ell_o n)^{1/2} \tag{25}$$

Therefore from eqs 23~25, the maximum point q_m and the maximum intensity $I(q_m)$ can be written as

$$q_m = \pi^{3/4}(C_o G_o \ell_o)^{1/2} n^{1/2} h^{-1/2} X_m(h) \tag{26}$$

$$I(q_m)/n \propto R_o [1 - h\exp(-X_m(h)^2/4)]/X_m(h) \tag{27}$$

Particularly for h = 1, the numerical calculation gives

$$X_m(1) = 2.2418.. \tag{28}$$

Since C_o, G_o and ℓ_o are constants, in this case we have from eqs 25~27, the qualitative results

$$R_o \propto n^{-1/2} \tag{29}$$

$$q_m \propto n^{1/2} \tag{30}$$

$$I(q_m)/n \propto n^{-1/2} \tag{31}$$

In ordinary polymers with single carbon bonds in the main chain the monomer length is $\ell_o = 2.5$ Å. Using this value and eq 28 in eq 26, and also assuming $C_o = G_o = 1$, q_m for h = 1 becomes

$$q_m = 8.36\, n^{1/2}\ \text{Å}^{-1} \tag{32}$$

These results are compared with the experimental results of small-angle neutron scattering of deuterated sodium poly(styrenesulfonate) solutions without added salts by Nierlich and others(1). In Figures 4 and 5 the straight lines show the

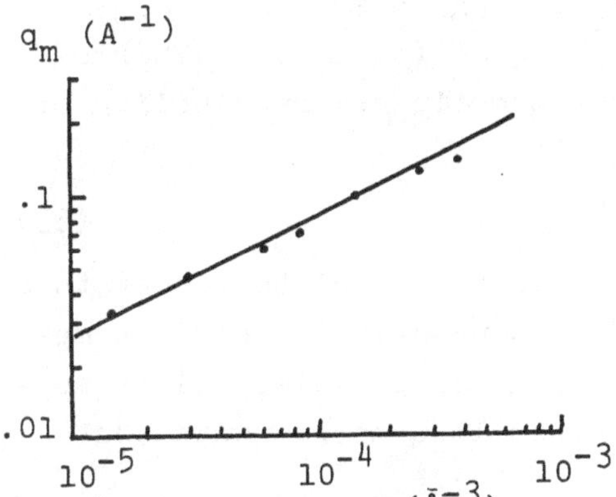

Fig. 4

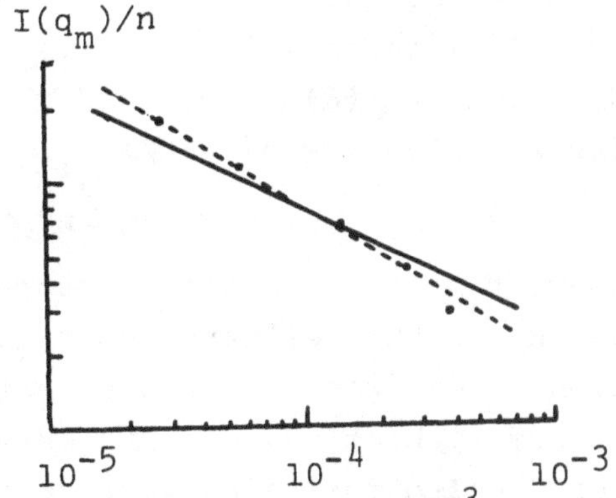

Fig. 5

calculated values by eqs 32 and 31 respectively, and the small dots show the experimental values.

Results

Figure 4 shows that eqs 30 and 32 agree well with the experimental results. The former qualitative result has been obtained by Hayter and others(9), but eq 32 gives an absolute value.

In polyelectrolyte solutions without added salts, the electrostatic repulsion between polymer chains is strong, and R_o can be assumed very large compared with nonelectrolyte polymer solutions. Therefore from eqs 15 and 19 we can assume $h \approx 1$ irrespectively of n, and I(q) has a maximum as obtained in the calculation. However it is considered that the added salts ions screen the electrostatic forces between polyelectrolyte chains, and this effect probably decreases R_o. So in the solutions with added salts we can assume a smaller value of h by eq 15. Figure 2 and 3 show that as h decreases I(q) increases at small aq, and the maximum disappears at $h \approx 0.85$. And for h smaller than this, I(q) is monotonically decreasing with q. Therefore this calculation can explain the experimental result (1~4).

The assumption $h \approx 1$ for very strong interaction means that I(q) has a very small value at $q \sim 0$ (eq 18). Thermodynamically I(0) is related to the osmotic compressibility of the solution $\kappa = n^{-1}(\partial n/\partial \pi)_T$

$$I(0) \propto k_B T n^2 \kappa \qquad (33)$$

where π is the osmotic pressure, T is the absolute temperature and k_B is the Boltmann constant. Consequently this assumption agrees with the known fact that κ of polyelectrolyte solutions without added salt is very small compared with that for nonelectrolyte polymer solutions(10).

References

1. Nierlich, M.,Williams, C.E., Boue, F., Cotton, J.P., Daoud, M., Farnoux, B., Jannink, G., Picot, C., Moan, M., Wolf, C., Rinaudo, M., de Gennes, P.G.:J. Phys.(Paris) 40, 701-704 (1979).
2. Plestil, J., Mikes, J., Dusek, L.: Acta Polym. 30, 29-32 (1979).
3. Ise, N., Okubo, T., Yamamoto, K., Kawai, H., Hashimoto, T., Fujimura, M., Hiragi, Y.,: J. Am. Chem. Soc. 102, 7901-7906 (1980).
4. Patokowsky, A., Gulari, E., Chu, B.: J. Chem. Phys. 73, 4178-4184 (1980).
5. Benmouna, M., Weill,G., Benoit, H., Akcasu, A.Z.: J. Phys. (Paris) 43, 1679-1685 (1982).
6. Koyama, R.,: Physica B 120, 418-421 (1983).
7. Koyama, R.,: Macromolecules (to be published 1984).
8. Koyama, R.,: J. Phys. Soc. Jpn. 34, 1029-1038 (1973), 36, 1409-1417 (1974).
9. Hayter, J., Jannink, G., Brochard-Wyart, F., de Gennes, P. G.: J. Phys.(Paris), Lett. 41, L-451-454 (1980).
10. Lifson, S., Katchalsky, A.: J. Polym. Sci. 13, 43-55 (1954).

LIGHT-INDUCED CONFORMATIONAL CHANGES OF MACROMOLECULES IN SOLUTION AS DETECTED BY FLASH PHOTOLYSIS IN CONJUNCTION WITH LIGHT SCATTERING MEASUREMENTS

Masahiro Irie
Institute of Scientific and Industrial Research, Osaka University, Osaka, Japan

Wolfram Schnabel
Hahn-Meitner-Institut für Kernforschung Berlin, Bereich Strahlenchemie, D-1000 Berlin 39, Federal Republic of Germany

Introduction

Recently, flash photolysis in conjunction with the light scattering detection method has been applied for investigations concerning the dynamics of macromolecules in solution. The dynamics of disentanglement diffusion were studied by measuring the rate of the change of the light scattering intensity (LSI) after very fast main-chain scission. In this case the diminution of the average molecular weight gave rise to a decrease of the LSI after irradiation of the polymer solution with a 20 ns flash [1]. Flash photolysis in conjunction with the LS-detection method is also applicable to measure the rate of conformational changes of macromolecules at constant chain length. As can be seen from the Debye equation (1) [2]

$$\frac{K\,c}{R_\vartheta} = \frac{1}{\overline{M}_w} + \frac{16\pi^2\ \langle s^2\rangle}{3\lambda_0^2\ \overline{M}_w}\sin^2(\vartheta/2) + 2A_2 \qquad (1)$$

the light scattering intensity R_ϑ (Rayleigh ratio) is correlated to the weight average molecular weight $\overline{M}_w$, the mean square radius of gyration $\langle s^2\rangle$ and the second virial coefficient A_2. Here, $K = (2\pi^2 n_0^2/N_A \lambda_0^4)\ (dn/dc)^2$, c is the polymer concentration, n_0 is the refractive index of the solvent, dn/dc is the specific refractive index increment, λ_0 is the wavelength of the incident light and N_A is Avogadro's number. Expansion of the polymer chain leads to an increase in $\langle s^2 \rangle$ and A_2 which causes a decrease in the LSI, whereas upon shrinkage of the chain, leading to a decrease in $\langle s^2 \rangle$ and A_2, the LSI increases. Advantage was taken of this behavior by measuring the rate of the photo-induced conformational change of a polyamide having backbone azobenzene groups [3]:

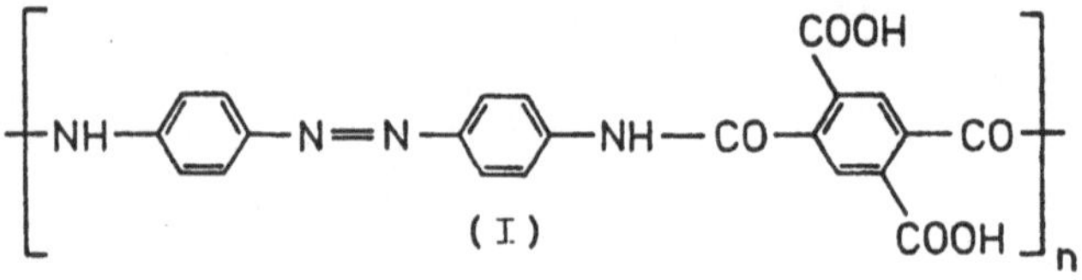

(I)

Upon irradiation with UV light (350 nm < λ < 410 nm) the extended conformation corresponding to all azogroups being in the trans-form is transformed to a compact conformation corresponding to a major fraction of azogroups being in the cis-form.

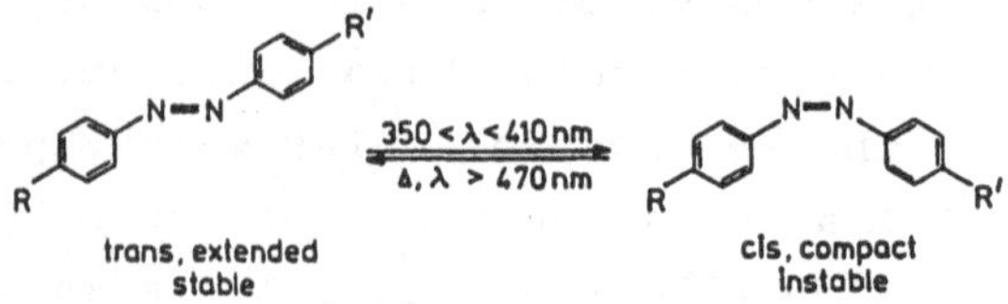

(2)

The reverse cis → trans reaction can be achieved either by heating or by irradiation with light of λ > 470 nm. It causes unfolding of the compact chain as indicated by a decrease of the light scattering intensity due to an increase of the radius of gyration. The rate of unfolding of polyamide I (M = 7×10^3) was measured after pretreatment by UV irradiation for trans → cis isomerization. In these experiments, dilute polymer solutions (0.3 g/l) in various solvents were irradiated with 15 ns flashes of 530 nm light at 22°C. The halflife of the decrease of the LSI in N,N-dimethylformamide (DMF) was 5.7×10^{-4} s. Additional data are given in Table 1.

Table 1. Conformational Relaxation of Polyamide Chains Subsequent to cis → trans isomerization of Azobenzene Groups in the Backbone

Solvent	$\tau_{1/2}$(s) [a)]
N,N-dimethylacetamide	5.8×10^{-4}
N,N-dimethylformamide (DMF)	5.7×10^{-4}
4:1[b)] DMF-ethanol	7.2×10^{-4}
3:2[b)] DMF-ethanol	11.0×10^{-4}
4:1[b)] DMF-water	8.8×10^{-4}

a) Halflives of the decrease of the light scattering intensity at 514 nm and 22°C.

b) Volume ratio.

From the results obtained with mixtures of DMF and ethanol it was concluded that upon worsening the solvent quality the rate of unfolding was retarded. In the following a new experiment concerning light-induced coil contraction of macromolecules containing pendant azobenzene groups will be presented. Moreover, an experiment concerning coil expansion of a polyelectrolyte due to a light-induced pH jump will be described.

Results

1) Coil contraction of polystyrene having pendant azobenzene groups. Coil contraction due to trans → cis isomerization was studied recently [4] with copolymers containing pendant azogroups: These copolymers,consisting of styrene and 4-(methacryloylamino) azobenzene (see Table 2),

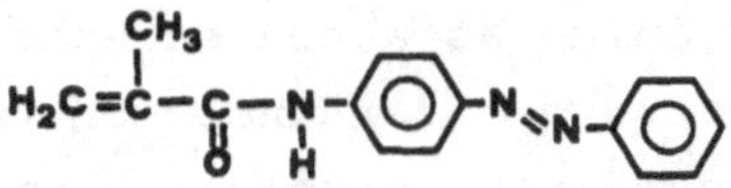

were found to precipitate in dilute cyclohexane solution at temperatures above the critical miscibility temperature upon irradiation by UV light [5] . Because the optical absorption spectra of the two isomers differ appreciably (λ_{max} of the

Table 2. Characterization of copolymers of styrene and 4-(methylacryloylamino)azobenzene

Sample	Content of Azobenzene Groups (mol %)	M_w a)
PS-A-2.2	2.2	3.1×10^4
PS-A-4.3	4.3	2.7×10^4
PS-A-5.6	5.6	2.0×10^4
PS-A-6.5	6.5	1.8×10^4

a) Estimated from GPC using a calibration curve for polystyrene.

trans form: 353 nm, λ_{max} of the cis form: 440 nm) the trans → cis isomerization induced by a 15 ns flash of 347 nm light could readily be detected by absorption measurements. It occurred during the flash, i.e. with a rate constant greater than $10^8 s^{-1}$. From the very fast trans → cis isomerization of pendant azo groups it was inferred that neighbouring phenyl groups did not interact strongly with trans azobenzene groups. <u>Light scattering measurements</u> revealed the following:
With all copolymers except PS-A-2.2, an increase of the LSI after the flash with halflives of several hundred µs was observed. A typical oscillogram demonstrating the LSI change is shown in Fig. 1, trace (a). It can be seen that the LSI decreased initially. This decrease is due to the concurrent increase of the optical absorption at 514 nm (see oscillogram (b)) and of the decrease of dn/dc as a consequence of trans → cis isomerization. The refractive index increment was measured with copolymer PS-A-2.2 in cyclohexane solution at 28°C and 546 nm: dn/dc = 0.125 ml/g before and dn/dc = 0.100 ml/g after trans → cis isomerization. Due to the low content of azo-

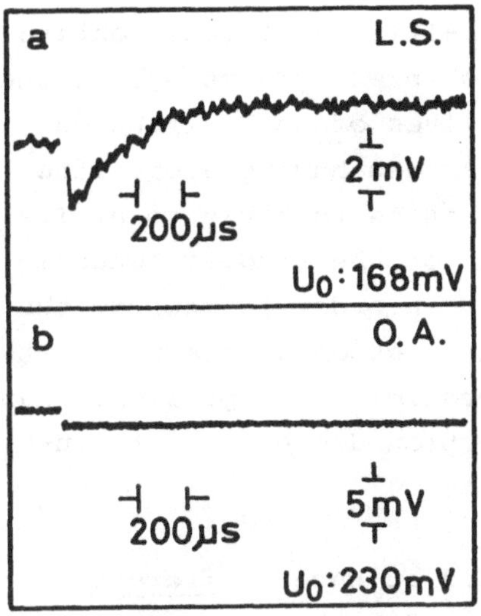

Fig.1. Oscillograms illustrating the increase in the light scattering intensity (a) and in the optical absorption (b) both at 514 nm after irradiation of copolymer PS-A-4.3 in cyclohexane solution (0.13 g/l) with a 15 ns flash of 347 nm light at 25°C.

benzene groups copolymer PS-A-2.2 did not precipitate after irradiation.

The relative slow increase in the LSI is considered to reflect the conformational change involving a decrease in both the radius of gyration of the coils and the second virial coefficient [6,7]. The halflives of the LSI increase for the various copolymers are compiled in Table 3. The fact that no increase in the LSI after the laser flash was observed with copolymer PS-A-2.2 is consistent with a result obtained by stationary irradiations: precipitation did not occur when the content of azobenzene units was less than 3 mol %[5].

Table 3. Conformational Relaxation of Polystyrene with Pendant Azobenzene Groups

Sample	$\tau_{1/2}$ [a] (s)
PS-A-2.2	--
PS-A-4.3	2.1×10^{-4}
PS-A-5.6	4.0×10^{-4}
PS-A-6.5	8.5×10^{-4}

a) Halflives of the increase in the light scattering intensity at 514 nm and 25°C.

It is interesting to note that the halflives of coil contraction of polystyrene with pendant azobenzene groups are of the same order of magnitude as the halflives of coil expansion measured for coil expansion of polyamides having azobenzene groups in the main chains. However, it is to early to arrive at general conclusions at this stage of the studies. Quantitative parameter studies are planned in order to measure the rate of conformational changes as a function of chain length, content of azobenzene groups etc. Preliminary experiments have shown that the halflife coil contraction decreases with increasing temperature.

Experiments carried out with PS-A-4.3 at three <u>different polymer concentrations</u> (0.11, 0.076 and 0.055 g/l) yielded the same halflife of LSI change. This result is clear evidence that the process observed in the range of several hundred μs is an intramolecular reaction. The fact that the isomerization of only a few azobenzene groups per chain induces significant conformational changes is attributed to a severe perturbation of the balance of polymer-solvent and polymer-polymer interaction by trans → cis isomerization of pendant azobenzene groups. In a former paper [5], it was pointed out that the dipole moment increases from 0.5 to 3.1 D upon trans → cis isomerization of the azo groups. It is feasible, therefore, that cis-azobenzene groups are more prone to interact with styrene base units than trans-azobenzene groups because of dipole-induced interactions. The phenomena observed with the copolymers dissolved in cyclohexane can be, then, interpreted as follows: trans-azobenzene groups interact more favorably with solvent molecules than with styrene base units, whereas the reverse situation is true for cis-azobenzene groups which interact less strongly with solvent molecules than with styrene base units. The enhanced capability of cis-azobenzene groups of interacting with other segments of the chain is considered to give rise to a shrinkage of the polymer chain in the initial stage after the trans → cis isomerization, i.e. at a time when this interaction is purely intramolecular (in very dilute solution). When different macromolecules collide at a later stage, intermolecular interaction will also become operative. As a consequence, the polymer chains will aggregate and precipitation will be observed. These processes are schematically illustrated in Fig. 2. This model is substantiated by the fact that the increase in the LSI after the flash was not inhibited by low molecular weight azobenzene compounds that were added to the polymer solution. The idea that interaction between cis-azo groups causes coil contraction can be discarded, therefore.

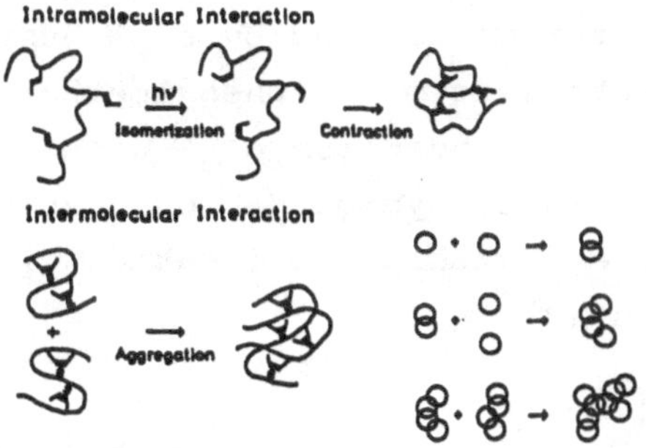

Fig.2. Schematic illustration of chain contraction and precipitation of polystyrene with pendant azobenzene groups.

Experiments with solutions of copolymer PS-A-4.3 in <u>different solvents</u> yielded results that are in accordance with this model. With decahydronaphthalene solutions similar effects were observed as with cyclohexane solutions. With benzene and methylene chloride solutions, on the other hand, no slow change in the LSI (in the μs range) was observed, although the trans → cis isomerization proceeded to the same extent and at the same rate as in cyclohexane solution and caused a very rapid decrease in the LSI during the flash. This result correlates very well with the findings that photo-stimulated precipitation occurred in cyclohexane and decahydronaphthalene but not in benzene and methylene chloride solutions [5].

Actually, these results were expected because benzene and methylene chloride are envisaged to interact with polar cis azobenzene groups. Therefore, this kind of interaction will predominate over the interaction of styrene base units with cis-azobenzene groups in this case. On the other hand, decahydronaphthalene cannot be polarized and will not interfere with the interaction between styrene base units and cis-azo groups.

2) Coil expansion of poly(methacrylic acid) induced by rapid neutralization[8]. The method employed to measure the rate of coil expansion of polyelectrolytes is based on a rapid change in the degree of ionization induced by a pH-jump which is achieved by flash photolysis of triphenylmethane leuco hydroxide derivates 1. Upon irradiation of 1 with a 20 ns laser flash of ultraviolet light, hydroxide ions can be produced "instantaneously", i.e., within the duration of the flash according to reaction (3)

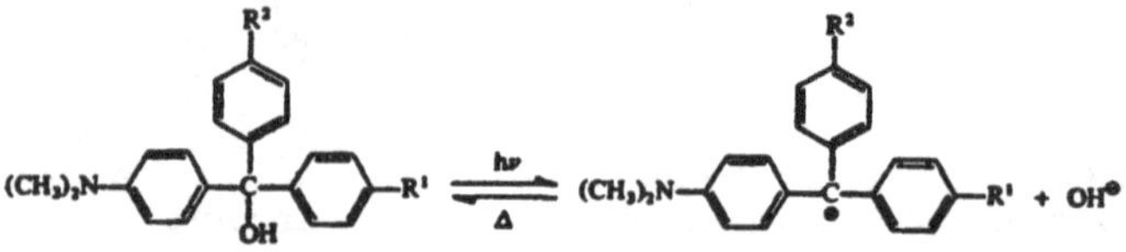

1

(3)

	1a	1b	1c
R^1	$N(CH_3)_2$	$N(CH_3)_3$	$N(CH_3)_2$
R^2	H	H	SO_3

The importance of the photo-stimulated ionic dissociation of triphenylmethane leuco oompounds was extensively reviewed by Bertelson [9]. Recently [10], it was pointed out that reaction (3) can be applied to induce pH-dependent chemical and physical reactions. It was shown that upon continuous irradiation of aqueous solutions containing 1, the H^+-concentration decreased almost instantaneously by several orders of magnitude after the light was switched on. It is noteworthy, in connection with the present work, that the pH decreased again with a halflife of several minutes, i.e., very slowly, after the light was switched off. Poly(methacrylic acid) (PMA) was chosen for primary tests, because it dissolves readily, at room temperature, in water and in methanol. The degree of ionization was adjusted by neutralization with alkali.

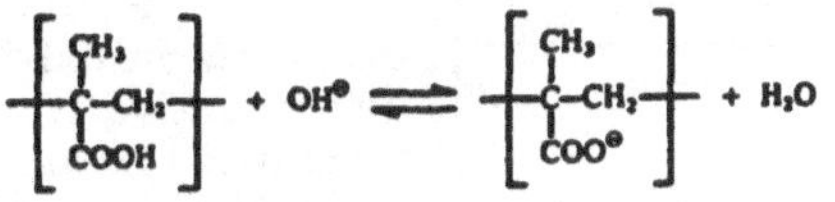

(4)

All experiments were carried out in methanolic solution containing 1a as hydroxide emitter. 1a is only slightly soluble in water. It turned out that PMA expands in methanolic solution in a similar manner as in aqueous solution. This was inferred from viscosity measurements. As is shown in Fig. 3, η_{sp}/c increased significantly upon increasing [NaOH] from 10^{-5} to 10^{-4} mol/l. The increase of η_{sp}/c with increasing

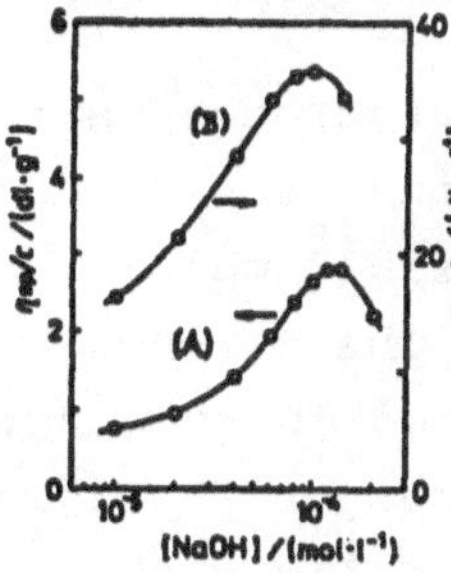

Fig.3. Coil expansion of poly (methacrylic acid) in methanolic solution at 30°C. Dependence of η_{sp}/c on [NaOH]. (A): PMA-1 ($\bar{M}_w$ = 9.6×10^4):8.0×10^{-2} g/l; (B): PMA-2 ($\bar{M}_w$ = 1.3×10^6): 8.4×10^{-2} g/l.

[NaOH] is due to intramolecular electrostatic repulsion between ionized carboxylic groups, whereas the decrease of η_{sp}/c at [NaOH] > 10^{-4} mol/l is indicative of a screening effect due to increased ionic strength.

When a methanolic solution of PMA containing 1a was continuously irradiated by light of λ > 270 nm, η_{sp}/c increased rapidly as is shown in Fig.4. After the irradiation was stopped, the viscosity dropped slowly and approached a value slightly higher than the initial value. The latter effect reflects that 1a is a weak base and dissociates slowly in a dark reaction, which is, however, much less pronounced than the photo-effect. The reversible change of the viscosity indicates that PMA coils expand as a consequence of the light-induced increase in the hydroxide ion concentration.

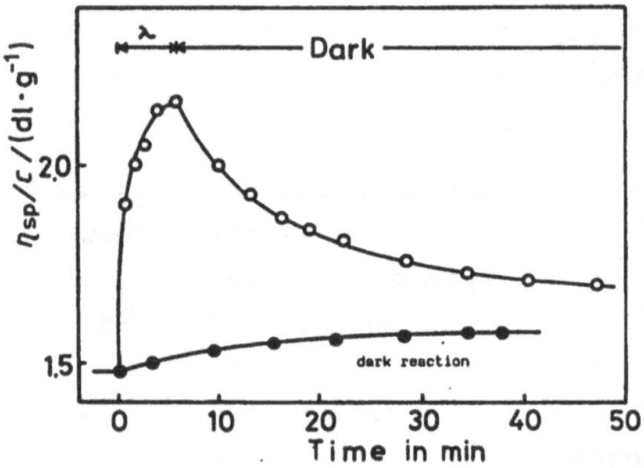

Fig.4. Photo-stimulated coil expansion of poly-(methacrylic acid) in methanolic solution at 23°C in the presence of triphenylmethane leuco hydroxide 1a (1.6×10^{-4} mol/l) during continuous irradiation with light of a Xenon lamp (Osram XBO, 450 W) at λ_{inc}>270 nm. Concentration of PMA-1: 9.6×10^{-2} g/l. (o): irradiated sample; (•): unirradiated sample, dark reaction.

The rate of this conformational change was measured by a flash photolysis experiment. Methanolic PMA solutions (8.4×10^{-2} g/l) containing 1a (2.0×10^{-5} mol/l) and NaOH (3.0×10^{-5} mol/l) were irradiated with 20 ns flashes of λ = 265 nm. The flashes were produced by a neodymium/YAG laser operated in conjunction with two frequency doublers. The absorbed dose per flash was ca. 4.5 J/l. As can be seen from Fig.5(a), the light scattering intensity (LSI) decreased after the flash according to two modes, a very rapid one and a slow one. The rapid mode which corresponds to the very first, unresolved, portion of the trace in Fig. 5(a) reflects the increase in the optical absorption due to triphenylmethyl cations and probably in the refractive index increment due to the ionic dissociation of 1a according to reaction (3). The increase in the optical absorption at λ = 514 nm, i.e. the wavelength of the analytical light used at the LSI measurements, is separately shown in Fig. 5 by trace (b). As a consequence of the increase in the absorption of the analytical light, the LSI is decreased.

The slow mode of the LSI decrease is ascribed to coil expansion following neutralization according to reaction (4). This interpretation was substantiated by the fact that LiBr ($1x10^{-3}$ mol/l) strongly suppressed the slow mode of the LSI decrease.

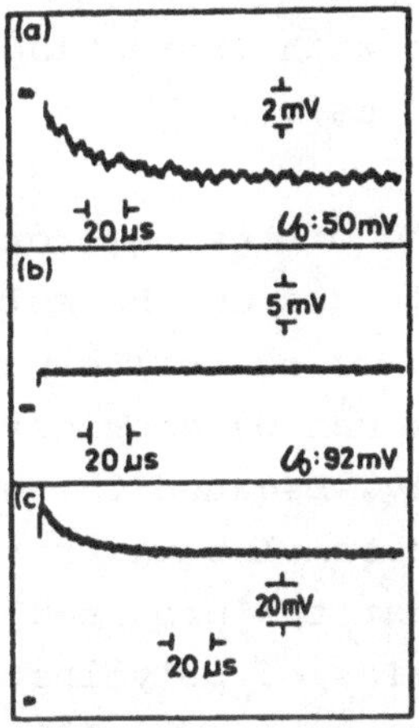

Fig.5. Flash photolysis of PMA-2 ($8.4x10^{-2}$ g/l) in methanolic solution containing 1a ($2.0x10^{-5}$ mol/l) and NaOH ($3.0x10^{-5}$ mol/l) at 23°C. Oscilloscope traces depicting changes in (a): light scattering intensity at 514 nm, (b): optical absorption at 514 nm, and (c): electrical conductivity at 40 V, AC, after the flash (20 ns, λ_{inc} = 265 nm).

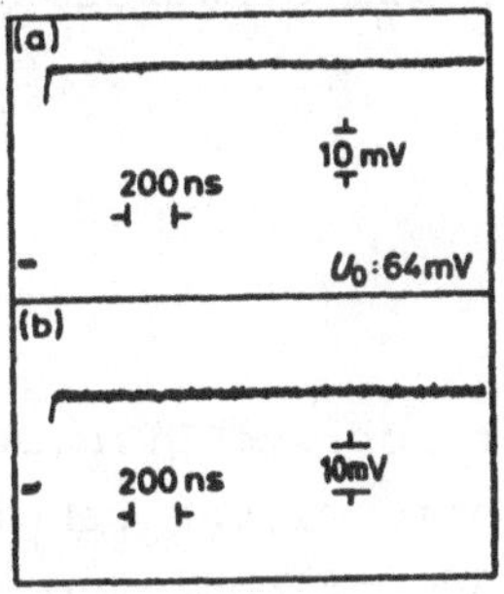

Fig.6. Photo-stimulated ionic dissociation of 1a in methanolic solution ($2.0x10^{-5}$mol/l) at 23°C. Oscilloscope traces depicting the increase in (a): the optical absorption at 522 nm and (b): the electrical conductivity at 40 V, AC after the flash (20 ns, λ_{inc} = 265 nm, absorbed dose: ca. 4.5 J/l ($1x10^{-5}$einstein/l)

Fig. 6 shows that the ionic dissociation of 1a according to reaction (3) indeed is very fast. These oscilloscope traces were recorded with a much higher time resolution than the traces in Fig. 5. They demonstrate the very rapid build-up (during the flash) of the optical absorption of triphenylmethyl cations and the simultaneous increase in the electrical conductivity due to the generation of ions [11].

It is interesting to note that the electrical conductivity decreased slightly after the flash in the presence of PMA as shown by trace (c) in Fig. 5. The partial decrease in the conductivity reflects the conversion of OH^- ions to less mobile macro ions possessing pendant carboxylate ions according to reaction (4). The decay rate increases with increasing polymer concentration (halflife: about 10 μs).

The rate of coil expansion was found to be of the same order of magnitude, but definitely lower. It depended on the molecular weight of the polymer. Typical halflives of coil expansion range from 10 to 30 μs. These values can be considered as rough order of magnitude estimates only, because the molecular weight distribution of both polymer samples was rather broad ($\overline{M}_w/\overline{M}_n > 5$). It appears, however, that the proposed method for the investigation of the dynamics of polyelectrolyte expansion gives satisfactory results and is capable of revealing reliable quantitative data with polyelectrolytes of narrow molecular weight distribution.

Acknowledgments

The authors are grateful to Dr. W. Wunderlich and Dr. M. Stickler, Röhm GmbH, Darmstadt, who provided one polymer sample (PMA-1) and determined the molecular weight of both samples by GPC. PMA-2 was prepared by Miss M. Kajander and Mr. R. Zuch at the Hahn-Meitner-Institut. The valuable assistance of Dr. G. Beck during the flash photolysis experiments is gratefully acknowledged.

References

1. (a) Schnabel, W.: Developments in Polymer Degradation, N. Grassie, ed., Applied Science Publ. Vol. 2, p. 35, London 1979;
(b) Tagawa, S., Schnabel, W.: Macromolecules 12, 663 (1979), Macromol. Chem. 180, 2371 (1979).

2. Tanford C.: Physical Chemistry of Macromolecules, Wiley, New York 1967.

3. Irie, M., Schnabel, W.: Macromolecules 14, 1246 (1981).

4. Irie, M., Schnabel, W.: Macromolecules,in print.

5. Irie, M., Tanaka, H.: Macromolecules 16, 210 (1983).

6. Krigbaum, W.R., Carpenter, D.K.: J.Phys.Chem. 59, 1166 (1955).

7. Štěpánek, P., Koňák, C., Sedláček, B.: Macromolecules 19, 1214 (1981).

8. Irie, M., Schnabel, W.: Makromol. Chem. Rapid Commun. 5, 413 (1984).

9. Bertelson, R.C.: Technique of Chemistry, Vol. 3, G.H.Brown, ed., Wiley, New York 1971, p. 294.

10. Irie, M.: J. Am. Chem. Soc. 105, 2078 (1983).

11. Brown, R.G., Cosa, J.: Chem. Phys. Lett. 45, 429 (1977).

PHOTON CORRELATION SPECTROSCOPY OF POLYMETHACRYLIC ACID

Presentation of a new type of correlator and some preliminary results on the conformational changes in PMA

Ph. Brak[+] and A. Persoons, Laboratory of Chemical and Biological Dynamics, Celestijnenlaan 200 D, B-3030 Leuven

Introduction

Photon correlation spectroscopy is a well established technique to measure the diffusion coefficient of macromolecules. It is also a fruitful technique to obtain an approximate value of M_w/M_n by means of the cumulant analysis.

Experimental

As compared to most commercial correlators our correlator (UAC 32 K A.T.N.E.) has some special features : delay increment from 10 ns up to 0.99 seconds, the number of delay increments up to 32752. The autocorrelation function is calculated in 128 channels, divided over the total delay time according to :

$$t_{i+1} = t_i + S \times 2^{(\mathrm{int}[(i - 16)/16])}$$

where S is the sample time and t_i the delay time of the ith channel. Therefore it is possible to use much shorter sample times without loss of data points. This means that we can obtain a maximum of signal [1] since the measured function is

$$G_k^{(2)}(\tau) = \langle n\rangle\langle n_k\rangle[1 + f(A)f(B)(1 + k)/(1 + \langle n\rangle) \, |g^{(1)}(\tau)|^2] \quad (1)$$

Physical Optics of Dynamic Phenomena and Processes in Macromolecular Systems

where $g^{(1)}(\tau)$ is the scattered field correlation function, $\langle n \rangle$ and $\langle n_k \rangle$ are the mean number of counts and the mean number of clipped counts respectively. f(A) is a multiplicative factor between zero and one, depending on the scattering geometry. f(A) is approximately one for the detection area up to one coherence area and approches 1 / M when the detector area is M times larger than the coherence area. f(B) is the correction factor for the finite sampling time effect and can be made very close to unity by choosing the sampling time to be much less than the characteristic correlation time. This means that we can maximize the preexponential factor since we are able to use much smaller sample times. The theoretical predicted baseline can be calculated from the monitor channels of the correlator. The PMA used in these preliminary measurements was prepared by anionic polymerisation of trimethylsilyl methacrylate in tolueen at 203° K and 1-1' diphenyl 3-methyl penthyllithium as initiator. This yields isotactic PMA (see NMR - spectrum Fig. 1) while in THF one should obtain syndiotactic PMA.

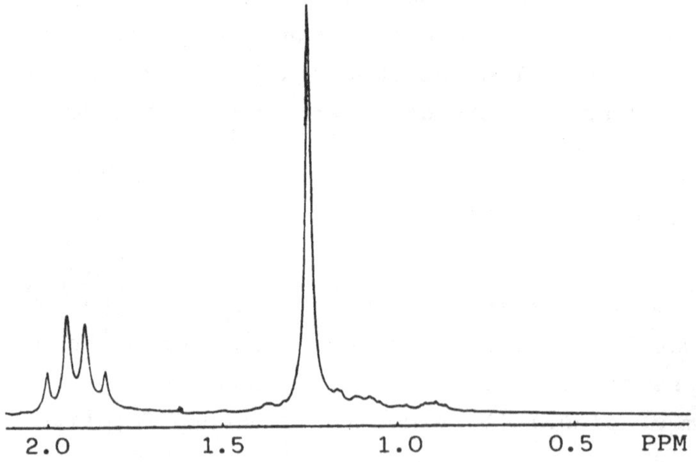

Fig. 1. NMR of isotactic PMA on a 250 MHz NMR in NaOD.

Results and discussion

Measurements were performed as a function of pH. The results are shown in Figure 2. The peculiar form of this plot can be completely explained considering the following formula

$$D = kT/\rho \; \psi(v+1)\,[\,1 + (\partial \ln(\psi(v+1))/\partial \ln n)] \tag{2}$$

where ψ is the osmotic coefficient, v the number of ions per molecule, ρ the hydrodynamic resistance and n the concentration.

At low ionisation degree, our polymer is insoluble and therefore no measurements can be made in this region. However, atactic PMA, which is soluble in this region, shows a very sharp increase of D, due to the increase in the number of counter ions which exert a strong drag effect on the coiled molecules [2]. In this pH range the molecules are hypercoiled and their hydrodynamic resistance is very low [3]. From pH 5.3 - 7 the molecules extend almost completely, by virtue of the intermolecular electrostatic repulsions so that ρ increases more rapidly than $v[(\psi + (\partial\psi/\partial \ln n)]$. This causes a decrease in D as would be expected from (2). After the initial uncoiling no further

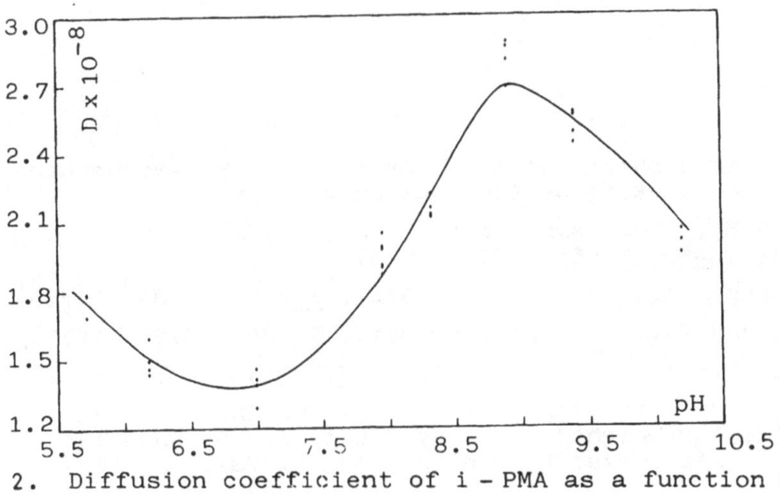

Fig. 2. Diffusion coefficient of i - PMA as a function of pH.

large change in ρ may be expected. The further increase in D can be understood by considering the further increase in charge on the macromolecule. Around pH 9 the decrease can be explained by a decrease in ψ while the product $v[\psi + (\partial\psi/\partial \ln n)]$ remains more or less constant.
The first and the second moment from the distribution function, obtained by the method of cumulants [4], gives us an approximate value for the polydispersity [5] :

$$M_w/M_n = 1 + \mu_2/\alpha^2 \langle\Gamma\rangle^2 \tag{3}$$

where μ_2 is the second moment and $\langle\Gamma\rangle$ $K^2 D_z$ (K the scattering vector), α is about 1/2 for Gaussian coils and 1/3 for spheres. When we calculate M_w/M_n by this formula we obtain values between 2.35 and 3. These are approximated values, the real values are expected to be 30 % or more smaller [5].

Conclusion

With our correlator the signal level reaches the theoretical predicted maximum. We showed that PCS is a good technique to study conformational changes in polyelectrolytes, and gives us a good estimate of the value of M_w/M_n.

References

+ Ph. Brak is a bursar of the "Instituut voor Wetenschappelijk Onderzoek in de Landbouw en de Nijverheid"

1. Chen, S.H., Veldkamp, W.B., Lai, C.C.: Rev. Sci. Instrum., 46, 1356 - 1367 (1975).
2. Katchalsky, A.: J. Polymer Sci., 13, 321 - 334 (1955).
3. Katchalsky, A.: J. Polymer Sci., 7, 393 - 412 (1951).
4. Koppel, D.E.: J. Polymer Sci. 57, 4814 - 4820 (1972).
5. Pusey, P.N.: in "Industrial Polymers Characterization by Molecular Weight", ed. by Green, J.H.S. and Dietz, R., (Transcripta Books, London) 26 - 36 (1973).

PHOTON CORRELATION SPECTROSCOPY OF MACROMOLECULAR SOLUTIONS UNDER HIGH PRESSURE

Bo Nyström and Jaan Roots
Department of Chemistry, University of Oslo
P.O.Box 1033, Blindern, 0315 Oslo 3, Norway.

Introduction

Recently the importance of the effect of hydrostatic pressure on dynamical properties of solutions of both synthetic flexible polymers (1,2) and proteins (3,4,5) has been recognized by photon correlation spectroscopy (PCS) investigations, and many features of current interest have been reported. It is well established that the PCS technique is a versatile and powerful tool in the study of diffusion processes, and the probing of the particle size of macromolecules in solution.

The response of many biological systems to high hydrostatic pressure has been studied from various angles of approach since Bridgman's pioneering work (6). It is now established that pressure constitutes a means of perturbing the environment of a molecule in a continuous controlled way (7). The conventional so-called denaturants are temperature and/or chemical composition of the medium, but these variables have some inconveniences (8). While pressure exclusively is correlated to effects which depend upon the volume, temperature alterations simultaneously produce changes in both volume and thermal energy in a way which makes it difficult to separate these two effects. In addition, it has often been observed that proteins in solution are temperature sensitive, in that even modest increase in temperature can produce irreversible configurational changes in the form of e.g.

Physical Optics of Dynamic Phenomena and Processes in Macromolecular Systems

aggregates. Environmental changes can also be achieved by variation in solvent composition or addition of various chemical species, but such changes are almost always ill defined and the results are difficult to interpret.

The molecules in a system respond to pressure by packing themselves more efficiently, and/or undergoing conformational changes via rotations around single bonds. The inter- and intramolecular interactions which may be perturbed by pressure less than about 10,000 atm include electrostatic interactions, hydrophobic interactions and hydrogen bonds, all of which play an important role in the structure, function and stabilization of most biological molecules.

The pressure dependences of the hydrodynamic radius, R_H, and the diffusion coefficient, D, are the two major results of the work presented here on macromolecular solutions. Although, special reference is paid to the behaviour of protein molecules at elevated pressures some comments on the behaviour of other polymer/solvent systems will also be given.

Experimental

Narrow distribution polystyrene (PS) (M_W = 1.11 x 10^5) and sodium polystyrene sulphonate (NaPSS) (M_W = 7.8 x 10^5) both from Pressure Chemical Co., and lysozyme and sulfhydryl blocked bovine serum albumin (BSA) monomer (standard protein powders) from Miles Laboratories were all used without further purification. The systems investigated were (I) BSA/tris-buffer (pH=7.4, I=0.04M), (II) lysozyme/acetic acid-acetate buffer (pH=4.0, I=0.1M), (III) NaPSS/0.1 M NaCl and (IV) PS/toluene; the solute concentrations were determined either by weighing or by spectrophotometry. All solutions were filtered, in an atmosphere of filtered air, through 0.22 µm Millipore filters directly into precleaned specially designed quartz cuvettes.

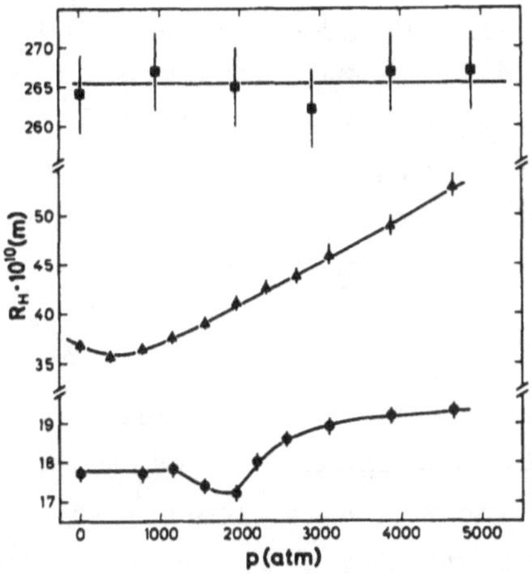

Figure 1. Hydrodynamic radius, R_H, versus pressure, p, for the systems (see Experimental): (I) ▲; (II) ●; (III) ■.

A portable vessel (9) for optical measurements under pressure was used; the pressure could be determined with ± 2.5% inaccuracy. The measurements were carried out at 25°C. The equipment used in the PCS measurements has been described previously (2); a fix scattering angle of 90° was used. Normalized photon counting correlation functions $g^{(2)}(\tau)$ were determined at 120 delay times τ. The analysis of data was based on the method of cumulants (10); a quadratic polynom in τ was fitted to the data to provide an average decay rate $\bar{\Gamma}$ and a normalized variance Q. The diffusion coefficient, D, and the hydrodynamic radius, R_H, were evaluated using $\bar{\Gamma}=D\cdot q^2$, where q is the magnitude of the scattering vector, and the Stokes-Einstein equation, respectively.

Results and Discussion

The pressure dependence of the effective hydrodynamic radius,

R_H, for the protein systems <u>viz</u>. bovine serum albumin (BSA) in a tris-buffer (pH=7.4 and ionic strength I=0.04M) and lysozyme in an acetic acid/acetate buffer (pH=4.0 and I=0.1M), and for the polyelectrolyte system sodium polystyrene sulphonate (NaPSS)/water in the presence of salt (I=0.1 M) is depicted in Figure 1. It is evident that R_H exhibits different features for these three systems.

Let us first discuss the protein systems. For the BSA system (I), R_H passes through a shallow minimum at moderate pressures (1000 atm) and then increases progressively with pressure, the net increase is approximately 31% over the pressure interval studied. In this context we may mention that PCS measurements at elevated pressures have been carried out previously on the system BSA in an acetic acid/acetate buffer at pH=4.7 (the isoelectric point) and at an ionic strength of 0.12M (4). The curve representing the pressure dependence of R_H was found to be superimposable on that reproduced in Figure 1 for the BSA system. These results indicate that the swelling process during denaturation of BSA is insensitive to even significant changes of the environmental conditions. For the lysozyme system (II), R_H passes through a pronounced minimum located at approximately 2000 atm, and upon further increase in pressure R_H increases, but gradually reaches a plateau value; the net increase is about 9%. The observed divergence in behaviour of the protein systems can probably be traced to intrinsic structural differences between the BSA and lysozyme molecules.

Upon releasing the pressure from moderate pressures (up to 2500 atm) the return of R_H was for both protein systems reversible, but from higher pressures the return exhibited an increasing feature of irreversibility. Values of R_H monitored successively during slow decompression from a reached pressure of about 4500 atm were significantly larger than the corresponding values obtained during the compression phase. Even

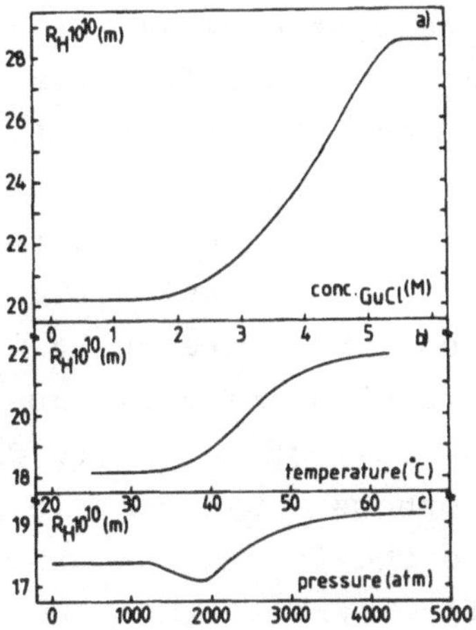

Figure 2. Effect of various denaturants on the hydrodynamic radius of lysozyme; a) guanidine hydrochloride (13); b) temperature (14); c) pressure (5).

when the measurements were repeated a long time after decompression the value of R_H was constant and significantly larger than that measured originally at atmospheric pressure. These findings indicate that the protein molecules do not return to their well-defined native conformation. In the spirit of a two state model one would expect the protein molecules to exist either as native or unfolded molecules. Analysing the present results within the framework of this model the findings indicate that the solution consists of a mixture of folded and unfolded species. However, an alternative interpretation of these data, without adopting the two state model, is that the molecules are statistically distributed over an array of (partially unfolded) conformational states. This interpretion is also compatible with the observed features. It is not unlikely that a possible existence of perturbed conformations may be due to an intricate situation where the transition to the energetically most favourable state is kinetically hindered.

For the polyelectrolyte system (III), R_H is independent of pressure, and the characteristic features of R_H observed for the protein systems (I and II) are absent. The intricate behaviour of R_H for the proteins, as opposed to that for the polyelectrolyte, is probably related to the difference in chemical structure of these molecules. In contrast to synthetic polyelectrolytes, which are built up of repeating monomer units, protein molecules are built up of a sequence of different amino acids, some of which are displaying a hydrophilic character and others a hydrophobic character. The forces which determine the folded structure are of two kinds, the network of specific intramolecular bonds such as hydrogen bonds, van der Waals forces and electrostatic interactions and a class of less understood protein-solvent interactions grouped together as hydrophobic interactions. In a naive macroscopic picture (11) the pressure induced denaturation of proteins may be visualized as a penetration of solvent into the folded structure, and thereby exposing the hydrophobic residues buried in the interior of the protein to the aqueous solvent. This conditional change may perturb the delicate balance of forces stabilizing the protein (12), and the net stability may decrease. The ultimate global effect may be a conformational change.

Lysozyme, which is a small protein with a single polypeptide chain, is frequently used in studies of denaturation phenomena because of its structural simplicity as compared with many other proteins. In order to compare the extent of denaturation induced by various denaturants, the hydrodynamic radius, obtained from PCS measurements on aqueous solutions of lysozyme, is plotted (see Figure 2) as a function of concentration of chemical agent (13), temperature (14), and pressure (5), respectively. Except for the minimum of the curve representing pressure denaturation, similar features of R_H are observed for all denaturants, viz. sigmoid-shaped curves with plateau regions below and above the transition region. However, the net change in R_H is different for the various denaturants. In the case of the chemical agent R_H increases by 45%, while the increase in R_H is 18% and 9% when temperature and pressure are used as perturbing variables, respectively. This finding seems to indicate that various denaturants cause different levels of conformational perturbation, with pressure being the weakest denaturant.

Works (15) on free energy changes in lysozyme denaturation have shown that native lysozyme is completely denatured by the addition of guanidinium chloride, but that partially denatured conformations, i.e. conformations which retain some of the native secondary and tertiary structures, appeared when temperature and other denaturants were used. The need for intermediate states between the native and the fully unfolded form was dictated by differences in the numerical changes observed for the various perturbing parameters studied. The present results support the conjecture of "intermediate states"; the transition curves for lysozyme under the denaturing conditions of temperature and pressure may be representative of transitions to partially unfolded conformations.

It is generally believed that conformational transitions of biopolymers are second-order phase transitions, and usually the transition features are rationalized in the spirit of a two state model. The transition curves observed for lysozyme under the denaturation conditions used here do not contradict the classical picture (16,17) that conformational transitions of proteins may be visualized as a globule-coil transition.

A notable feature of the curve in Figure 2 representing pressure denaturation (cf. also Figure 1) is the minimum before the transition begins. A similar tendency, but less pronounced, has also been observed for the BSA system (3) (cf.

Figure 1). The protein molecules thus appear to be contracted at moderate pressures. This "pretransition" phenomenon is probably due to a change in the delicate balance of forces stabilizing the native structure, or the existence of local structural fluctuations which may diminish with pressure. It should be noted that this contraction behaviour seems to be unique for pressure; no behaviour of this kind has been observed in PCS measurements on protein solutions when e.g. temperature has been used as a variable. However, a possible "pretransition" contraction effect in temperature studies may be obscured or cancelled due to thermal expansion (18,19) of the native protein molecules. As a matter of fact, in a recent PCS study (18) on the thermally induced denaturation of ribonuclease, a small globular protein of the same type as lysozyme, the particle size was even found to gradually increase in the pretransition region.

Figure 3 illustrates the pressure dependence of the ratio D^p/D^1 (the superscripts 1 and p denote atmospheric and considered pressure, respectively) at different concentrations for the systems I, II and IV. It is evident that for the system PS/toluene (IV) (good, nearly athermal solvent), the pressure dependence of D, both for dilute and semidilute solutions, is directly correlated to the change in solvent viscosity with pressure (η^1/η^p; solid curve), and the expression $D^p = D^1\eta^1/\eta^p$ is found to hold (2). For the protein systems, on the other hand, the change in solvent viscosity (solid curve) can only partially account for the pressure dependence of D. For the protein systems the ratio D^p/D^1 falls off more rapidly with pressure than the corresponding change in solvent viscosity. Furthermore the decrease in the ratio D^p/D^1 is more pronounced for the higher concentrated protein solutions; this trend is more accentuated for the BSA system than for the lysozyme system. In general, at higher concentration we expect a more intricate behaviour due to the feature of enhanced intermolecular interactions. For a solution of inter-

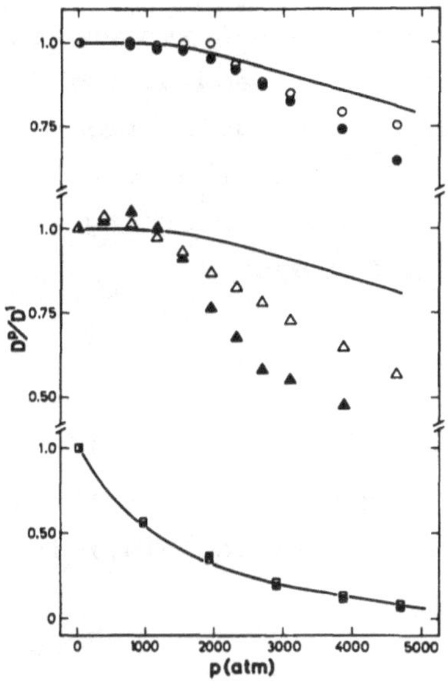

Figure 3. The ratio D^p/D^1 versus pressure, p, for the systems (see Experimental) and concentrations C (kg cm^{-3}):(I) △ 4.70; ▲ 73.2; (II) ○ 9.20; ● 80.1; (IV) □ 4.43; ■ 53.0. The solid curves indicate the relative change (η^1/η^p) in solvent viscosity with pressure.

acting Brownian particles the diffusion motion is frequently described by introducing a wave vector, q, and a concentration dependent effective diffusion coefficient, D, which is given by $D = D_o/S(q)$, where D_o is the "free-particle" diffusion coefficient and S(q) is the static structure factor. This factor characterizes the interparticle interaction of the system. For a system of non-interacting particles S(q) equals 1. The lysozyme system, for which the electrolyte concentration is fairly high, may represent a weakly interacting or noninteracting system. In this case the interaction situation should be quite insensitive to conditional changes (20) and it is not unlikely that the structure factor even at high concentration is more or less unaffected by pressure. For the BSA system, on the other hand, where no simple electrolyte was added, the interparticle interaction balance may be more delicate; it is possible that an external perturbation in the form of e.g. pressure may change the structure factor governing the long-range particle correlations in the system.

Acknowledgement

A research grant (J.R) from the Royal Norwegian Council for Scientific and Industrial Research (NTNF) is gratefully acknowledged.

References

1. Hecht, A.M., Geissler, E., Chosson, A.: Polymer 22, 877-881 (1981).
2. Roots, J., Nyström, B.: Macromolecules 15, 553-556 (1982).
3. Nyström, B., Roots, J.: Chem.Phys.Lett. 91, 236-240 (1982).
4. Nyström, B., Roots, J.: J.Chem.Phys. 78, 2833-2837 (1983).
5. Nyström, B., Roots, J.: Makromol. Chem. 185, 1441-1447 (1984).
6. Bridgman, P.W., J.Biol.Chem. 19, 511-512 (1914).
7. Heremans, K.: Ann.Rev. Biophys. Bioeng. 11, 1-21 (1982).
8. Jaenicke, R.: Ann.Rev. Biophys. Bioeng. 10, 1-67 (1981).
9. Claesson, S., Malmrud, S., Lundgren, B.: Trans.Faraday Soc. 66, 3048-3052 (1970).
10. Koppel, D.E.: J.Chem.Phys. 57, 4814-4820 (1972).
11. Weber, G., Drickamer, H.G.: Quart.Rev.Biophys. 16, 89-112 (1983).
12. Grütter, M.G., Hawkes, R.B.: Naturwissenschaften 70, 434-438 (1983).
13. Dubin, S.B., Benedek, G.B., Biochemistry 12, 714-720 (1973).

14. Nicoli, D.F., Benedek, G.B.: Biopolymers 15, 2421-2437 (1976).

15. Ahmad, F., Contaxis, C.C., Bigelow, C.C.: J.Biol.Chem. 258, 7960-7963 (1983).

16. Lifshitz, I.M., Grosberg, A. Yu., Khokhlov, A.R.: Rev. Mod. Phys. 50, 683-713 (1978).

17. Williams, C., Brochard, F., Frisch, H.L.: Ann.Rev.Phys. Chem. 32, 433-451 (1981).

18. Wang, C.-C., Holland Cook, K., Pecora, R.: Biophys. Chem. 11, 439-442 (1980).

19. Jøssang, T., Feder, J., Rosenqvist, E.: Scattering Techniques Applied to Supramolecular and Nonequilibrium Systems, Edited by Chen., S.-H., Chu,B., Nossal, R. Plenum 1981 pp.831-837.

20. Nyström, B., Johnsen, R.: Chemica Scripta 22, 82-84 (1983).

INTRAMOLECULAR MOTILITY IN PIG IMMUNOGLOBULIN G STUDIED BY NEUTRON SPIN ECHO TECHNIQUE

Sándor Borbély, Yuriy M. Ostanevich
Joint Institute for Nuclear Research, Dubna, USSR

László Cser, Béla Faragó
Central Research Institute for Physics, Hungarian Academy of Sciences, Budapest, Hungary

Ferenc Mezei
Institute Laue-Langevin, Grenoble, France

Frantisek Franek
Institute for Molecular Genetics, Czechoslovak Academy of Sciences, Prague, Czechoslovakia

Introduction

Present knowledge of the three-dimensional structure of immunoglobulin G is based on X-ray diffraction studies (1) and on scattering analyses of the molecular shape of IgG in solution (2). Both these approaches yield static pictures of the investigated structures. Since the IgG molecule consists of relatively independent parts connected by loose sections of polypeptide chains, the notion of intramolecular motility in IgG appears very plausible. A powerful tool for studying intramolecular motility recently became available through the development of the neutron spin echo technique. This method provides information not only on the time dependence, but also on the momentum dependence of the process. The flexibility of the parts of the molecule may be demonstrated by the effective diffusion coefficient D_{eff} as a function of the transferred momentum Q.

Physical Optics of Dynamic Phenomena and Processes in Macromolecular Systems

Experimental

All experiments were performed using pig anti-dinitrophenyl antibody of the precipitating type (3), which was dissolved in a 50 mM sodium phosphate buffer containing 100 mM sodium chloride. The buffer was prepared in deuterium oxide. The viscosity of the solvent and solutions were determined by using a capillary viscosimeter. The translational diffusion coefficient D^o_{tr} was calculated according to the Svedberg equation, and corrected to the experimental conditions (14°C and deuterium oxide solvent).

Results

From the time dependence of the spin echo signals effective diffusion coefficients $D_{eff}(Q)$ were evaluated (for the theory see 4). These values are given in Fig.1 for different solutions as a function of the momentum transfer Q. $D_{eff}(Q=0)$, that is equal to the translational diffusion coefficient D_{tr}, could not be determined from the measured curves, since the accuracy and resolution were inadequate in the range of $Q < 1/R_g$. The values of $D_{eff}(Q=0)$ were therefore extracted from hydrodynamic parameters using the Stokes-Einstein formula. The concentration dependence of $D_{eff}(Q=0)$ was assumed to have been introduced by using solution viscosity instead of solvent viscosity. For the interpretation of the experimental data two types of model were constructed. In the first model the molecule has a rigid T-shaped form (rigid model) and only permitted motion -apart from the translational diffusion- is the rotational diffusion of the global molecule. In the second model a wagging or conelike wobbling motion of the F_{ab} arms around the hinge region should be considered (flexible model). The rotational diffusion coefficient of the whole molecule D^T_r, the rotational diffusion coefficient of one F_{ab} arm around the hinge region D_r and the maximum tilting angle of F_{ab} arms

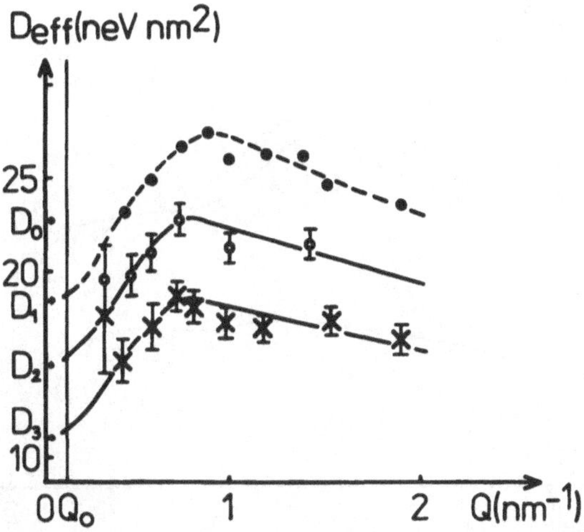

Fig. 1. Experimental values of D_{eff} as a function of Q at different concentrations of IgG: 7.33% (w/v) (crosses), 3.67% (w/v) (circles), extrapolated values to zero concentration of IgG (full circles). Q_o is the reciprocal value of R_g (radius of gyration). D_o is the translational diffusion coefficient of IgG molecule as calculated from the sedimentation data, D_1, D_2 and D_3 are the same for the concentrations of zero, 3.67% and 7.33% respectively.

were considered as independent changeable parameters (in detail see 5). The model calculations were restricted to the solution of 7.33%, since these data are the most accurate and were obtained over a wide Q-range.

According to our concept the $D_{eff}(Q) - D_{tr}$ values correspond either to the rotational motion of the molecule as a whole, or a wobbling type motion of the F_{ab} arms. To check the validity of above assumptions and to make a choice between them the values obtained by model calculations were compared with the experimental ones. As may be seen from Fig.2 a satisfying fit could be achieved by the flexible model with the tilting angle $\beta_{max} = 50^{o}$ and diffusion coefficient $D_r = 1/1220\ ns^{-1}$. These values are in good agreement with those of Valentine and Green (6) and Hanson et al. (7) who showed that the deviation of F_{ab} arms from the T-shaped form may reach a value of 60° and the diffusion coefficient of the wobbling-type motion of F_{ab} arms determined by the nanosecond fluorescence polari-

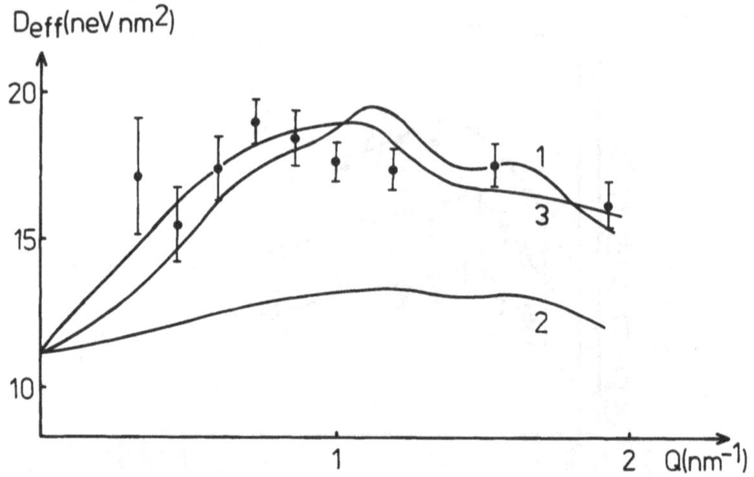

Fig. 2. Computed plots of D_{eff} vs. Q for molecular models of IgG. 1, rigid model (D_r^T = 1/1600 ns^{-1}); 2, rigid model (D_r^T = 1/6000 ns^{-1}); 3, flexible model (β_{max} = 50°, D_r = 1/1220 ns^{-1})

zation technique is D_r = 1/1050 ns^{-1} . To achieve a good approximation in the case of rigid model we had to choose the value D_r^T = 1/1600 ns^{-1} , which is much higher than that obtained from hydrodynamic radius (1/6000 ns^{-1}) . Therefore we could not accept the rigid model as a realistic solution.

References

1. Deisenhofer, J., Colman, P.M., Huber, R., Haupt, M., Schwick, G.: Hoppe-Seyler's Z. Biol. Chem. 357, 435-445 (1976).
2. Cser, L., Gladkih, I.A., Franěk, F., Ostanevich, Yu. M.: Colloid and Polymer Sci. 259, 625-640 (1981).
3. Franěk, F., Doskočil, J., Šimek L.: Immunochemistry 11, 803-809 (1974).
4. Neutron Spin Echo (Mezei, F., Ed.), Lecture Notes in Physics, Vol. 128, Springer Verlag, Heidelberg 1980.
5. Alpert, Y., Cser, L., Faragó, B., Franěk, F., Mezei, F., Ostanevich, Yu. M.: Eur. J. Biochem., in press.
6. Valentine, R.C., Green, N.M.: J. Mol. Biol. 27, 615-617 (1967).
7. Hanson, D,C., Yguerabide, J., Schumaker, V.N.: Biochemistry 20, 6842-6852 (1981).

THE CONFORMATION OF A DNA-PROTEIN COMPLEX

M.A. Scheerhagen, H. van Amerongen, M.E. Kuil, R. van Grondelle, Joh. Blok

Department of Biophysics, Physics Laboratory of the Free University, De Boelelaan 1081, 1081 HV Amsterdam, The Netherlands

Introduction

The protein encoded by gene 32 of bacteriophage T4(GP32) is a so-called "helix destabilizing protein". It has an essential function in the replication of T4 DNA (1). It binds tightly and cooperatively (binding site between 7 and 10 nucleotides) to single-stranded (ss) DNA (2), inducing a characteristic change in the circular dichroism spectrum of DNA (3). In order to obtain information on the structure of DNA in the complex with GP32, in particular on the axial increment (ax) i.e. the projection of the base-base distance on the helix axis, we measured the rotational and translational diffusion coefficient (D_R, D_T) of short, homodisperse single-stranded DNA complexed with GP32, using electric field induced birefringence and quasi-elastic light scattering (QELS).

Results

1) Electric birefringence. The binding of GP32 with a 145-base DNA fragment leads to a marked change in the observed birefringence. As is shown in Figure 1 (i) the signal of the complex is large in comparison with those of the separate components, (ii) the birefringence of the complex is positive in contrast to that of single- and double-stranded DNA, (iii) in comparison with the dimensions of the uncomplexed protein (MW = 33.5 kD) its field free relaxation is slow, an indication that free GP32 tends to aggregate into clusters. To represent the field-free decay of the complex at least two

Physical Optics of Dynamic Phenomena and Processes in Macromolecular Systems

exponentials are required (Fig. 1). After correction to standard conditions (20^{0}C, water) we obtained the following relaxation times for the 145b-GP32-complex:
$\tau_1^{20,w} = 0.88 \pm 0.07$ µs (35%) and $\tau_2^{20,w} = 4.2 \pm 0.2$ µs (65%). The biphasic decay cannot be ascribed to heterodispersity of the single-stranded DNA as was checked by gel electrophoresis. Moreover mono-exponential decays with the correct (4) relaxation time were obtained with the 145 b.p. double-stranded DNA from which the ss material was prepared by heating. We cannot yet present a satisfactory interpretation of τ_1. In complexes with still shorter fragments (5) the fast decay components were also observed with about equal relative contribution. The value of τ_1 depended on the length of the complex. Since the value of τ_2 does not decrease by lowering the complex concentration, τ_2 probably corresponds to an unhampered end-to-end rotation of the complex with $D_R^{20,w} = 4.0 \times 10^4\ s^{-1}$.

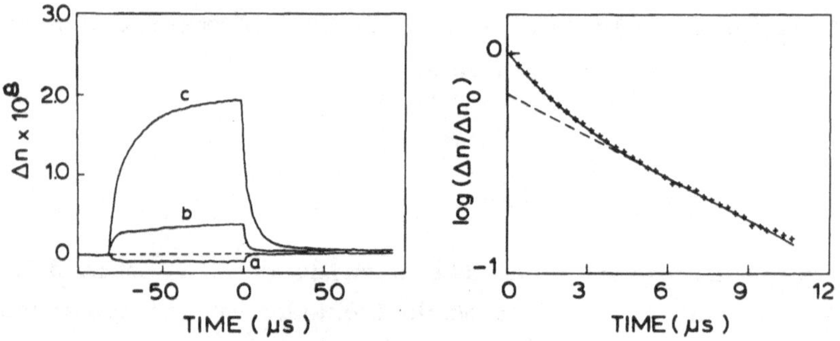

Fig. 1. Left: Transient electric birefringence. Field on at t = -80 µs and switched off at t = 0 a) 145-base ss DNA, 39.5 µM (P) b) GP32, 8.1 µM c) DNA-GP32 complex, concentrations as for a and b. Temperature 10^{0}C. Buffer 1 mM Na_2HPO_4 + 0.1 mM Na_2 EDTA + 4 mM NaCl, pH = 7.7.
Right: Semilogarithmic plot of field-free decay of DNA-GP32 complex.

2) Quasi elastic light scattering. Homodyne QELS was used to determine the value of D_T for the DNA-GP32-complex (145 b). Figure 2 shows the autocorrelation function, obtained after

removal of dust particles by a combination of filtration and centrifugation, leading to $D_T = (1.95 \pm 0.05) \times 10^{-7}\ cm^2\ s^{-1}$ at 50 mM NaCl irrespective of the scattering angle between 30 - 150°. D_T was only slightly dependent on salt concentration: $D_T = (1.78 \pm 0.05) \times 10^{-7}\ cm^2\ s^{-1}$ at 10 mM NaCl and $D_T = (2.05 \pm 0.05) \times 10^{-7}\ cm^2\ s^{-1}$ at 200 mM NaCl. The former value gives a lower estimate since interaction between the complexes must be expected at such low salt concentrations.

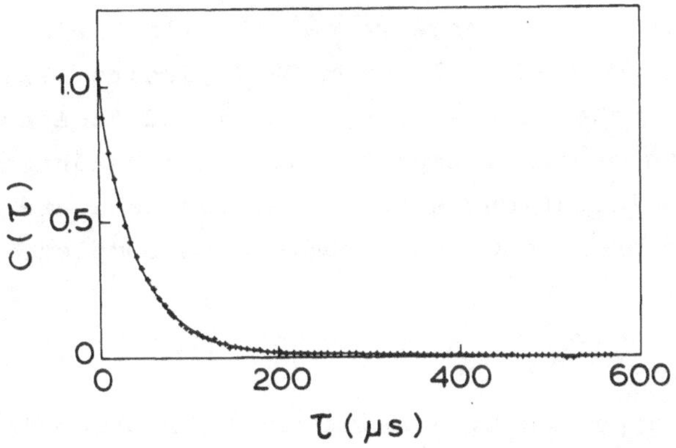

Fig. 2. Autocorrelation function of intensity variations in the light scattering at 90° from a solution of the DNA-GP32 complex. [DNA] = 73.8 μM (P), [GP32] = 23.4 μM, T = 20.4°C, buffer 1 mM Na_2HPO_4 + 0.1 mM Na_2 EDTA + 50 mM NaCl, pH = 7.7.

3) <u>Calculation of dimensions</u>. Approximating the complex as a rigid rod D_R and D_T can be written as a function of length and diameter of the rod (6, 7). Combination of both diffusion coefficients leads to a diameter of 47 Å and a length, corresponding to an axial increment of 4.5 Å for the DNA-GP32-complex of 145 bases. The axial increment for short RNA fragments (76b, 83b) in complex with GP32 is even as large as 5.5 Å (5).

4) <u>Calculations of circular dichroism (CD)</u>. At first sight the large value of ax seems contradictory to the rather intense CD spectrum of the DNA in the complex.
However calculations for a particular polynucleotide (poly rA)

using the method of Cech (8) show that the characteristic CD spectrum can be roughly predicted with conformations characterized by (i) a significant tilt of the bases with respect to the helix axis and also (ii) a large number of base pairs per helical turn or a position of the bases close to the helix axis or both (5).

Conclusion

The hydrodynamic measurements reveal that the axial increment of DNA increases dramatically upon GP32 binding. The positive birefringence of the complex may be assigned to a contribution from the protein probably together with a substantial tilt of the bases. CD calculations show that these data are not in conflict with the observed CD spectrum of the complex.

Acknowledgement

This work was supported by a grant from the Netherlands Organisation for the Advancement of Pure Research (Z.W.O.). The authors are indebted to Mrs. C.A. Vlaanderen for preparation of the GP32-samples.

References

1. Breschkin, A., Mosig, G.: J. Mol. Biol. 112, 279 (1977).
2. Newport, J.W., Lonberg, N., Kowalczykowski, S.C., Von Hippel, P.H.,: J. Mol. Biol. 145, 105 (1981).
3. Anderson, R.A., Coleman, J.E.: Biochemistry 14, 5485 (1975).
4. Elias, J.G. Eden, D.,: Macromolecules 14, 410 (1981).
5. Scheerhagen, M.A., Blok, Joh., Van Grondelle, R.: Publication in preparation.
6. Mandelkern, M., Elias, J.G., Eden, D., Crothers, D.M.: J. Mol. Biol. 152, 153 (1981).
7. Tirado, M., Garcia de la Torre, J.: J. Chem. Phys. 71, 2581 (1979).
8. Cech, C.M.: Ph.D. Thesis, University of California, Berkeley (1975).

ELECTROOPTICAL KERR EFFECT STUDIES OF HEMOGLOBIN*

Zdzisław Błaszczak
Institute of Physics, Adam Mickiewicz University
Grunwaldzka 6, 60-780 Poznań, Poland

Michael Witrh, Bengt Nordén
Physical Chemistry Dept., Chalmers University of Technology
S-412 96 Göteborg, Sweden

Introduction

One of the electrooptical phenomena is the Kerr effect /KE/ (1). In this phenomenon the optical anisotropy, arising in the initially isotropic transparent medium under the influence of the electric field applied, is measured by the induced optical birefringence. In liquids KE is mainly due to the effect of molecules orientation in result of the interaction of permanent and induced dipole moments of the molecules with the field. Studies of the Kerr effect provide valuable information about structure, conformation and aggregative properties of both small and big molecules.

Experimental

The KE studies of aqueous solution pose several experimental problems. High-voltage pulse of short duration must be applied to the Kerr cell to minimize the heating of the sample caused by its conductance. The Kerr cell should be designed so that its conductivity, capacitance and volume were minimum.

*Work supported in part by Research Project MR I.5.10.04

Physical Optics of Dynamic Phenomena and Processes in Macromolecular Systems

These criteria were partly fulfilled by using Teflon for the body of the cell and stainless steel for the electrodes /4.8 cm in length with 0.2 cm gap/. The high voltage pulse was applied to the cell from a high-voltage pulse generator, laboratory made, according to the scheme described in (2). Single pulses of 2 to 5 kV amplitude and 5 to 500 μs duration could be used. As a light source we used He-Ne or argon laser made by Spectra Physics. The Kerr cell was arranged in the usual way, with the electric field applied at an angle of 45° to the crossed polarizer and analyzer. Initially, in the absence of the electric field,no light passing through the crossed optical system could be detected by means of a photomultiplier. The electric field pulse and the light pulse, stimulated by the former, appearing behind the analyzer were recorded by a digital oscillograph. Knowing the amplitudes of both pulses we can calculate the Kerr constant B of the medium, defined by the formula: $\delta = 2\pi B \times l \times E^2$, where δ is the relative retardation between two plane-polarized light waves having electric vectors parallel and perpendicular to the direction of the applied electric field E. The optical path length is denoted by l.

We measured KE of water solution of human hemoglobin. As we know, hemoglobin /Hb/ which is contained in red blood cell , serves as the oxygen carrier in blood and plays a vital role in the transport of carbon dioxide and hydrogen ions. The Hb molecule is nearly spherical with a diameter of 5.5 nm. Its four polypeptide chains, held together by noncovalent interactions, are packed in a tetrahedral array. As far as we know KE investigations of human Hb have not been performed, however such studies for horse Hb were performed in the sixties (3). Therefore the studies of human hemoglobin were undertaken to supplement those former data as well as because the hemoglobin molecule is particulary interesting as it is more α-helical than most other globular proteins.

We used the crystallized and lyophilized Hb of Sigma production, solved in bidistilled water to obtain oxy-Hb. From this form the deoxy- and methemoglobin were prepared according to standard procedure.

Results

At first we measured the Kerr constant B_w for pure water, with different values of the applied field intensity and duration, and then the Kerr constant of hemoglobin-water solution as a function of Hb concentration and electric field strength. All the measurements were carried out at a temperature of 393 K. Some examples of the pulses observed on the oscillograph screen are presented in Fig. 1. Part of the obtained results are summarized in Table 1.

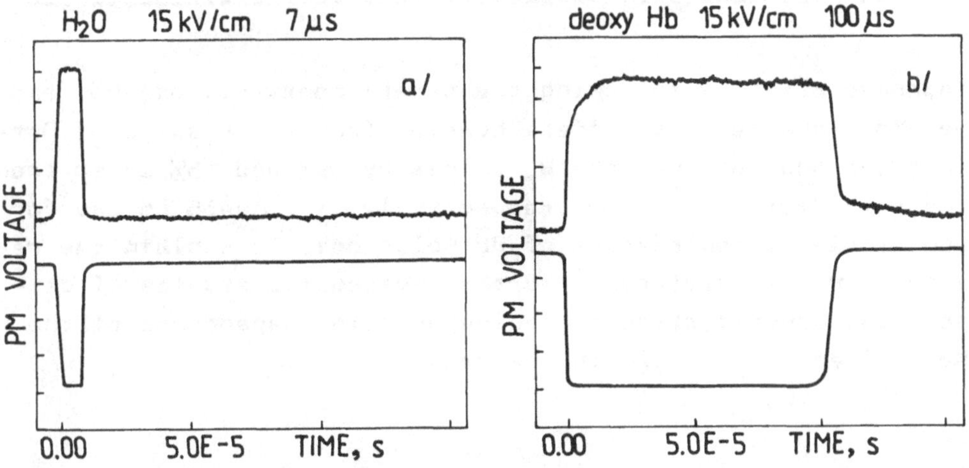

Fig.1. Typical example of KE signal /up/ and electric field pulse /down/ of water a/ and water-deoxyHb solution b/

Discussion

The Kerr constant values obtained for water are in a good agreement with the literature data (4). As follows from Table 1, the solution-solvent Kerr constants B_r are higher then 1, and

Table 1. Some average B and relative $B_r=B/B_w$ values of Kerr constants for λ=514.5 nm of the liquides studied. Data for 0.89 a/ and 8.90 b/ µmol of Hb concentration. Pulse duration: water - 20 µs, Hb - 100 µs

Liquid	$B \times 10^{14}$ /m V^{-2}/	B_r
water	1.63 /for 632.8 nm/	
	2.52	1
	3.86 /for 457.9 nm/	
deoxy-Hb	2.63 a/	1.04
	3.31 b/	1.31
oxy-Hb	2.70 a/	1.07
	3.58 b/	1.42
met-Hb	2.78 a/	1.10
	3.62 b/	1.44

increase slightly following the order: deoxy-Hb, oxy-Hb, and met-Hb. This results differs however from the results of Orttung (3) who obtained the B_r values by 39% and 15% lower than those for met- and oxyHb, respectively, for λ=456 nm and for comparable concentrations of Hb solutions. To explain the reasons for this difference further systematic studies of dispersion, concentration and pulse duration dependence of the Kerr effect in hemoglobin are required.

References

1. Kerr, J.: Phil.Mag. S.5., 50, 337-446, 446-458 1875, S.5., 8, 85-102, 229-245 1879, S.5., 13, 153-262 1882 .
2. Yamaoka, K., Matsuda, K.: J.Sci. Hiroshima Univ. ser. A, 43, 185-203 1980 .
3. Orttung, W.H.: J.Amer.Chem.Soc. 87, 924-926 1965, J.Phys. Chem. 73, 418-423, 2908-2915 1969 .
4. Orttung, W.H., Meyers, J.A.: J.Phys.Chem. 67, 1905-1910 1963; Beevers, M.S.,Khanarian, G.: Aust.J.Chem. 32, 262-269 1979 , 33, 2585-2595 1980 .

MOLECULAR MOBILITY IN LIQUID POLYMERS AND POLYMERIZATION PROCESS INVESTIGATED BY ACOUSTICAL METHODS

P. Hauptmann

Technische Hochschule Leuna-Merseburg, Sektion Physik
GDR - 42 Merseburg

Introduction

The ultrasonic field acting on liquid polymers such as solutions or dispersions can detect the molecular motional behaviour or the particle growth during the polymerization. It will be reported about the acoustical relaxation compared with conformational calculations, and the observation of polymerization processes with ultrasound.
Acoustical methods are well known methods in the investigation of relaxation processes in different materials. Contrary to the optical methods other interactions between the incident waves and the molecules exist. The applied frequency range is relatively broad. The influence of the solvent on the ultrasonic absorption in poly(vinyl acetate) solutions was studied. A fundamental advantage of acoustic waves is the ability to pass through turbid media. In the following an example will be shown for the application of the ultrasonic field in the detection of the VAC polymerization.

Theoretical

The excess absorption $\Delta\alpha$ in discrete relaxation processes can be described in the following manner

$$\Delta\alpha/f^2 = A/[1 + (f/f_1)^2] + B/[1 + (f/f_2)^2] + C \qquad (1)$$

Physical Optics of Dynamic Phenomena and Processes in Macromolecular Systems

f_1, f_2 are the relaxation frequencies, A, B the relaxation strengths. C is the classical absorption and f the measuring frequency. Equation (1) describes a two-relaxation process. In the case of a broad relaxation spectrum the equations

$$\alpha/f^2 \sim \int_0^\infty [H(\tau)/(1+\omega^2\tau^2)]\,dt \qquad (2)$$

or

$$\alpha/f^2 \sim f^{-n} \qquad (3)$$

are valid with $H(\tau)$ the distribution function and n an exponent which can be varied. The experimental results were fitted with the equations (1) to (3) and can be explained with molecular models.

For the process control during the polymerization the velocity is more suitable. It can be shown that c of the composite medium is a definite function of the several components i

$$c = c_0 + \sum_i (\Delta c/\Delta k)^i\, k^i \qquad (4)$$

c_0 is the velocity of the dispersion medium, $(\Delta c/\Delta k)^i$ is the velocity-concentration coefficient of the i-th component and k^i is the corresponding concentration.

Experimental

The ultrasonic absorption was measured with different methods: reverberation 200 kHz - 3 MHz (1); differential 2 - 20 MHz; pulse technique 20 - 150 MHz (2).

The velocity was determined with an automatic pulse-travelling method of high accuracy connected with a microprocessor.

Poly(vinyl acetate) was the investigated polymer.

Results

The excess absorption of PVAC in different solvents is shown in Fig. 1. The results are summarized in Table 1.

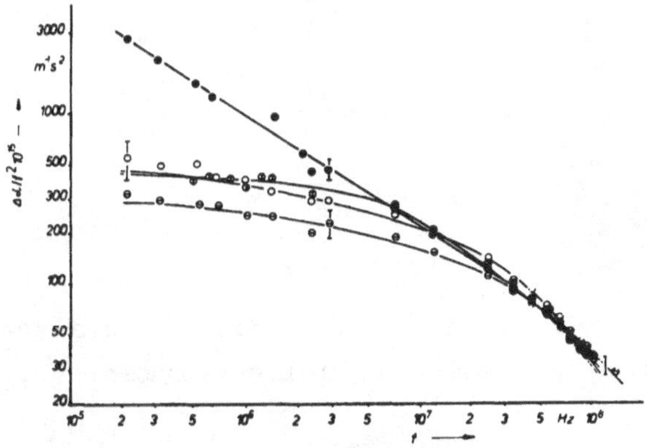

Fig. 1. $\Delta\alpha/f^2$ as a function of measuring frequency (20°C): PVAC-methanol (k = 13.5 %) ●, PVAC-acetone (k = 5%) ⊖, PVAC-toluene (k = 5%) ○, PVAC-cyclohexanone (k = 5%) ⦶

Table 1. Relaxation parameters after eq. (1)

PVAC in solvent	A $10^{-15}m^{-1}s^2$	f_1 MHz	B $10^{-15}m^{-1}s^2$	f_2 MHz	C $10^{-15}m^{-1}s^2$
Toluene	140	13	100	60	6
Cyclohexa-none	228	14.3	85	72.9	1.4
Acetone	117	15	81	70	9
Methanol	2040	0.8	100	53	12

In Fig. 2 the dependence of the ultrasonic velocity on the polymerization time is represented. At every moment c provides an information about the polymerization state.

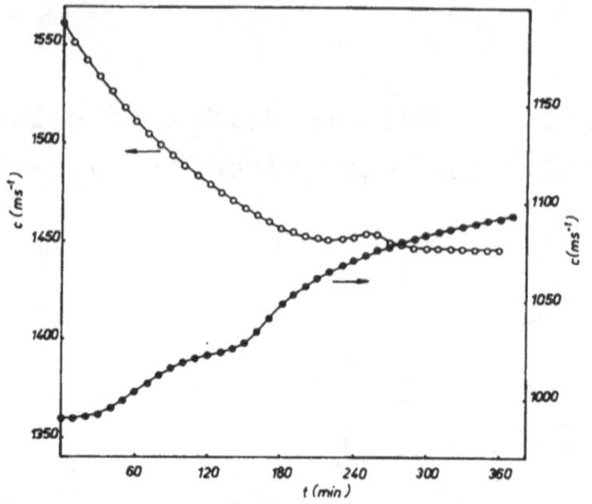

Fig. 2. Velocity c as a function of time t in a semicontinuous-polymerization (○) and a solution-polymerization of VAC (●).

Discussion

The acoustic behaviour of PVAC in various solvents (except methanol) can be explained with a two-relaxation model. A very localized HF-process and a more cooperative LF-process cause the attenuation. PCILO-calculations (3) of units similar this polymers gave rise to the assumption that the transitions are in the range of simple trans-gauche transitions according the single-bond-motion (4). The influence of the solute on the conformational energy is less than 20 %. PVAC dissolved in methanol shows in the low frequency range a broad spectrum caused by normal mode processes or even entanglements. The same sample but in dispersion was investigated during the polymerization. The velocity provides important information on the conversion, the distribution of monomer, the influence of stabilizer on the properties and others (5). The application of this method provides a very good tool in detecting the polymerization regime and its wanted controlling.

References

1. Hauptmann, P., Alig, I., Säuberlich, R.: Acustica 50, 189 (1982).

2. Säuberlich, R., Hauptmann, P.: Exp. Techn. Physik 29, 21 (1981).

3. Alig, I.: Dissertation (Merseburg) 1983.

4. Helfand, E., Wassermann, Z. R., Weber, T. A.: Macromolecules 13, 526 (1980).

5. Barett-Gültepe, M. A., Gültepe, M. A., Yeager, E. B.: J. Phys. Chem. 87, 1039 (1983).

THEORY OF LIGHT SCATTERING FROM SYSTEMS OF INTERACTING SPHERICAL MACROMOLECULES IN SOLUTION

Rudolf Klein, Gerhard Nägele and Walter Hess

Fakultät für Physik der Universität Konstanz
D-7750 Konstanz, Federal Republic of Germany

I. Introduction

Suspensions of charged spherical particles have in recent years been investigated by static and quasielastic light scattering [1] and by neutron scattering [2,3,4]. These experiments were carried out on a number of macromolecules, which are of interest for biophysics, on colloids and on micellar systems. These systems can for many purposes be characterized as charged hard spheres surrounded by a cloud of counter-ions and of additional small ions if salt has been added to the dispersion. The macroions are large compared to the small ions and to the molecules of the solvent. The role of the solvent is to provide a dielectric constant ε for the charged particle interactions and a friction for the motion of the macroions.

The first part of this paper deals with static properties of charged suspensions. The local order, as it is reflected in the static structure factor and the radial distribution function, is investigated. Here, our objective is to take the finite size of the small ions into account and to explore the effects of various types of polydispersity of the macroions on the static structure factor for dispersions which might be typical for micellar systems. In the second part some results for dynamical properties, such as the dynamic structure factor $S(k,t)$, are reviewed. It will be shown that a monodisperse system of interacting spherical macromolecules has a non-exponential $S(k,t)$ outside the extreme forward direction, which arises from the viscoelastic behavior of the system.

Physical Optics of Dynamic Phenomena and Processes in Macromolecular Systems

II. Static Properties of a System of Charged Spherical Particles in Solution

A frequently used model for static and dynamic properties of dispersions is the so-called one-component macrofluid (OCM) [5]. Within this model the effects of the small ions are summarily taken into account by introducing a screening for the Coulomb potential between macroions, leading to the famous Derjaguin-Landau - Verwey-Overbeek (DLVO) potential, whose repulsive part is given by

$$U_{DLVO} = \frac{(Ze)^2}{\varepsilon(1+\kappa_D d/2)^2} \frac{e^{-\kappa_D(r-d)}}{r} \tag{1}$$

where d is the macroion diameter, Ze its charge, and κ_D is the Debye screening parameter

$$\kappa_D^2 = \frac{4\pi\beta e^2}{\varepsilon} \sum_i n_i z_i^2 \tag{2}$$

Here, n_i is the number density, $z_i e$ the charge of the i-th particle species and the sum is over all species of counterions and salt ions, ε is the dielectric constant of the solvent and $\beta = (kT)^{-1}$. Eq. (1) is the asymptotically exact solution of the linearized Poisson-Boltzmann equation under the assumption of point-like salt ions.

Within the Mean-Spherical-Approximation (MSA) Hayter and Penfold [5] were able to derive analytic expressions for the static structure factor S(k) and the radial distribution function g(r) for the OCM as long as the concentration is not too small. An improvement, which also covers low concentrations and high charges, was introduced by Hansen and Hayter [6], called the rescaled MSA. In order to get information about the distribution of the small ions around a given central macroion and the influence of the finite size of the small ions, one has to treat all ions on the same footing. For this reason we have used the primitive model (PM) of electrolyte theory for our highly asymmetric system. All species are treated as charged hard spheres with number densities n_i, valency z_i and diameter d_i; the solvent is only taken into account by providing a dielectric constant ε. Then the interaction potentials are, instead of (1), pure Coulomb potentials

$$u_{ij}(r) = \begin{cases} \infty & , \quad r < \dfrac{d_i+d_j}{2} \\ \dfrac{z_i z_j e^2}{r} & , \quad r > \dfrac{d_i+d_j}{2} \end{cases} \qquad (i,j = 1,\ldots,m) \tag{3}$$

Another reason for choosing the PM is the possibility to treat mixtures of macroions in order to understand the effects of polydispersity in scattering experiments. There are different kinds of polydispersity which have different effects on the measured static structure factor, namely scattering polydispersity, size polydispersity and charge polydispersity. The treatment presented here is restricted to systems of moderate to high macroion concentrations and rather low macroion charges where rescaling in the sense of Hansen and Hayter is unnecessary.

In the m-component system under study there are m(m+1)/2 (partial) radial distribution functions $g_{ij}(r)$ and an equal number of partial structure factors $S_{ij}(k)$. The basis for their calculation are the Ornstein-Zernike equations

$$h_{ij}(r) = c_{ij}(r) + \sum_{k=1}^{m} n_k \int d^3r' \; c_{ik}(|\underline{r}-\underline{r}'|) \; h_{kj}(r') \tag{4}$$

where $h_{ij}(r) = g_{ij}(r) - 1$, and $c_{ij}(r)$ denotes the direct correlation functions. The MSA consists in the following approximation for the direct correlation function:

$$c_{ij}^{MSA}(r) = -\beta \, u_{ij}(r) \quad ; \quad r > d_{ij} = \frac{1}{2}(d_i+d_j) \tag{5}$$

Since $g_{ij}(r) = 0$ for $r < d_{ij}$, eq. (4) together with (5) determines c_{ij} for $r < d_{ij}$ and, more important, g_{ij} and h_{ij} for $r > d_{ij}$. From the solution of the system (4) and (5) one gets the partial structure factors by

$$S_{ij}(k) = \delta_{ij} + (n_i n_j)^{1/2} \, h_{ij}(k) \tag{6}$$

where $h_{ij}(k)$ is the Fourier transform of $h_{ij}(r)$.

The $S_{ij}(k)$ determine the intensity $I(k)$ of scattered light, x-rays or neutrons through

$$I(k) \propto \sum_{i,j=1}^{m} (N_i N_j)^{1/2} f_i(k) f_j(k) S_{ij}(k) \tag{7}$$

where N_i is the total number of particles of species i, and

$$f_i(k) = f_i(0) b_i(k) \tag{8}$$

denotes the scattering amplitude of a particle of species i. The form factor $b_i(k)$ for a homogeneous spherical particle is

$$b_i(k) = \frac{3j_1(kd_i/2)}{kd_i/2} \tag{9}$$

where j_1 is the first spherical Bessel function.

The procedure to solve the equations (4) and (5) in MSA follows earlier work by Blum [7], Blum and Hoye [8] and Hiroike [9] and will be published elsewhere together with a systematic representation of the results. Here, some of them will be presented and discussed.

(a) Influence of the finite size of the small ions

$S_{ij}(k)$ and $g_{ij}(r)$ are calculated from a three-component PM. The first species are the macroions of valency z_1, number concentration n_1 and hard-sphere diameter d_1. The counterions and the ions of an additional 1-1 electrolyte form the other two components. We assume for simplicity that all small ions have the same diameter d_{sm} (which is varied in the calculation) and that they carry one elementary charge ($|z_2|=|z_3|=1$).

First we consider the case without salt. Fig. 1 shows the $g_{ij}(r)$ for a situation which might be typical for a micellar system. The correlations between the macroions are given by g_{11} and the counterion distribution around a central macroion is described by g_{12}. Due to the attractive interaction the counterions accumulate in the extreme neighborhood of the central macroion, the maximum of g_{12} being at $r/d_1 = 0.55$ as expected if

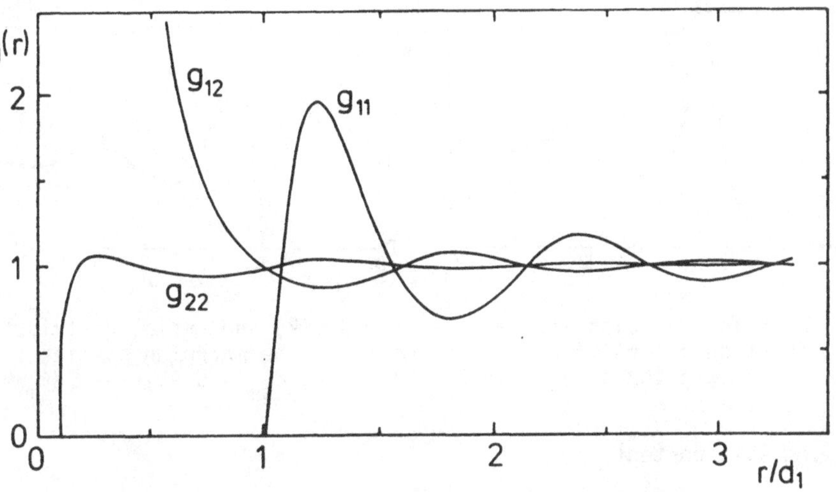

Fig. 1. Macroion-macroion (g_{11}), macroion-counterion (g_{12}) and counterion-counterion (g_{22}) radial distribution functions of a two-component PM with number concentration $n_1 = 4.6 \times 10^{18} cm^{-3}$, valency $z_1 = -20$ and hard-sphere diameter $d_1 = 50$ Å of the macroions. The counterions diameter d_{sm} is set equal to 5 Å.

$d_{sm} = 5$ Å . A shallow minimum of g_{12} appears at the position of a "shell" of nearest neighbor macroions, which is followed by a smaller maximum arising from the correlations between the macroions.

In Fig. 2 the structure factor $S_{11}(k)$ and the radial distribution function $g_{11}(r)$ of the macroions are shown for point-like counterions and for $d_{sm} = 5$ Å . As expected, S_{11} and g_{11} exhibit somewhat stronger structure for $d_{sm} = 5$ Å, since finite size small ions are not as effective in screening the interaction among macroions as point-like ions. This excluded volume effect is less pronounced at smaller number densities n_1 of the macroions.

The influence of the finite size of the small ions is of particular importance if one adds salt. At the rather large total volume concentrations of the systems in Figs. 1 and 2 the hard-core part of the interaction more and more determines S_{11} and g_{11} . Fig. 3 shows the effect of added salt relative to the salt-free case. The structure in $S_{11}(k)$ is being reduced and the increase at k=0 reflects the increased osmotic compressibility with

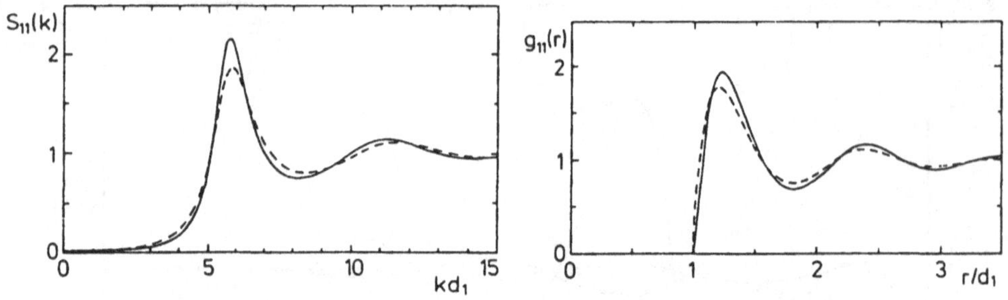

Fig. 2. Macroion-macroion structure factor $S_{11}(k)$ and radial distribution function $g_{11}(r)$ of a two component PM. The macroion species parameters are the same as in Fig. 1. (——) d_{sm} = 5 Å; (---) d_{sm} = 0 Å.

increasing salt content.

(b) Binary mixtures of macroions

So far the scattered radiation was assumed to arise from a single species of macroions (denoted by the index 1). It is an advantage of the multicomponent primitive model to generalize easily to the case of several species of macroions scattering radiation with different scattering powers f_α . The measured structure factor is of the form of eq. (7) and can be defined as

$$S^M(k) = \frac{1}{N\overline{f^2}} \sum_{\alpha,\beta} (N_\alpha N_\beta)^{\frac{1}{2}} f_\alpha f_\beta S_{\alpha\beta}(k) \tag{10}$$

where

$$\overline{f^2} = \frac{\sum_\alpha N_\alpha f_\alpha^2}{\sum_\alpha N_\alpha} \tag{11}$$

The normalization in (10) is chosen such that $S^M(k) = 1$ for noninteracting particles. For the case of pure scattering polydispersity one assumes that all macroions behave identically with regard to their structural properties but that there are two types of particles differing in their scattering power ($f_1 \neq f_2$). Then $g_{\alpha\beta} = g$ for all α and β and (10) reduces to

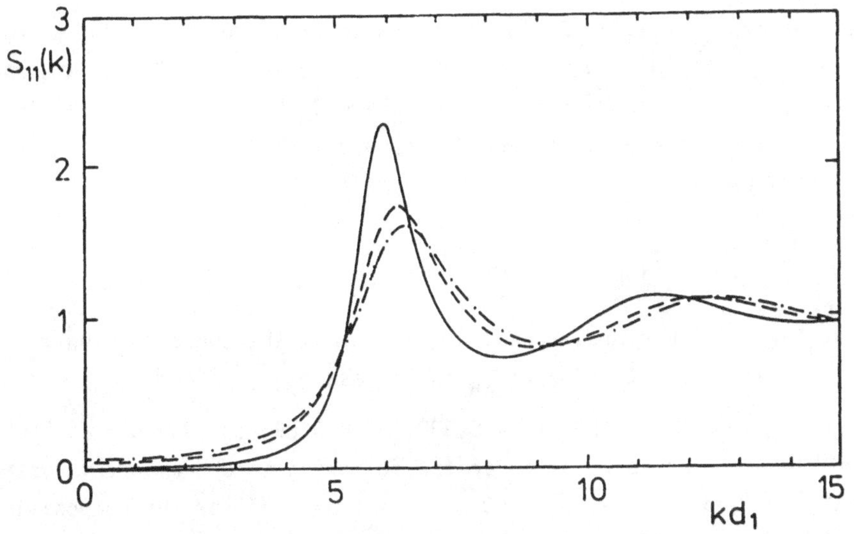

Fig. 3. Structure factor $S_{11}(k)$ of the macroions for a three-component PM, which models a system with added salt. $n_1 = 5.0\times10^{18}cm^{-3}$, $z_1 = -20$, $d_1 = 50$ Å and $d_{sm} = 5$ Å . n_3 is the cation respectively anion number concentration of the added 1-1 electrolyte. (——) $n_3/n_1 = 0$, that is without added salt; (---) $n_3/n_1 = 50$; (-··-) $n_3/n_1 = 100$.

$$S^M(k) = y\, S^{id}(k) + (1-y) \equiv S^D(k) \qquad (12)$$

where

$$y = \bar{f}^2 / \overline{f^2} \qquad \text{and} \qquad \bar{f} = (\sum_\alpha N_\alpha f_\alpha)/(\sum_\alpha N_\alpha) \qquad (13)$$

$S^{id}(k)$ in (12) denotes the structure factor of an effective one-component system. Whereas the simple result (12) is exact for pure scattering polydispersity it has been proposed to use it also as an approximation for size polydispersity [10]. The usefulness of this decoupling approximation for charged systems is now being tested.

In the following we restrict ourselves to a binary mixture and assume both species to have equal refractive indices. As long as d_i is small compared to the wavelength of light we have f_α = const x d_α^3 with the same constant for both species (α = 1,2). Defining the ratio of diameters $\lambda = d_1/d_2$ with $d_1 \leqslant d_2$ and $x_i = N_i/(N_1 + N_2)$, the quantity $y(x_2)$ in eq. (13) has its minimum at the concentration $x_2^* = \lambda^3/(1+\lambda^3)$, where it has the value $y(x_2^*) \equiv y^* = 4\lambda^3/(1+\lambda^3)^2$. The ratio of incoherent to coherent static scattering is given from eq. (12) as

$$I_{inc}/I_{coh} = R/S^{id}(k) \tag{14}$$

with $R = (1-y)/y$. Therefore, R has its maximum at the concentration $x_1^* = 1 - x_2^*$, so that the ratio I_{inc}/I_{coh} is largest at x_1^*.

We have used $\lambda = 0.8$, so that $x_1^* = 0.339$, $y^* = 0.896$ and R = 0.116. The expression (10) is then calculated for the following three-component systems: $n_1 = 3\times10^{18}cm^{-3}$, d_1 = 40 Å, d_2 = 50 Å, $z_1 = z_2 = -15$ for the compositions n_1/n_2 = 1; 0.3 and 0.05. In addition there are corresponding numbers of univalent counterions with d_{sm} = 4 Å. These results are compared with $S^D(k)$, eq. (12), where $S^{id}(k)$ is the structure factor $S_{11}(k)$ of a two-component system, which consists of macroions of number concentration (n_1+n_2), valency z = - 15 and a volume fraction equal to the volume fraction of the two species of macroions in the real system, that is $\eta = \pi(n_1d_1^3+n_2d_2^3)/6$. The diameter $\bar{d}$ of the macroions in the effective system is then given by $(n_1+n_2)\bar{d}^3 = (n_1d_1^3+n_2d_2^3)$. The second component of the effective system consists of neutralizing counterions with d_{sm} = 4 Å.

It is found that $S^D(k)$ is a very good approximation to $S^M(k)$ near the main maximum of the structure factor. But near k=0 large discrepancies are found in particular at higher relative concentrations n_2/n_1; $S^D(0)$ overestimates the true structure factor $S^M(0)$. Fig. 4 shows the results for $n_2 = n_1$. This failure of the decoupling approximation at small wavevectors has already been noticed by Vrij [11] and by Pusey, Fijnaut and Vrij [12] within the Percus-Yevick approximation for hard-core systems. We have included the corresponding hard-sphere results in Fig. 4, since in this case the MSA treatment is identical to the Percus-Yevick approximation. It is seen that the slight difference between $S^D(k)$ and $S^M(k)$ for the hard-core system near k_{max} is reduced by charging up the particles.

For the investigation of charge polydispersity effects we have taken

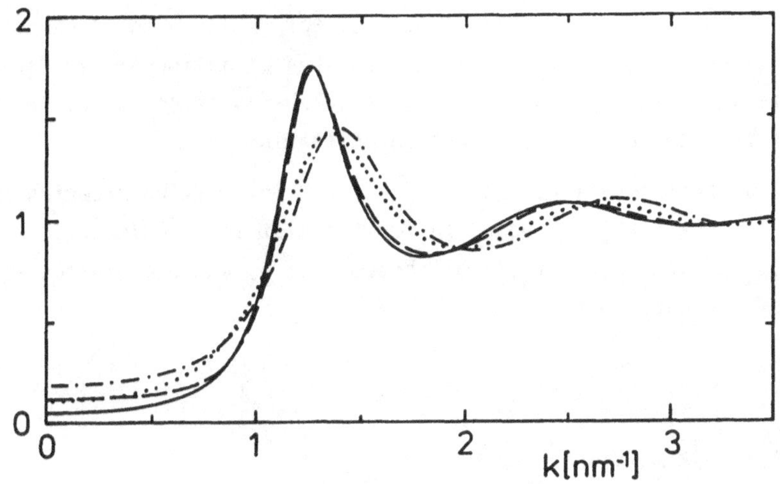

Fig. 4. Structure factors $S^M(k)$ and $S^D(k)$ of a binary mixture of macroions of composition $n_2/n_1=1$, together with the corresponding functions of the equivalent uncharged system. The parameters are $n_1 = 3\times10^{18}$ cm^{-3}, $d_1 = 40$ Å, $d_2 = 50$ Å, $z_1=z_2= -15$ and $d_{sm} = 4$ Å. (——) $S^M(k)$; (···) $S^M(k)$ of the equivalent uncharged system; (---) $S^D(k)$; (-··-) $S^D(k)$ of the equivalent uncharged system.

$n_1 = n_2$, $z_1 = -15$, a fixed diameter ratio $\lambda = 0.8$ and have varied the charge ratio from $z_2/z_1 = 0.66$ to 1.33 and have compared the results for $S^M(k)$ and $S^D(k)$ for effective systems with $\bar{z} = (z_1+z_2)/2$. It is found that $S^D(k)$ for k values near k_{max} is still an acceptable approximation to $S^M(k)$ within the chosen parameter intervals. With regard to the long-wavelength region the gross discrepancy remains; in addition $S^M(0)$ changes from 0.075 to 0.030 in varying z_2/z_1 within the above interval, whereas $S^D(0) \simeq 0.11$ for all charges $\bar{z}$.

III. Results for Dynamical Properties

Photon correlation experiments on polystyrene spheres [13,14,15] detect fluctuations of the system on the scale which is of the same order as the average distance between macroions. Therefore, the scattering experiments probe the system on the length scale of its short-range structure and detect the temporal changes of this structure. For this reason it cannot be expected that macroscopic hydrodynamic equations of motion are sufficient for a theoretical description of the dynamical behavior of macromolecular suspensions as seen in light and neutron scattering.

The basic transport equation, appropriate for N interacting Brownian particles, is the Fokker-Planck equation for the distribution function $f(\Gamma;t) \equiv f(\underline{p}_1,\ldots,p_N,\underline{r}_1,\ldots,\underline{r}_N;t)$ of the momenta $\underline{p}_i$ and coordinates $\underline{r}_i$, $i=1,\ldots,N$ of the macroions

$$\frac{\partial}{\partial t} f(\Gamma;t) = \hat{\Omega} f(\Gamma;t) \tag{15}$$

where the Fokker-Planck operator is given by

$$\hat{\Omega} = -\sum_i \{\frac{1}{m} \underline{p}_i \cdot \frac{\partial}{\partial \underline{r}_i} + \underline{F}_i \frac{\partial}{\partial \underline{p}_i}\} + \zeta_o \sum_i \frac{\partial}{\partial \underline{p}_i} \cdot (k_B T \frac{\partial}{\partial \underline{p}_i} + \frac{1}{m} \underline{p}_i). \tag{16}$$

Here, $\underline{F}_i$ is the force acting on particle i by all other macromolecules and ζ_o is the one-particle friction coefficient of the macroions. Hydrodynamic interactions can be neglected for the highly charged and rather dilute systems of polystyrene spheres. This is not justified for more concentrated and less charged systems.

Various collective and one-particle properties have been calculated on the basis of eq. (15) by Hess and Klein [16]. Here, we will only mention some of the results, which are of particular importance for dynamic light scattering. The dynamic structure factor $S(k,t)$ or its Laplace transform $\tilde{S}(k,z)$ can be calculated with the help of the projection operator formalism of Mori and Zwanzig [17]. In the experimentally accessible time regime of the correlator one obtains

$$\tilde{S}(k,z) = \frac{S(k)}{z + \dfrac{c_T^2(k)k^2 m}{\tilde{\zeta}(k,z)}} \tag{17}$$

Here, $S(k) = S(k,t=0)$ is the static structure factor, $c_T^2(k) = k_B T/m\, S(k)$ and $\tilde{\zeta}(k,z)$ is a generalized dynamic friction function, which generalizes the friction coefficient ζ_o of a single macroion to a frequency and wave-vector dependent function $\tilde{\zeta}(k,z)$. This quantity is found to consist of two parts, namely the one-particle friction due to the presence of the solvent and a dynamical part which arises from the direct interactions between the moving macroions:

$$\tilde{\zeta}(k,z) = \zeta_o + \frac{k^2}{c} \tilde{\eta}(k,z) \tag{18}$$

Here, c is the concentration of macroions and $\tilde{\eta}(k,z)$ is the Laplace transform of the longitudinal dynamic viscosity of the interacting macroions.

One of the most interesting results of the above mentioned light scattering experiments on nearly monodisperse samples is the fact that outside the extreme forward direction the dynamic structure factor S(k,t) cannot be represented by a single exponential function of time. From our general result (17) it is evident that this behavior is due to the z-dependence of $\tilde{\zeta}(k,z)$; the deviations of S(k,t) from a single exponential therefore arise from the viscoelastic properties of the suspension.

This has been checked by calculating the dynamic longitudinal viscosity function $\eta(k,t)$. The Mori-Zwanzig formalism provides an expression for $\eta(k,t)$ in the form of a time-dependent correlation function. This function has been evaluated [16] by using the mode-coupling approximation [17]. The result is

$$\eta(k,t) = \frac{2\beta c^2}{(2\pi)^3} \int d^3k' \, \mathfrak{F}[S(k')] \, S(\tfrac{1}{2}\,\underline{k}+\underline{k}',t) \, S(\tfrac{1}{2}\,\underline{k}-\underline{k}',t) \tag{19}$$

Here, $\mathfrak{F}[S(k)]$ is a known function of the static structure factor and S(k,t) is the dynamic structure factor, whose Laplace transform is expressed in terms of $\tilde{\eta}(k,z)$ in eq. (17) and eq. (18). Therefore, eqs. (17) to (19) form a closed set of equations which was solved numerically, where one takes S(k) either from a static light scattering experiment or from a theory of the type described in the first part of this paper.

For a comparison with experiment we have proceeded in the following way: From the experimentally determined S(k,t) one can define a mean-relaxation time through

$$\tau(k) = \frac{1}{S(k)} \int_0^\infty S(k,t)\, dt \qquad (20)$$

If $\mu_1(k) = D_o k^2/S(k)$ denotes the first cumulant of $S(k,t)$, where D_o is the diffusion coefficient of a non interacting particle, then $\Delta(k) = \{\mu_1(k) - \tau^{-1}(k)\}/\mu_1(k)$ is an appropriate measure of the deviation of $S(k,t)$ from a single exponential in time. On the other hand, from eqs. (17), (18) and (20) $\tau(k)$ can also be expressed as

$$\tau(k) = [\zeta_o + \frac{1}{c}\, \eta(k)\, k^2]/(c_T^2\,(k)\, k^2)$$

where

$$\eta(k) = \int_0^\infty \eta(k,t)\, dt$$

is the (k dependent) longitudinal viscosity. Using the numerical solution of the self-consistent set of eqs. (17) and (19), one obtains a theoretical result for $\Delta(k)$. The agreement with the light scattering data of Grüner and Lehmann [15] is within 20 % , see ref. [16]. It should be noted that this agreement is obtained without the fitting of an adjustable parameter.

From this discussion it becomes clear why the measured correlation functions in QELS are non-exponential at finite k. The concentration fluctuations, which give rise to light scattering, are coupled to stress fluctuations of the highly correlated system of macroions. The stress fluctuations reflect the non-trivial viscoelastic behavior of the system, as described by $\eta(k,t)$. If $S(k,t)$ were a single exponential, $\tau(k)$ would be equal to the inverse of the first cumulant and $\Delta(k)$ would vanish.

The fact that the suspension exhibits viscoelasticity, leads to elastic behavior, if the system is probed on a sufficiently short time scale. The corresponding "high frequency elastic constants" have been calculated [18]. It is found that the longitudinal constant varies as the square of the concentration c, whereas the shear elastic constants increases with $c^{4/3}$. Finally it is pointed out that various other physical properties characterizing the dynamical behavior of suspensions of charged spherical macromolecules have been calculated on the same basis (16,18,19,20).

IV. Conclusion

The aim of this paper has been to demonstrate that structural and dynamical properties of suspensions of charged spherical macromolecules can be described and understood on a microscopic basis. It has been shown that in fairly concentrated suspensions like micellar systems the role of the small ions should be taken into account for the calculation of S(k). Furthermore, various polydispersity effects have been investigated for static properties of charged hard spheres. With regard to the dynamical structure factor it was found that even a monodisperse system of charged hard spheres can in general not be described by a single exponential function, since at finite scattering vectors the viscoelastic stress fluctuations give rise to a more complicated behavior.

References

1. Pusey, P.N., Tough, R.J.A., in: Dynamic Light Scattering and Velocimetry: Applications of Photon Correlation Spectroscopy, R. Pecora, ed., Plenum, New York 1984.
2. Hayter, J.B., Penfold, J., J. Chem. Soc. Faraday Trans. 1, 77,1851 (1981).
3. Kalus, J., Hoffmann, H., Reizlein, K., Ulbricht, W., Ibel, K., Ber. Bunsenges. Phys. Chem. 86, 37 (1982).
4. Cebula, D.J., Goodwin, J.W., Jeffrey, G.C., Ottewill, R.H., Parentich, A., Richardson, R.A., Faraday Disc. No. 76, 37 (1983).
5. Hayter, J.B., Penfold, J., Mol. Phys. 42, 109 (1981).
6. Hansen, J.P., Hayter, J.B., Mol. Phys. 46, 651 (1982).
7. Blum, L., Mol. Phys. 30, 1529 (1975).
8. Blum, L., Hoye, J.S., J. Phys. Chem. 81, 1311 (1977).
9. Hiroike, K., Mol. Phys. 33, 1195 (1977).
10. See the contribution by Pusey, P.N., in Faraday Dis. No. 76, 93 (1983).
11. Vrij, A., J. Chem. Phys. 69, 1742 (1978).
12. Pusey, P.N., Fijnaut, H.M., Vrij, A., J. Chem. Phys. 77, 4270 (1982).
13. Brown, J.C., Pusey, P.N., Goodwin, J.W., Ottewill, R.H., J. Phys. A 8, 664 (1975).
14. Dahlberg, P.S., Boe, A., Strand, K.A., Sikkeland, T., J. Chem. Phys. 69, 5473 (1978).

15. Grüner, F., Lehmann, W., J. Phys. A 12, L 303 (1979); 15, 2847 (1982).

16. Hess, W., Klein, R., Adv. Phys. 32, 173 (1983).

17. See, for instance, Boon, J.P., Yip, S., Molecular Hydrodynamics, McGraw Hill, New York, 1980.

18. Klein, R., Hess. W., Faraday Disc. No. 76, 137 (1983).

19. Hess, W., Klein, R., J. Phys. A 15, L 669 (1982).

20. Klein, R., Hess, W., in: Ionic Liquids, Molten Salts and Polyelectrolytes, K.H. Bennemann, F. Brouers, D. Quitmann, eds., Lecture Notes in Physics, vol. 172, Springer-Verlag, Berlin, 1982.

COOPERATIVE GROWTH OF MOLECULAR AGGREGATION IN SEMI-CONCENTRATED PBLG SOLUTIONS INVESTIGATED BY THE DYNAMIC ELECTRO-OPTICAL METHOD

Hiroshi WATANABE, Tatsuo NAKANO, and Youjiro FUKUDA

Department of Chemistry, College of Arts and Sciences,
University of Tokyo, Komaba, Meguroku, Tokyo, JAPAN

In order to elucidate the mechanism of transition from semi-concentrated solution to a lyotropic liquid crystalline phase, molecular aggregation of PBLG in semi-concentrated solutions was investigated by dynamic electro-optical methods. Molecular weights of PBLG used are (2.1, 10.3, 18.6, and 28.0)x10000 (sample I, II, III, and IV) and the solvents used are m-cresol(good), 1,2-dichloroethane(fairly good), dioxane(poor), and benzene(very poor). Electric fields applied are rectangular pulses with amplitude up to 20 KV and pulsed sine waves with amplitude 4 KV(p-p). The frequency range available in this experiment was from 200 Hz to 20 KHz.

For the DC fields (rectangular pulse) and the low frequency AC fields, the electric birefringence is positive and oscillates with a frequency of 2ω where ω is the angular frequency of the AC field. On increasing the ω, the amplitude of the oscillating signals decreases as expected in the permanent dipole orientation. However, the sign of the static component of the birefringence signal changes from positive to negative on further increase of the ω, and the absolute value of the signals increases on increasing value of ω. A

Physical Optics of Dynamic Phenomena and Processes in Macromolecular Systems

typical example is shown in Fig. 1, in which the ○ are the amplitude of oscillation and the ▲ are the static values. The negative birefringence appears beyond some critical concentration which depends on the molecular weight and the nature of solvents. Such an inversion of the birefringence signals from the positive to the negative in semi-concentrated PBLG solutions in m-cresol has already been reported by Mori et al(1) by using weak fields. In this work we applied high field which enabled us to observe the field free relaxation after the removal of the pulsed fields. A birefringent plate, its optical axis parallel to the field direction, was placed in-between the polarizer and the analyzer in order to check the sign of the signals. A typical birefringence signal observed in high frequency region is shown in Fig. 2a. The signal is positive just after the application of a high frequency AC pulse but goes down slowly to negative and reaches to a steady state. Just after the annihilation of the AC pulse, the signal down shoots first and then

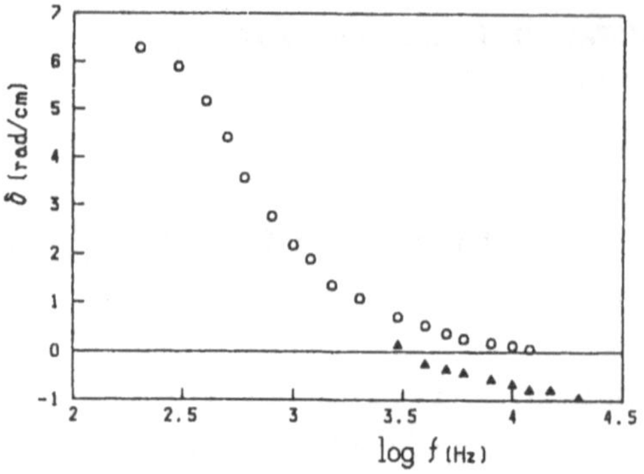

Fig. 1. Angular frequency dependencies of AC (○) and DC (▲) components of EB.

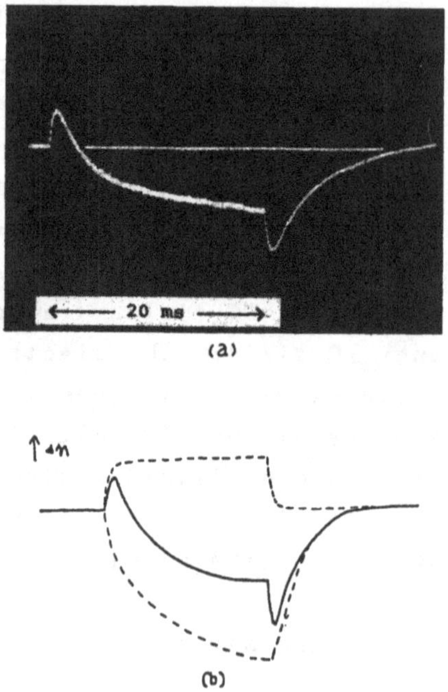

Fig. 2. An EB signal and its decay for 20 KHz AC pulse, and (b) its decomposition into positive and negative components.

gradually decays. The relaxation is very slow. Fig. 2b presents how the complicated birefringence signal in Fig. 2a can be interpreted as a superposition of a fast positive signal and a slow negative signal.

The remarkable feature of the negative birefringence in the high frequency is a very slow relaxation after the removal of the AC field. The field free relaxation of the negative birefringence expressed as the normalized retardation δ/δ_0, where δ_0 is the phase difference between normal and abnormal light and δ is the value at t=0, are compared with the ones of positive birefringence for DC and 200 c/s AC fields in Fig. 3a. The decay for AC field is obtained by a sudden annihilation of the field at a peak of oscillating signal. In Fig. 3b the relaxations of positive birefringence for several values of ω are compared. It is clear that the relaxation is the faster the larger the value of ω.

These results strongly suggest that a sort of large aggregation is formed in the semi-concentrated solutions and the aggregates orient under the high frequency AC field.

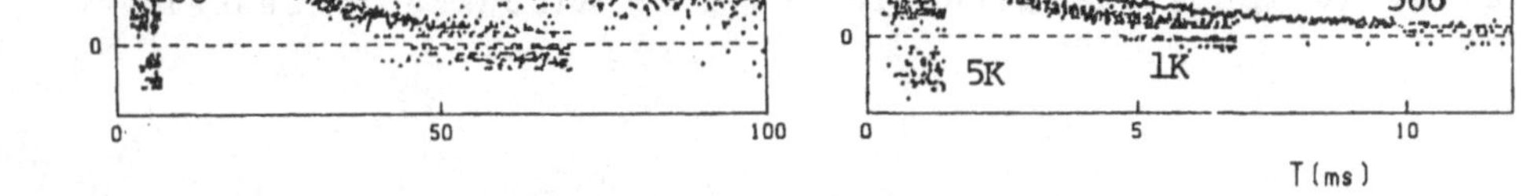

Fig. 3. Normalized decay curves of the EB in PBLG(IV)/EDC solution. c=70g/dm^3. Frequencies of the AC field and DC are denoted on each curve. EB is negative for 20 KHz and the others are positive. Note that the time scale in (a) and (b) is different. Inset in (a) is a close observation of the initial part of negative EB by a fast sampling.

The orientation mechanism is of induced dipole moment type and the direction of the orientation is perpendicular to the field, as suggested by the negative birefringence. The perpendicular orientation of PBLG helix is confirmed by the measurement of the electric dichroism. That is, the sign of dichroism at 257nm, the middle of absorption band of side chain benzyl group, is negative at the DC and low frequency AC fields but positive at high frequency AC fields. The absolute value of the electric dichroism is very small, however, indicating that the movement of benzyl groups on the side chains is relatively free with respect to the back bone helix.

The negative birefringence in high frequency region obeys the Kerr's law in the range of observation. From the concentration dependence of the Kerr constant of the negative birefringence for sample IV/m-cresol solutions we could estimate the threshold concentration of this specific polymer-solvent system is about $14g/dm^3$ beyond that the negative birefringence appears.

The results suggest a cooperative growth of the aggregates in pre-transition region of semi-concentrated PBLG solutions, consequently a new mechanism for lyotropic liquid crystalline phase formation which is different from the relevant explanations. The volume of the aggregates estimated from the Kerr constant and the decay and possible origins of the induced dipole moment which orients the helix perpendicular to the field direction will be discussed elsewhere.

References

1. Mori, Y., Ookubo, N., Hayakawa, R., Wada, Y.:J. Polym. Sci., Polym. Phys. Ed., 20, 2111 (1982)

LIGHT SCATTERING ANOMALY OBSERVED IN DILUTE SOLUTIONS OF POLY (VINYL ALCOHOL) AGED AT HIGH CONCENTRATION

Libuše Mrkvičková, Čestmír Koňák, Blahoslav Sedláček

Institute of Macromolecular Chemistry, Czechoslovak Academy of Sciences, 162 06 Prague 6, Czechoslovakia

Introduction

When studying the ageing of poly(vinyl alcohol) by integral light scattering (ILS) we observed an anomaly consisting in a strong decrease in ILS intensity at lower angles. This phenomenon was first reported by Doty and Steiner (1). It was recognized to be characteristic of the strict "liquid-like type" structure (2). Under such conditions, the effect is caused by ordering of particles (leading to a considerably large excluded volume) and not by the aggregation itself - this would indicate an opposite effect. Therefore, we decided to try to elucidate this problem by using also other LS methods.

Experimental

Samples. After dissolution (5 hours at 80°C) of poly(vinyl alcohol), PVA (BHD Chemical LTD., England; $M_w = 22\ 300$), solutions of concentrations c = 1.6, 6.4, 10, 16% w/v were left to age at 25°C. Before measurements the samples were diluted considerably (≲ 1% w/v). Aqueous 0.1 N KCl was used as solvent.

Methods. ILS measurements were performed with a Fica 40 000 apparatus (vertically polarized primary beam of wavelength $\lambda = 546.1$ nm, angle interval 30-150°). For quasielastic light scattering (QELS) measurements a homodyne spectrometer with 96 channel correlator (He-Ne laser, $\lambda = 632.8$ nm, interval 18-

Physical Optics of Dynamic Phenomena and Processes in Macromolecular Systems

105°) was used. Turbidity was measured with a Perkin-Elmer-Hitachi UV-VIS spectrometer 340 (λ = 435.8, 546.1, 684,3 nm). For the characterization of supermolecular particles (SMP), both integral (ITR) and differential (DTR) turbidity ratio methods were applied (3). The solution were purified with a Beckman L8-55 ultracentrifuge (rotor SW.1, 15 000 r.p.m., 1 hour).

Results and Discussion

The minimum on the reciprocal particle scattering function $P^{-1}(\Theta)$ was observed by ILS if the solution was aged at a higher concentration than the critical one, $c = 3.3\times10^{-2}$ gcm^{-3}, i.e., when the space was filled with molecularly dispersed PVA (4). The magnitude and position of this extreme are practically stable in time and depend on the given concentration after dilution ($\lesssim$1% w/v)(Fig.1). For this reason the dissymmetry of

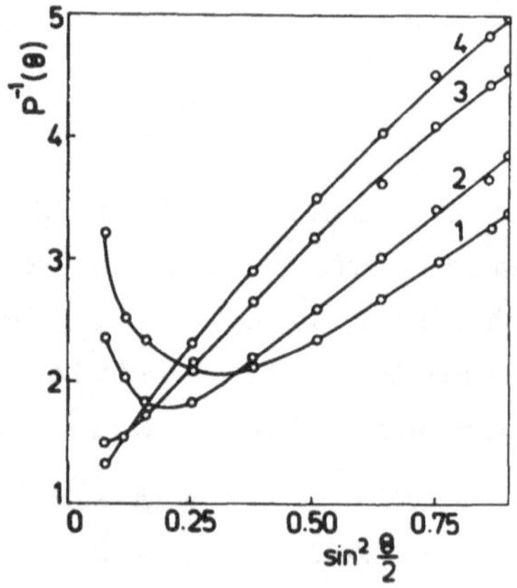

Fig. 1. Function $P^{-1}(\Theta)$ at concentration 1 - 0.0128, 2 - 0.0064, 3 - 0.0032, 4 - 0.0008 gcm^{-3}. PVA aged for 3 weeks in 16% solution.

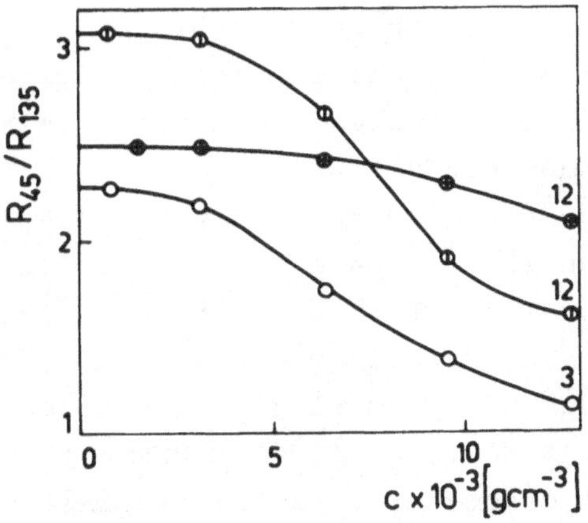

Fig. 2. Dissymmetry R_{45}/R_{135} vs. concentration for aged PVA in o 16, ⊕ 10, ⊕ 6.4% solutions. Curve labels correspond to the time of ageing in weeks.

reduced intensity of scattered light (R_Θ) is also concentration dependent (Fig.2). The higher the concentration during ageing, the deeper the minimum on the angular dependence of Kc/R_Θ

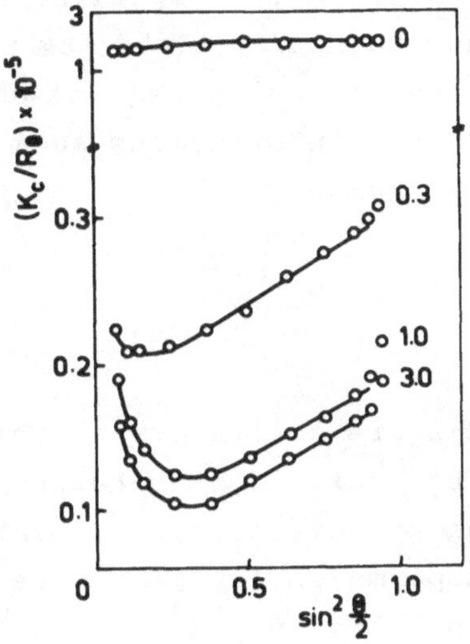

Fig. 3. Angular dependence of Kc/R_Θ values for aged PVA in 16% solution after 0, 0.3, 1, 3 weeks; c = 0.0128 g cm^{-3}.

values (Fig.1) which are shifted to higher angles. The dependence on the ageing time manifests itself similarly (Fig.3). This suggests an interparticle interference in the partially ordered system under study.

During the ageing mass of dispersed particles (M_w^p) and their radius of gyration $(\bar{S}^2)_z^{1/2}$ have increased, i.e., strong aggregation took place. The QELS method characterized the aggregates thus formed from hydrodynamic viewpoint (R_H - hydrodynamic radius) as a system of SMP moving in the molecularly dispersed PVA in aqueous 0.1N KCl. The ITR and DTR methods have proved that only a small fraction of particles presented in the system are responsible for the effect observed. For example, the concentration of SMP formed in a 16% solution after 3 weeks was approximately a hundred times lower than the total concentration in a diluted solution. Parameters of PVA in the system are: $M_w^p = 2.4\times10^6$, $(S^2)_z^{1/2} = 75$ nm, $(R_H)_z = 62$ nm, $M_w^p/M_n^p = 1.5$. The ratio $(R_H)_z/(\bar{S}^2)_z^{1/2} = 0.84$ corresponds to the soft spheres value (5). No minimum could be recorded on the angular dependence of the collective diffusion coefficient (D_c - from the short time limit data) within the experimentally accessible range (18-105°). This behaviour probably results from dynamic interactions between the diffusing SMP and chains of molecularly dispersed PVA. These interactions seem to predominate over those between SMP themselves.

Conclusion

Probably two mechanisms are active during the ageing of poly (vinyl alcohol): 1. Attractive forces between the individual macromolecules, caused by inital crystallization (6,7), lead to the formation of supermolecular particles by hydrogen bonds (6). 2. A small part of these particles (i.e., the largest and the most compact) are able, above a certain concentration, to order their surroundings due to repulsive forces (created probably by their rather hydrophobic surface). This brings about a strong external interference leading, especially at low angles, to an opposite deformation of the P(Θ) function than expected in the case of aggregation.

References

1. Doty, P., Steiner, F.: J. Chem. Phys. 20, 85 (1952).
2. Porod, G.: Monatsch. Chem. 103, 395 (1972).
3. Sedláček, B., Koňák, Č.: J. Coll. Interface Sci. 90, 60 (1982).
4. Aharoni, S.M.: J. Macromol. Sci. - Phys. B15, 347 (1978).
5. Burchard, W.: Adv. Polym. Sci. 48, 1-125 (1983).
6. Pritchard, J. G.: Poly(vinyl alcohol): Basic Properties and Uses, Gordon and Breach, London 1970.
7. Mrkvičková, L. et al: will be published.

LIGHT-SCATTERING FROM CARRAGEENAN SOLUTIONS

Chantal De Jonghe, Harry Reynaers

Laboratorium voor Macromoleculaire Chemie, KULeuven,
B-3030 Heverlee, Belgium

Kees Bloys van Treslong, Frans Varkevisser

Gorlaeus Laboratoria der Rijksuniversiteit Leiden,
Leiden, The Netherlands

Introduction

κ- and ι-carrageenan are sulphated polysaccharides of algal origin; they are alternating copolymers of 1,3-linked β-D-galactose and 1,4-linked 3,6-anhydro-α-D-galactose. For both polymers the hydroxylgroup at C(4) of the 1,3-linked residue is sulphated and in the case of ι-carrageenan sulphation occurs as well at C(2) of the 1,4-linked residue. Chemical heterogeneity however occurs along the polymer chain : the ideal structures being interrupted by replacement of some of the 3,6-anhydrogalactose residues by galactose-2,6-disulphate, galactose-6-sulphate, galactose-2-sulphate or galactose (1). Both polysaccharides exhibit strong gelation behaviour in the presence of K^+-ions. The physical crosslinking of the gel-network has been ascribed to a two coils-double helix thermo-reversible transition on the basis of an increase in optical rotation at the gel-setting temperature and X-ray evidence (2,3). Multiple attempts to explore the nature of the aggregation process by light-scattering failed to give an unequivocal answer with respect to the postulated coil-helix transition (4-7).
Light-scattering measurements are presented on native carrageenans under gelling and non-gelling conditions as well as on samples degraded to segments by the combined Smith periodate oxidation-alkaline borohydride techniques.

Physical Optics of Dynamic Phenomena and Processes in Macromolecular Systems

Results

All measurements reported in the tables were obtained either by a FICA 50 (λ = 436 nm) or the Chromatix KMX-6 (λ = 633 nm) instruments. Solutions were cleared from dust by filtration through Nuclepore (0.45 and 0.22 μm) filters directly in the light scattering cell. The whole filtration procedure was performed at 80°C to eradicate thermal history of the samples. In contrast with literature information light scattering from carrageenan solutions under non-gelling conditions (0.1M NaCl or 0.1M LiCl) exhibit the normal concentration dependence and in the case of κ-carrageenan undistorted Zimm plots are obtained down to 30° , pointing to the absence of microgel particles in solution. With respect to the angular dependence of the scattered light, the solutions appeared to be stable for at least five days. Inspection of Table 1 clarifies nevertheless some scattering in the results as to the absolute value of the molecular weights of our samples. This has been attributed to some uncertainty as to the state of aggregation in our samples and the values of dn/dc as already reported by Jones (5) and Norton (7). An interesting result of the present study comes from light scattering observations on κ-carrageenan segmented by the Smith periodate oxidation (Table 2). Although the molecular weights are seemingly not very different after 8 h or 24 h of hydrolysis, the latter sample is highly polydisperse as observed from osmotic pressure measurements (8). Analogous measurements on samples of ι-carrageenan (Table 3 and 4) reveal the strong tendency of this material to aggregate.

In order to get more insight in the molecular behaviour of κ-carrageenan under gelling conditions above the gel-setting temperature (Tg) and at Tg (53.5°C for the present system) a solution was prepared of segmented κ-carrageenan ($\langle M \rangle_w$ = 10 700) in 0.1M KCl solution. After filtration at 80°C the angular dependence of the light scattering was followed as a function of time and temperature. At the highest temperature no angular dependence is observed except at the smallest angles. Lowering of the temperature induces formation of microgel, the high angle part remaining however unaffected. At Tg (53.2°C) there is a strong increase in total light scattering at all angles at once, which means that all molecules participate simultaneously to the aggregation process prior to reticulation.

Table 1 : Light-scattering data for κ-carrageenan

Experimental conditions 25°C	dn/dc (ml/g)	$<M>_w$	$<A_2>$ (mol ml/g^2)	$<R_G^2>^{1/2}$ nm
0.1M NaCl-KMX-6	0.157	354 000	4.6×10^{-3}	
0.1M NaCl- "	0.157	430 000	3.7×10^{-3}	
0.1M NaCl- "	0.157	302 000	3.2×10^{-3}	
0.1M NaCl- "	0.140	450 000	3.9×10^{-3}	
0.1M LiCl- "	0.157	265 110	3.6×10^{-3}	
0.1M LiCl- "	0.142	324 200	2.9×10^{-3}	
0.1M LiCl- "	0.142	421 000	3.9×10^{-3}	
0.1M NaCl-FICA	0.160	278 400	3.5×10^{-3}	71.6

Table 2 : Light-scattering data for segmented κ-carrageenan

Experimental conditions 25°C	dn/dc (ml/g)	$<M>_w$	$<A_2>$ (mol ml/g^2)
0.1M NaCl-KMX-6	0.157 (0.138)	10 700	1.5×10^{-2}
0.1M NaCl "	0.117	110 000	3.9×10^{-3}
0.1M NaCl "	0.117	101 500	4.0×10^{-3}

Table 3 : Light-scattering data for ι-carrageenan

Experimental conditions 25°C	dn/dc (ml/g)	$<M>_w$	$<A_2>$ (mol ml/g^2)	$<R_G^2>^{1/2}$ (nm)
0.1M NaCl-KMX-6	0.114	1 496 000	7.23×10^{-4}	
0.1M NaCl-FICA	0.163	570 000	7.96×10^{-4}	123.2

Table 4 : Light-scattering data for segmented ι-carrageenan

Experimental conditions 25°C	dn/dc (ml/g)	$\langle M\rangle_w$	$\langle A_2\rangle$ (mol.ml/g^2)
0.1M NaCl, KMX-6	0.100	550 000	2.65×10^{-3}
0.1M NaCl, "	0.125	540 000	1.18×10^{-3}
0.1M NaCl, "	0.114	903 000	1.21×10^{-3}
0.1M NaCl, "	0.100	290 000	1.16×10^{-3}

Conclusions

Under non-gelling conditions stable solutions of κ-carrageenan and κ-carrageenan segments can be prepared. In contrast with literature information these solutions exhibit normal concentration and angular dependence (Zimm plot). ι-carrageenan solutions appear to be strongly aggregated and seem actually less suitable for light scattering studies. Studies of the light scattering of κ-carrageenans under gelling conditions indicate a strong tendency of the solutions to form microgel even above Tg, the gel-setting temperature. At Tg, all molecules participate in the aggregation process prior to reticulation.

References

1. Rees, D.A., Williamson, F.B., Frangou, S.A. & Morris, E.R. : Eur.J.Biochem., 122, 71-79 (1982).
2. Rees, D.A. : Biochem.J., 126, 257-273 (1972).
3. Arnott, S., Scott, W.E., Rees, D.A. & McNab, C.G.A. : J.Mol.Biol., 90, 253-267 (1974).
4. Rochas, C. : Thesis, Université de Grenoble (1982).
5. Jones, R.A., Staples, E.J., & Penman, A. : J.Chem.Soc., Perkin Trans.II, 1608-1612 (1973).
6. Morris, E.R., Rees, D.A., & Robinson, G. : J.Mol.Biol., 138, 349-362 (1980).
7. Norton, I.T., Goodall, D.M., Morris, E.R., & Rees, D.A. : J.Chem.Soc., Faraday Trans., 79, 2475-2488 (1983).
8. De Jonghe, Chantal : Thesis, University of Leuven (1984).

STUDIES OF PROTEIN - NUCLEIC ACID INTERACTIONS BY PHOTON CORRELATION SPECTROSCOPY. II. tRNA - BSA INTERACTIONS AT INTERMEDIATE AND HIGH IONIC STRENGTH+

Adam Patkowski

Quantum Electronics Laboratory, Institute of Physics,
A. Mickiewicz University, Grunwaldzka 6, 60-780 Poznań, Poland

Ben Chu

Department of Chemistry, State University of New York at
Stony Brook, Stony Brook, N.Y. 11794, USA

Introduction

The importance of transfer RNA - protein interaction studies has long been recognized for their implications to enzymatic reactions in a living cell. By means of biochemical and physicochemical methods tRNA-protein interactions have been characterized in terms of association constants, interaction sites and strength, thermodynamic parameters and internal structure changes of interacting macromolecules. In our previous paper (1) we have shown that photon correlation spectroscopy can provide us with such parameters as diffusion coefficients of reactants and reaction products as well as reaction rate constants characterizing tRNA-BSA interactions at low ionic strength. Since the previously obtained results (1) suggested an electrostatic nature of binding, we have studied the effect of counterions on this reaction in order to check this hypothesis.

+Work supported in part by Research Project MR I.5.10.04

Physical Optics of Dynamic Phenomena and Processes in Macromolecular Systems

Theoretical

The spectrum of light scattered from a chemically reacting system has been studied theoretically for some simple reactions (2,3,4). The first order correlation function for the reaction:

$$A \underset{k_b}{\overset{k_a}{\rightleftharpoons}} B \tag{1}$$

where k_a and k_b are the forward and backward reaction rate constants, is given by the formula (5) :

$$g^{(1)}(\tau) = B_+ \exp(S_+\tau) + B_- \exp(S_-\tau) \; , \tag{2}$$

where

$$S_\pm = -\tfrac{1}{2}(\gamma_1 + \gamma_2) \pm \tfrac{1}{2}\left[(\gamma_1 - \gamma_2)^2 + 4k_a k_b\right]^{1/2}, \tag{3}$$

$$\gamma_1 = K^2 D_1 + k_a, \qquad \gamma_2 = K^2 D_2 + k_b, \qquad K = \frac{4\pi}{\lambda_o} n_o \sin\frac{\Theta}{2} \tag{4}$$

$$B_\pm = \pm(S_+ - S_-)^{-1}\Big[\alpha_1^2\, c_1^o\, (S_\pm + \gamma_2) + \alpha_1 \alpha_2 (c_1^o k_a + c_2^o k_b) + \alpha_2^2\, c_2^o\, (S_\pm + \gamma_1)\Big] \; . \tag{5}$$

Here the quantities D_1, D_2, α_1, α_2 and c_1^o, c_2^o are the translational diffusion coefficients, optical polarizabilities and equilibrium concentrations of A and B, respectively; K is the scattering vector which depends on the refractive index of the solvent, n_o, wavelength of the incident radiation, λ_o, and the scattering angle Θ. $k_a = 4k_a' c_1^o$ for dimerization.

Results and Discussion

Our experimental setup, the data fitting procedures and the characteristics of tRNA and BSA have been described earlier (1). tRNA-BSA interactions were studied in the following solvents: (i) 0.01 M sodium cacodylate, pH 6.0; (ii) 0.01 M sodium cacodylate, 0.1 M NaCl, pH 6.0 and (iii) 0.01 M sodium cacodylate, 1 mM $MgCl_2$, pH 6.0. In all these solutions BSA concentration was constant and amounted to 3.5 mg/ml, while tRNA concentration was in the range of 0.2 to 40.0 mg/ml.

1. Aggregate composition.

For tRNA-BSA solutions in 1 mM sodium cacodylate, pH 6.0 two kinds of aggregates were found (1) of the following composition: $tRNA(BSA)_2$ and $tRNA(BSA)_4$. In solvents (i) and (ii) only the smaller aggregates were observed. For solvent conditions (iii) the concentration of aggregates was to low to allow for any reliable estimates by light scattering.

2. Reaction rate constants.

At low ionic strength (1mM sodium cacodylate) (1) the aggregation rate constant amounted to $8x10^7$ $M^{-1}s^{-1}$. The theoretically estimated rate constant for a diffusion controlled reaction of these macromolecules amounts to $7x10^9$ $M^{-1}s^{-1}$ and is two orders of magnitude higher than the experimental one. The value of the dissociation rate constant was below the experimental error. At higher ionic strength (solvent (i)) the aggregation rate constant increased to $3x10^8$ $M^{-1}s^{-1}$ and the dissociation rate constant increased to about 10^3 s^{-1}thus the stability and the concentration of aggregates decreased. Further increase of the ionic strength of tRNA-BSA solutions (solvent (ii)) and presence of 1 mM $MgCl_2$ (solvent (iii)) caused a strong decrease of the aggregate stability and concentration making in this way any estimation of the reaction rate constants impossible.

3. Nature of binding.

On the basis of the above presented data we can propose the following interpretation of tRNA-BSA aggregation: The aggregates are formed due to Coulombic local attraction between fragments of tRNA and BSA of opposite electric charge since the aggregates are most stable at the lowest ionic strength. For longer distances between tRNA and BSA electrostatic repulsion prevails due to equal sign of the net electric charge of these macromolecules. This repulsion causes a marked decrease of the aggregation rate constant in comparison with the one estimated for a diffusion controlled reaction at the lowest ionic strength. An increase of the ionic strength of solutions causes a more effective screening of the long range repulsive interactions which results in an increase of the aggregation rate constant. On the other hand the local attraction of tRNA and BSA fragments decreases due to a partial neutralization of the macromolecular charge by counter-ions and thus the stability of the aggregates decreases. From the fact that the tRNA-BSA aggregation almost disappears in solutions containing 1 mM magnesium ions one can conclude that only the expanded tRNA structure appearing at low ionic strength with no magnesium ions present (5) can take part in the aggregation process. Partial folding of tRNA tertiary structure caused by high NaCl concentration (solvent (ii)) or by presence of magnesium ions (solvent (iii)) (5) decreases tRNA-BSA aggregation. This leads to the conclusion that to allow for the aggregation some parts of the expanded tRNA structure must be accessible for interactions with BSA. The appearence of bigger aggregates, $tRNA(BSA)_4$, only at the lowest ionic strength may result from geometrical reasons, i.e. from the fact that tRNA structure in these conditions is the most extended one (5) and thus one tRNA molecule is accessible for interactions with four BSA molecules.

References

1. Patkowski, A., Chu, B.: J. Chem. Phys. 73, 3082-3087 (1980).
2. Benbasat, J.A., Bloomfield, V.A.: Macromolecules 6, 593-597 (1973).
3. Berne, J., Giniger, R.: Biopolymers 12, 1161 (1973).
4. Berne, J., Pecora, R.: Dynamic Light Scattering, Wiley, New York 1976.
5. Patkowski, A., Chu, B.: Biopolymers 18, 2051-2072 (1979).

SOLVATION OF PROTEIN MOLECULES STUDIED BY COMBINED QELS AND SAXS INVESTIGATIONS

Klaus Gast, Dietrich Zirwer, Paul Plietz, Jürgen J. Müller
Gregor Damaschun and Heinz Welfle

Department of Molecular Biophysics, Central Institute of Molecular Biology of the Academy of Sciences of the GDR, 1115 Berlin, GDR

Introduction

Quasi-elastic light scattering (QELS) is an efficient tool to study proteins in solution. However, there is one problem which cannot be solved from QELS data alone: How do solvation and non-spherical shape contribute to the ratio of frictional coefficients? Small-angle X-ray scattering (SAXS) yields detailed information about the structure of the protein but not about protein solvation. The combination of both methods helps to overcome this problem and represents a pertinent way to estimate the outer solvation shell. Experimental results for the two proteins 11S globulin from Vicia faba and 7S globulin from Phaseolus vulgaris are presented.

Results

1) Determination of structural parameters by QELS

The primary data is the normalized autocorrelation function

$$G^{(1)}(\tau) = 1/N \int_0^\infty c(M)\, M \cdot P\,(q\, R(M))\, \exp\,(-q^2\, D(M)\tau)\, dM$$

N - normalization factor, c - weight concentration, M - molecular mass, P - scattering form factor, q - scattering vector, D - translational diffusion coefficient.

The absence of protein aggregates can be easily checked looking at

a) The correlation function profile. $\ln(|G(\tau)|)$ must be nearly a straight line and/or $\delta_z = K_2 / K_1^2$ (K_1, K_2 - first and second cumulant, respectively) must be very small.
b) The angular independence of the calculated diffusion coefficients.
These conditions are well satisfied for our two proteins. Therefore the diffusion coefficients and all related parameters can be obtained with high accuracy. The following molecular parameters can be derived from the first cumulant $K_1 = q^2 D$. Extrapolation to zero concentration and standard conditions yields the diffusion coefficient $D^o_{20,W}$ and the frictional coefficient $f = K T/D$. Via the Stokes-Einstein relation we can calculate the Stokes' radius R_s. Knowing the sedimentation coefficient s and the partial specific volume $\bar{v}$ we can calculate the molecular mass $M_{s,D} = R T s/D (1 - \bar{v} \varrho)$. From R_s, $M_{s,D}$, and $\bar{v}$ the ratio of frictional coefficients $f/f_o = R_s/R_o$ can be calculated, where $R_o = (3 M \bar{v}/4 \pi N_A)^{1/3}$ is the radius of an exactly spherical unsolvated molecule.

2) Detailed structural characterization by SAXS
The experimental scattering curves of the proteins and the methods of data-handling are given in refs. (1) and (2). A summary of the molecular parameters determined by SAXS and by QELS is given in Table 1. The experimental SAXS curves were compared with theoretical scattering curves of oblate ellipsoids of revolution with different axial ratios b/a. The best agreement with the experimental curves was found with axial ratios b/a of 0.7 and 0.3 for the 11S and the 7S globulin, respectively. Quaternary structure models which fit the SAXS data are presented in Fig. 1.

3) Combination of data from SAXS and QELS - an efficient approach to the study of protein solvation
The parameters necessary for these calculations are the Stokes

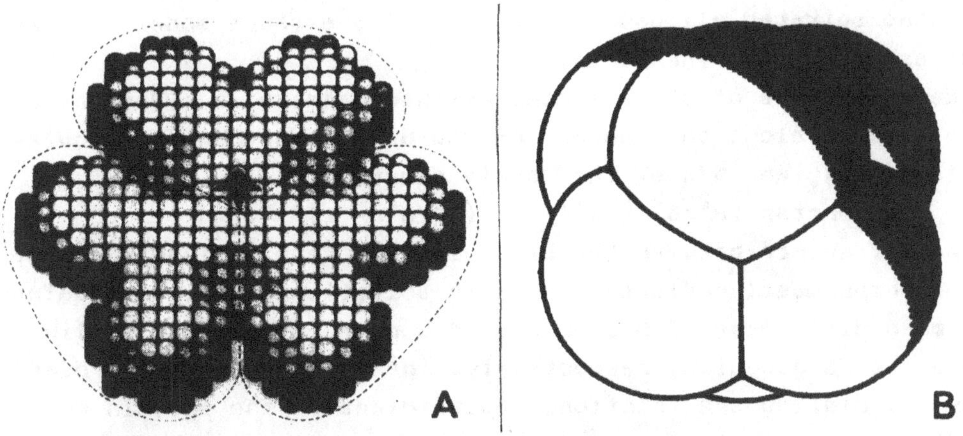

Fig. 1. Quaternary structure models for A) the 7S globulin from Phas. vulg. and B) the 11S globulin from Vicia faba

Table 1. Structural parameters of the proteins

Parameter	7S globulin	11S globulin
Molar mass (g/mol)	$1.4 - 1.5\times10^5$	3.5×10^5
Radius of gyration (nm)	4.1	4.4
Maximum dimension, L (nm)	13.0	13.0
Intersection length, d_c (nm)	3.3	6.4
Surface, S (nm^2)	365	426
Stokes' radius, R_s (nm)	4.7	6.32
Translational diffusion coefficient, $D^0_{20,W}$ (cm^2/s)	4.5×10^{-7}	3.38×10^{-7}
Ratio of frictional coefficients, f/f_o	1.36	1.36
Partial specific volume, $\bar{v}$ (ml/g)	0.729	0.729
Sedimentation coefficient, s (S)	7.3	13.0
Solvation (g H_2O/g protein)	0.5	0.54

radius from QELS and the shape model from SAXS (an ellipsoid of revolution with the axial ratio b/a). The following assumptions are made. The thickness of the solvation shell is nearly constant all around the molecule and the solvation shell is small compared to the molecular dimensions, so that the axial ratios

of the solvated ellipsoid b'/a' and of the shape model b/a can be assumed to be identical.
The dimensions of the solvated ellipsoid which is hydrodynamically equivalent to a sphere of radius R_s can then be calculated from R_s and b/a by the Perrin equation
$$b' = R_s \cdot \arctan\,(a^2/b^2 - 1)^{1/2} \,/\, (a^2/b^2 - 1)^{1/2}.$$
The difference between the dimensions of this ellipsoid and of the shape model reflects the outer solvation shell. We obtained a mean difference of 0.6 nm and of 0 nm for the 11S globulin and the 7S globulin, respectively. The method can be improved by calculating the frictional coefficient of the protein molecule, if a detailed structure model is available from X-ray scattering data (3).

Conclusions

The mean difference of 0.6 nm between the semiaxes of the two ellipsoids in the case of the 11S globulin points to an outer solvation shell of corresponding thickness. In the case of the 7S globulin the ellipsoids have identical dimensions. This is attributed to a rough surface structure of this protein (see also the model in Fig. 1) allowing a lot of solvent to be localized within the volume of the shape model.

References

1. Plietz, P. et al.: Int. J. Biol. Macromol. 5, 356-359 (1983).
2. Plietz, P. et al.: Biochim. Biophys. Acta 784, 140-146 (1984).
3. Müller, J.J. et al.: studia biophysica 93, 39-46 (1983).

LIGHT SCATTERING STUDIES OF BOVINE EYE LENS α_L-CRYSTALLIN AT HIGHER CONCENTRATIONS : SHORT- AND LONG-RANGE ORDER

Christiane Andries, Marc Van Laethem, Julius Clauwaert
University of Antwerp, Department of Biochemistry, Biophysics Research Group, Universiteitsplein 1, B-2610 Antwerpen, Belgium.

Introduction

In order to understand the light scattering behaviour of the eye-lens cytoplasm, we are studying the different α_L-crystallin species present in the different parts of the eye lens and the solution behaviour of α_L-crystallins in different solvent conditions. The emphasis on this class of proteins seems justified by the fact that these proteins, as a result of their larger size and high concentration, offer by far the largest contribution to the light scattering scattering behaviour of the cytoplasm. We report on our studies of the calf and bovine eye lens protein α_L -crystallin with photon correlation spectroscopy and statical light scattering in the concentration range up to 200 g/l at lower (I = 0.08) and higher (I = 0.32) ionic strength conditions.

Results

1) Isolation and concentrating of α_L-crystallins.
The preparation of α_L-crystallins has been described (1). For the higher concentration studies, the material from two zonal centrifugations had to be collected and concentrated to a final volume of about 1 ml. The quality of the final solutions has always been controlled by diluting a small amount of this solution to a concentration of about 1 mg/ml and measuring the diffusion coefficient : this was always in the expected range of $(2.30 \pm 0.05) \times 10^{-7}$ cm^2 s^{-1} for cortical bovine α_L-crystallin and $(2.50 \pm 0.05) \times 10^{-7}$ cm^2 s^{-1} for cortical calf α_L-crystallin.

Physical Optics of Dynamic Phenomena and Processes in Macromolecular Systems

The quality factor Q of these correlation functions was also excellent.

2) Photon correlation spectroscopy and absolute intensity measurements of scattered light.

The apparatus and the precautions which have been taken during photon correlation spectroscopy and light scattering measurements, have been described (1).

Because the correlation function at higher protein concentration is no longer single-exponential, the Malvern Type 7023 correlator has been extended to 96 channels. Photon correlation spectra have been measured at 45°, 90° and 135° and T = 298°K.

Due to the fact that the light scattering becomes angle-dependent at high α_L concentrations, this has been measured in the angle range 30° to 140° at 10° intervals. In the low concentration range (up to 50 g/l), the α_L-crystallin solution behaviour can be described by an extended hard-sphere potential or an equivalent shielded Coulomb potential; the characters of these potentials do depend, and in the proper way as expected for electrostatic interactions, on the ionic strength of the solution.

At higher concentrations (C $\rangle$ 70 g/l), the scattering behaviour is quite complicated, resulting in non-exponential correlation functions.

3) Data analysis.

Four methods have been used for the analysis of the correlation functions.

The cumulant analysis (2) showed very clearly that in the highly concentrated solutions of α_L-crystallin, a complex pattern of diffusion modes is present. This method does not permit to deduce the exact distribution function of the decay times because only a limited number of cumulants can be accurately determined and a distribution is not unambiguously characterized by its cumulants. But the cumulant analysis offers an a posteriori control of the distribution function which can be calculated by other methods.

The distribution function G (Γ,K), where $\Gamma = D\bar{K}^2$ and $\bar{K}$ the scattering vector, can be determined in one way by postulating a form of G ($\Gamma,\bar{K}$). A discrete sum of deltafunctions

$G(\Gamma,\bar{K}) = \sum_i A_i \; \delta[\Gamma - \Gamma_i]$ can in our conditions even be simplified to a two exponential fit where the correlation function can be described by

$$|g^1(\tau)|^2 = |A_1 e^{-\Gamma_1 \tau} + A_2 e^{-\Gamma_2 \tau}|^2 + B$$

This fitting procedure has been used, in combination with several tests (3,4), for α_L-crystallin solutions at a concentration c of proteins $>$ 100 g/l because only in these conditions the ratio between the two amplitudes A_2/A_1 and the difference between Γ_2 and Γ_1 was large enough to allow a meaningful application of this procedure.

Most methods which aim at a direct inversion of the integral $g^1(\tau) = \int G(\Gamma\bar{K}) \, e^{-\Gamma\tau} \, d\Gamma$ are badly conditioned :noise and the limited measuring range cause instability of the solution (5). Recently an exponential sampling method was proposed based on the eigenvalues and -functions of the Laplace transform, which should be very well suited for our problem (6). The distribution function is expanded in δ-functions

$$G(\log(\Gamma)) = \sum_{n=1}^{N} a_n \; \delta[\log \Gamma - \log \Gamma_n]$$

with

$$\Gamma_{n+1} = \Gamma_n \, e^{\pi/w_m}$$

w_m determines the resolution and depends on the quality of the measurements; a_n is determined by at least squares method.

In another approach, we have used the CONTIN program to analyse the experimental correlation functions (7). CONTIN is a very flexible fully portable Fortran IV package for solving linear integral equations of the first kind and systems of possible ill-conditioned algebraic equations. It is capable of estimating the unknown function $S(\lambda)$ in

$$y_k = \int_a^b F_k(\lambda) \; S(\lambda) \, d\lambda + \sum_{i=1}^{N_L} L_{ki} \; \beta_i + \varepsilon_k$$

where y_k are the observed data
$F_k(\lambda)$ unknown functions
ε unknown noise components

The optional sum over the N_L known L_{ki} and N_L unknown β_i permits, for example, a constant background term β to be included by setting $N_L = 1$ and all the $L_{k1} = 1$. The algorithm and the several possible options are described in detail by Provencher (7). We have used the user-orientable option for photon correlation spectroscopy to analyse the diffusion distribution function and the polydispersity of the α_L-crystallin solutions at high concentrations.

4) Light scattering.

The Rayleigh ratios of the α_L-crystallin solutions have been calculated from the experimental intensities I_{α_L} and I_{Ben}, after application of corrections for the difference in volume seen at different angles, the dead time of the photomultiplier, the dark current:

$$R_{\alpha_L} = (I_{\alpha_L}/I_{Ben}) x (n_0/n_{Ben})^2 x R_{Ben} x 2/(1+\rho_v)$$

Discussion

The analysis of the correlation spectra of the highly concentrated α_L-solutions with the 4 different methods resulted in almost identical conclusions ; these solutions contain two scattering entities : a smaller unit with a larger diffusion coefficient D_1 and a quite narrow distribution, and a larger unit with a smaller diffusion coefficient D_2 but showing a rather broad distribution.

The smaller scattering unit can be identified as the single α_L-crystallin. The concentration dependence of D_1 at these high concentrations can be predicted from the extrapolation of the experimental data obtained at low concentration $D_m = D_0(1 + \alpha\phi)$; D_1 shows no angle dependence and must be interpreted as the collective relaxation of the free but interacting α_L -crystallins.

Both the angular dependence of the scattered intensity as the angular dependence of the amplitude A_2 of the second class of scattering units prove that larger scattering units are formed and a detailed comparison of the results of the various analysis techniques at various measuring times, leads to the conclusion that a broad range of large scattering units is present.

We can calculate an estimate for the size of these scattering units from

the Stokes-Einstein relation and introducing the diffusion coefficient D_2 although it is uncertain what kind of viscosity η has to be introduced in this situation (8). If we assume that the scattered intensity is a superposition of the light scattered by the "free" α_L-crystallins and the large units, and that no interference occurs between the light scattered by the two types, the angle-dependent intensity is related to the large scattering units. Using the Guinier approximation, the radius of gyration R_G can be calculated.
These calculations result in a mean content of about 10^3 individual α_L-crystallins for these large scattering units.
Biochemical studies have excluded the formation of α_L-aggregates in these conditions (9). Different observations favour the hypothesis of cluster formation due to a secondary minimum in the potential as a net effect of electrostatic repulsion and Van der Waals attraction (8). These clusters can be described as the formation of correlated parts in the solution and in this way the existence of local areas of order over large distances is suggested.

References

1. Andries, Chr., Guedens, W., Clauwaert, J., Geerts, H. : Biophys. J. 43, 345-354 (1983).
2. Koppel, D.E. : J. Chem. Phys. 57, 4814-4820 (1972).
3. Laiken, S.L., Printz, M.P. : Biochemistry 9, 1547-1553 (1970).
4. Mathiez, P., Weisbuch, G., Mouttet, C. : Biopolymers 18, 1465-1478 (1979).
5. Pike, E.R. : In Scattering Techniques Applied to Supramolecular and Non-Equilibrium Systems. Chen, S.-H., Chu, B., Nossal, R. ed. : 179-200, Plenum Press, New York 1981.
6. Ostrowsky, N., Sornette, D., Parker, P., Pike, E.R. : Optica Acta 28, 1059-1070 (1981).
7. Provencher, S.W. : Comput. Phys. Commun. 27, 229-242 (1982).
8. Bauer, G.B. : J. Phys. Chem. 84, 1592-1598 (1980).
9. Siezen, R.J., Owen, E.A. : Biophys. Chem. 18, 181-194 (1983).

MAGNETO-OPTICS OF COLLOIDS

R. V. Mehta
Sardar Vallabhbhai Regional College of Engineering and Technology, Surat-395007, India

Abstract

Magnetically induced birefringence, dichroism, extinction, scattered intensity and degree of polarization are defined in terms of scattering matrix. Certain special cases are analysed. Experimental techniques to interpret the data are briefly described.

I. Introduction

In 1845 W.R.Grove (1) demonstrated that the transparency of a suspension containing magnetite particles in water increases when a magnetic field is applied parallel to the light beam. Magneto-optical effects in truly colloidal solutions were first studied by Majorana (2). He observed that certain specimens of aged iron-oxide colloids became birefringent and dichroic when placed in a magnetic field transverse to the light beam. The effects were found to be (a) proportional to square of the magnetic field and (b) in the colloid the linearly polarized light which was retarded more was also absorbed to a greater extent. Schmauss attributed this effect to the orientation of suspended particles (3) and the ultramicroscopic nature of the particles was first demonstrated by Cotton and Mouton (4). Langevin gave the quantitative explanation for the origin of electro and magneto birefringence in pure liquids and the same theory is applicable to the magneto birefringence in colloids

Physical Optics of Dynamic Phenomena and Processes in Macromolecular Systems

provided the colloidal particles are small compared to the wavelength of incident light (5). Bitter's discovery - that the colloidal magnetite can be used to study the domain structure on the ferromagnetic surface (6) - has attracted attention of many workers towards magneto-optical effects (7,8,9). Elmore gave a theory for the optical transmission under a magnetic field. Ewel (10) and Heller (11) observed the reversal in sign of the birefringence of iron oxide colloids upon increasing the field and temperature. Heller has systematically studied this effect and attributed this effect to the reversal in the orientation of the $Fe_2O_3.H_2O$ crystallites (12). An alternative explanation for this effect is suggested by Naik and Desai (13). Magnetically induced transmission changes in colloidal graphite were investigated by Stott (14). Average size and shape of the particles were deduced using a theory based on geometrical shadow approximation. Subsequently it was found that it is more appropriate to use dipole scattering theory to interpret the observed polarization dependence of extinction coefficients in this colloid (15). The dipole scattering theory was successfully applied to interpret magneto--optical effects of many other colloidal solutions (16-20). This theory has been extended for larger particles using anomalous diffraction theory by Meeten (21-25). Recently interest in the study of magneto-optical effect has been stimulated because of the many possible commercial applications of stable magnetic fluids (also known as ferrofluids). These fluids are the concentrated colloidal solutions of magnetite or cobalt in various dispersing media (26). In order to prevent aggregation due to van der Waals' and magnetic attraction the particles in the fluid are coated with a layer of surfactant. Measurements of the induced birefringence, dichroism, scattering etc. are carried out by many workers and possible explanations are given (27-46). Practical applications of these effects are also developed by several workers (47-53). Though the use of the magneto-optical techniques in colloids is known since last 40 years, a general account which incorporates various types of magneto-optical effects, their causes, treatment of data etc. is not available in a single paper. An attempt in this direction is made here. In Section II basic theory of magneto-optical effects is summarized. Certain new expressions for intensity, degree of polarization and scattering coefficient for longitudinal configuration are also given. Treatment of data and inferences that can be drawn is discussed in Section III. Section IV describes experimental set-up. Certain experimental results are discussed in Section V.

II. Theory

The colloidal solution may exhibit magnetically induced anisotropy if the dispersed particles are anisotropic and show preferred orientation in space. The anisotropy of the particles may be either due to their shape or inherent optical anisotropy. Preferred orientation may occur if the particles either possess permanent magnetic moment or anisotropic (para- or dia-) magnetic susceptibilities or both. The colloid then resembles an uniaxial crystal with its optic axis along the direction of the applied magnetic field. It can be shown that (54) the induced birefringence, dichroism and extinction are given by*

$$\Delta n = 2\pi k^{-3} n_m [\int_n \sum_i \text{Im}\, S_1(0)\, f(n)\, dn - \int_n \sum_i \text{Im}\, S_2(0)\, f(n)\, dn] \tag{1}$$

$$\Delta n' = n_m k^{-1} [(C_{ext})_{\ell} - (C_{ext})_r] \tag{2}$$

and

$$(C_{ext})_{\ell,r} = 4\pi k^{-2} n_m [\int_n \sum_i \text{Re}\, S_{1,2}(0)\, f(n)\, dn] \tag{3}$$

*For definition see List of symbols

The intensity and degree of polarization of the scattered light are given by

$$I = Tr\zeta^{s} \quad (4)$$

$$P = 1 - 4|\zeta^{s}|/Tr\zeta^{s} \quad (5)$$

The detailed expression for the coherency matrix of the medium is given in an earlier paper (55,59). The orientation distribution function f(n)dn is to be calculated from the magnetic properties of the particles. In all the cases it is necessary to calculate elements of the matrix $S(\theta,\phi)$. Though the rigorous solution for scattering by spheroids of arbitrary size, shape and refractive index is now available (56), so far as the magneto-optics of colloids is concerned only few special types of the scatterers are considered, viz. anisotropic dipole scatterers, ellipsoids with small phase shift, conducting spheroids and very large particles. A summary of expressions for the above parameters will be given here. Detailed derivations are given in previous papers (39,52-59).

1. Dipole Scatterers

The case that has been extensively studied and applied is that of anisotropic dipole scatterers. When the magnetic field is applied transverse to the direction of the incident light, then it has been shown that (16,39)

$$\Delta n = 2n_m \; N \; Re \; \frac{(\beta_1 - \beta_2)}{3} \; (1 - 3L_1) \quad (6)$$

$$(C_{ext})_{\ell}^{T} = - 4\pi kN \; Im \; [\beta_1 + (\beta_1 - \beta_2) \; \phi(h)] \quad (7)$$

$$(C_{ext})_{r}^{T} = - 4\pi kN \; Im \; [(\frac{\beta_1 + \beta_2}{2}) - (\frac{\beta_1 - \beta_2}{2}) \; \phi(h)] \quad (8)$$

Here $(h) = 1 - 2L_1$.

Further the change in scattered intensity is given by

$$(\frac{\Delta I}{I_o})_U^T = \frac{2N}{15} k^{-4} R^{-2} \, [7X + 5Y - 15L_1(2X + Y) + 45XL_2 - \text{Sin}^2\theta\{3X + 5Y - 15L_1(2X + Y) + 105XL_2\}] \quad (10)$$

$$P^T = \frac{X + Y - L_1(4X + 3Y) + 5XL_2 - A\,\text{Sin}^2\theta}{X + Y + 2\beta_2\beta_2^* - L_1(2X + Y) + 3XL_2 - B\,\text{Sin}^2\theta} \quad (11)$$

where

$$X = (\beta_1 - \beta_2)(\beta_1 - \beta_2^*)$$

$$Y = \beta_1\beta_2^* - 2\beta_2\beta_2^* + \beta_2\beta_1^*$$

$$A = \beta_1\beta_1^* - L_1(5X + 2Y) + 12XL_2$$

$$B = \beta_1\beta_1^* - 2L_1(2X + Y) + 7XL_2$$

The parameter $\beta_j (= \alpha_j - \frac{2}{3} ik^3|\alpha_j|^2)$ includes the contribution due to pure scattering (19). Expressions for the functions $\theta(h)$ and $\psi(h)$ are given elsewhere (57,60). Function L_2 is related to $\psi(h)$ by

$$\psi(h) = 1 - 4L_1 + L_2 \quad (12)$$

Zero field value of the extinction is given by

$$(C_{ext})_o = - 4\pi kN \, I_m \, (\frac{2\beta_1 + \beta_2}{3}) \quad (13)$$

This leads to

$$(C_{ext})_\ell^T - (C_{ext})_0 = 2\,[(C_{ext})_0 - (C_{ext})_r^T] \quad (14)$$

The total scattering coefficient is obtained by integrating intensity over the surface of sphere with radius R. Hence

$$(C_{sca})_\ell = \frac{16\pi N k^4}{3} \, [3\beta_2\beta_2^* + L_1(2X + Y) + 2X\,L_2] \quad (15)$$

$$(C_{sca})^T_\ell = \frac{16\pi N\,k^4}{3}\,[\beta_1\beta_1^* - 2Y\,L_1 - 7X\,L_2] \tag{16}$$

$$(C_{sca})^T_U = \frac{8\pi N\,k^4}{3}\,[4\beta_2\beta_2^* + X + Y + L_1\,(2X + Y) - 5XL_2] \tag{17}$$

Similarly for longitudinal configuration we get

$$(\Delta n)^L = 0 \tag{18}$$

$$(C_{ext})^L_r = (C_{ext})^L_1 = (C_{ext})^T_r \tag{19}$$

$$(\frac{\Delta I}{I_o})^L_U = \frac{2N}{15}\,k^{-4}R^{-2}\,[10Y(3L_1 - 1) + 8X(15L_2 - 1) - \sin^2\theta\{10X\,L_2 - 15L_1(2X - Y) - (2X + 5Y)\}] \tag{20}$$

$$P^L = \frac{\sin^2\theta\,[L_1(X - 4Y) - 16X\,L_2 - 8\beta_2\beta_2^*]}{8[\beta_2\beta_2^* + L_1Y + 4L_2X - \frac{1}{2}\sin^2\theta\{\beta_2\beta_2^* - L_1(2X - Y) + 12XL_2\}]} \tag{21}$$

and

$$(C_{sca})^L_U = \frac{8\pi Nk^4}{3}\,[\beta_2\beta_2^* + L_1(X + Y)] \tag{22}$$

2. Rayleigh-Gans Particles

When the colloidal particles satisfy the conditions for R-G scattering (54) all the above parameters can be modified by incorporating a suitable form factor $R(\Theta,\phi)$. This factor will be different for 'T' and 'L' configurations. Thus

$$(\frac{\Delta I}{I_o})^{T,L}_{U,R-G} = R(\theta,\phi)^{T,L}\,(\frac{\Delta I}{I_o})^{T,L}_{U,dipole} \tag{23}$$

Similarly one can define other parameters. It is straightforward to show that the relation (14) will be satisfied in this case while the equation (19) will not be obeyed (46).

3. Small Conducting Spheroids

When particles are metalic and are much smaller then the wavelength of light the scattering may not be pure electric dipole scattering. In the limit when $m = \infty$, the scattering consists of electric and magnetic dipole radiation (54). In this case it has been shown that (20)

$$(C_{ext})_1^T = (C_{ext})_1^e + (C_{ext})_r^m \tag{24}$$

$$(C_{ext})_1^T = (C_{ext})_r^e + (C_{ext})_1^m \tag{25}$$

$$(C_{ext})^L = (C_{ext})_r^e + (C_{ext})_r^m \tag{26}$$

where $(C_{ext})_1^{e(m)}$ is the extinction coefficient due to electric (magnetic) dipole when the symmetry axis of the particle is parallel to the electric vector of the incident light. It is to be noted that now

$$(C_{ext})^L \neq (C_{ext})_r^T$$

and

$$(C_{ext})_1^T - (C_{ext})_0 \neq 2[(C_{ext})_0 - (C_{ext})_r^T]$$

Recently scattering by small magnetic sphere is studied by Kerker et al (63). Here also electric and magnetic dipole radiation dominates. If their analysis is extended to the case of small magnetic spheroids, then similar conclusions can be arrived.

4. Large Particles

When the size of colloidal particles is more than 20 times the wavelength of light, the scattered light arises due to contributions of reflection, refraction and diffraction. The diffraction part depends only on the form and size of the particles while the former two parts depend on the state of polarization as well. Detailed expressions for the induced intensity taking into account all the three aspects has not been worked out. However variation in cattered intensity in case of a ferrofluid placed in a magnetic field is interpreted in terms of diffraction effects (27). Extinction can easily be calculated. For diamagnetic disk shaped particles (62)

$$(C_{ext})^T = \frac{Nh}{E_h} \left| \frac{4A}{\pi} E + B\,\Phi \right|$$

$$(C_{ext})^L = \frac{Nh}{E_h} \left| A\,\Phi + \frac{4B}{\pi} E_\gamma \right|$$

where A is the area of cross section of the particle perpendicular to the symmetry axis and B is that containing the symmetry axis

$$h = [(\chi_2 - \chi_1)V_H]^{1/2}/2k_BT\ , \quad E_h = \int_0^h e^{-x^2}\,dx$$

$$\Phi = \frac{1}{h^2}\,[1 - \exp(-h^2)]\,, \quad E_\gamma = \int_0^{\pi/2} \exp(-h^2 \operatorname{Cos}^2\gamma)\,\operatorname{Sin}^2\gamma\, d\gamma$$

Similar expressions for other types of particles are also derived.

5. Composite Systems

In some cases particles in a dispersion are heterodispersed i.e. it contains a mixture of fine and coarse particles. Magnetically induced extinction in such a dispersion exhibit an interesting behaviour. In the transverse configuration when the incident light is polarized with the electric vector perpendicular to the field, extinction coefficient first increases with the field, reaches to a maximum, then decreases to zero and at a certain field reappears with reversal in its sign. Field strength at which the maximum and reversal occur depends on the ratio of number of large to small particles (64). This reversal is purely due to scattering phenomena i.e. the reversal occurs because the scattering matrix of large and small particles under orientation is quite different. Reversal also occurs when a dispersion either contain two different types of particles (e.g. polar and diamagnetic) or contains particles possessing two types of moments. The technique to ascertain the type of reversal is fully discussed in a previous paper (61). It has been shown that when the induced extinction exhibit reversal only in one state of polarization of incident light, viz. $(C_{ext})_r$, then the system definitely contains mixture of small, and large particles. On the other hand if the induced extinction attains maximum as well as reversal in both the states of linear polarization then it is necessary to study wavelength dependence of reversal in birefringence as well as dichroism. If the field strengths at which effects become maximum (H_M) and reverse their sign (H_N) are independent of wavelength of the incident light then the system must contain particles which are acted upon by two orthogonal torques. If H_M and H_N are different for birefringence and dichroism then the system contains two types of particles, one of which are particles oriented by permanent magnetic moment and others are oriented by torque due anisotropic susceptibility.

6.Other Types of Particles

Induced optical anisotropy for a system containing scatterers whose size falls in anomalous diffraction region are theoretically studied by several workers (65,23,66). Meeten has used A.D. theory to interpret the observed effects in kaolinite suspensions. Using Stevenson approach Stoimenova and Stoylov have derived expressions for induced birefringence and scattered intensity (65). Experimental analysis based on this theory is not yet available.

III. Experimental Setup

Several arrangements are described for the measurement of the various magneto-optical parameters (20,37,44). A monochromatic light beam is obtained either from a monochromator, laser or interference filter and is collimated by a suitable lens. The beam is made linearly polarized by polarizer (Polaroid or Glan prism). The linearly polarized light is allowed to pass through a measuring cell which is kept between the two poles of an electromagnet. For measurements in the longitudinal configuration these pole pieces are bored through. The transmitted or scattered light is then either allowed to pass through a second polarizer or is received directly by a photomultiplier. For the measurement of birefringence a waveplate is sometimes introduced either between the first polarizer and the cell or between the cell and the analyzer. Detection may be done either by a d.c. measuring device or by a lock-in amplifier or by a C.R.O. Setting up of the first polarizer with respect to the direction of the magnetic field may be carried out by a magneto-optical effect called bimagnetic rotation (48). Measurement of the field free relaxation time which in turn measures average particle size - can also be carried out by incorporating a simple exponential time base in C.R.O. (67). Details of the various setups are described in the above references.

With the above experimental setup the following investigation can be carried out:
(i) Variation of induced extinction (or dichroism) as a function of applied field, (ii) variation of induced extinction (or dichroism) as a function of wavelength, (iii) birefringence variation, (iv) wavelength dependence of the induced birefringence, (v) polarization dependence of induced extinction, (vi) variation of scattered intensity as a function of field strengths and wavelength.

To supplement the data sometimes temperature dependence (17,61) and concentration dependence (27) of the effects are also studied.

Treatment of the Data

The experimental study of the various magneto-optical effects mentioned above, together with the appropriate theoretical expressions described earlier, are used to deduce the following inferences regarding the colloidal solutions:

i) Nature of orienting torque, viz. polar, anisotropic or combination of the two (17)
ii) Size and size-distribution (15)
iii) Shape factor of the particles (14)
iv) Optical anisotropy (13)
v) Magnetic anisotropy (19)
vi) Aggregate formation (46)
vii) Spatial ordering in the medium (40).

To arrive at the quantitative deduction one has first to decide about the nature of the orienting torque. This is usually done by least-square fitting of theoretical and experimental curves. Various theoretical curves are given in previous papers. Ratio of optical polarizabilities α_1/α_2 is determined from saturation values of an observed parameter, e.g. $(C_{ext})_{l\infty}$ and $(C_{ext})_{r\infty}$. In certain cases it is possible to determine the saturation values by plotting the parameter versus $1/H^2$. In other cases experimental curve of logarithm of the observed parameter (e.g. induced phase difference δ) versus logarithm of the magnetic field is matched with the theoretical curve of logarithm of the appropriate function versus logarithm of h. To match these curves, coordinates of the experimental curve are suitably shifted from the actual values. The ordinate shift is shown to be equal to the saturation value of the parameter, while abscissa shift yield the value of torque due to magnetic property from which magnetic properties, viz. polar moment, anisotropy in susceptibility etc., can be calculated (19).

From the saturation values ratio of optical polarizability is calculated. If the particles are optically isotropic then one can calculate shape factor from the depolarization values. The method is fully discussed elsewhere. If the particles are optically anisotropic then saturation values obtained from the extinction study and the birefringence study will be different. In case when the colloidal particles are polydisperse then the two curves mentioned above will not match over the entire range. Matching at low and high field yields two moments of the distribution curve. Assuming an appropriate distribution one can plot size or volume distribution curves (15,62).

Magnetostatic attraction between individual particles may lead to formation of aggregates in magnetic colloids (46). This can be ascertained by extinction measurements in the transverse and the longitudinal configurations. From the theoretical investigation it can be shown that if

$$(C_{ext})_1^T - (C_{ext})_o = (C_{ext})_o - (C_{ext})_r^T \quad \text{and} \quad (C_{ext})_r^T = (C_{ext})^L$$

then no aggregates are formed. If the 1st relation is satisfied while second one is not satisfied then aggregates formation has taken place but aggregates are within Rayleigh-Gans region. If both the relations do not hold then large aggregates are present in the colloid.

IV. Results

Table 1 summarizes some of our published results.
Recently the effects in a commercial magnetic colloid EMG-707 were studied by us. The colloid was obtained from Ferrofluidics corp. (USA). 0.2 ml of the colloid was diluted with distilled water to 150 ml. Figure 1 shows the variation in extinction with the applied field for the transverse configuration. In the longitudinal case transmission changes were found to be

Table 1. Magneto-optical characteristics of some colloids

No.	Colloid	Source	Torque	Size µm	Shape D/T	$m\times10^{16}$ emu/cm³	$e\times10^{4}$ emu/cm³	Ref.
1	Graphite[a]	AC	DMA	0.73	6.5	-	-	15
2	Oil dag	AC	DMA	0.68	5	-	-	15
3	Graphite[b]	DC	DMA	0.55	5.3	-	-	-[d]
4	Graphite	JDJ	DMA	10	-	-	-	62
5	Graphite	Mix	DMA	-	-	-	-	64
6	Bentonite	EM	DMA	0.06	4.5	-	0.16	18 19
7	αFe_2O_3	Lab	PMA	-[c]	-	-	2.0	13
8	Fe_3O_4	Lab	PMM	0.014	-	-	-	16
9	$CdO.Fe_2O_3$	Lab	OTPA	0.055	-	2.8	0.8	17
10	$NiO.Fe_2O_3$	Lab	OTPA	0.05	-	1.4	0.5	17
11	Molybdenite	AC	DMA	0.06	-	-	-	20

[a]Aquadag, [b]in water. AC Acheson Colloid Co., DC Dixon Crucibles (UK), JDS J.D. Jones Co. (India), Mix mixture of 3 & 4, EM Evans Medical (UK), Lab synthesized in the Laboratory. DMA diamagnetic anisotropy, PMA paramagnetic anisotropy, PMM permanent magnetic moment, OTPA orthogonal torques due to polar (P) and anisotropic (A) susceptibility. [c]$V = 14.6\times10^{-18}cm^3$. [d]J. Optics 6, 1 (1977).

dependent on time (Fig.2). It is observed that the equation (20) is not satisfied by this colloid. Further $(C_{ext})^L$ increases with time. It was observed that upon removal of the field initial transmission was restored. This suggests aggregation is induced in the colloid which confirms the results of Hayes (27). It was reported that the particles in the colloid are polydisperse and follow log-normal distribution with $\sigma = 0.45$ and volume average size = 11 nm. Figure 3 shows log-log matching of the theoretical and the experimental curves. Since curves do not fit over the entire range, so particles in the fluid must be polydisperse. From the abscissa shift

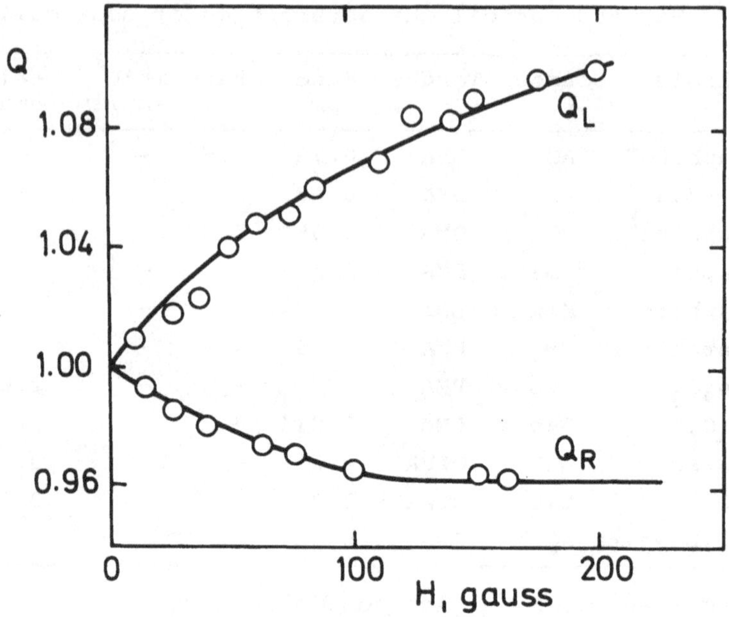

Fig. 1. Magneto-optical extinction for diluted ferrofluid EMG 707

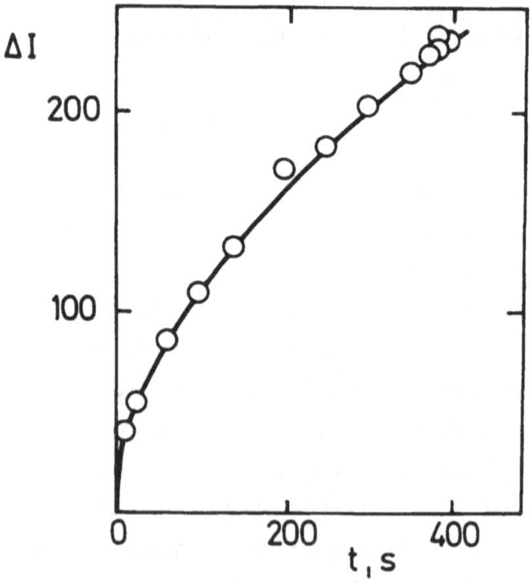

Fig. 2. Variation of induced intensity ΔI with time t for EMG 707

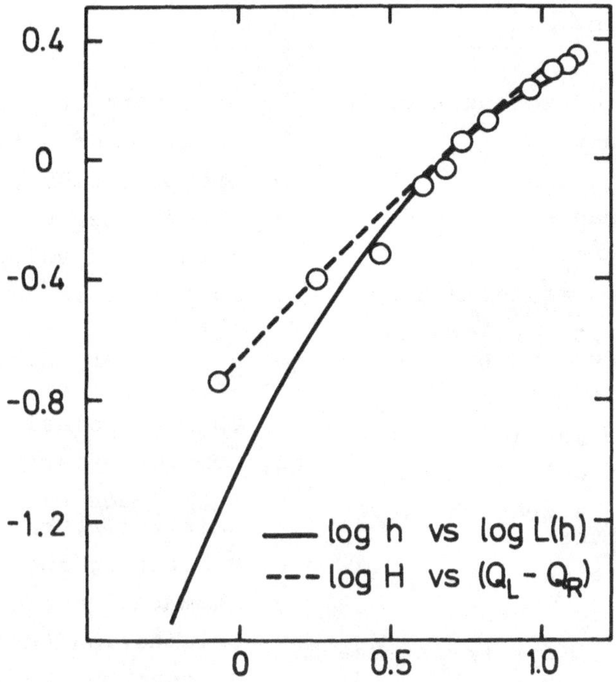

Fig. 3. Log-log matching of the theoretical and experimental curves.

we get m = 29.3×10^{-16} emu/cm³. This leads to I_s = 6600 Gauss. This is a good agreement with I_s = 6000 Gauss - the domain magnetization of Fe_3O_4.

Acknowledgement

Thanks are due to Dr. K. Raj of Ferrofluidics Corporation for providing samples of ferrofluids and certain data. Thanks are also due to Mr. J. M. Patel for his help in the experimental work.

List of Symbols

D	Diameter of the particle
E_h	Error function
E_γ	Error function
H	Applied magnetic field
I	Intensity of light
I_s	Saturation magnetization
L	Superscript for longitudinal configuration
$L_{1,2}$	Langevin function
N	Number of particles per cm^3
P	Degree of polarization
R	Distance between particle and detector
S	Scattering matrix
T	Temperature
T	Superscript for transverse configuration
V	Volume of particle
e	Superscript for optical polarizability due to electric field of light
h	Function of H
i	Imaginary part
k	Propagation constant
k_B	Boltzmann constant
n_m	Complex refractive index
m	Magnetization
Δ_n	Birefringence
0	Zero field value
S	Superscript for scattering
U	Suffix for unpolarized light
α_j	Optical polarizability
β_j	Complex polarizability
r	Angle made by symmetry axis w.r.t. the field
θ	Scattering angle
ϕ	Azimuthal angle
Φ	Orientation function
χ_i	Susceptibility
ζ	Coherency matrix
Σ	Summation
*	Complex conjugate
Re	Real part of
Im	Imaginary part of
Tr	Trace of

References

1. Makeehan, L.W.: Phys. Rev. 57, 1177 (1940).
2. (a) Majorana, Q.: Compt. Rend. 135, 159 (1902).
   (b) Majorana, Q.: Accad. Lincei, Atti. 11, 374, 531 (1902).
3. Schmauss, A.: Ann. Physik 12, 186 (1903).
4. Cotton, A., Mouton, H.: Compt. Rend 141, 317, 349 (1905).
5. Langevin, P.: Radium 7, 249 (1910).
6. Bitter, F.: Phys. Rev. 38, 1903 (1931).

7. Heaps, C.W.: Phys. Rev. 57, 528 (1940).
8. Elmore, W.C.: Phys. Rev. 60, 593 (1941).
9. Mueller, M., Shamos, M.S.: Phys. Rev. 62, 631 (1942).
10. Ewell, A.W.: Phys. Rev. 17, 292 (1903).
11. Heller, W.: Properties magneto-optiques des solutions colloidales, Herman, Paris (1939).
12. Heller, W., Quimfe, G., Ta, Y.: Phys. Rev. 62, 497 (1942).
13. Naik, Y.G., Desai, J.N.: Indian J. Pure Appl. Phys. 3, 27 (1965).
14. Stott, F.D.: Proc. Phys. Soc. B-62, 418 (1949).
15. Shah, H.S., Desai J.N., Naik, Y.G.: Indian J. Pure Appl. Phys. 6, 282 (1968).
16. Dave, M.J., Mehta, R.V., Naik, Y.G., Shah, H.S., Desai, J.N.: Indian J. Pure Appl. Phys. 6, 364 (1968).
17. Desai, J.N., Naik, Y.G., Mehta, R.V., Dave, M.J.: Indian J. Pure Appl. Phys. 7, 534 (1969).
18. Mehta, R.V., Shah, H.S., Dave, M.J.: J. Colloid Interface Sci. 35, 41 (1971).
19. Mehta, R.V.; J. Colloid Interface Sci. 42, 165 (1973).
20. Mehta, R.V., Shah, H.S.: J. Phys. D., Appl. Phys. 7, 2483 (1974).
21. Meetan, G.H.: J. Chem. Soc. Faraday Trans. 75, 1406 (1979).
22. Meetan, G.H.: J. Colloid Interface Sci. 87, 407 (1982).
23. Meetan, G.H.: J. Colloid Interface Sci. 73, 38 (1980).
24. Meetan, G.H.: J. Colloid Interface Sci. 74, 181 (1980).
25. Meetan, G.H.: J. Colloid Interface Sci. 84, 235 (1981).
26. Rosensweig, R.E.: Advances in Electronics and Electron Physics 48, 239 (1979).
27. Hayes, C.F.: J. Colloid Interface Sci. 52, 239 (1975).
28. Hayes, C.F., Hwang, S.R.: J. Colloid Interface Sci. 60, 443 (1977).
29. Martinet, A.: Rheol. Acta 13, 260 (1974).
30. Moiorov, M.M.: Magnetohydrodynamics 3, 39 (1977).
31. Neitzol, U., Barner, K.: Phys. Lett. 63A, 327 (1977).
32. Mehta, R.V.: Thermomechanics of Magnetic Fluids (Berkovsky, B., Ed.) Hemisphere 139 (1978).
33. Bibik, E.E., Lavrov, I.S.: Kolloid Zh. 26, 391 (1964).
34. Bibik, E.E., Lavrov, I.S., Merkushev, O.M.: Kolloid Zh. 28, 631 (1966).

35. Bibik, E.E., Lavrov, I.S., Merkushev, O.M.: Chem. Abs. 70 (1969).

36. Bibik, E.E., Simonov, A.A., Korenev, A.D., Lavrov, I.S., Lreniva, M.A.: Kolloid Zh. 33, 648 (1977).

37. Skibin, Y.N., Chekanov, V.V., Raikher, Y.L.: Sov. Phys. JETP 45, 496 (1977).

38. Scholten, P.C.: IEEE Trans. Magn. MAG-11, 1400 (1975).

39. Mehta, R.V.: IEEE Trans. Magn. MAG-16, 203 (1980).

40. Scholten, P.C.: IEEE Trans. Magn. MAG-16, 221 (1980).

41. Davies, H.W., Llewellyn, J.P.: J. Phys. D. Appl. Phys. 13, 2327 (1980).

42. Davies, H.W., Llewellyn, J.P.: J. Phys. D. Appl. Phys. 12, 311 (1979).

43. Davies, H.W., Llewellyn, J.P.: J. Phys. D. Appl. Phys. 12, 1357 (1979).

44. Llewellyn, J.P.: J. Phys. D. Appl. Phys. 16, 95 (1983).

45. Bacri, J.C., Salin, D.: J. Physique-Letters 432, 771 (1982).

46. Mehta, R.V.: J. Magn. Magn. Mat. 39, 64 (1983).

47. Martinet, A.: Nouv. Rev. D'optique Applique 2, 283 (1977).

48. Mehta, R.V., Dave, M.J., Desai, J.N.: Appl. Opt. 10, 2786 (1971).

49. Goldberg, P., Hansford, J., Von Herrden, P.J.: J. Appl. Phys. 42, 3874 (1971).

50. Muray, J.J.: Appl. Opt. 4, 1011 (1965).

51. Skibin, Y.N., Chekanov, V.V.: Magnetohydrodynamics 2, 137 (1977)

52. Taketomi, S.: Japan J. Appl. Phys. 22, 1137 (1983).

53. Chung, D.Y., Hickman, T.R., De Paula, R.P., Cole, J.H.: J. Magn. Magn. Mat. 39, 71 (1983).

54. Van de Hulst: Light Scattering by Small Particles, John Willey and Sons (1957).

55. Parke, N.G.: J. Math. Phys. 28, 2 (1949).

56. Asano, S., Yamamoto, G.: Appl. Opt. 14, 29 (1975).

57. Mehta, R.V., Shah, H.S., Desai, J.N.: J. Colloid Interface Sci. 36, 80 (1971).

58. Mehta, R.V., Desai, R.V.: J. Colloid Interface Sci. 46, 285 (1974).

59. Mehta, R.V., Patel, J.M.: J. Optics 6, 5 (1977).

60. Kielich, S.: Acta Physica Polonica A 37, 447 (1970).

61. Mehta, R.V., Bhagat, D.M.: IEEE Trans. Mag. MAG-16, 258 (1980).

62. Bhagat, D.M., Mehta, R.V., Shah, H.S.: J. Colloid Interface Sci. 81, 14 (1981).

63. Kerker, M., Wang, D.S., Giles, C.L.: J. Opt. Soc. Am. 73, 765 (1983).

64. Mehta, R.V., Shah, H.S., Bhagat, J.B., Bhagat, D.M.: IEEE Trans. Magn. MAG-16, 1324 (1980).

65. Stoimenova, M., Stoylov, S.: J. Colloid Interface Sci. 76, 502 (1980).

66. Sokerov, S., Stoimenova, M.: J. Colloid Interface Sci. 46, 94 (1974).

67. Mehta, R.V., Dave, M.J.: Rev. Sci. Instrum. 42, 1074 (1971).

CHARACTERIZATION OF INVERSE POLYACRYLAMIDE LATICES BY QUASI-ELASTIC LIGHT SCATTERING

Françoise Candau, Christine Holtzcherer
Centre de Recherches sur les Macromolécules (CNRS)
6,rue Boussingault - 67083 STRASBOURG Cedex (France)
Sauveur Candau
Laboratoire d'Acoustique Moléculaire - Université Louis Pasteur
4,rue Blaise Pascal - 67070 STRASBOURG Cedex (France)

Introduction

The quasi-elastic light scattering technique tends to become a routine technique for the characterization of latices. The experimentally measured parameter is the translational diffusion coefficient of the particles which, in the infinite dilution limit, is inversely proportional to their hydrodynamic radii. For hard spheres, the hydrodynamic radius is equal, within the experimental accuracy to the geometric radius.

Recently, we have developed an original method of preparation of inverse polyacrylamide latices using a microemulsion polymerization procedure (1). The acrylamide was polymerized inside water-swollen micelles stabilized by Aerosol OT and dispersed in toluene. The structure of these systems is rather complex and their characterization by classical light scattering arises some difficulties (2).

The purpose of this paper is to show that quasi-elastic light scattering can still provide a measurement of the particle size. We also probe by means of the same technique the evolution of the particle size when a seeding procedure is applied to the systems in order to increase the polyacrylamide concentration . The principle of the latter procedure is analogous to that commonly used in normal emulsion polymerization.

Physical Optics of Dynamic Phenomena and Processes in Macromolecular Systems

Results

1 - One stage polymerization. The procedure of the polymerization has been described elsewhere (1). Fig. 1 shows the variation of the average diffusion coefficient D with the volume fraction of the dispersed phase ϕ for 3 microemulsions differing by their acrylamide content and for the corresponding latices formed after polymerization of acrylamide (Tab.1). Two features can be observed.

Table 1. Characteristics of the systems

Systems	Composition (wt/wt)				R_1 (Å)	R_2 (Å)
	Tol.	AOT	Water	AM	(before pol.)	(after pol.)
N	67.80	16.90	10.22	5.08	50	215
A	62.20	17.30	10.10	3.40	38	160
M 4512 e	76.37	13.47	7.99	2.17	37	250

First, the absolute values of the slope of D(ϕ) are much larger for the microemulsions than for the latices, the higher the acrylamide content, the larger the value of the slope. Addition of acrylamide to the AOT/water/toluene system leads to an enhancement of the attractive interaction forces between particles due to the localization of the monomer at the W/O interface. After polymerization, the polyacrylamide is strongly collapsed within the particles and the attractive forces are strongly reduced.

Second, the polymerization process produces a considerable decrease in the diffusion coefficient. This indicates a growth of the particle size (Tab.1). On the other hand, the variance of the autocorrelation function of the scattered light, which is associated with the particle size distribution, is not strongly affected by the polymerization (Fig.2). As a consequence of the particle growth, the AOT emulsifier required to stabilize the polymer latex particles is in excess and therefore small AOT micelles (d $\sim$ 30 Å) are formed in the medium which coexist with latex particles (2). The presence of these small micelles scattering scarcely light cannot be detected in the auto-correlation function which thus provides a true measurement of the radius of the latex particles.

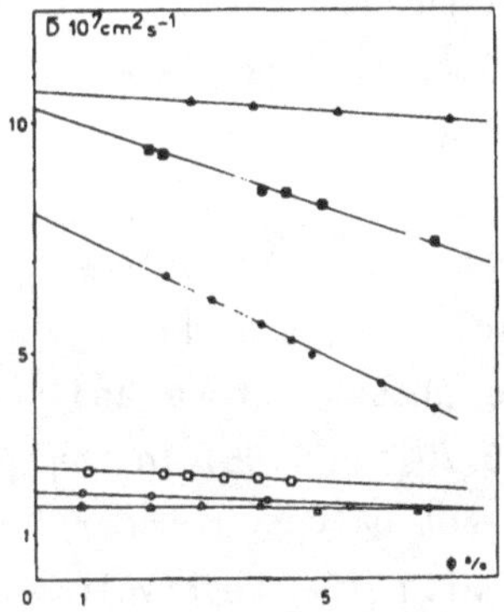

Fig. 1. Variation of D with Ø

microemul. ▲ M 4512 e
■ A; ● N

latices △ M 4512 e
□ A; o N

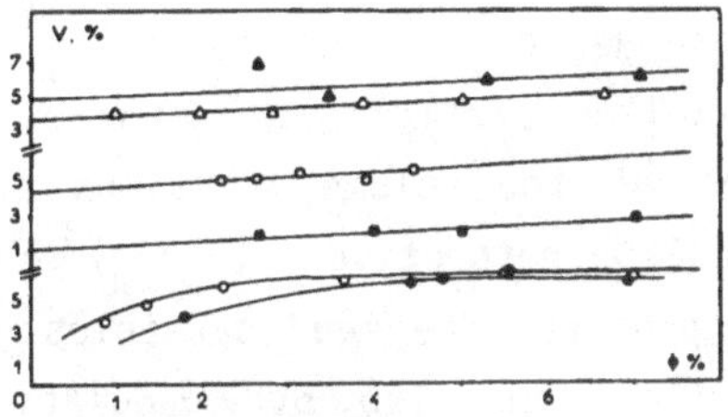

Fig. 2. Variation of the variance with Ø for the corresponding systems

2 - Seeding process . The seeding procedure consists in swelling pre-polymerized latex particles (seed system) with monomer followed by a second step polymerization (3). This allows to prepare latexes with higher polymer contents. In conventional emulsion polymerization, the stabilization of the system is achieved by addition of emulsifier to the seed latex. In our case, the emulsifier can be provided by the AOT micelles present in the medium. As a matter of fact, it is possible to dissolve into the seed system acrylamide powder at concentrations up to 11%(wt/wt). Acrylamide obviously dissolves into the latex particles rather than in the AOT micelles as demonstrated by the large decrease in the diffusion coefficient (cf.Fig.3b). This could be inferred from the Kelvin equation, considering the large difference in the chemical potentials of monomer in the two classes of particles.

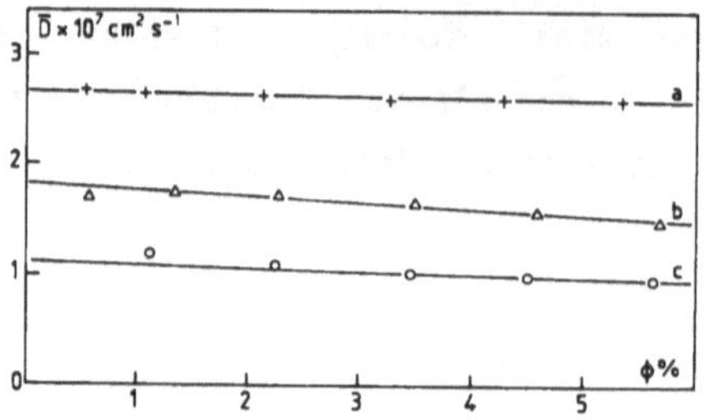

Fig. 3. Variation of D with the volume fraction a) seed latex, R_a = 150 Å b) swollen seed latex, R_b = 215 Å; c) final latex R_c = 350 Å

The swelling degree of the particles that is the ratio of the volume of the swollen particle over that of the seed particle, is much larger than that of the volume fractions of the dispersed phases before and after swelling respectively ($(R_b/R_a)^3 = 2.95$ whereas $(\phi_b/\phi_a) = 1.06$). In the swelling process, the seed particles grow presumably not only by monomer incorporation but also by coagulation and fusion with the reservoir micelles. It can also be observed that the swelling with acrylamide is also accompanied by an enhancement of the attractive forces, as shown by the negative slope of the D(Ø) variation (Fig.2b).

The polymerization of acrylamide in the seed again produces a further growth of the particles as inferred from the results of Fig. 3c, since the D values are considerably lowered by the polymerization process. The variance is not strongly affected indicating that the distribution of the particle size remains relatively narrow.

Acknowledgements: We wish to thank Dr. Y.S. Leong for providing some of the samples.

References

1) Leong, Y.S., Candau, F. : J.Phys.Chem. 86, 2269 (1982).
2) Candau, F., Leong, Y.S., Pouyet, G., Candau S. : J.Coll.Int.Sci. (in press).
3) Smith, W.V. : J.Amer.Chem.Soc. 70, 3695 (1948).

INTERACTION BETWEEN MICELLES OF POLY(STYRENE-b-HYDROGENATED BUTADIENE-b-STYRENE) IN 1,4-DIOXANE/n-HEPTANE MIXTURES

Čestmír Koňák, Petr Štěpánek, Zdeněk Tuzar
Institute of Macromolecular Chemistry, Czechoslovak Academy of Sciences, 162 06 Prague 6, Czechoslovakia

Introduction

Quasielastic light scattering (QELS) from systems of interacting particles provides the collective diffusion coefficient, D_c. In dilute solutions of particles, D_c can be represented by a virial expansion in the particle volume fraction ϕ or in the concentration c: $D_c/D_o = 1 + k_D\phi = 1 + k_D'c$, where D_o is the diffusion coefficient at infinite dilution, k_D and k_D' are interaction parameters; $k_D = k_D'/\bar{v}$, $\bar{v}$ being the partial specific volume of the particles. The simplest model system for theoretical calculation of k_D is a dispersion of hard spheres. Two types of interactions are considered in the calculation of k_D: static ones, proportional to the second virial coefficient, A_2, and hydrodynamic ones, which are responsible for the scatter of calculated k_D values (cited, e.g., in (1)) ranging from -6 to +3. Most of the model systems for experimental testing (lattices, colloid dispersions of metals, etc.) are unsuitable due to electrostatic interactions. For silica particles with hydrocarbon chains on their surfaces (so far the best approximation of hard spheres) in cyclohexane, $k_D = 1.3 \pm 0.2$ (2).
In this study, block copolymer micelles, for which geometrical and hydrodynamical dimensions coincide (3), were used as an experimental model for hard spheres. A_2 and k_D were measured by integral and quasielastic light scattering and compared with theoretical values.

Physical Optics of Dynamic Phenomena and Processes in Macromolecular Systems

Experimental

Micellar system: Spherical micelles with aliphatic cores and polystyrene shells of three-block copolymer Kraton G-1650 (Shell product) in mixtures 1,4-dioxane/0-30 vol.-% n-heptane. Methods: Quasielastic light scattering (QELS), Ar laser $\lambda = 514.5$ nm, $\Theta = 90°$, 96 channel digital correlator. D_c values were evaluated from autocorrelation functions (in all cases by the perfect single-exponential fit, indicating that micelles were practically monodisperse). Hydrodynamic radius, R_H, was calculated from the Stokes formula $R_H = kT/6\pi\eta D_c$, where η is the solvent viscosity. Integral light scattering (ILS): Sofica instrument, $\lambda = 546$ nm, $\Theta = 30° - 150°$. M_w and A_2 were evaluated from $K'c/R_o = 1/M_w + 2A_2c$, where K' is the optical constant and R_o is the Rayleigh ratio at $\Theta \to 0$.

Results

QELS and ILS data on block copolymer micelles (Figs. 1a and 1b) provide values of basic parameters like relative micellar mass, M_w, hydrodynamic radius, R_H, and interaction parameters, A_2 and k_D (Table 1).

Table 1: Light scattering data for Kraton micelles in 1,4-dioxane/n-heptane mixtures at 26°C. For the definitions of measured and calculated quantities see text.

n-Heptane	Measured values				Calculated values	
	$M_w \times 10^{-6}$	$A_2 \times 10^6$	R_H	k_D'	$A_2^{HS} \times 10^6$	k_D
vol.-%		$cm^3\,g^{-2}mol$	nm		$cm^3\,g^{-2}mol$	
0	4.80	2.1	19.5	4.02	3.07	1.09
10	4.44	4.2	22.3	6.63	5.66	1.05
20	4.20	7.5	23.2	7.54	7.19	1.01
30	3.77	12.7	24.9	11.95	10.97	1.15

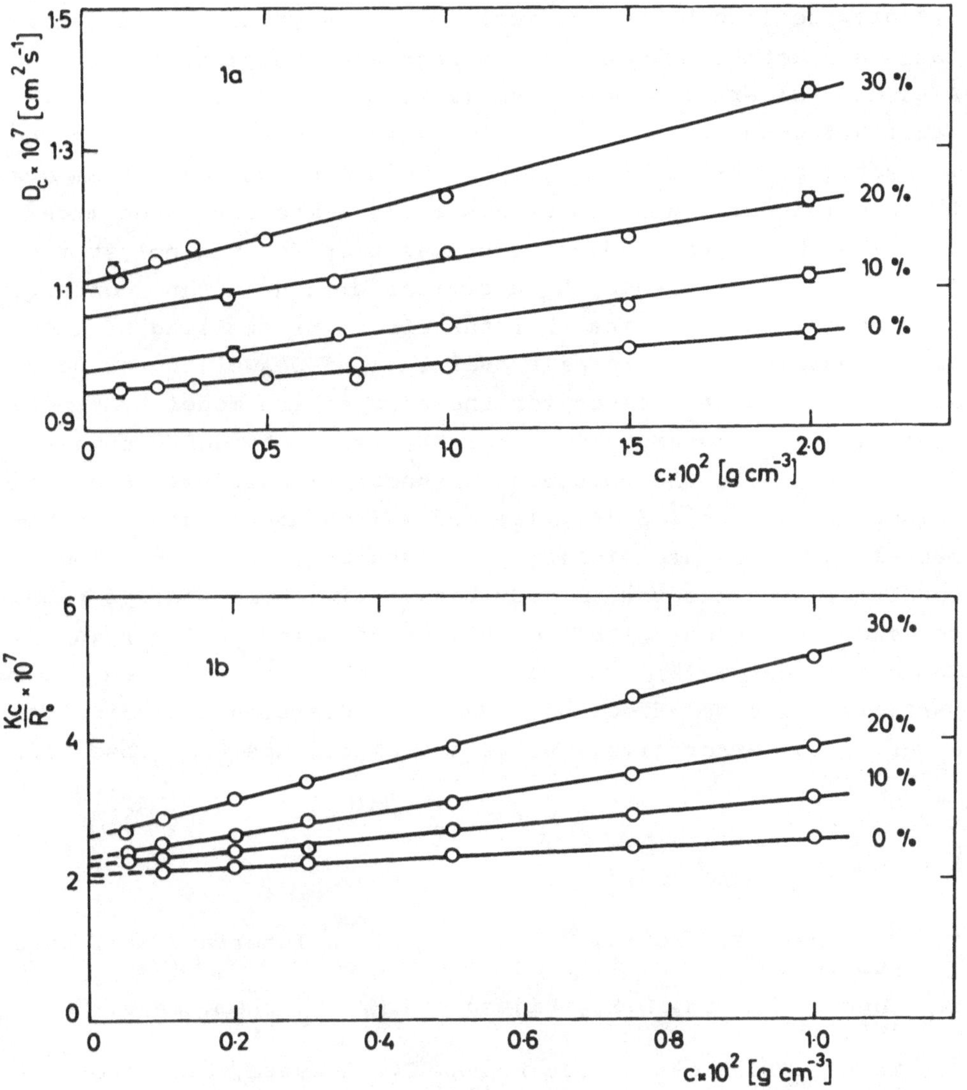

Fig. 1. Concentration dependences of D_c (1a) and Kc/R_o (1b) for Kraton micelles in 1,4-dioxane/n-heptane solvent mixtures at 26°C. Curves are labelled by vol.-% of n-heptane.

1,4-Dioxane is a better solvent for polystyrene, while n-heptane is a better solvent for hydrogenated polybutadiene. Although solvent mixtures used (i.e., up to 30 vol.-% n-heptane) are good solvents for polystyrene and poor ones for the aliphatic blocks, an increase of the n-heptane content improves the thermodynamic quality of a solvent mixture towards micellar aliphatic cores and deteriorates that for the polystyrene shell. As a consequence, M_w decreases with n-heptane content. The increase of R_H means that the effect of swelling of micellar aliphatic cores prevails over that of deswelling of polystyrene shells. A_2 values for the hard sphere model system is given (4) by $A_2^{HS} = 4N_A V/M_w^2$, where N_A is the Avogadro number and $V = 4\pi R_H^3/3$ is the volume of a spherical particle. Since the values of A_2 from ILS (Fig.1a) and A_2^{HS} obtained using experimental R_H values are similar, the assumption that the block copolymer micelles under study behave like hard spheres seems to be correct. The calculated values of $k_D (= k_D' N_A V/M_w)$ in Table 1 (1.01 - 1.15) are smaller than that for modified silica particles (2). No effect of solvent composition exceeding the experimental error (i.e., solvent quality) has been observed.

References

1. Štěpánek, P., Koňák, Č.: Adv. Colloid Interface Sci. to be published.
2. Kops-Werkhoven, M.M., Fijnaut, H.M.: J. Chem. Phys. 74, 1618 (1981).
3. Tuzar, Z., Pleštil, J., Koňák, Č., Hlavatá, D., Sikora, A.: Makromol. Chem. 184, 2111 (1983).
4. Yamakawa, H.: Modern Theory of Polymer Solutions, Harper and Row, New York 1980.

PROPERTIES OF BLOCK COPOLYMER MICELLES NEAR THE C.M.C. and C.M.T.

Zdeněk Tuzar, Petr Štěpánek, Čestmír Koňák

Institute of Macromolecular Chemistry, Czechoslovak Academy of Sciences, 162 06 Prague 6, Czechoslovakia

Introduction

Micellization of block copolymers in selective solvents (i.e. good solvents for one block, which are poor solvents for the other block) obeys the model of closed association (1). This model is characterized by an equilibrium between unimer (molecularly dissolved copolymer) and spherical micelles (having a core formed of the insoluble blocks and a shell containing the soluble blocks), with a sharp mass and size distribution. Closed association model also assumes the existence of a critical micelle concentration (c.m.c.), below which only unimer and above which also micelles can be detected by a given method. In analogy to c.m.c., a critical micelle temperature (c.m.t.) can be defined, above which, at a given copolymer concentration, no micelles would be present. The objective of this study has been to detect and describe block copolymer micelles near c.m.c. and c.m.t. by light scattering methods.

Experimental

Copolymer: Fraction of a three block copolymer poly(styrene-b-hydrogenated butadiene-b-styrene) (Kraton G-1650, Shell product), free of homopolystyrene ($M_w = 7x10^4$, 29 wt.-% styrene). In selective solvents used, 1,4-dioxane and 1,4-dioxane/30 vol.-% n-heptane mixture, micelles are formed with aliphatic core and polystyrene shell (2).

Physical Optics of Dynamic Phenomena and Processes in Macromolecular Systems

Methods. Quasielastic light scattering (QELS): homodyne spectrometer, 96 channel digital correlator, Ar laser (λ = 514 nm), $\Theta = 90°$. Collective diffusion coefficient, D_c, was evaluated from the autocorrelation function $G(\tau) = Ae^{-\Gamma\tau} + B$, where $\Gamma = 2D_cK^2$, K being the scattering vector, τ time delay and A,B constants. - Integral light scattering (ILS): Sofica goniometer, λ = 546 nm, $\Theta = 30°-150°$. Processed data provide the average molar mass $M_w^{(av)} = M^{(m)}w^{(m)} + M_w^{(u)}w^{(u)}$, w being the mass fraction, (m) and (u) pertain to micelles and unimer, respectively.

Results

Micelles near c.m.c.

C.m.c. for any block copolymer micelles can be expected much smaller than for, e.g., soap micelles; this has not been determined by light scattering yet. Assuming the closed association model, the ILS data should follow the idealized plot on Fig.1 (1).

Fig.1. Schematic plot $1/M_w^{(av)} = f(c)$ for closed association.

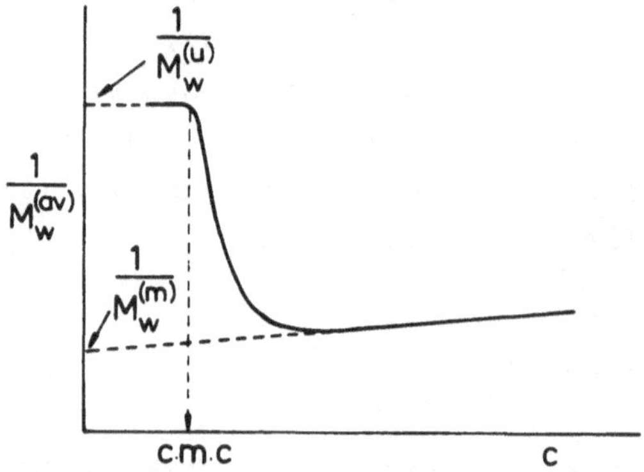

Experimental data (Fig.2) show a different behavior of micelles at low c. Both QELS and ILS failed to record c.m.c. The perfect single exponential fit in QELS measurements indicates practically monodisperse micelles even at the lowest measured concentration. Instead of an increase of D_c and $1/M_w^{(av)}$ with dilution, a decrease of both values below $c = 1x10^{-3}$ g cm^{-3} is

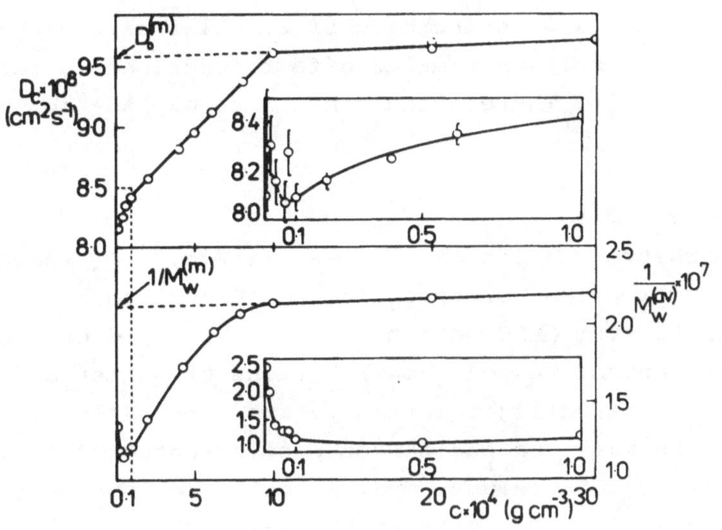

Fig. 2. Concentration dependences of D_c and $1/M_w^{(av)}$ for micellar system Kraton/1,4-dioxane at 26°C.

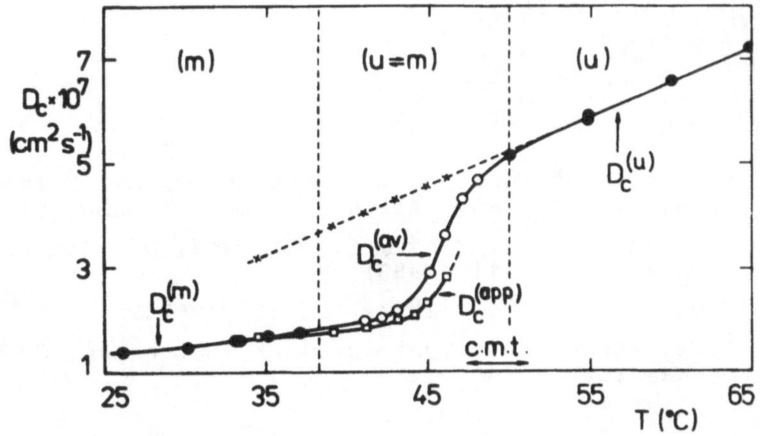

Fig. 3. Temperature dependence of D_c for Kraton solution ($c = 5 \times 10^{-3}$ g cm^{-3}) in 1,4-dioxane/30 vol.-% n-heptane. For explanation of symbols see text.

observed, implicating an increase of micellar size and mass. This finding resembles a similar effect described in the literature (cf. 3). There is no conclusive explanation of this effect.

Micelles near c.m.t.

Three temperature regions can be seen in Fig.3. D_c values in the two outer regions corresponding to unimer and micelles (●) were evaluated from QELS data by a perfect single exponential fit. In the central region, contributions of unimer and micelles to the scattered light intensity are comparable. $D_c^{(av)}$ values (o) are based on the forced single exponential fit of poor quality. $D_c^{(app)}$ values, evaluated by the four-parameter two-exponential fit (□) using extrapolated values (×) for the unimer contribution, must not be identified with $D_c^{(m)}$; $D_c^{(app)} \geq D_c^{(m)}$. Unfortunately, above 46°C in the vicinity of the c.m.t. micelles contribute to the total intensity of scattered light so little that $D_c^{(app)}$ cannot be evaluated. The complex problem of the evaluation of $D_c^{(u)}$ and $D_c^{(m)}$ in the central region is the object of a further study.

References

1. Elias, H.-G., in: "Light Scattering from Polymer Solutions", Huglin, M.B., ed., Academic Press, London 1972
2. Tuzar, Z., Pleštil, J., Koňák, Č., Hlavatá, D., Sikora, A.: Makromol.Chem. 184, 2111 (1983)
3. Lally, T.P., Price, C.: Polymer 15, 325 (1974)
4. Berne, B.J., Pecora, R.: Dynamic Light Scattering, J.Wiley and Sons, New York 1976, p.41

LATERAL DIFFUSION OF MICELLES MEASURED BY FLUORESCENCE RECOVERY AFTER PHOTOBLEACHING

Preliminary results on micelles labeled with a solubilized fluorescent dye

Werner Van De Sande (°), André Persoons
Laboratory for Chemical and Biological Dynamics
University of Leuven, Celestijnenlaan 200 D, B-3030 Leuven

Introduction

Fluorescence Recovery After Photobleaching (FRAP) is a well established technique to measure the lateral diffusion of fluorescent molecules and macromolecules labeled with a fluorophore (1,2,3). Our preliminary results show that the FRAP technique is also applicable to micellar solutions. The labeling of the micelles is achieved by solubilizing a fluorescent dye.

Materials and Methods

We used the cationic surfactant N,N-dimethyldodecylammonium chloride (DDAC). The preparation and purification is dis-cribed in (4). The fluorescent dye was an N(10)-alkylated acridine orange derivative (3,6-bis(dimethylamino)-10-dodecyl acridinium bromide, Fig. 1). The preparation is dis-cribed in (5). A new and simple purification method yielded a purity of 99% (to be compared to 93% obtained earlier (6)). The dye is very slightly soluble in water. Solutions were prepared by mixing aqueous solutions of the three components (surfactant, NaCl and dye).

Physical Optics of Dynamic Phenomena and Processes in Macromolecular Systems

Fig.1. The fluorescent dye 3,6-bis(dimethylamino)-10-dodecyl acridinium bromide (λ_{max} = 492 nm).

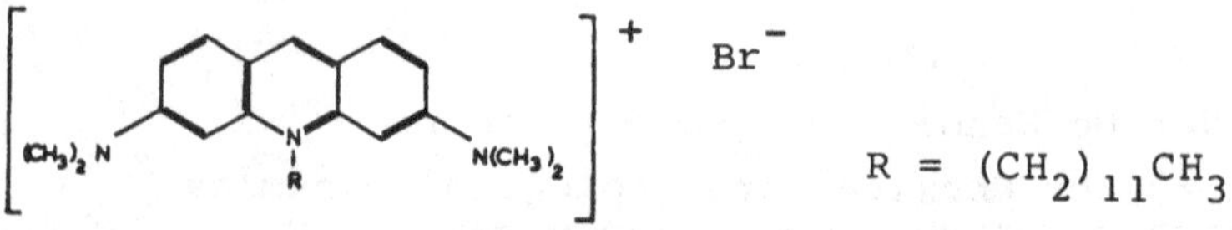

The main features of the FRAP apparatus were as follows. An argon-ion laser was operated at 488 nm. The laser intensity of the bleaching pulse was about 200 mW. To monitor the fluorescence recovery this intensity was optically attenuated by a factor 10^4. Typical bleaching times were 5 ms The radius of the uniform circular beam, focussed through a microscope objective, was 6 µm.

Theory and Data Analysis

The experimental curve F(t) was normalized in the form of the fractional fluorescence recovery curve f(t):

$$f(t) = \frac{F(t) - F(0)}{F(\infty) - F(0)} \qquad (1)$$

The fractional recovery curve was fitted to the exact closed formula (7):

$$f(t) = \exp(-x) \times \{ I_0(x) + I_1(x) \} \qquad (2)$$

I_0 and I_1 are modified Bessel functions, $x = 2\tau_D/t$, $\tau_D = \omega^2/4D$, ω is the radius of the circular beam, D is the diffusion coefficient.
The value of the intensity F(0), immediately after bleaching, was obtained by fitting the initial part of F(t) to a poly-

nomial of second degree. The initial points after the bleaching pulse were omitted.
The value of the intensity $F(\infty)$, after full recovery, was calculated as the mean fluorescence intensity before the bleaching pulse. If the decrease in the total sample concentration due to photobleaching can be neglected, this assumption is valid.
We also want to mention the possibility of accumulating several normalised fluorescence curves, thereby improving the signal to noise ratio.

Results

A typical fractional fluorescence recovery curve (eq.1) and the corresponding fit (eq.2) is shown in Fig.2. The measured diffusion coefficient 3.7×10^{-7} cm^2 s^{-1} is comparable to the diffusion coefficient measured by quasi-elastic light scattering on the same solution without added dye: 1.7×10^{-7} cm^2 s^{-1}.

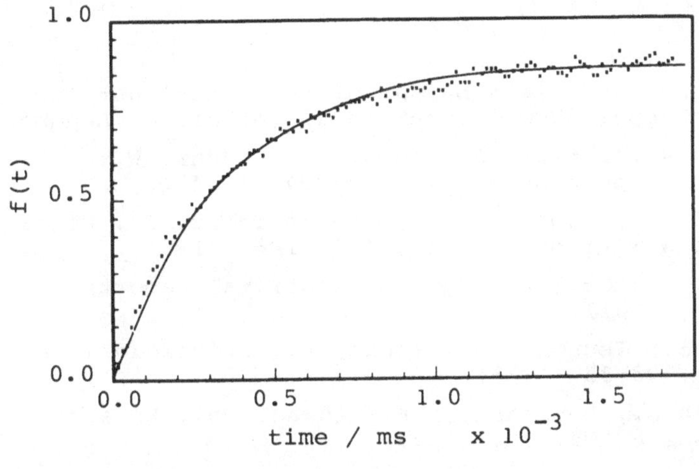

Fig.2. Fractional fluorescence recovery curve (x) and fitted curve (solid line) of 2×10^{-2} M DDAC in 2 M NaCl and 10^{-5} M dye.

The FRAP technique offers the possibility of extrapolating the fluorophore concentration to infinite dilution while the micellar "background" concentration can be kept constant. One thus obtains the tracer diffusion coefficient of the micelles. QELS measures the mutual diffusion coefficient (8). By comparison of the results of both techniques one can thus experimentally investigate the dependence of both diffusion coefficients on several parameters.

Acknowledgements

The authors want to thank the Research fund of the University for grant OT/X/32 and the National Fund for Scientific Research for grant FKFO No. 2 0074 83. Dr. Y. Engelborghs is acknowledged for his helpfull comments.

References

(°) W. Van De Sande is a bursar of the "Instituut voor Wetenschappelijk Onderzoek in Nijverheid en Landbouw".

1. Peters, R., Peters, J., Tews, K.H., Bähr, W.: Biochim.Biophys.Acta 367, 282-294 (1974).
2. Axelrod, D., Koppel, D.E., Schlessinger, J., Elson, E., Webb, W.W.: Biophys.J. 16, 1055-1069 (1976).
3. Peters, R.: Cell Biology International Reports 5, 733-759 (1981).
4. Ozeki, S., Tsunoda, M., Ikeda, S.: J.Colloid Interface Sci. 64, 28-35 (1978).
5. James, A.D., Robinson, B.H.: Advan. Mol. Relaxation Processes 8, 287-304 (1976).
6. Yamagishi, A., Masui, T., Watanabe, F.: J.Phys.Chem. 85, 281-285 (1981).
7. Soumpasis, D.M.: Biophys.J. 41, 95-97 (1983).
8. Phillies, G.D.J.: J.Colloid Interface Sci. 86, 226-233 (1982).

MOLECULAR DYNAMICS SIMULATION OF THE N-ALKANES ROTATOR PHASE

M.A.Mazo, E.F.Oleynik
Institute of Chemical Physics, USSR Academy of Sciences, ul. Kosigyn 4, Moscow, 117977, USSR
N.K.Balabaev, L.V.Lunevskaya
Computer Research Centre, USSR Academy of Sciences, Pushchino, 142292, USSR
A.G.Grivtsov
Institute of Physical Chemistry, USSR Academy of Sciences, Leninskii Prosp. 31, Moscow, 117312, USSR

Abstract

The method of molecular dynamic simulation has been applied for the investigation of structure and dynamics of chain molecular system (C_{10}-n-alkane). The behaviour of individual molecules as well as condensed system of 31 molecules at the temperature 150-275 K has been analyzed. The influence of chemical defects, such as methylene group or chlorine atom, on the dynamics of the system has been considered.

Introduction

The structure and the properties of amorphous polymers in the glassy state attract an ever increasing attention of investigators /1,2/. Emphasis is placed on the nature of chain mobility in a solid phase, on the geometry and the energy of moving entities and on the relation between mobility and concentration of different defects. Successful application of the method of molecular dynamic (MD) simulation to polymeric systems /3÷5/ seems to make it promising for the investigation of solid state of polymers.

To ascertain what part various types of defects play in the formation of amorphous polymeric structures, we attempted earlier /6-7/ to investigate this problem from a model of the internal region of a polyethylene single crystal into which various physical and chemical defects could be introduced. Due to computer limitations we could work with the

Physical Optics of Dynamic Phenomena and Processes in Macromolecular Systems

chains of C_{10}-type. It turned out that in the temperature range 155-275 K the model corresponded to the "rotator" or "hexagonal" phase of n-alkanes which was subjected to intensive study last years /8-13/. It has been recently found that along with the rotational mobility of molecules this phase exhibits intensive dynamics of backbone torsional angles /8,10/. This fact makes it difficult to interpret some experimental data and requires more careful investigations of the behaviour of the n-alkane rotator phase /14-16/.

In this paper we present some results of the MD simulation of structure and dynamics of C_{10}-n-alkane based on the previously reported structural cell /6,7/. Part of the results have been recently published in /17/. We believe this analysis is the first step in the MD investigation of amorphous polymers in the solid state.

The structural model and simulation method

A structural cell for the computer simulations had the shape of rectangular parallelepiped which contained 31 molecules of $C_{10}H_{20}R_2$ (Fig.1). The long axes of molecules were aligned along the cell axis Z. 14 molecules (arranged on the boundaries of the structural cell, Fig.1) were motionless during simulation experiments. The configurations of these 14 molecules were always kept in the form of a planar zig-zag with parameters of orthorhombic n-alkane crystal. The rest (17 chains) were moved during the simulations in accordance with interatomic potential.

The atoms of each molecule were represented by material points. All chemical bonds were rigid. Valence angles were not fixed and were determined by the special potentials of atom interactions /7/. All chemically non-bonded atom-atom interactions were introduced by Lennard-Jones (6-12) potential. Chain-end atoms R of mobile chains were bonded with the boundary planes (Z-planes) by quadratic potential.

In most of simulations chain-end atoms R of mobile chains were bounded with the boundary planes (z-planes) by potential

$$U_b(z)=C_b(Z_b-z)^2,$$

where z is z-coordinate of atoms R, Z_b is the coordinate of Z-plane and C_b = 1 kcal $Å^{-2}$.

Chemical defects were introduced into the system by the substitution of $-CH_3$ (Me) or -Cl for H-atom in the centre of the cell. The group $-CH_3$ was the material point with the corresponding parameters. The values of the parameters used are given in /7,17/.

Equations of motions for the 17 mobile molecules were integrated by the known algorithm /18/. The time step for the integration was 2×10^{-3} ps. The fluctuations of the total energy of the system over the whole run (about 10 ps) were smaller than 0.2%. The temperature of the system was determined from the average kinetic energy per degree of freedom.

Initially, all the molecules were arranged in the cell to make an ideal orthorhombic polyethylene crystal /17/. The parameters of this cell were: a=7.25 Å, b=4.95 Å, c=2.53 Å and setting angles φ=45° (Fig.1). The atoms of the 17 mobile chains were given random momenta corresponding to 370 K, and the motions of all the atoms over the period of 10 ps. Then a special run of 10^4 time steps was performed to obtain the equilibrium parameters of the cell to be agreed with the chosen interaction potentials, temperature and to eliminate external pressure. Unlike the other runs, the parameters a, b and c of the boundary chains varied slowly with the internal pressure. Over the run the total energy of the system slowly reduced.

At some moments the dimensions of the cell were fixed, and the obtained coordinates and velocities of atoms were taken as initial for further simulation. The value of the parameters a, b and c, density of the system and temperature of the performed runs are given in Table 1.

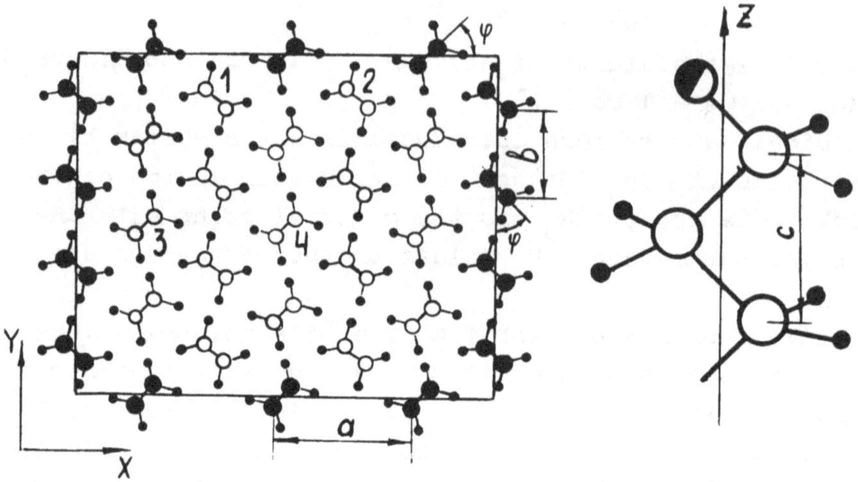

Figure 1. The structural cell for the computer simulation. ○, ● - C-atoms, • - H-atoms, ◐ - end group R. Chains with ● - marked C-atoms were kept immobile during the simulation (boundary chains). Their location corresponds to the orthorhombic lattice. Chains with ○ - marked C-atoms were mobile. Their positions shown are regular, whereas in the simulation they usually vary.

Table 1. Lattice parameters, density and temperature over the MD runs

Defect	Temperature K	Lattice parameters, Å			a/b	Density g cm^{-3}
		a	b	c		
-	155	7.08	4.38	2.517	1.62	1.20
-	200[a]	7.21	4.45	2.485	1.62	1.17
-	210	7.21	4.45	2.492	1.62	1.17
-	235	7.22	4.46	2.491	1.62	1.16
-	275	7.34	4.52	2.463	1.63	1.15
Methyl	275	7.32	4.54	2.466	1.61	1.14
Cl	275	7.32	4.54	2.460	1.61	1.14

a - In this case the potential U_b was not used.

Results

1. Crystal structure. Configurations of 17 mobile chains in their positions inside the structural cell were determined by the total energy of the system and the boundary conditions. Therefore in the computer simulation it was necessary to check the arrangement of the molecules in the cell and to compare them with the ideal lattice parameters.

The analysis showed that the system was crystalline over all the runs. This is clearly seen from Fig.2, where is shown the distribution of projection of the C-atoms of mobile chains onto crystal axes a, b, c at 155 K. The location of all the carbon atoms is practically of a plane-type, that reflects the existence of a crystalline lattice.

At a higher temperature one can observe essential broadening of the maxima and lowering their heights on the projection to c-axis. The maxima do not disappear completely even at 275 K. We have calculated the average projections of the

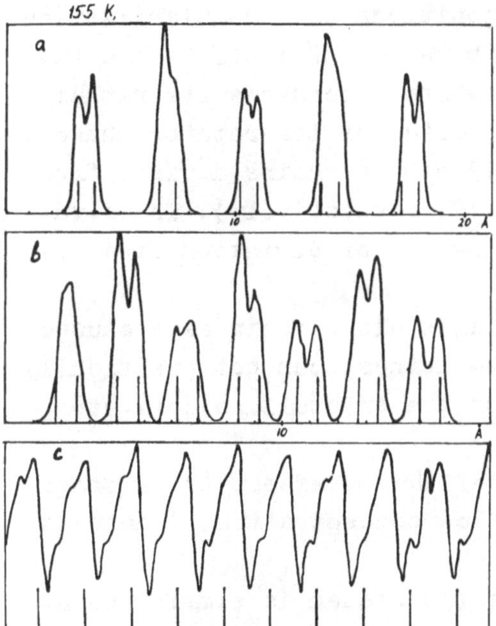

Figure 2. Distribution projections of the C-atoms of mobile chains onto crystalline axes at 155 K. The perpendiculars to x--axes show the position of C-atom location of boundary chains.

centres of masses of molecules (c.m.) onto X,Y plane and their standard deviations. Such a projection for the 275 K is shown in Fig.3. It has been revealed that the projections

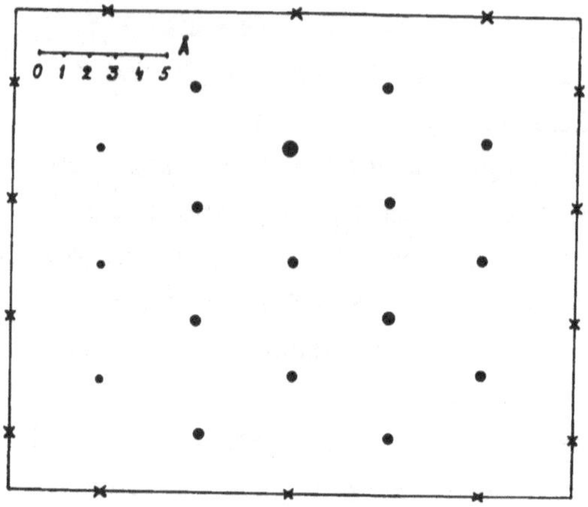

Figure 3. Average projections of the centres of masses (c.m.) of molecules onto (x,y) plane at 275 K: x – c.m. of motionless boundary molecules; • – c.m. of mobile molecules. The radii of the circles are proportional to standard deviation. The scale is shown in the left corner.

of c.m. of the chains fluctuate with small amplitudes (≤ 0.2 Å) around the average positions of a regular lattice with parameters close to a and b values of Table 1. The values a, b, a/b and their temperature dependence are rather close to the X-ray experimental data for the rotator phase of n-alkanes /9/ (a and b for 17 mobile chains deviate from experimental values by 4% and 14%, respectively). The c-value is normal for n-alkane chains (1% of deviation from experiment).

Restricted mobility of molecules along their axes caused by interactions with boundary z-plaines does not practically influence the mean parameters of the crystalline cell at least at 200 K.

Chemical defects within the cell do not affect the average sizes of the cell due to their low concentration (4 defects per 1000 carbon atoms).

Thus in the above parameters the system is similar to n-alkanes in the rotator phase.

2. Molecular conformations. As is found, all the backbone torsional angles (TA) fluctuate around the trans-positions with variable amplitudes. In Fig.4 is shown the dynamics of TA of the central molecule at 275 K. In Fig.5 are given the histograms of TA of all the mobile molecules for some runs. It is clear that at the temperature higher than 210 K one can observe fluctuations of TA from ±60° to ±120°. This may be termed as gauche conformations in the chains. Lifetime of the gauche conformations reached 3.5 ps at 275 K.

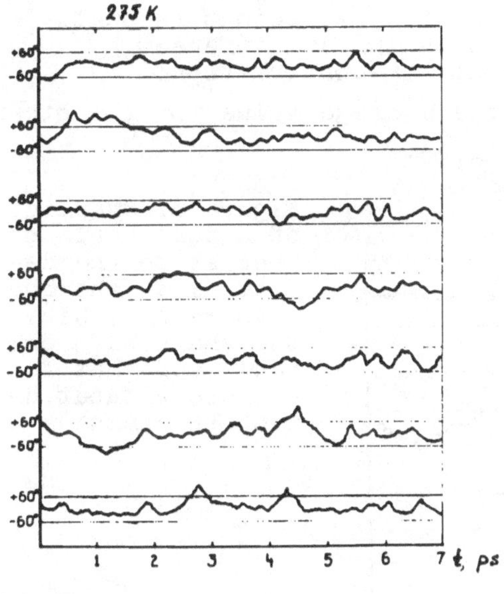

Figure 4. Dynamics of the backbone torsional angles. Each line corresponds to a single TA of the central chain at 275 K.

The careful analysis of these gauche conformations existing at 275 K for a period longer than 0.5 ps has revealed single $g^{\pm}$ conformations in 40% of all the cases. The conformations like $g^{\pm}tg^{\mp}$ were observed in 22% of all the cases. We have also found the $g^{\pm}g^{\pm}$ (7%), $g^{\pm}g^{\pm}g^{\pm}$ (15%) and $g^{\pm}g^{\pm}g^{\pm}g^{\pm}$(5%) conformations.

The chemical defects affect substantially the dynamic behaviour of the TA. The skeleton chemical bonds nearest to the defect appear only in trans-conformations and have TA

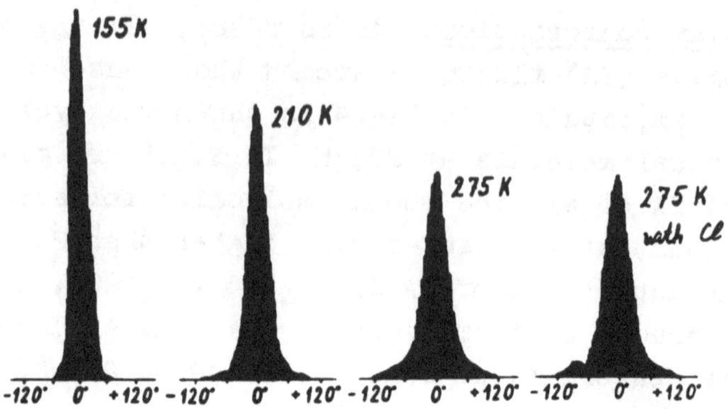

Figure 5. Distribution of the backbone torsional angles of all the mobile chains at different runs. 0° corresponds to trans-conformation.

amplitudes much smaller than the average value for the other TA at the same temperature (Fig.6).

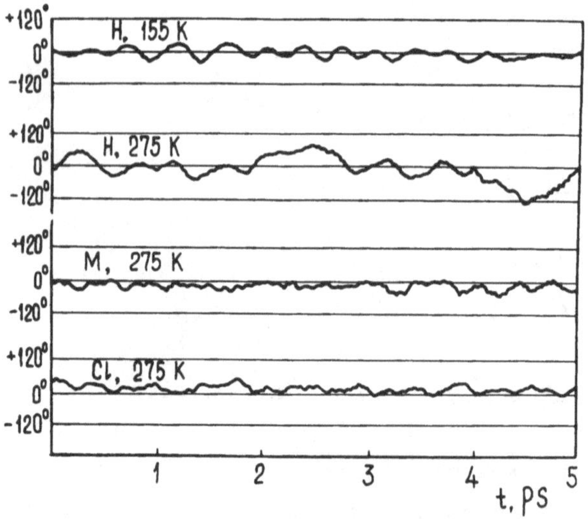

Figure 6. Dynamics of a torsional angle, nearest to the defect (M is the chain with methyl, Cl is the chain with chlorine), and that for a chain without defect (H-chains).

3. Valence angles. The potential (1) is chosen such that the equilibrium value of the CCC-angles is identically 109.5°. But the average CCC-angles in all the mobile chains

measured over each run were equal to 111.5±1°, which is in good agreement with experimental data /19/. The valence angles are independent of the chain position within the structural cell and of the temperature, though we can observe small systematic fluctuations (±0.3°) along the chain.

4. Rotating mobility of molecules. Since all the molecules in the cell in average exist in planar zig-zag conformations we can determine the momentary plane of the carbon atoms for each molecule. In Fig.7 are given histograms of distribution of angles between x-axis and these chain planes

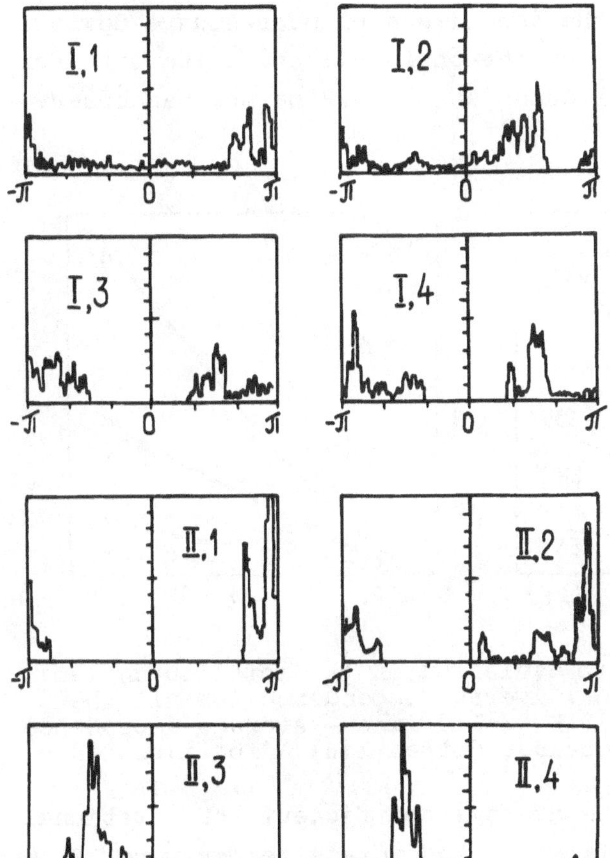

Figure 7. Distribution of angles between x-axis and chain planes. I - 275 K, II - 155 K; arabic numerals designate molecules according to Fig.1.

for 4 molecules at 155 K and 275 K.

All simulations have shown a wide spectrum of motions of the planes changing from a rather fast rotation to weak vibrations around equilibrium. The intensity of these rotations increases with the temperature.

No correlation between rotational mobility of the molecules, their average orientation and the position in the structural cell has been observed. However, at 155 K the system tends to the orthorhombic phase: some molecules located close to the boundary vibrate near the positions corresponding to the regular orthorhombic lattice of n-alkanes.

In Fig.8 is shown the dependence of mean-square deviation of the rotation angles of the chain planes on the time for different runs. In all cases at $t > 0.2$ ps one can observe a diffusion regime.

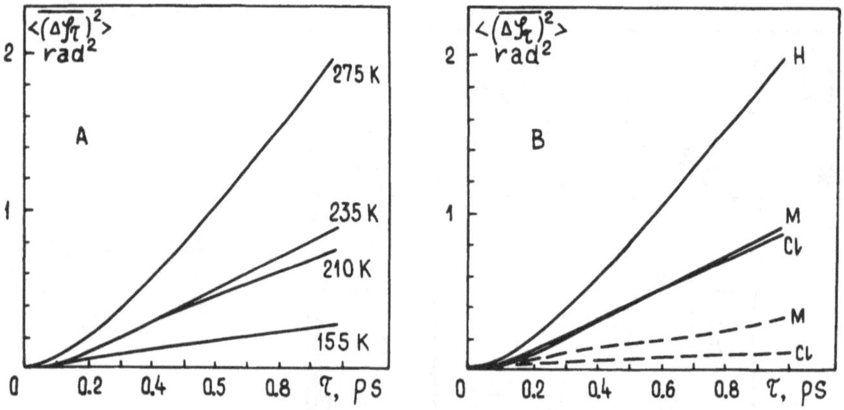

Figure 8. Mean square deviation of the rotation angles for different runs. A. Average dependence for all the mobile chains. B. 275 K. Solid line - average dependence for all defectless chains; dotted line - for isolated defective molecule.

Fig.9 represents the average coefficients of rotational diffusion D_r for mobile molecules versus temperature. D_r was determined as a ratio of mean square deviation of the rotation angles to time, here the time is equal to 1 ps. It is interesting, that the order of magnitude of D_r value is close to experimental data /20/.

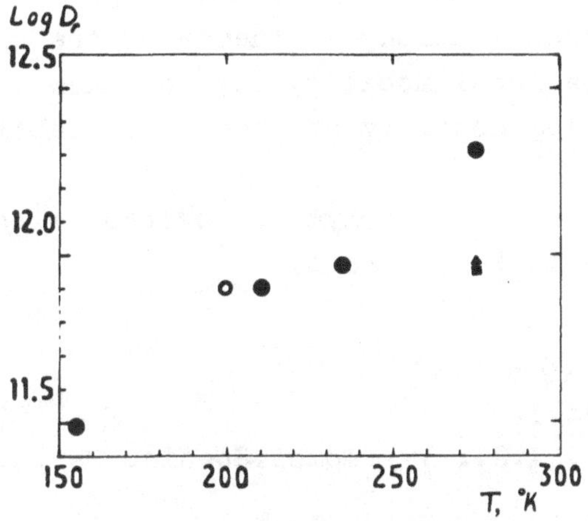

Figure 9. Logarithms of coefficients (D_r) of rotational diffusion of n-alkane chains vs temperature.
● - run without chemical defect;
○ - run without potential U_b;
▲ - run with the methyl--defect;
■ - run with Cl defect.

The temperature and chemical defects influence markedly the dynamics of rotational mobility: the rotational mobility of the defective chain and of surrounding chains is lower. The deceleration of the rotation is accounted for by the increase in both the average inertia momentum of the defective chain and the interchain friction due to the defect volume. The value of D_r for surrounding chains drops when introducing defects but is practically independent of the defect mass.

The examination of various factors characterizing the structure and the dynamics of single molecules in the model has shown that they do not depend on their position in the structural cell. This indicates that the boundary conditions are not too bad, though the potential U_b may considerably influence molecular dynamics above 200 K.

Qualitative analysis of the rotational mobility of molecules proved it more complex than was previously assumed /14-16/. For example, no sudden rotations of single molecules or groups of adjacent molecules through large angles (90° or 180°) were observed. We have neither found explicit relation between longitudinal diffusion of molecules and their rotational mobility. It is clear that this problem needs further investigation.

It should be noted that such a strong influence of the chemical defect upon the rotational mobility may be taken into consideration when analysing mobility of chemically labelled systems.

The study of the dynamics of solid polymeric systems with higher defect concentrations is in progress.

References

1. Faraday Disc. Chem. Soc. 68 (1979).
2. Ann.N.Y.Acad.Sci. 371 (1981).
3. Balabaev N.K. and Grivtsov A.G.: Vysokomol.Soedin. Ser.B. 23, 121 (1981).
4. Gotlib Yu.Ya., Balabaev N.K., Darinskii A.A. and Neelov I.M.: Macromolecules 13, 602 (1980).
5. Grivtsov A.G., Malenkov G.G. and Abaturov L.V.: Mol.Biol. 17, 577 (1983).
6. Balabaev N.K., Grivtsov A.G., Lunevskaya L.V., Mazo M.A., Topolcaraev V.A. and Oleynik E.F.: Prepr. 12th Eur.Conf. on Macrom.Phys.Molecul.Mobility in Polym.Syst., Leipzig, Sept. 1981, p.36.
7. Balabaev N.K., Bulatov V.V., Grivtsov A.G., Lunevskaya L.V., Mazo M.A., Oleynik E.F. and Topolcaraev V.A.: Mathematical Methods for Study of Polymers. All Union.Conf., p.71, Pushchino, USSR, 1982.
8. Ewen B., Strobl G.R. and Richter D.: Faraday Disc.Chem. Soc. 69, 19 (1980).
9. Ungar G.: J.Phys.Chem. 87, 689 (1983).
10. Snyder R.G., Maroncelli M., Strauss H.L., Elliger C.A., Cameron D.G., Casal H.L. and Mantsch H.H.: J.Am.Soc. 105, 133 (1983).
11. Taylor M.G., Kelusky E.C., Smith I.C.P., Casal H.L. and Cameron D.G.: J.Chem.Phys. 78, 5108 (1983).

12. Doucet J., Denicolo I., Graivech A.F. and Germain C.: J. Chem.Phys. 80, 1647 (1984).
13. Forsman H. and Andersson P.: J.Chem.Phys. 80, 2804 (1984).
14. Hoffman J.D.: J.Chem.Phys. 20, 541 (1952).
15. McClure D.W.: J.Chem.Phys. 49, 1830 (1968).
16. Ishinabe T.: J.Chem.Phys. 72, 353 (1980)
17. Mazo M.A., Balabaev N.K., Lunevskaya L.V., Grivtsov A.G. and Oleynik E.F.: Dokl.Akad.Nauk SSSR, 277, 412 (1984).
18. Balabaev N.K.: The motion simulation of molecules with rigid links, Prepr., Research Computer Centre Acad.Sci. USSR, Pushchino, 1981.
19. Wunderlich B.: Macromolecular Physics, v.1, London, Acad.Press, 1973.
20. Barnes J.D.: J.Chem.Phys. 58, 5193 (1973).

THE KERR EFFECT RELAXATION IN HIGH ELECTRIC FIELDS

Hiroshi Watanabe and Akio Morita

Department of Chemistry, College of Arts and Sciences,
University of Tokyo, Komaba, Meguroku, Tokyo 153, Japan

I. Introduction

Since the theoretical and the experimental work of Benoit(1), it has been well established that the dynamic electro-optical methods provides us useful information for the electrical and the hydrodynamical properties of macro-molecules in solution. The dynamic electro-optical effects (DEOE) is directly related with the Brownian motion under the influence of the external electric fields. In the electro-optical measurements of macromolecular solutions, it is not rare that the condition of $V/k_BT>1$ where V is the potential energy with respect to the molecular orientation and k_BT is the thermal energy is realized.

In the further developments of the theory of DEOE, analytical expressions for the rapidly reversing field(2) and the sinusoidal field(3,4) were obtained. A perturbation method was introduced for the step-up and the rapidly reversing fields to extend the theory to the non-linear region(5,6).

In this report we shall present some recent developments

Physical Optics of Dynamic Phenomena and Processes in Macromolecular Systems

of the analytical and numerical treatments of DEOE(7-13). They are the following: (i) A new method for calculating the DEOE at the sudden change of a static electric field. (ii) A new method for calculating the DEOE at a sudden application of time varying electric fields. (iii) A new method of calculating the DEOE for the extremely high fields will be presented. The problem will be discussed in the light of the rotational Brownian motion under the extremely high field.

The problem of DEOE is to obtain the orientation factor given by

$$\Phi(t) = \frac{\int_0^{\pi} P_2(\cos\theta) f(\theta, t) \sin\theta \, d\theta}{\int_0^{\pi} f(\theta, t) \sin\theta \, d\theta} \qquad (1)$$

where $P_2(x) = (3x^2 - 1)/2$ is the Legendre polynomial of degree 2, t is the time, and $f(\theta, t)$ is the angular distribution function.

Upon the assumption that the macromolecule is axially symmetric and non-interacting each other, f may be given by the rotational diffusion equation

$$\frac{\partial f(\theta, \tau)}{\partial \tau} = \frac{1}{\sin\theta} \frac{\partial}{\partial \theta} \left[\sin\theta \left(\frac{\partial f}{\partial \theta} + \frac{1}{k_B T} \frac{\partial V}{\partial \theta} f \right) \right] \qquad (2)$$

where $\tau = Dt$ is the reduced time, D being the rotational diffusion constant about the transverse axis of the molecule, $k_B T$ is the thermal energy, and V is the potential function. The potential function is given by

$$V = -\mu E(t) \cos\theta - \tfrac{1}{2}(\alpha_1 - \alpha_2) E^2(t) \cos^2\theta \qquad (3)$$

where μ is the permanent dipole moment, α_1 and α_2 are the electric polarizabilities parallel and perpendicular to the symmetry axis, respectively, and E(t) is the external electric field.

II. DEOE following the sudden change of a homogeneous electric field.

In order to obtain $\Phi(\tau) = \langle P_2(\cos\theta(\tau)) \rangle$ in Eq. (1), we expand $f(\theta, t)$ in a series of Legendre polynomials,

$$f(\theta,t)=\sum_{n=0}^{\infty}a_n(t)P_n(\cos\theta) \tag{4}$$

Putting Eq. (4) into Eq. (2), we find the recurrence relation

$$\frac{da_0(\tau)}{d\tau}=0 \tag{5.a}$$

$$\begin{aligned}\frac{1}{n(n+1)}\frac{da_n(\tau)}{d\tau}&=-\left[1-\frac{2g(\tau)}{(2n-1)(2n+3)}\right]a_n(\tau)\\&+e(\tau)\left[\frac{1}{2n-1}a_{n-1}(\tau)-\frac{1}{2n+3}a_{n+1}(\tau)\right]\\&+2g(\tau)\left[\frac{n-1}{(2n-3)(2n-1)}a_{n-2}(\tau)-\frac{n+2}{(2n+3)(2n+5)}a_{n+2}(\tau)\right]\\&(n=1,2,3,\ldots)\end{aligned} \tag{5.b}$$

where $e=\mu E/k_BT$, $g=\Delta\alpha E/2k_BT$ ($\Delta\alpha=\alpha_1-\alpha_2$). Dividing both side of Eq. (5. b) by a_0 and using the relation

$$\frac{a_n(\tau)}{a_0}=(2n+1)\langle P_n(\cos\theta)\rangle=(2n+1)q_n(\tau) \tag{6}$$

we find that

$$\begin{aligned}\frac{dq_n(\tau)}{d\tau}&=-d_n(\tau)q_n(\tau)+e(\tau)w_n'[q_{n-1}(\tau)-q_{n+1}(\tau)]\\&+2g(\tau)[v_n'q_{n-2}(\tau)-y_n'q_{n+2}(\tau)]\end{aligned} \tag{7}$$

where

$$d_n(\tau)=n(n+1)\left[1-\frac{2g(\tau)}{(2n-1)(2n+3)}\right] \tag{8.a}$$

$$w_n'=\frac{n(n+1)}{2n+1} \tag{8.b}$$

$$v_n'=\frac{(n-1)n(n+1)}{(2n-1)(2n+1)} \tag{8.c}$$

$$y_n'=\frac{n(n+1)(n+2)}{(2n+1)(2n+3)} \tag{8.d}$$

For the sudden application of a homogeneous field E at $\tau=0$ to a system in which a Boltzmann distribution has been established by another homogeneous field E^2, we obtain a matrix equation

$$\begin{aligned}\mathbf{Q}(\tau)&=\sum_{l=0}^{\infty}\mathbf{T}\left[\frac{(-\tau)^l\Lambda^l}{l!}\right]\mathbf{T}^{-1}\mathbf{Q}^0-\sum_{l=0}^{\infty}\mathbf{T}\left[\frac{(-\tau)^{l+1}\Lambda^l}{(l+1)!}\right]\mathbf{T}^{-1}\mathbf{C}\\&=\mathbf{T}\exp(-\tau\Lambda)\mathbf{T}^{-1}\mathbf{Q}^0+\mathbf{T}(\mathbf{I}-\exp(-\tau\Lambda))\Lambda^{-1}\mathbf{T}^{-1}\mathbf{C}\end{aligned} \tag{9}$$

where

$$\mathbf{Q}^0 = \begin{bmatrix} q_1^0 \\ q_2^0 \\ q_3^0 \\ \vdots \end{bmatrix} \tag{10}$$

with

$$q_n^0 = \frac{\int_{-1}^{1} P_n(u)\exp(\quad e_0 u + g_0 u^2)\,du}{\int_{-1}^{1} \exp(\quad e_0 u + g_0 u^2)\,du} \tag{11}$$

and

$$\mathbf{C} = \begin{bmatrix} ew_1' \\ 2gv_2' \\ 0 \\ 0 \\ \vdots \end{bmatrix} \tag{12}$$

T is the eigen matrix of

$$\mathbf{B} = \begin{bmatrix} d_1 & ew_1' & 2gy_1' & 0 & 0 & 0 & \cdots \\ -ew_2' & d_2 & ew_2' & 2gy_2' & 0 & 0 & \cdots \\ -2gv_3 & -ew_3' & d_3 & ew_3' & 2gy_3' & 0 & \cdots \\ 0 & -2gv_4' & -ew_4' & d_4 & ew_4' & 2gy_4' & \cdots \\ \cdots & \cdots & \cdots & \cdots & \cdots & \cdots & \cdots \end{bmatrix} \tag{13}$$

and Λ is the diagonal matrix whose elements are the eigen-values of B, so that

$$\exp(-\tau\Lambda) = \begin{bmatrix} \exp(-\lambda_1\tau) & 0 & 0 & \cdots \\ 0 & \exp(-\lambda_2\tau) & 0 & \cdots \\ 0 & 0 & \exp(-\lambda_3\tau) & \cdots \\ \cdots & \cdots & \cdots & \cdots \end{bmatrix} \tag{14}$$

Letting $E_0=0$, and therefore $Q^0=0$, we have the rise processes. Then Eq. (9) reduces to

$$\mathbf{Q}^b(\tau) = \mathbf{T}[\mathbf{I} - \exp(-\tau\Lambda)]\,\Lambda^{-1}\mathbf{T}^{-1}\mathbf{C} \tag{15}$$

For $\tau \to \infty$, we have

$$\mathbf{Q}(\infty) = \mathbf{T}\Lambda^{-1}\mathbf{T}^{-1}\mathbf{C} = \mathbf{B}^{-1}\mathbf{C} \tag{16}$$

and Eq. (15) can be written

$$\mathbf{Q}^b(\tau) = \mathbf{T}[\mathbf{I} - \exp(-\tau\Lambda)]\mathbf{T}^{-1}\mathbf{Q}(\infty) \tag{17}$$

Letting $E=-E_0$ in Eq. (9), we have the transients for the reversing fields. With

$$Q^{+} = B^{-1}(E)C(E) = Q(\infty) \tag{18.a}$$

$$Q^{-} = B^{-1}(-E)C(-E) \tag{18.b}$$

the transients for the reversing fields can be expressed as

$$Q^{r}(\tau) = T\exp(-\tau\Lambda)T^{-1}(Q^{+} - Q^{-}) + Q^{+} \tag{19}$$

Letting E=0, therefore C=0, we have the decay processes as

$$Q^{d}(\tau) = \exp(-\tau B^{0})Q^{0} \tag{20}$$

where

$$\exp(-\tau B^{0}) = \begin{bmatrix} \exp(-1\cdot 2\tau) & 0 & 0 & \cdots \\ 0 & \exp(-2\cdot 3\tau) & 0 & \cdots \\ 0 & 0 & \exp(-3\cdot 4\tau) & \cdots \\ \cdots & \cdots & \cdots & \cdots \end{bmatrix} \tag{21}$$

Fig. 1 show the unnormalized buildup(a) and reversing(b) for $R=e^2/2g=-2$. It is noted that the increase of the field strength enhances the contribution of the induced dipole moment compared with that of the permanent dipole moment. The value of the field strength is shown as e^2 in the figure caption.

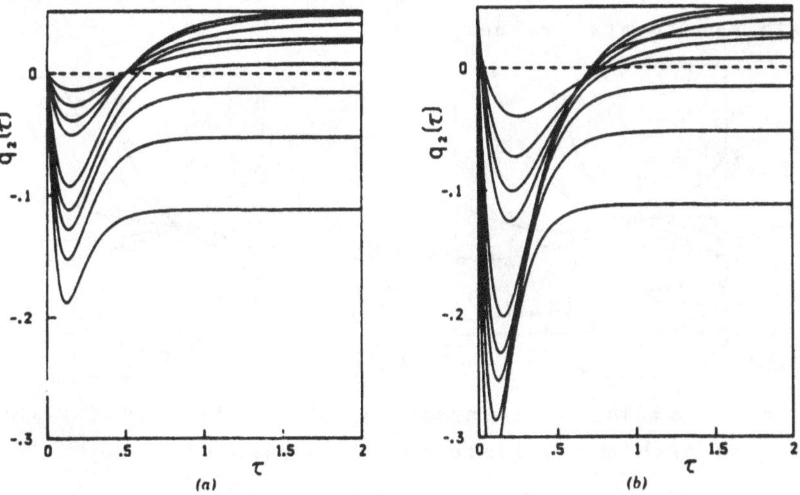

Fig. 1 Unnormalized buildup(a) and reversing(b) birefringence signals for R=-2. From the top to bottom in the initial part: e^2=1, 2, 3, 4, 8, 10, 12, 15, 20. The sign of the stationary state value reverses in the interval $10<e^2<12$.

The areas surrounded by the normalized transients of the birefringence denoted by $(S^b)_2$, $(S^r)_2$, and $(S^d)_2$ are functions of R and provides us an additional method for determining the electrical parameters of the molecules from the transient birefringence(14, 15). From Eq. (17) and Eq. (18), we have

$$(\mathbf{S}^b)_m = \int_0^\infty \left\{1 - \left[\frac{[\mathbf{Q}^b(\tau)]_m}{(\mathbf{Q}^+)_m}\right]\right\} d\tau = \frac{[\mathbf{B}^{-1}(E)\mathbf{Q}^+]_m}{(Q^+)_m} \quad (22)$$

and from Eq. (20) we have

$$(\mathbf{S}^r)_m = \int_0^\infty \left\{1 - \left[\frac{[\mathbf{Q}^r(\tau)]_m}{(\mathbf{Q}^-)_m}\right]\right\} d\tau = \frac{[\mathbf{B}^{-1}(-E)(\mathbf{Q}^+ - \mathbf{Q}^-)]_m}{(Q^-)_m} \quad (23)$$

where [column matrix]$_m$ represents the m-th element of the column matrix, and S^b and S^r are the column matrices whose second elements correspond to the normalized birefringence. The area surrounded by the decay curves $(S^d)_m$ are, from Eq. (21)

$$(\mathbf{S}^d)_m = [m(m+1)]^{-1} \quad (24)$$

which does not depend on the field strength. Figures 2(a) and (b) show the dependence of $(S^b)_2$ and $(S^r)_2$ on the field strength for several values of R.

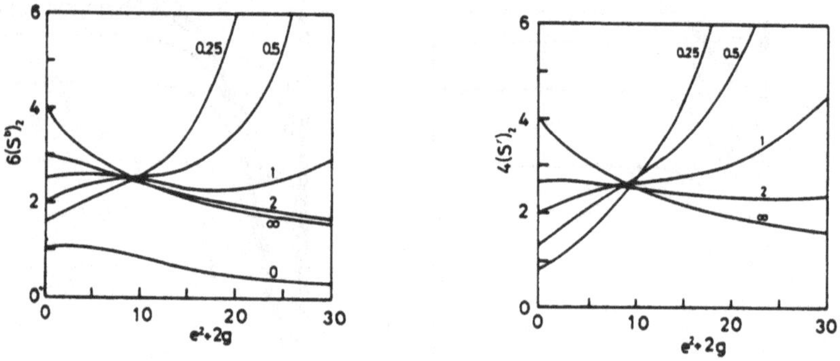

Fig. 2. Field strength dependence of (a) $(S^b)_2$ and (b) $(S^r)_2$. Numerals on each curve represent the value of R.

III. DEOE for time-varying electric fields.

In order to obtain analytical expressions for DEOE for time-varying electric field, we write Eq. (7) in the following matrix form, noting that $q_0(\tau)=1$:

$$\mathbf{Q}(\tau) = \mathbf{D}(\tau)\mathbf{Q}^0 + \int_0^\tau e(\tau)\mathbf{W}(\tau-\tau')\mathbf{Q}(\tau')\,d\tau'$$

$$+2\int_0^\tau g(\tau')\mathbf{G}(\tau-\tau')\mathbf{Q}(\tau')\,d\tau' + \mathbf{R}(\tau) \tag{25}$$

where

$$\mathbf{D}(\tau) = \begin{bmatrix} D_1(\tau) & 0 & 0 & \cdots \\ 0 & D_2(\tau) & 0 & \cdots \\ 0 & 0 & D_3(\tau) & \cdots \\ \cdots & \cdots & \cdots & \cdots \end{bmatrix} \tag{26}$$

$$\mathbf{W}(\tau) = \begin{bmatrix} 0 & -w_1' D_1(\tau) & 0 & 0 & \cdots \\ w_2' D_2(\tau) & 0 & -w_2' D_2(\tau) & 0 & \cdots \\ 0 & w_3' D_3(\tau) & 0 & -w_3' D_3(\tau) & \cdots \\ \cdots & \cdots & \cdots & \cdots & \cdots \end{bmatrix} \tag{27}$$

$$\mathbf{G}(\tau) = \begin{bmatrix} u_1 D_1(\tau) & 0 & -y_1 D_1(\tau) & 0 & 0 & \cdots \\ 0 & u_2 D_2(\tau) & 0 & -y_2 D_2(\tau) & 0 & \cdots \\ v_3' D_3(\tau) & 0 & u_3 D_3(\tau) & 0 & -y_3 D_3(\tau) & \cdots \\ \cdots & \cdots & \cdots & \cdots & \cdots & \cdots \end{bmatrix} \tag{28}$$

$$\mathbf{R}(\tau) = \begin{bmatrix} w_1' \int_0^\tau e(\tau') D_1(\tau-\tau')\,d\tau' \\ 2v_2' \int_0^\tau g(\tau') D_2(\tau-\tau')\,d\tau' \\ 0 \\ 0 \\ \vdots \end{bmatrix} \tag{29}$$

in which

$$D_n = \exp[-n(n+1)\tau] \tag{30}$$

If we write

$$\mathbf{K}(\tau,\tau') = e(\tau')\mathbf{W}(\tau-\tau') + 2g(\tau')\mathbf{G}(\tau-\tau') \tag{31}$$

and

$$\mathbf{P}(\tau) = \mathbf{D}(\tau)\mathbf{Q}^0 + \mathbf{R}(\tau) \tag{32}$$

Eq. (24) becomes an integral equation

$$\mathbf{Q}(\tau) = \int_0^\tau \mathbf{K}(\tau,\tau')\mathbf{Q}(\tau')\,d\tau' + \mathbf{P}(\tau) \tag{33}$$

The solution of Eq. (33) is

$$Q(\tau)=\int_0^{\tau}K(\tau,\tau_1)P(\tau_1)\,d\tau_1+\int_0^{\tau}\int_0^{\tau_1}K(\tau,\tau_1)K(\tau_1,\tau_2)P(\tau_2)\,d\tau_1 d\tau_2$$
$$+\int_0^{\tau}\int_0^{\tau_1}\int_0^{\tau_2}K(\tau,\tau_1)K(\tau_1,\tau_2)K(\tau_2,\tau_3)P(\tau_3)\,d\tau_1 d\tau_2 d\tau_3+\cdots \qquad (34)$$

For extremely low value of E(τ), by taking terms up to $E^2(\tau)$ for q (τ), we find that Eq. (34) gives

$$q_2(\tau)=q_2(0)\exp(-6\tau)\left[1+\frac{2}{7}\int_0^{\tau}g(\tau_1)\,d\tau_1\right]$$
$$+q_1(0)w_2'\int_0^{\tau}\exp[-6(\tau-\tau_1)]\exp(-2\tau_1)e(\tau_1)\,d\tau_1$$
$$+w_1'w_2'\int_0^{\tau}\int_0^{\tau_1}\exp[-6(\tau-\tau_1)]\exp[-2(\tau_1-\tau_2)]\,e(\tau_1)e(\tau_2)\,d\tau_1 d\tau_2$$
$$+2v_2'\int_0^{\tau}\exp[-6(\tau-\tau_1)]\,g(\tau_1)\,d\tau_1 \qquad (35)$$

Analytical expressions for DEB for three types of the time-varying electric fields were obtained.

(i) Continuous reversing square pulses with a period of 2T

The analytical expression of DEB for the continuously reversing field with period 2T as defined in Fig. 3 is given by

$$\frac{q_2(\tau)}{e/15}=\left\{\frac{3}{2}\frac{(1-e^{-2\tau'})^2}{1+e^{-4\tau'}}e^{-2m\tau'}\left[e^{-4m\tau'}-(-1)^m\right]\right.$$
$$+\frac{(1-e^{-2\tau'})^2(1-e^{-6m\tau'})}{1+e^{-2\tau'}+e^{-4\tau'}}-\frac{3}{2}(-1)^m\frac{1-e^{-2\tau'}}{1+e^{-2\tau'}}e^{-2m\tau'}$$
$$\left.+\frac{2-e^{-2\tau'}}{1+e^{-2\tau'}}\right\}e^{-6\Delta\tau}+\frac{3}{2}\frac{(-1)^m(1-e^{-2\tau'})e^{-2m\tau'}-2}{1+e^{-2\tau'}}e^{-2\Delta\tau}$$
$$+1+\frac{2g}{e^2}(1-e^{-6\tau}) \qquad (36)$$

In order to show how the system reaches the stationary state, $q_2(\tau)/q_2^0$ versus $\Delta\tau$ in case of $g_0=0$ and $\tau'=1$ is plotted in Fig. 4. It is evident in this particular case of that m=3 already satisfies the condition for the steady state.

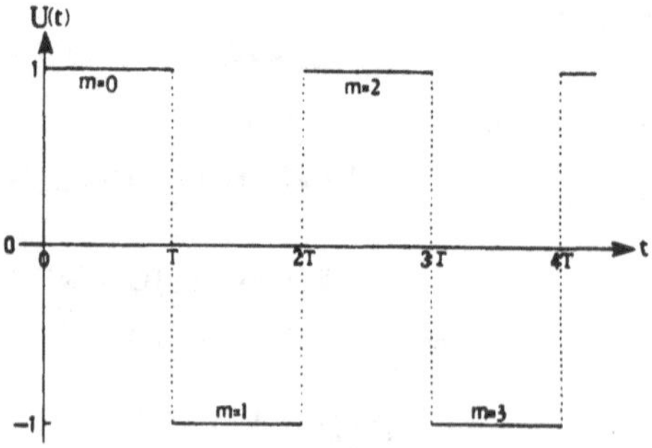

Fig. 3 Definition of U(t).

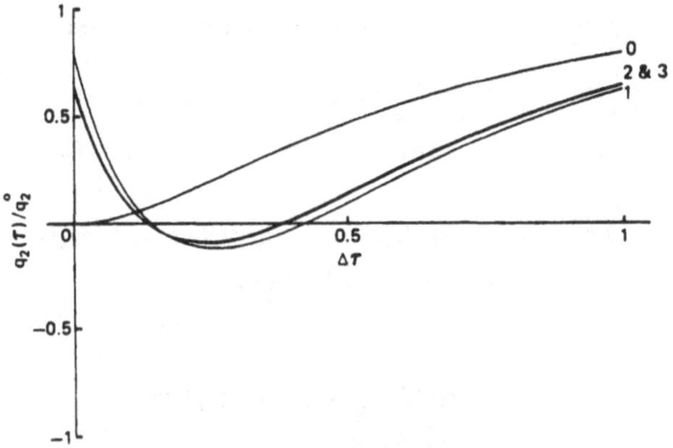

Fig. 4. Plots of $q_2(\tau)/q_2^0$ obtained from Eq. (35) versus $\Delta\tau$ for various values of m, τ', and g_0=0. The numerals on the right of the curves represent values of m.

(2) Exponentially rising and decaying electric fields.

For the exponentially rising and decaying fields $E(t)=E_0[1-\exp(kt)]$ (field 1) and $E(t)=E_0\exp(-kt)$ (field 2), we obtain from Eq. (35) that

$$q_2(\tau)=\frac{4}{5}e_0^2\left\{\frac{1-e^{-6\tau}}{12}-\frac{(4-k')(e^{-k'\tau}-e^{-6\tau})}{2(2-k')(6-k')}+\frac{k'(e^{-2\tau}-e^{-6\tau})}{8(2-k')}\right.$$
$$\left.-\frac{k'[e^{-(2+k')\tau}-e^{-6\tau}]}{2(4-k')(2-k')}+\frac{e^{-2k'\tau}-e^{-6\tau}}{2(2-k')(3-k')}\right\}$$
$$+\frac{4}{5}g_0\left[\frac{1-e^{-6\tau}}{6}-\frac{2(e^{-k'\tau}-e^{-6\tau})}{6-k'}+\frac{e^{-2k'\tau}-e^{-6\tau}}{2(3-k')}\right] \tag{37}$$

and

$$q_2(\tau)=q_2(0)e^{-6\tau}\left[1+\frac{2g_0}{7k'}(1-e^{-2k'\tau})\right]+q_1(0)\frac{6e_0}{5(4-k')}[e^{-(2+k')\tau}-e^{-6\tau}]$$
$$+\frac{4e_0^2}{5}\left[\frac{e^{-2k'\tau}-e^{-6\tau}}{2(2-k')(3-k')}-\frac{e^{-(2+k')\tau}-e^{-6\tau}}{(2-k')(4-k')}\right]$$
$$+\frac{2g_0}{5(3-k')}(e^{-2k'\tau}-e^{-6\tau}) \tag{38}$$

for field 1 and field 2, respectively.

(3) The Sinusoidal Electric Field E(t)=E cos t

Putting $E(\tau)=\cos\omega'\tau$ where $\omega'=\omega/D$ and τ=Dt in Eq. (35), we obtain after integration,

$$q_2(\tau)=\frac{4}{5}\frac{e_0^2}{4+\omega'^2}\times\left\{\frac{1}{6}(1-e^{-6\tau})-\frac{2}{16+\omega'^2}[4e^{-2\tau}\cos\omega'\tau+\omega'e^{-2\tau}\sin\omega'\tau-4e^{-6\tau}]\right.$$

$$+\frac{1}{36+4\omega'^2}\left[(6-\omega'^2)\cos 2\omega'\tau+5\omega'\sin 2\omega'\tau+(\omega'^2-6)e^{-6\tau}\right]\Big\}$$
$$+\frac{2g_0}{5}\left\{\frac{1}{6}(1-e^{-6\tau})+\frac{1}{36+4\omega'^2}\left[6\cos 2\omega'\tau+2\omega'\sin 2\omega'\tau-6e^{-6\tau}\right]\right\} \quad (39)$$

Taking terms up to the order of E^4, we find that Eq. (35) gives

$$q_2(\tau)=2v_2'I_1(\tau)+w_1'w_2'I_2(\tau)-2y_2'w_1'w_2'I_3(\tau)-w_1'^2w_2'^2I_4(\tau)$$
$$-2w_2'w_3'v_2'I_5(\tau)-w_1'w_2'^2w_3'I_6(\tau) \quad (40)$$

In Eq. (40)

$$I_1(\tau)=\int_0^\tau D_2(\tau,\tau_1)g(\tau_1)\,d\tau_1 \quad (41)$$

$$I_2(\tau)=\iint_{\tau\geq\tau_1\geq\tau_2\geq 0} D_2(\tau,\tau_1)D_1(\tau_1,\tau_2)e(\tau_1)e(\tau_2)\,d\tau_1 d\tau_2 \quad (42)$$

$$I_3(\tau)=\iiint_{\tau\geq\tau_1\geq\tau_2\geq\tau_3\geq 0} D_2(\tau,\tau_1)D_1(\tau_1,\tau_2)D_2(\tau_2,\tau_3)$$
$$\times e(\tau_1)e(\tau_2)g(\tau_3)\,d\tau_1 d\tau_2 d\tau_3 \quad (43)$$

$$I_4(\tau)=\iiiint_{\tau\geq\tau_1\geq\tau_2\geq\tau_3\geq\tau_4\geq 0} D_2(\tau,\tau_1)D_1(\tau_1,\tau_2)D_2(\tau_2,\tau_3)D_1(\tau_3,\tau_4)$$
$$\times e(\tau_1)e(\tau_2)e(\tau_3)e(\tau_4)\,d\tau_1 d\tau_2 d\tau_3 d\tau_4 \quad (44)$$

$$I_5(\tau)=\iiint_{\tau\geq\tau_1\geq\tau_2\geq\tau_3\geq 0} D_2(\tau,\tau_1)D_3(\tau_1,\tau_2)D_2(\tau_2,\tau_3)$$
$$\times e(\tau_1)e(\tau_2)g(\tau_3)\,d\tau_1 d\tau_2 d\tau_3 \quad (45)$$

$$I_6(\tau)=\iiiint_{\tau\geq\tau_1\geq\tau_2\geq\tau_3\geq\tau_4\geq 0} D_2(\tau,\tau_1)D_3(\tau_1,\tau_2)D_2(\tau_2,\tau_3)D_1(\tau_3,\tau_4)e(\tau_1)$$
$$\times e(\tau_2)e(\tau_3)e(\tau_4)\,d\tau_1 d\tau_2 d\tau_3 d\tau_4 \quad (46)$$

where

$$D_n(\tau,\tau')=\exp\left[-\int_{\tau'}^{\tau}d_n(\tau'')\,d\tau''\right] \quad (47)$$

The explicit expression for the steady state birefringence were obtained(13). Field strength dependence of the steady state birefringence for the special case of R=-2 calculated by the equation obtained is given in Fig. 5.

The DEB for the larger value of E, the set of the first order differential equations,Eq. (7) was solved numerically. An example for a special case of R=-2 is given in Fig. 6.

Typical birefringence signals observed on a semi-concentrated PBLG/benzene solution (5%) for 50 Hz and 1K Hz electric field are presented in Fig. 7.

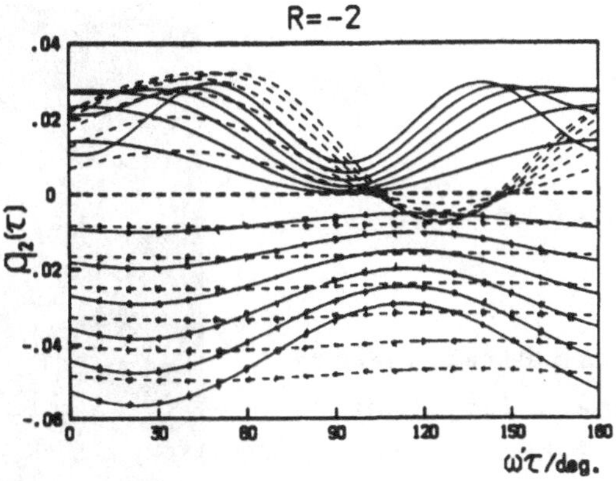

Fig. 5. Field Strength dependence of the steady state birefringence for R=-2 obtained from Eq. (41) vs. $\omega'\tau$ for various values of ω and e^2. From the bottom to top, $e^2=0.5$, 1, 1.5, 2, 2.5, 3 and $\omega'=$ 0.1 (———); 1 (- - - - - -), 10 (—•—•—), 100 (-•-•-•-).

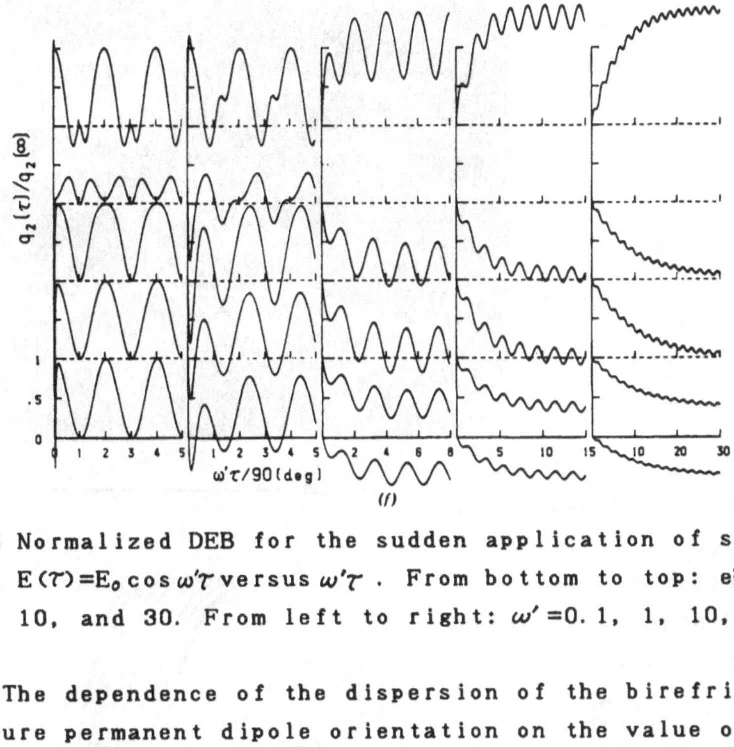

Fig. 6 Normalized DEB for the sudden application of strong field $E(\tau)=E_0 \cos\omega'\tau$ versus $\omega'\tau$. From bottom to top: $e^2=0.01$, 1, 3, 10, and 30. From left to right: $\omega'=0.1$, 1, 10, 30, 100.

The dependence of the dispersion of the birefringence for pure permanent dipole orientation on the value of e^2 obtained by the numerical calculations is demonstrated in Fig. 8. It is remarkable that the dispersion shifts to the higher value of ω' with increasing value of e^2. This may be understood to be due to the increase of orienting force causing faster reorientation of the molecule under the alternating field.

(a)

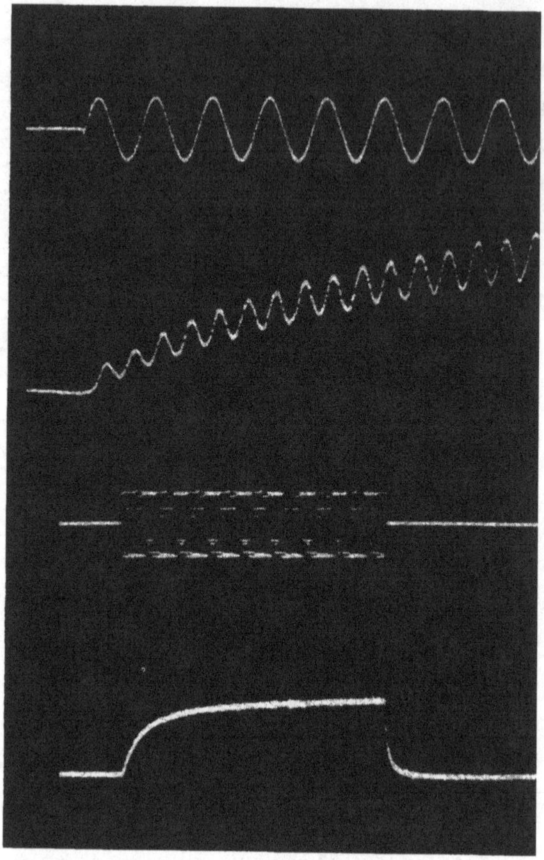

(b)

Fig. 7. Birefringence signals for 50 Hz (a) and 1 KHz (b) AC field of a PBLG/benzene solution. c=50g/dm³.

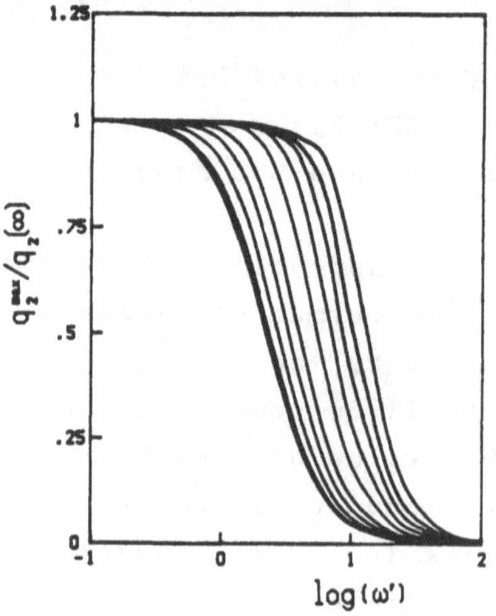

Fig. 8. Normalized dispersion curves of the electric birefringence for pure permanent dipole orientation. From the left to right: e= 0.0001, 1, 2, 3, 4, 6, 8, 10, 12, and 15.

IV. DEOE in infinitely high, time varying field.

For an infinitely high field, the random torque in the stochastic-differential equation may become negligible compared with the torque due to the external field. Therefore, the angular motion of the symmetrical molecule can be described by the deterministic differential equation

$$\omega_x(t) = \frac{1}{I_x\beta_x} N_x(\theta, \phi, \psi) \tag{48.a}$$

$$\omega_y(t) = \frac{1}{I_y\beta_y} N_y(\theta, \phi, \psi) \tag{48.b}$$

$$\omega_z(t) = \frac{1}{I_z\beta_z} N_z(\theta, \phi, \psi) \tag{48.c}$$

where $\omega_i(t)$ (i=x, y, z) is the angular velocity about the particle axis labeled i, and the N $(\theta \phi \psi)$ is the external torque in which θ, ϕ, ψ are the Eularian angles specifying the orientation of the molecule. Then the continuity equation for the distribution function f

$$\frac{\partial f}{\partial t} + \frac{\partial}{\partial M_x}(\dot{M}_x f) + \frac{\partial}{\partial M_y}(\dot{M}_y f) + \frac{\partial}{\partial M_z}(\dot{M}_z f) = 0. \tag{49}$$

where $M_y = \sin\theta\cos\phi$, $M_x = \sin\theta\sin\phi$, and $M_z = \cos\theta$ give rise to

$$\frac{\partial f}{\partial t} + i \sum_{j=x,y,z} L_j \frac{N_j f}{I_j \beta_j} = 0 \tag{50}$$

in which

$$L_x = \frac{1}{i}\left(M_z \frac{\partial}{\partial M_y} - M_y \frac{\partial}{\partial M_z}\right) = \frac{1}{i}\left(\cos\psi \frac{\partial}{\partial\theta} + \frac{\sin\psi}{\sin\theta}\frac{\partial}{\partial\phi} - \cot\theta \sin\psi \frac{\partial}{\partial\psi}\right) \tag{51.a}$$

$$L_y = \frac{1}{i}\left(M_x \frac{\partial}{\partial M_z} - M_z \frac{\partial}{\partial M_x}\right) = \frac{1}{i}\left(-\sin\psi \frac{\partial}{\partial\theta} + \frac{\cos\psi}{\sin\theta}\frac{\partial}{\partial\phi} - \cot\theta \cos\psi \frac{\partial}{\partial\psi}\right) \tag{51.b}$$

$$L_z = \frac{1}{i}\left(M_y \frac{\partial}{\partial M_x} - M_x \frac{\partial}{\partial M_y}\right) = \frac{1}{i}\frac{\partial}{\partial\psi} \tag{51.c}$$

are the angular operators. It should be noted that the diffusion terms $D_j L_j f$ (j=x, y, z) are missing in Eq. (50). For the symmetrical molecule, we have from Eq. (50) that

$$\frac{\partial f}{\partial \tau} = -\frac{\partial}{\partial u}\{[e(\tau)(1-u^2)+2g(\tau)(1-u^2)u]f\} \tag{52}$$

where $u = M_z = \cos\theta$, while the differential equation for the directional cosine is

$$\frac{du}{d\tau} = e(\tau)(1-u^2)+2g(\tau)(1-u^2)u \tag{53}$$

Equations (52) and (53) can be solved analytically for special case of (1) $e \neq 0$, $g=0$ (the pure permanent dipole orientation) and (2) $e=0$, $g \neq 0$ (the pure induced dipole orientation). The solution for Eq. (52) with the initial conditions

$$f(u,0) = C_1 \exp(e_0 u) \tag{54}$$

for case 1 and

$$f(u,0) = C_2 \exp(g_0 u^2) \tag{55}$$

for case 2 are given by

$$f(u,\tau) = C_1 \frac{4x(\tau)}{[(1-u)x(\tau)+1+u]^2} \exp(e_0) \exp\left[\frac{2e_0(u-1)x(\tau)}{1+u+(1-u)x(\tau)}\right] \tag{56}$$

and

$$f(u,\tau) = C_2 \frac{y(\tau)}{[u^2+(1-u^2)y(\tau)]^{3/2}} \exp\left[g_0 \frac{u^2}{u^2+(1-u^2)y(\tau)}\right] \tag{57}$$

respectively, where C_1 and C_2 are constants determined from the normalization condition for $f(u,0)$. The analytical expressions of $\langle u^2 \rangle$ for case 1 and 2 are given in ref. 13 and the further development on this subject will be published elsewhere.

References

1. Benoit, H.: Anal. Phys. **6**, 561 (1951).
2. Tinoco, I., Yamaoka, K.: J. Phys. Chem. **63**, 423 (1959).
3. Ogawa, S., Oka, S.: J. Phys. Soc. Japan **15**, 658 (1960).
4. Thurston, G.B., Bowling, D.L.: J. Colloid Interface Sci. **30**, 34 (1969).

5. Matsumoto, M., Watanabe, H., Yoshioka, K.: J. Phys. Chem. **74**, 2181 (1970).

6. Koopmans, G., de Bore, J., Greve, J.: Electro-optics and Dielectrics of Macromolecules and Colloids, B.R. Jennings Ed., Plenum, New York, 1979.

7. Morita, A., Watanabe, H.: J. Chem. Phys. **70**, 4708 (1979).

8. Watanabe, H., Morita, A.: J. Chem. Phys. **73**, 5884 (1980).

9. Watanabe, H., Morita, A.: J. Chem. Phys. **75**, 379 (1981).

10. Morita, A., Watanabe, H.: J. Chem. Phys. **75**, 1320 (1981).

11. Morita, A., Watanabe, H.: J. Chem. Phys. **77**, 1193 (1982).

12. Watanabe, H., Morita, A.: J. Chem. Phys. **78**, 5311 (1983).

13. Watanabe, H., Morita, Adv. Chem. Phys. **54**, in press.

14. Yoshioka, K., Watanabe, H.: Nippon Kagaku Zasshi (J. Chem. Soc. Japan) **84**, 626 (1964).

15. Yoshioka, K., Watanabe, H.: Physical Principles and Techniques of Protein Chemistry, Part A, A.S.J. Leach Ed., Academic Press, New York, 1969.

DIELECTRIC RELAXATION OF A FLEXIBLE MOLECULE

Akio Morita and Hiroshi Watanabe

Department of Chemistry, College of Arts and Sciences,
University of Tokyo
Komaba, Meguro-ku, Tokyo 153, Japan

In an earlier letter,[1] by setting up stochastic differential equations of the Langevin type and by putting them into Smoluchowski equation, we have obtained an analytical expression for the Laplace transform of the time dependent electric polarization in terms of infinite continued fractions for a flexible molecule having two equivalent rigid arms with an intramolecular interaction.

In this letter, we report some interesting new features in the Cole-Cole plot obtained from this result. It is shown that the presence of the interaction gives rise to a significant deviation from the Debye semicircle, furthermore, this theoretical result leads to similar shapes of the Cole-Cole plots to those considered previously by empirical relations such as the Cole-Cole,[2] Davidson-Cole[3] and Williams-Watts types.[4]

On assuming that the potential energy function V_i due to the intramolecular interaction is given by

$$V_i = \frac{\gamma}{2} k_B T \cos^2\alpha, \qquad (1)$$

where k_B is the Boltzmann constant, T is the absolute temperature and 2α is an angle between the two equivalent rigid rod-like rotating arms, we found that[1]

Physical Optics of Dynamic Phenomena and Processes in Macromolecular Systems

$$\frac{P(\tau)}{P(0)} = \sum_{k=0}^{\infty} q_{2k}(\tau)\,\frac{r_{2k}}{r_0}\,, \tag{2}$$

where $P(\tau)$ is the electric polarization after the sudden removal of an applied electric field at t=0 and it is expressed as a function of τ=Dt (D is the rotational diffusion constant of one of the arms and t is the time). In Eq. (2), r_{2k} is defined by the integral:

$$r_{2k} = \int_0^1 u^{2k+3}\,\exp\left(-\frac{\gamma u^2}{2}\right) du,$$

and $q_{2k}(\tau)$ is a function satisfying the recurrence relation:

$$\frac{dq_{2k}}{d\tau} = -\left[(2k+1)^2 + 1 + (2k+1)\gamma\right] q_{2k} + \left[(2k+2)(2k+3) - 1\right] q_{2(k+1)}$$

$$+ (2k-1)\gamma\, q_{2(k-1)}\,,$$

where

$$q_{2k}(0) = \delta_{k,0}\,,$$

in which $\delta_{m,n}$ is Kronecker's delta. The Laplace transform of $q_{2k}(t)$ with respect to time t was expressed in terms of infinite continued fractions.

The linear response theory relates the Laplace transform of P(t)/P(0) to the complex dielectric constant $\varepsilon(\omega) = \varepsilon'(\omega) - i \times \varepsilon''(\omega)$ at a frequency $\omega/2$ by the relation

$$\frac{\varepsilon(\omega) - \varepsilon_\infty}{\varepsilon_s - \varepsilon_\infty} = X'(\omega) - iX''(\omega) = 1 - i\omega \int_0^{\infty} \frac{P(t)}{P(0)}\, e^{-i\omega t}\, dt \tag{3}$$

where ε_s and ε_∞ represent $\varepsilon'(\omega)$ at infinitely small and large frequencies, respectively. We have carried out numerical calculations whose results are shown graphically in Figs. 1-6.

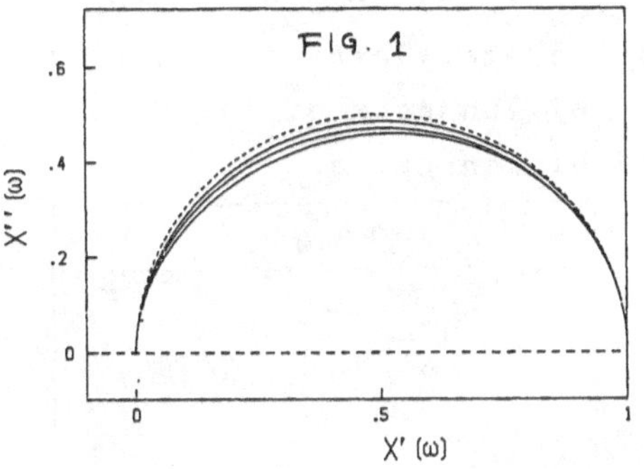

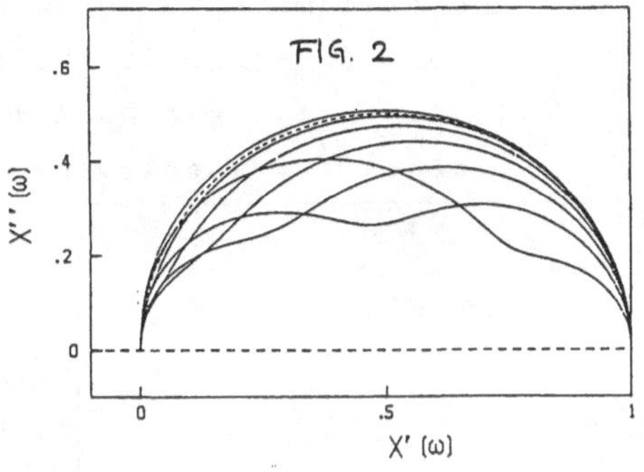

Fig. 1: Cole-Cole plot for $\Upsilon>0$. From the top to the bottom $\Upsilon=0$ (Debye semicircle is represented by the broken curve), 20, 10 and 4.

Fig. 2: Cole-Cole plot for $\Upsilon<0$. The curve above the Debye semicircle (broken curve) is for $\Upsilon=-1$. The other curves are for $\Upsilon=-2$, -3, -3.5, -4 and -4.5 from the top to the bottom, if they are viewed from the right hand side near $X'(\omega)=0.8$.

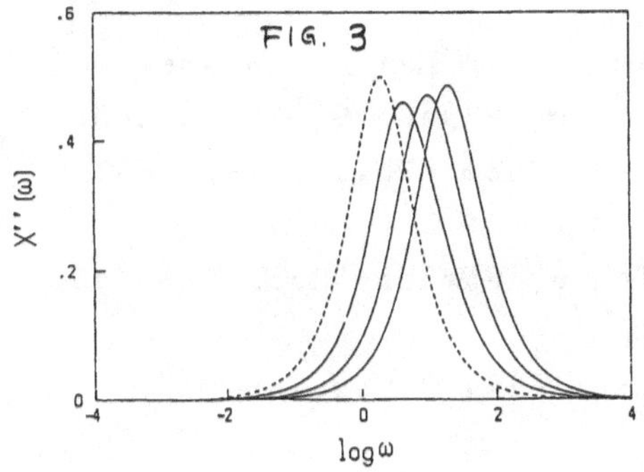

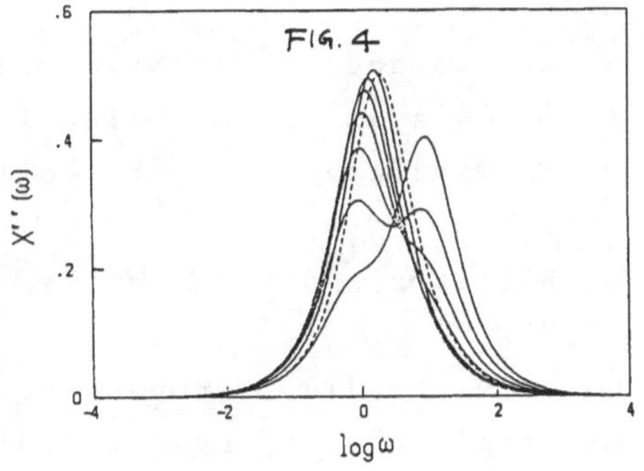

Fig. 3: Plots of $X''(\omega)$ vs. $\log\omega$ for $\Upsilon > 0$. From the top to the bottom; $\Upsilon = 20$, 10, 4 and 0, if the curves are viewed from the right hand side near $\log\omega = 2$.

Fig. 4: Plots of $X''(\omega)$ vs. $\log\omega$ for $\Upsilon < 0$. From the top to the bottom; $\Upsilon = 0$ (broken curve), −1, −2, −2.5, −3, −3.5, −4, −4.5, if the curves are viewed from just a little bit right hand side of the peak of the Debye process represented by the broken curve.

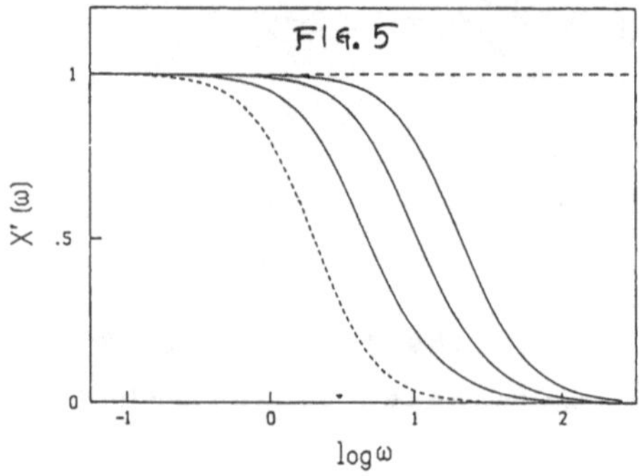

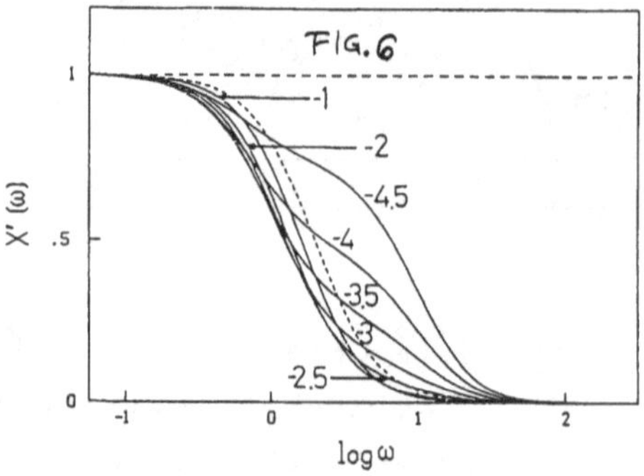

Fig. 5: Plots of $X'(\omega)$ vs. $\log\omega$ for $\Upsilon > 0$. From the left to the right sides; $\Upsilon = 0$ (broken curve), 4, 10, 20.

Fig. 6: Plots of $X'(\omega)$ vs. $\log\omega$ for $\Upsilon < 0$. The broken curve is for $\Upsilon = 0$, and the numbers indicate the values for Υ.

References

1. A. Morita and H. Watanabe, Chem. Phys. Letts. in press
2. K. S. Cole and R. H. Cole, J. Chem. Phys., **9**, 341 (1949)
3. D. W. Davidson and R. H. Cole, J. Chem. Phys., **19**, 1484 (1951)
4. G. Williams and D. C. Watts, Trans. Faraday Soc., **66**, 80 (1970)
5. B. K. P. Scaife, "Complex Permittivity", English University Press, London (1971)

PHOTON STATISTICS IN LIGHT SCATTERING

Jan Peřina
Laboratory of Optics, Palacký University, 771 46 Olomouc, Czechoslovakia

Introduction

In this contribution we present a quantum dynamical theory of the photon statistics of light scattered by random media. The photon statistics of such light are usually described phenomenologically by means of the Mandel photodetection equation in which the integrated intensity is modulated in an appropriate way, e.g. with the use of the negative exponential probability distribution in gaussian media, or the log-normal probability distribution for some regimes of fluctuations in the turbulent atmosphere, etc. Using the coherent state technique, we can provide a more rigorous theory of the photon statistics of scattered light involving the quantum treatment of the radiation field while the random medium is described through the fluctuating permittivity. To do this we can adopt the Heisenberg picture making use of the Heisenberg - Langevin equations and quantum characteristic functions, or the Schrödinger picture using the master equation and the generalized Fokker - Planck equation. Finally we can obtain the measurable photocount distribution and its factorial moments. We show that this approach describes scattered light as a generalized superposition of coherent and chaotic fields in a quite rigorous way and the photocount statistics are expressed in terms of the Laguerre polynomials. Such an approach proves to be useful also in quantum and nonlinear optics in treating optical fields having no classical ana-

Physical Optics of Dynamic Phenomena and Processes in Macromolecular Systems

logues and exhibiting antibunching of photons, sub-poissonian statistics and squeezing.

Theory

We start with the hamiltonian of radiation in the form

$$\hat{H}_{rad} = \frac{1}{8\pi} \int_V [\hat{\underline{E}}^2(\underline{x},t) + \hat{\underline{H}}^2(\underline{x},t)]\, d^3x + \frac{1}{8\pi}\int_V \varepsilon'(\underline{x},t)\hat{\underline{E}}^2(\underline{x},t)d^3x$$
$$= \sum_l \hbar\omega_l(\hat{a}_l^+\hat{a}_l + \tfrac{1}{2}) - \sum_{l,m} \hbar K'_{lm}(t)(\hat{a}_l^+ - \hat{a}_l)(\hat{a}_m^+ - \hat{a}_m) \qquad (1)$$

where $\underline{x}$ and t are space and time variables, $\hat{\underline{E}}$ and $\hat{\underline{H}}$ are the electric and magnetic strength operators of the electromagnetic field in a volume V, ε' represents the fluctuating part of the permittivity $\varepsilon = 1 + \varepsilon'$ (the mean value is $\langle\varepsilon'\rangle = 0$), $\hat{a}_l$ and $\hat{a}_l^+$ are the annihilation and creation operators of a photon in the mode l of the frequency ω_l and

$$K'_{lm}(t) = \frac{(\omega_l\,\omega_m)^{1/2}}{16\,\pi\, c^2} \int_V \varepsilon'(\underline{x},t)\underline{u}_l(\underline{x})\cdot\underline{u}_m(\underline{x})\, d^3x \qquad (2)$$

is the quantity fluctuating in the random medium expressed as the matrix element of ε' in terms of the modal functions $\underline{u}_l(\underline{x})$ in the volume V (c being the light velocity).

Losses are then described by an additional hamiltonian of a reservoir system coupled to radiation modes in the form

$$\hat{H}_{add} = \sum_{j,l} \hbar\psi_l^{(j)}\hat{c}_l^{(j)+}\hat{c}_l^{(j)} + \sum_{j,l}[\hbar\varkappa_{jl}\,\hat{c}_l^{(j)+}\,\hat{a}_j + h.c.] \quad (3)$$

where $\hat{c}_l^{(j)}$ and $\hat{c}_l^{(j)+}$ are the annihilation and creation operators of a reservoir harmonic oscillator (phonon) in the mode l coupled to the radiation mode j having a frequency $\psi_l^{(j)}$; the first term in (3) represents the free hamiltonian of the reservoir and the second term is the interaction hamiltonian with the coupling constant $\varkappa_{jl}$ (h.c. denotes the hermitian conjugate term).

From the total hamiltonian $\hat{H} = \hat{H}_{rad} + \hat{H}_{add}$ we can obtain the Heisenberg - Langevin equations of motion for radiation modes,

$$\frac{d\hat{a}_j}{dt} = -(i\,\omega_j \mathcal{K}_j^{1/2}\,\mathrm{ch}\,\frac{\varphi_j}{2} + \frac{\gamma_j}{2})\hat{a}_j + i\,\omega_j \mathcal{K}_j^{1/2}\mathrm{sh}\,\frac{\varphi_j}{2}\;\hat{a}_j^{+} + \hat{\mathcal{L}}_j \quad (4)$$

where $\mathcal{K}_j = 1 + 4\,K'_{jj}/\,\omega_j$, $\varphi_j = \ln \mathcal{K}_j$, $\hat{\mathcal{L}}_j$ is the Langevin fluctuating force ensuring the validity of the boson commutation rules $[\hat{a}_j(t),\hat{a}_k^{+}(t)] = \delta_{jk}$ for all times t and γ_j is the damping constant related to the coupling constant $\varkappa$.

The equation (4) can be solved in the form

$$\hat{a}_j(t) = u_j(t)\hat{a}_j + v_j(t)\hat{a}_j^{+} + \sum_{l} w_{jl}(t)\hat{c}_l^{(j)} \quad (5)$$

where $\hat{a}_j$ and $\hat{a}_j^{+}$ are the initial values of the annihilation and creation operators respectively and

$$u_j(t) = \exp(-\gamma_j t/2)\left[\cos(\omega_j \mathcal{K}_j^{1/2}t) - i\,\mathrm{ch}(\varphi_j/2)\sin(\omega_j \mathcal{K}_j^{1/2}t)\right]$$

$$v_j(t) = i\,\exp(-\gamma_j t/2)\,\mathrm{sh}(\varphi_j/2)\,\sin(\omega_j\,\mathcal{K}_j^{1/2}t)\;; \quad (6a)$$

the time dependent functions $w_{jl}(t)$ are more complex, but for the further use only the expression

$$\sum_{l} |w_{jl}(t)|^2 \approx 1 - \exp(-\gamma_j t) \quad (6b)$$

is needed.

Now we can calculate the antinormal quantum characteristic function of parameters $\{\beta_j\}$

$$C_{\mathcal{A}}(\{\beta_j\},t) = \mathrm{Tr}\left\{\hat{\rho}(0)\prod_j \exp[-\beta_j^{*}\hat{a}_j(t)]\,\exp[\beta_j\hat{a}_j^{+}(t)]\right\} \quad (7)$$

$\hat{\rho}(0)$ and $\hat{a}_j(t)$ being the initial density matrix and the annihilation operator in the Heisenberg picture, Tr denotes the trace of the operator. Substituting (5) to (7), the characteristic function is obtained in the form

$$C_{\mathcal{A}}(\{\beta_j\},t) = \Big\langle \prod_j \exp\Big\{ -[B_j(t)+1]|\beta_j|^2 + \frac{1}{2} C_j^*(t)\,\beta_j^2 + \frac{1}{2} C_j(t)\,\beta_j^{*2} + \xi_j^*(t)\,\beta_j - \xi_j(t)\beta_j^*\Big\}\Big\rangle \tag{8}$$

where

$$B_j(t) = |v_j(t)|^2 + \langle n_{dj}\rangle\,(\,1 - \exp(-\gamma_j t)\,)$$

$$C_j(t) = u_j(t)v_j(t)$$

$$\xi_j(t) = u_j(t)\,\xi_j + v_j(t)\,\xi_j^* \quad ; \tag{9}$$

here $\langle n_{dj}\rangle$ represents the mean number of chaotic phonons coupled to the radiation mode j , $\xi_j(t)$ are the time dependent complex amplitudes of the radiation field and ξ_j are their initial values ; the radiation field is assumed to be in the initial coherent state $|\{\xi_j\}\rangle$, $\hat{a}_j\,|\{\xi_j\}\rangle = \xi_j\,|\{\xi_j\}\rangle$, and the angle brackets in (8) mean the average over initial fluctuations of ξ_j and over fluctuations of χ_j (φ_j).

Performing the Fourier transformation of (8), we arrive at the corresponding quasi-probability function for the complex amplitudes $\{\alpha_j\}$

$$\Phi_{\mathcal{A}}(\{\alpha_j\},t) = \int C_{\mathcal{A}}(\{\beta_j\},t) \prod_j \exp(\alpha_j\beta_j^* - \alpha_j^*\beta_j)\,\frac{d^2\beta_j}{\pi^2}$$

$$= \Big\langle \prod_j \frac{1}{\pi K_j^{1/2}(t)} \exp\Big[-\frac{B_j'(t)}{K_j(t)}\,|\alpha_j - \xi_j(t)|^2 + \frac{C_j^*(t)(\alpha_j - \xi_j(t))^2 + \text{c.c.}}{2K_j(t)}\Big]\Big\rangle \tag{10}$$

(the integrations are performed over the whole complex plane β_j, $d^2\beta_j = d(\mathrm{Re}\,\beta_j)d(\mathrm{Im}\,\beta_j)$) and $K_j(t) = B_j'^2(t) - |C_j(t)|^2$,

$B'_j(t) = B_j(t) + 1$. Thus we see that this process is described by the averaged shifted multifold gaussian probability distribution with correlated real and imaginary parts of the complex field amplitudes ; if the incident radiation field is coherent, then it is represented by the superposition of the coherent field $\xi_j(t)$ and the chaotic field $K_j(t)$.

Adopting the Schrödinger picture and starting with the equation of motion for the density matrix $\hat{\rho}$,

$$i\hbar \frac{\partial \hat{\rho}}{\partial t} = [\hat{H}, \hat{\rho}] \tag{11}$$

and using the q-c - number correspondence of the coherent-state technique, we can prove that the quasi-distribution (10) satisfies the generalized Fokker - Planck equation

$$\frac{\partial \phi_{\mathcal{A}}}{\partial t} = \Big\{ \sum_j \frac{\partial}{\partial \alpha_j}\Big[\Big(\frac{\gamma_j}{2} + i\omega_j \chi_j^{1/2} \operatorname{ch} \frac{\varphi_j}{2}\Big) \alpha_j - i\omega_j \chi_j^{1/2} \times \operatorname{sh} \frac{\varphi_j}{2} \alpha_j^*\Big] + \text{c.c.} + \gamma_j(\langle n_{dj}\rangle + 1) \frac{\partial^2}{\partial \alpha_j \partial \alpha_j^*} - \Big[\frac{i}{2} \omega_j \chi_j^{1/2} \operatorname{sh} \frac{\varphi_j}{2} \frac{\partial^2}{\partial \alpha_j^2} + \text{c.c.}\Big]\Big\} \phi_{\mathcal{A}} \tag{12}$$

(c.c. denotes the complex conjugate terms).

The generating function $\langle \exp(-\lambda W)\rangle$ of the integrated intensity W by a photodetector (λ is a parameter) can be obtained directly from the normal quantum characteristic function $C_{\mathcal{N}}(\{\beta_j\},t) = C_{\mathcal{A}}(\{\beta_j\},t) \exp(\sum_j^M |\beta_j|^2)$ by means of the formula

$$\langle \exp(-\lambda W)\rangle = (\pi\lambda)^{-M} \int \exp\Big[-\frac{1}{\lambda} \sum_j^M |\beta_j|^2\Big] \times C_{\mathcal{N}}(\{\beta_j\},t)\, d^2\{\beta_j\} \tag{13}$$

M being the number of degrees of freedom (modes). This formula provides that

$$\langle \exp(-\lambda W)\rangle = \Big\langle \left[1+\lambda(E-1)\right]^{-M/2} \left[1+\lambda(F-1)\right]^{-M/2} \times \exp\left[-\frac{\lambda A_1}{1+\lambda(E-1)} - \frac{\lambda A_2}{1+\lambda(F-1)}\right]\Big\rangle \quad (14)$$

where

$$E-1 = B-|C| = |v|(|v|-|u|) + \langle n_d\rangle (1-\exp(-\gamma t))$$

$$F-1 = B+|C| = |v|(|v|+|u|) + \langle n_d\rangle (1-\exp(-\gamma t))$$

$$A_{1,2} = \frac{1}{2}\left\{\sum_j^M |\xi_j(t)|^2 \mp \frac{1}{2|C|}\left[C^*\sum_j^M \xi_j^2(t) + c.c.\right]\right\} ; \quad (15)$$

here and in the following we assume uniform fluctuations and the same losses in various radiation modes. The expression in the angle brackets of (14) is just the product of the generating functions for the Laguerre polynomials L_n^M ; consequently the measurable photocount distribution p(n) and its factorial moments $\langle W^k\rangle$, k = 1,2,..., can be expressed in the form

$$p(n) = \frac{1}{n!}\frac{d^n}{d(-\lambda)^n}\langle \exp(-\lambda W)\rangle\Big|_{\lambda=1} = \Big\langle \frac{(1-1/F)^n}{(EF)^{M/2}} \times \exp\left(-\frac{A_1}{E}-\frac{A_2}{F}\right)\sum_{k=0}^{n}\frac{1}{\Gamma(k+M/2)\Gamma(n-k+M/2)}\left(\frac{1-1/E}{1-1/F}\right)^k \times L_k^{M/2-1}\left(-\frac{A_1}{E(E-1)}\right) L_{n-k}^{M/2-1}\left(-\frac{A_2}{F(F-1)}\right)\Big\rangle \quad (16a)$$

$$\langle W^k\rangle = \frac{d^k}{d(-\lambda)^k}\langle \exp(-\lambda W)\rangle\Big|_{\lambda=0} = k!\,\Big\langle (F-1)^k \times \sum_{l=0}^{k}\frac{1}{\Gamma(l+M/2)\Gamma(k-l+M/2)}\left(\frac{E-1}{F-1}\right)^l L_l^{M/2-1}\left(-\frac{A_1}{E-1}\right)$$

$$\times L_{k-1}^{M/2-1}\left(-\frac{A_2}{F-1}\right)\Big\rangle \qquad (17a)$$

where Γ is the gamma function.

The quantities E-1 and F-1 in (14), (16a) and (17a) play the role of the "mean numbers of chaotic photons" (the subtraction of unity eliminates the contribution of the physical vacuum), while the quantities $A_{1,2}$, related to the incident field, play the role of the "mean numbers of coherent photons"; E-1 and F-1 represent the quantum noise.

The deviation of the expression within the angle brackets in (14) from the Poisson generating function $\exp(-\lambda W(t))$, $W(t) = \sum_j^M |\xi_j(t)|^2$, reflects the change of the photon statistics caused by the dynamics of the process, whereas the average over the initial complex amplitudes, represented by the angle brackets, leads to an additional change of the photon statistics with respect to the initial state of light. The character of the dynamical change of the photon statistics depends on the sign of the "mean number of chaotic photons" $E-1 = B-|C|$ ($F-1 = B+|C|$ is always non-negative). If it is positive, as in the case of scattering of light by fluctuating random media, the uncertainty (bunching) of photons is higher than for the Poisson statistics corresponding to the coherent field, i.e. the photon statistics are super-poissonian. If it is negative, this uncertainty is reduced below that for the coherent field and non-classical fields exhibiting antibunching of photons, sub-poissonian statistics and squeezing may occur, as is typical in nonlinear optical processes. For the mean number of scattered photons we obtain

$$\langle n(z/c)\rangle = \langle |u|^2 + |v|^2\rangle \langle n_0\rangle + M\langle |v|^2\rangle + M\langle n_d\rangle (1-\exp(-\gamma z/c)) \qquad (18)$$

z being the distance travelled by the radiation in the random medium; here the first term represents the response to the incident radiation with the mean number of photons $\langle n_0\rangle$, the second term represents the self-radiation of the medium, which is in general non-zero even when $\langle n_o \rangle = 0$ and the third term is the reservoir contribution. If the incident radiation is strong enough, the quantum effects may be neglected ($|v| = 0$) and $E-1=F-1 \approx \langle n_d \rangle (1-\exp(-\gamma z/c))$. In this case the simpler expressions for the statistics of the signal $\langle n_c \rangle = A_1 + A_2 = W(t)$ and noise $\langle n_{ch} \rangle = M \langle n_d \rangle (1-\exp(-\gamma z/c))$ are obtained,

$$p(n) = \left\langle \frac{1}{\Gamma(n+M)} \left(1+\frac{\langle n_{ch} \rangle}{M}\right)^{-M} \left(1+\frac{M}{\langle n_{ch} \rangle}\right)^{-n} \exp\left(-\frac{\langle n_c \rangle M}{\langle n_{ch} \rangle + M}\right) \right.$$

$$\left. \times\ L_n^{M-1}\left(-\frac{\langle n_c \rangle M^2}{\langle n_{ch} \rangle (\langle n_{ch} \rangle + M)}\right) \right\rangle \qquad (16b)$$

$$\langle W^k \rangle = \left\langle \frac{k!}{\Gamma(k+M)} \left(\frac{\langle n_{ch} \rangle}{M}\right)^k L_k^{M-1}\left(-\frac{\langle n_c \rangle M}{\langle n_{ch} \rangle}\right) \right\rangle \qquad (17b)$$

If moreover losses are neglected ($\gamma = \langle n_d \rangle = \langle n_{ch} \rangle = 0$), we obtain the generating function for the averaged Poisson process $\langle \exp(-\lambda W(t)) \rangle$ leading to the photocount distribution given by the Mandel photodetection equation with random modulation, in agreement with the phenomenological descriptions,

$$p(n) = \int_0^\infty \frac{(\mathcal{X} W_o)^n}{n!} \exp(-\mathcal{X} W_o)\, P(W_o) \bar{P}(\mathcal{X})\, dW_o\, d\mathcal{X}$$

$$= \int_0^\infty p_o(n, \mathcal{X})\, \bar{P}(\mathcal{X})\, d\mathcal{X} \qquad (19)$$

W_o being the integrated intensity of the incident light, $P(W_o)$ its probability distribution and $p_o(n,\mathcal{X})$ its photocount distribution ; $\bar{P}(\mathcal{X})$ is the probability distribution of the fluctuating quantity $\mathcal{X}$ having the form

$$\bar{P}(\mathcal{X}) = \frac{\exp(-\mathcal{X}/\langle \mathcal{X} \rangle)}{\langle \mathcal{X} \rangle}, \quad \langle W^k \rangle = k! \langle \mathcal{X} \rangle^k \langle W_o^k \rangle \qquad (20a)$$

in Gaussian media and

$$\bar{P}(\mathcal{K}) = \frac{1}{(2\pi)^{1/2}\sigma\mathcal{K}} \exp\left[-\frac{(\ln\mathcal{K})^2}{2\sigma^2}\right], \langle\mathcal{K}^k\rangle = \langle\exp(k\varphi)\rangle = \exp(k^2\sigma^2/2) \quad (20b)$$

in the turbulent atmosphere ($\sigma \in (0,3/2)$). We see that chaotic light scattered by a gaussian medium is super-chaotic, since $\langle W^k\rangle = (k!)^2 \langle\mathcal{K}\rangle^k \langle W_o\rangle^k$.

Conclusions

We have developed the quantum dynamical theory of the photon statistics including quantum fluctuations and losses in scattering media. If these effects are neglected, the phenomenological descriptions of the photon statistics of light scattered by random media are derived. In usual situations the quantum effects can be shown to be non-substantial . In general, one finds that the peak of the photocount distribution is shifted to lower values of the photoelectron number n and the photocount distributions are broadened with increasing level of fluctuations. For higher (saturated) levels of fluctuations the information about the spectrum of light is practically lost in the photocount statistics. This treatment allows to include small as well as strong, and slow as well as fast fluctuations.In application to scattering by a small number of particles N in the scattering volume, we can obtain for the mean number of particles

$$\langle N\rangle = \frac{1}{(\exp\sigma^2 - 2)^2}\left(\frac{\exp 3\sigma^2 - 6}{\exp\sigma^2 - 2} - 9\right) \quad (21)$$

from the deviations of the actual photocount statistics from the Bose - Einstein statistics.

The presented results can further be applied to light scattering by rotating ground glass disk, nematic liquid

crystals with voltage, turbulent atmosphere and to speckles.
For further details we refer the reader to works [1-4].

References

1. Peřina, J.. Progress in Optics, Vol. 18, Ed. E.Wolf, North-Holland, Amsterdam 1980, p.127.

2. Peřina, J.: Quantum Statistics of Linear and Nonlinear Optical Phenomena, D.Reidel, Dordrecht, Boston, Lancaster 1984, Ch. 8.

3. Koňák, Č., Křepelka, J., Peřina, J.: Opt.Acta 30, 1645-1654 (1983).

4. Křepelka, J., Peřina, J., Koňák, Č., Štěpánek, P.: Czech.J.Phys. B 34, in print (1984).

ATTENUATION OF A COHERENT FIELD IN A DENSE DISPERSION OF PARTICLES

Čestmír Koňák, Petr Štěpánek
Institute of Macromolecular Chemistry, Czechoslovak Academy of Sciences, 162 06 Prague 6, Czechoslovakia

Jaromír Křepelka, Jan Peřina
Laboratory of Optics, Palacký University
771 46 Olomouc, Czechoslovakia

Introduction

In recent years there has been an increased interest in the propagation characteristics of the coherent field in a dense distribution of scatterers. The intensity of light I_t transmitted through a turbid medium at a zero scattering angle generally consists of the coherent intensity I_c from the primary laser beam and the incoherent intensity I_{ch} from the chaotic scattered light. The integral intensity measurements of the attenuation constant α of the coherent light are usually restricted to experimental conditions when $I_{ch} << I_c$.
The purpose of this contribution is to demonstrate that α of the coherent light intensity can be adequately determined even under conditions when $I_{ch} \sim I_c$ by photocount distribution measurements of transmitted light.

Theory

The theoretical photocount distribution p(n,T) for a superposition of coherent and chaotic optical fields is given by the relation (ref.(1)):

Physical Optics of Dynamic Phenomena and Processes in Macromolecular Systems

$$p(n,T) = \frac{1}{\Gamma(n+M)} \left(1 + \frac{\langle n\rangle}{M(1+r)}\right)^{-M} \times \left(1 + \frac{M(1+r)}{\langle n\rangle}\right)^{-n} \times$$
$$\times \exp\left(- \frac{r\langle n\rangle M}{\langle n\rangle + M(1+r)}\right) L_n^{M-1} \left(- \frac{r(1+r)\langle n\rangle M^2}{\langle n\rangle(\langle n\rangle+M(1+r))}\right), \qquad (1)$$

where Γ is the gamma function, L_n^{M-1} are Laguerre polynomials, M is the number of degrees of freedom ($M = 2TS/\tau_c S_c(1+p^2)$, where T is the time detection interval, S the area of photocathode, τ_c the coherence time, S_c the coherence area and p the degree of polarization); $\langle n\rangle = \eta\langle I\rangle TS$ is the mean number of photocounts per T (where η is the photoefficiency and $\langle I\rangle$ the mean intensity of the incident light) and $r = \langle n_c\rangle/\langle n_{ch}\rangle$ (where $\langle n_c\rangle$ and $\langle n_{ch}\rangle$ are the mean numbers of photopulses corresponding to coherent and chaotic optical fields, respectively).

Experimental

Samples: Latex spheres (~1 µm in diameter) were suspended (concentration $c = 7.6\times10^{-4}$ g/ml) in spectrophotometric cells with different path lengths (L = 1,2,5,7, and 10 mm).
Method: The experimental set-up of the optical measuring system is shown in Fig. 1.

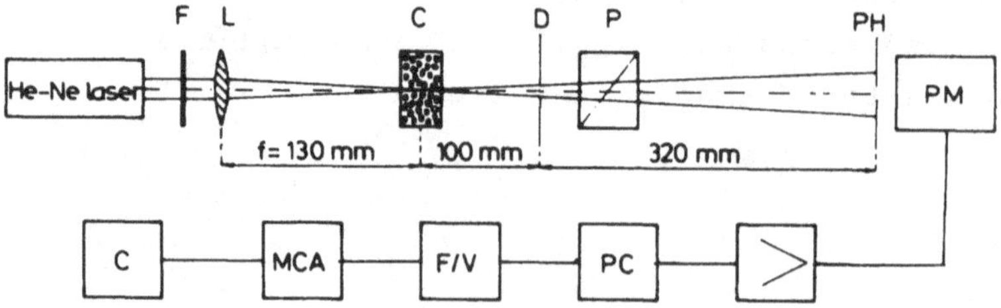

Fig. 1. A block diagram of the instrumentation used. F, neutral density filter; L, lens; C,cell; D, diaphragm (2 mm in diameter); P, polarizer; PH, pinhole (75 µm in diameter); PM, photomultiplier; PC, photon counting system; F/V, frequency-to-voltage convertor; MCA, multichannel analyzer; C, computer.

The experimental conditions were selected so that $\tau_c \gg T(=10^{-4}s)$ and $S_c > S(=4.5\times10^{-5}cm^2)$. The transmitted intensity was conveniently reduced by neutral density filters.

Results

The experimental photocount distributions of transmitted light are shown for different path lengths in Fig. 2.

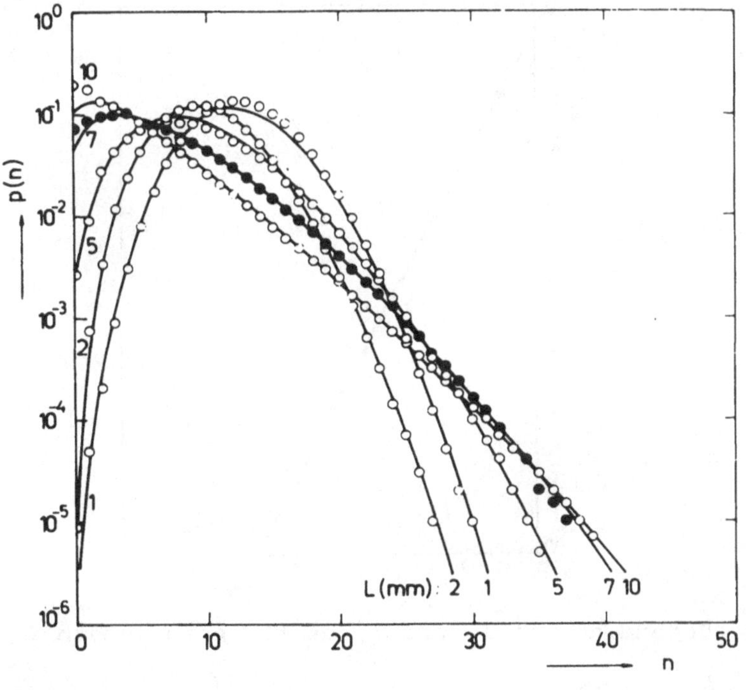

Fig. 2. Comparison of the experimental (points) and theoretical (full lines) photocount distributions.

The experimental p(n,T) distributions were fitted to the theoretical distribution (1) (see Fig.2). M was determined from p(n,T) for sample with L=10 mm where $I_{ch} \gg I_c$ ($\alpha \doteq 0$). The results of the analysis are given in Table 1.

Table I. The analysis of experimental p(n,T) (M=1.69 for all L).

L	r	n_t	n_c	I_t, photons/s	I_c, photons/s
10	≈0	4.74	≈0	4.74×10^4	-
7	1.77	6.48	4.08	3.24×10^5	2.04×10^5
5	8.07	9.00	8.00	9.00×10^6	8.01×10^6
2	92.7	9.62	9.52	2.40×10^9	2.38×10^9
1	$>10^6$	12.47	12.47	1.56×10^{10}	1.56×10^{10}
0	∞	15.75	15.75	9.84×10^{10}	9.84×10^{10}

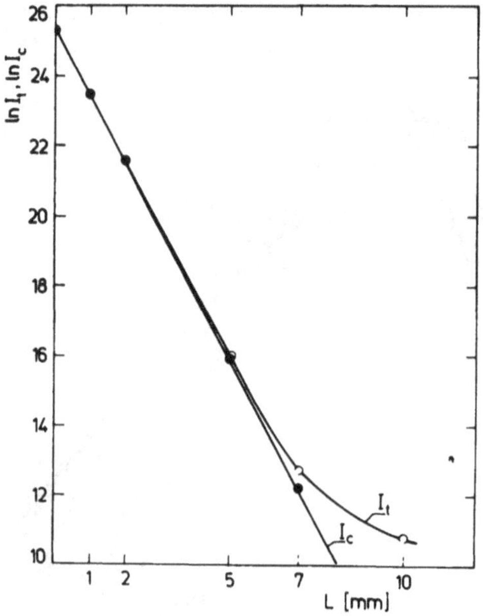

Fig. 3. Dependence of I_c and I_t on the path length L.

Values of I_c follow the Lambert-Beer's law even for $I_{ch} \sim I_c$ (see Fig.3) and α(=18.8 cm^{-1}) can be established with sufficient reliability for high values of αL. We suppose that the procedure can be extended to the case $I_{ch} > I_c$.

References

1. Peřina, J.: Quantum Statistics of Linear and Nonlinear Optical Phenomena, D. Reidel.Publ.Comp., Dordrecht 1984.

DETERMINATION OF POLYDISPERSITY INDEX BY QUASIELASTIC LIGHT SCATTERING

Petr Štěpánek, Zdeněk Tuzar, Čestmír Koňák

Institute of Macromolecular Chemistry, Czechoslovak Academy of Sciences, 162 06 Prague 6, Czechoslovakia

Introduction

Many efforts in polymer and colloid science have been devoted to the characterization of molecular weight distribution and its moments. The polydispersity can be determined either as a distribution curve, e.g., from GPC or the sedimentation velocity method, or as a ratio of two different moments of the distribution curve, usually the weight and number averages of molecular weight, M_w and M_n, as measured by, e.g., integral light scattering and osmometry. For both technical and physical reasons, polydispersity determined by those methods may be subjected to a considerable error, especially with samples of high M_w ($>10^6$) and a low polydispersity index M_w/M_n (<1.3). This is the case of polystyrene standards labelled often with polydispersity index values, determination of which is at the limit of possibilities of the method used.

We propose a new method based on QELS measurements that seems to be very sensitive to extremely low polydispersity index values and is at the same time independent of molecular weight.

Theory

For a monodisperse sample the autocorrelation function is a single exponential and the experimentally determined relaxation time τ_c should not depend on the sampling time of the correlator, ΔT, provided that τ_c spans over at least 10 chan-

Physical Optics of Dynamic Phenomena and Processes in Macromolecular Systems

nels and that the full curve covers at least one relaxation time (1).
For samples with even a very low polydispersity, non-constant values of τ_c are obtained by the forced single exponential fit, when varying ΔT.
To correlate changes in τ_c with the values of M_w/M_n, autocorrelation curves for solutions of polydisperse polymers were computer simulated, using the relation (unnormalized)

$$G(m\ \Delta T) = [\int f(M) \exp(-D(M) K^2 m\ \Delta T) dM]^2 ,$$

where m = 1,...,100 is the channel number, K is the scattering vector, $D(M) \sim M^{-\alpha}$ is the diffusion coefficient (α = 0.5-0.6, depending on solvent power (2)); f(M) is the molecular weight distribution function, either Gaussian or lognormal.

The simulated correlation curves were force fitted by a single exponential and apparent values of $\tau_c(\Delta T)$ were determined for various M_w/M_n. Provided α is constant throughout the range of M_w, τ_c may be normalized, e.g., by the value of τ_c^{10} (= 10 ΔT) to obtain a plot τ_c/τ_c^{10} vs. $\Delta T/\Delta T^{10}$ (ΔT^{10} corresponding to τ^{10}), which is independent of M_w (Fig.1).
In this simplest version, the method supposes $(KR_G)^2 \ll 1$ (R_G is the radius of gyration) and further a low concentration, $c \to 0$. Under these conditions, the additional contributions of internal modes, dissymmetry and interparticle interaction can be neglected.

Experimental

A homodyne QELS spectrometer was used with a He-Ne or Ar laser. The sample temperature was controlled at 26,2°C ± 0.1°C. The scattering angle was varied in the range 9° to 90°. The samples were polystyrene standards (see Table 1) dissolved in toluene or cyclopentane. They were optically cleaned by

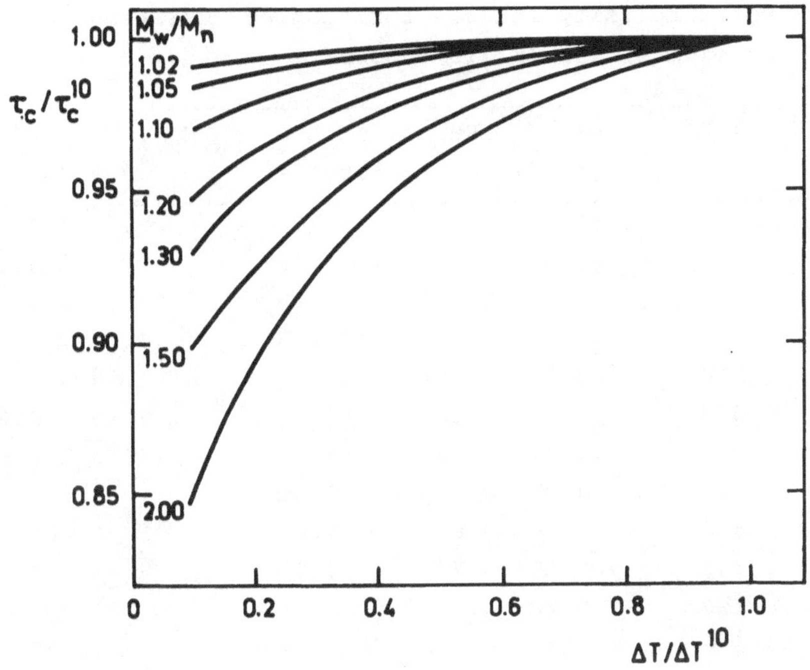

Fig.1. Dependence of the normalized relaxation time τ_c/τ_c^{10} on the normalized sampling time $\Delta T/\Delta T^{10}$ for $\alpha = 0.58$

filtration (0.22 μ) for $M_w < 1.2\text{x}10^6$ and by centrifugation for $M_w > 2\text{x}10^6$. Apparent relaxation times were measured at various sampling times and the normalized results were plotted (Fig.1). Polydispersity indexes were determined from comparison with theoretical curves (Table 1).

The influence of dissymmetry and internal modes was tested on the sample with $M_w = 1.26\text{x}10^6$, by performing the measurements at different angles. It was found that for this method, the condition $KR_G < 0.5$ is imperative, otherwise the experimental data cannot be well fitted and the estimated polydispersities are higher. E.g., for $KR_G = 0.82$ and 1.30 values of $M_w/M_n \doteq 1.25$ and 1.50 were obtained instead of $M_w/M_n = 1.08$ (see Table 1).

Table 1: Polydispersity indices for samples PC (Pressure Chemicals) and TS (Toyo Soda)

Sample	$M_w x10^3$	$cx10^3$ (g/cm^3)	KR_G	M_w/M_n labelled	M_w/M_n found
PC	9	50.0	0.037	1.06	1.03 ± 0.01
PC	20	20.0	0.060	<1.06	1.04 ± 0.03
PC	50	10.0	0.103	1.06	1.04 ± 0.03
PC	110	5.0	0.165	<1.06	1.05 ± 0.02
PC	390	0.8	0.351	<1.10	1.08 ± 0.02
PC	670	0.6	0.484	<1.15	1.20 ± 0.04
TS	1 260	0.34	0.320	1.05	1.08 ± 0.04
PC	2 000	0.30	0.525	<1.30	1.40*
TS	2 890	0.40	0.653	1.09	1.20*
TS	3 840	0.20	0.769	1.05	1.15*
TS	6 770	0.20	0.864	1.14	1.20*
TS	20 600	0.10	0.762	not given	1.35*

*Preliminary values obtained with $KR_G \gtrsim 0.5$.

Conclusion

The sensitivity of QELS to polydispersity is demonstrated. The method can be used for a quick determination of polydispersity index. The values of M_w/M_n determined for polystyrene standards confirm the labelled values. The lognormal distribution seems to be a reasonable approximation for the molecular weight distribution of the standards studied. The Gaussian distribution (more realistic for very narrow fractions) gives identical results with lognormal distribution for $M_w/M_n < 1.2$.

References

1. Oliver, C.J.: Photon Correlation and Light Beating Spectroscopy, p.151, H.Z. Cummins and E.R. Pike, Eds., Plenum Press, New York, London 1974
2. Flory, P.J.: Principles of Polymer Chemistry, Cornell University Press, Ithaca 1953

APPLICATION OF PICOSECOND LASER PULSES IN LIGHT SCATTERING STUDIES

Bruno Van Wonterghem (°), André Persoons
Laboratory for Chemical and Biological Dynamics
University of Leuven, Celestijnenlaan 200 d, B-3030 Leuven

Introduction

Until now correlation times in light scattering studies below 10 ns can only be measured by interferometric (1) or dispersive (2) methods. We are developing a method for measuring intensity fluctuation correlation functions on a very short time scale (10 ps to 10 ns) in the time domain, using ultrashort laser pulses. This time domain seems very interesting because it spans a whole range of local motions in solution (rotation) as well as in macromolecules and supramolecular structures. We shall denote the instantaneous scattering power of a sample by the function f(t), which takes into account both the polarizability tensor orientation and the location of the different scattering centers (3).

Theory

The method we propose is based on a CW injection-locked dye laser (Rh6G) with a cavity dumper. This system delivers a train of picosecond pulses (duration less then 2 ps) with a variable repetition rate (0 to 4 Mhz). The peak power of the pulses reaches up to 3 kW at a wavelength of 590 nm.

The principle of the apparatus is schematically depicted in Figure 1. The incoming light pulse is divided by a beam-

Physical Optics of Dynamic Phenomena and Processes in Macromolecular Systems

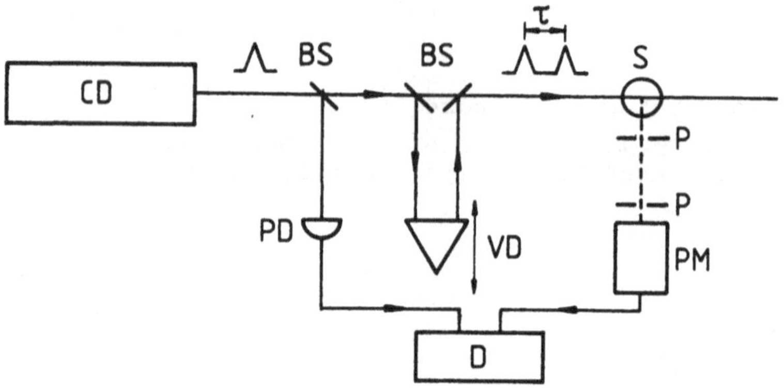

Figure 1: Principle of the apparatus.
(CD: cavity dumper, PD: photodiode, VD: variable delay line, BS: beam-splitter, S: scattering cell, P: pinhole, PM: photomultiplier, D: electronic data acquisition (see text))

splitter in two equal parts, one part being delayed by a variable delay time τ with respect to the other part. The instantaneous scattering power f(t) of the sample is therefore probed by two identical picosecond pulses separated by a time τ. Because of the slow response time of the photomultiplier (several ns) the response to the two scattered pulses is integrated into only one anode pulse. We shall mention two possible methods to extract $|g_f^{(1)}(\tau)|^2$ or $g_f^{(2)}(\tau)$, resp. the first and second order autocorrelation functions of the instantaneous scattering power f(t).

1) Anode pulseheight analysis. The photomultiplier output (anode) pulses are sent to a multichannel pulseheight analyzer which accumulates the charge distribution function of the integrated anode pulses. The photomultiplier used (Philips XP 2262) can resolve light pulses up to 5 photoelectrons. From this charge spectrum it is possible to calculate the counting distribution of the photoelectrons P(n) scattered by the respective pulse pairs. This is done by deconvolution with the single photoelectron

charge distribution (4). If we denote by I_p the intensity of the exciting pulses the total intensity scattered by a pulse pair I_s can be written as

$$I_s(t)=|f(t)|^2 I_p + |f(t+\tau)|^2 I_p \qquad (1)$$

by squaring and averaging over a sufficient number of pulse pairs, (1) becomes, assuming ergodicity of the function f(t),

$$<I_s^2>=2<I_p^2>\{<|f(t)|^2|f(t+\tau)|^2> + <|f(t)|^4>\} \qquad (2)$$

This expression equals per definition

$$<I_s^2>=2<I_p^2>\{g_f^{(2)}(\tau) + g_f^{(2)}(0)\} \qquad (3)$$

Using photoelectron statistics $<I_s^2>$ and $<I_p^2>$ can be related to factorial moments of P(n), respectively of the scattered pulses and the exciting pulses. The complete function $g_f^{(2)}(\tau)$ can be obtained by repeating this measurement for different delay times τ.

2) Pulse counting. The two-pulse scattering configuration (Fig. 1) allows us to determine the joint photon count probability function $P(n,m;\phi)_{n,m=0}$ (5) in an elegant way if the delay time τ is less then the photomultiplier response time. In our approach $P(0,0;\tau)$ is the probability that an exciting pulse pair did not cause an anode pulse. By simply counting the number of anticoincidences between the exciting pulses (monitored by a photodiode) and the discriminated photomultiplier output signal $P(0,0;\tau)$ is determined. Repeating this procedure for several delay times τ gives us the function $P(0,0;\tau)$ from which $|g_f^{(1)}(\tau)|^2$ can be calculated according to the formula (6, 7)

$$P(0,0;\tau)=\{1 + 2<n> + <n>^2(1 - |g_f^{(1)}(\tau)|^2)\}^{-1} \qquad (4)$$

where $\langle n \rangle$ is the mean number of photoelectrons detected per scattered pulse pair. Note that eq. 4 only applies to Gaussian scattering processes f(t). Obviously this method is much simpler then the first one.

Measurements have not yet been done but experimental work is in progress and results are soon to be expected.

Notes and references

° B.Van Wonterghem is a research assistent of the National Fund for Scientific Research (Belgium).

1. Alms, G.R., Bauer, D., Brauman, J.I., Pecora, R.: J. Chem. Phys. 58, 5570-5578 (1973).
2. Perrot, M., Bouachir, M., Lascombe, J.: Mol. Phys. 42, 551-564 (1981).
3. Berne, B.J., Pecora, R.: Dynamic Light Scattering, Wiley-Interscience, New York . London 1976
4. Arecchi, F.T.: IEEE J. Quant. Electr. QE-2, 341-350 (1966).
5. $P(n,m;\phi)$ means the probability that at time t n photoelectrons and at time $t+\phi$ m photoelectrons are detected. $P(0,0;\tau)$ thus means the probability that no photoelectron is detected at time t and time $t+\tau$.
6. Saleh, B.: Photoelectron Statistics, Springer Series in Optical Sciences, Volume 6, Springer-Verlag, Berlin - Heidelberg - New York 1978
7. Furcinitti, P., Kuppenheimer, J., Narducci, L.M., Tuft, R.A.: J. Opt. Soc. Am. 62, 792-796 (1972).

APPLICATION OF EXPERIMENTAL TECHNIQUES FOR OBTAINING PHOTON CORRELATION FUNCTIONS AT HIGH PRESSURE

Jaan Roots and Bo Nyström
Department of Chemistry, University of Oslo,
P.O.Box 1033, Blindern, 0315 Oslo 3, Norway.

Pressure variations are closely related to effects associated with volume changes, and should in principle be more amenable to theoretical interpretations, as opposed to temperature variations which induce simultaneous changes in volume and thermal energy in a way that makes it difficult to separate these two effects. A number of investigations of polymer solutions, using various experimental techniques, have been carried out at high pressure and different features have been illuminated. However, photon correlation spectroscopy on polymer solutions have been vastly neglected, probably due to considerable experimental difficulties.

Measurements at elevated pressures require elaborate techniques, and special equipment which in general is not commercially available. A design (1) of a portable vessel for optical measurements has been adopted, and the vessel has been adapted (2) to photon correlation spectroscopic experiments at pressures up to 5000 atm. The potential of experiments at high pressures is illustrated by results on a number of polymer/solvent systems of different complexity, viz.
(a) Synthetic polymer/organic solvent system for which the thermodynamic properties are nearly independent of pressure (2); the change in solvent viscosity will predominate the diffusion behaviour both in dilute and semidilute solutions. Pressure constitutes an excellent tool to continuously in-

Physical Optics of Dynamic Phenomena and Processes in Macromolecular Systems

crease the viscosity of organic solvents over a large range. It may be mentioned that this feature has recently been made use of in studies of intramolecular excimer formation (3,4). (b) Synthetic polymer having ionizable groups, in aqueous solution under conditions where electrostatic interactions effectively are screened (5). (c) Biologically produced, specific-sequence copolymers, whose monomer units are the naturally occuring amino acids (some displaying a hydrophilic, others a hydrophobic character), in aqueous solution (5,6). These biopolymers (globular proteins) are distinguished by their ability to undergo an induced transition from a form that executes, at most, small fluctuations about a well defined (native) conformation, to an unfolded form.

The pressure vessel used in these studies (PV in Figure 1) consists of a steel cylinder (a) with a central axial hole and a smaller channel, both of which can be closed at the ends with O-ring sealed pistons (b), supported by screws (c). The two side pistons and the bottom piston are channeled and provided with windows (d) which are tightened by the unsupported area technique. The solution to be studied is enclosed in a cylindrical quartz cuvette (e) having an optically flat bottom and two, opposite placed, optically flat windows. The cuvette is closed with a plunger equipped with O-rings; the filling port is sealed with an O-ring-fitted screw. The plunger isolates the sample solution from the pressure transmitting fluid. To apply pressure the top piston is pressed down by a steel rod (f) passing through a central hole in the upper supporting screw; the pressure is transmitted to the solution by the downward movement of the plunger in the cuvette. When the desired pressure in the cell has been reached, it is locked in by turning the top screw down against the top piston. The vessel can then be removed from the press and placed in the light-scattering instrument (see Figure 1), (or in another adapted experimental set-up).

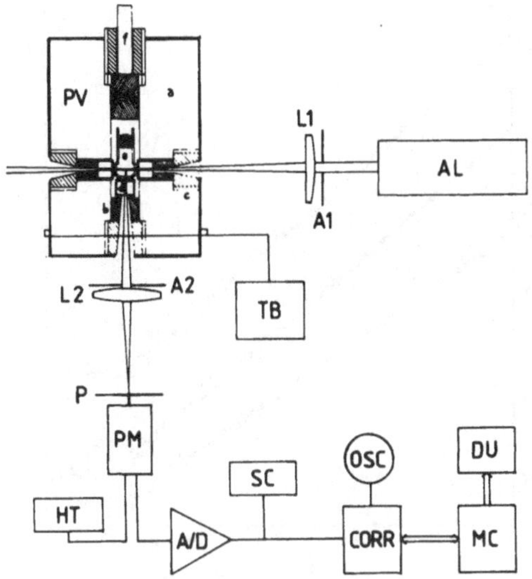

Figure 1. Instrumental set-up for photon-correlation experiments at high pressure. PV: cross-section of the pressure vessel and light-scattering cuvette; TB: thermoregulated bath with external circulation; AL: Argon ion laser operated in TEM_{00}-mode (and with intra-cavity etalon), λ=488nm; A1, A2: apertures, L1: plano-convex lens (f=150 mm); L2: biconvex lens (f=75 mm); P: pinhole (Ø=100 μm); PM: photomultiplier tube; HT: high tension for PM tube; A/D: amplifier/discriminator; SC: counter; CORR: 4-bit correlator, 128 channels; OSC: oscilloscope; MC: micro-computer on-line; DU:display unit/printer.

The solutions were filtered through 0.22 Millipore filters, in an atmosphere of filtered air, directly into carefully pre-cleaned cuvettes; the cuvettes were assembled under the same conditions. Before the pressure vessel was assembled and an experimental run started the passage of an unfocused laser-beam through the cuvette was visually examined with an eye-piece; the solution was discarded if light-speckles could be observed. An example of normalized photon correlation functions obtained at different pressures is shown in Figure 2.

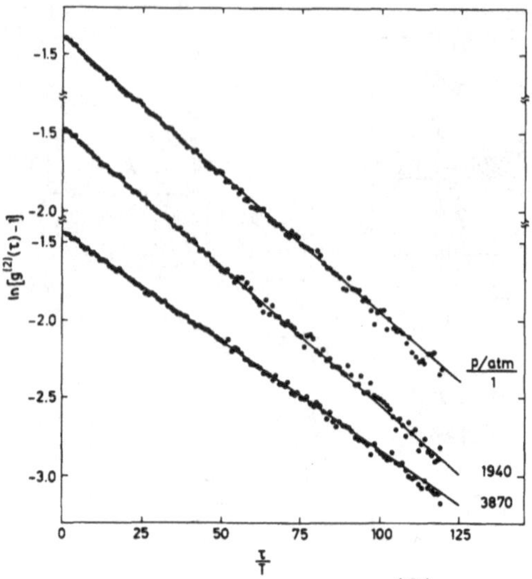

Figure 2. Semi-logarithmic plot of $g^{(2)}(\tau)$-t vs. channel number (τ/T) for lysozyme in acetic acid/acetate buffer (pH=4.0, ionic strength I=0.1 M) at a concentration of 9.20 kg m^{-3} (the value at atmospheric pressure) and at the pressures indicated, and the channel time T=0.2 μs (6). The solid lines represent the initials slope from a least-squares fit of a quadratic polynom in τ to the experimental data. The data points have been shifted by arbitrary amounts vertically to avoid overlap.

References

1. Claesson, S., Malmrud, R., Lundgren, B.: Trans.Faraday Soc. 66, 3048-3052 (1970).
2. Roots, J., Nyström, B.: Macromolecules 15, 553-556 (1982).
3. Fitzgibbon, P.D., Frank, C.W.: Macromolecules 14, 1650-1658 (1981).
4. Roots, J., Nyström, B.: Polymer Commun. 25, 166-169 (1984).
5. Nyström, B., Roots, J.: Chem.Phys.Lett. 91, 236-240 (1982).
6. Nyström, B., Roots, J.: Makromol.Chem. 185, 1441-1447 (1984).

BIMODAL ANALYSIS OF QELS DATA

R. Johnsen
Institute of Physical Chemistry,
University of Uppsala, Uppsala, Sweden

The relationship between the intensity autocorrelation function $g^{(2)}(\tau)$ and the electric field autocorrelation function $g^{(1)}(\tau)$ has been shown to be

$$g^{(2)}(\tau) = 1 + [g^{(1)}(\tau)]^2$$

It is often found that $g^{(1)}(\tau)$ is not a single exponential function. The cumulant method [1] has most often been used to determine the mean and variance of the distribution of Γ, the inverse relaxation time.

It has been found by some researchers to be advantageous to approximate $g^{(1)}(\tau)$ with a single exponential function plus an adjustable baseline, and this approach has been extended to the sum of several exponential functions plus an adjustable baseline. The origin of this adjustable baseline has variously been attributed to long-term laser drift or the presence of dust. In our experimental set-up, we have been unable to detect laser drift. We have also developed a solution-cleaning technique of continuous filtration resulting in consistently dust-free solutions at semidilute concentrations. We have therefore chosen to fix the baseline and fit a sum of exponentials to the autocorrelation function.

$$g^{(2)}(\tau) - 1 = \beta[A_1 \exp(-\Gamma_1\tau) + (1-A_1) \exp(-\Gamma_2\tau)]^2$$

β is an instrumental factor, with a positive value less than unity. A_1 is the fractional amount of "mode 1", (with time constant Γ_1), and $1-A_1$ is the fractional amount of "mode 2".

For calculation purposes, the process is described by the expression

$$Y(I) = P_1[P_3 \exp(-I\ P_2) + (1-P_3) \exp(-I\ P_4)]^2$$

Physical Optics of Dynamic Phenomena and Processes in Macromolecular Systems

where P_1 and P_3 are β and A_1 in the previous equation and I is the channel number. P_2 and P_4 are reciprocal "relaxation times" with relaxation time expressed in numbers of channels. Physically reasonable relaxation times would lie between one channel and infinity; thus P_2 and P_4 should be constrained to the interval 0-1, as should P_1 and P_3.

Although the analysis could be extended to include additional exponential terms, it is well known [2] that numerical separation of multiple exponents is extremely difficult, and can be achieved only with data of the highest precision.

The method of analysis is via an equally weighted non-linear regression procedure according to Marquardt [3]. The calculations were performed interactively on the same microcomputer as was used for data collection and reduction. The programs are written in FORTH for speed and compactness, and the microcomputer was fitted with an arithmetic processor for increased speed and accuracy.

In order to determine the reliability of the calculating procedure, it was used to evaluate a number of simulated experiments. When the calculating procedure was applied to real experiments (eg PS in THF in dilute solution) the residuals scattered randomly, and their distribution was apparently Gaussian. Fig.1 shows the residuals from a typical experiment, and Fig.2, the same residuals after sorting in ascending order. The smooth curve in Fig. 2 is the error function with the same standard deviation as the residuals. Thus normally distributed pseudo-random numbers, typical of the residuals in a real experiment, were added to a simulated experiment using parameters also typical of the real experiments. A value of 10^{-3} was used for the standard deviation of residuals (those for real experiments lie in the range $(0.5 - 1.5)\times10^{-3}$).

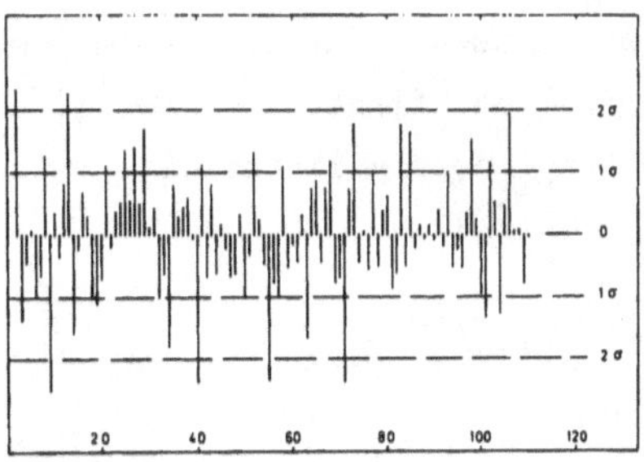

Fig. 1

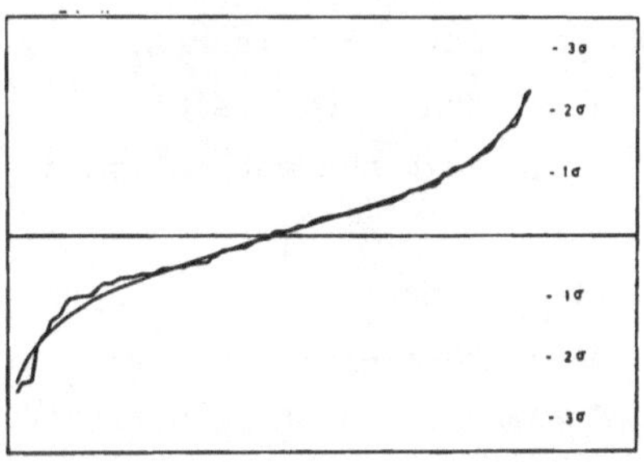

Fig. 2

Table 1 summarizes results from two sets of simulations. Each set comprises 24 simulated experiments, each with the same parameter values, but with different errors added. The first row in each set defines the ideal value of each of the four parameters. The second row of the table gives the mean value of the parameter estimates for each parameter. A measure of the scatter of the estimates is found in the standard deviation, in row number three.

Each experimental analysis yields not only an estimate of the four parameters, but an estimate of the "adjustability" of that parameter based on, among other things, the scatter of the residuals in that particular experiment. The "adjustability" is expressed as estimated standard deviation of that parameter. For each parameter, the average of the 24 estimates of these standard deviations is presented in the fourth row. Points to note in the table are: a) the agreement between rows 2 and 1 (the parameter estimates agree with the "true" values), and b) the agreement between rows 4 and 3 (the "adjustability" of a parameter estimate agrees with the "variability" of the estimates).

The upper half of Table 1 (Set I) tests a reasonable set of parameters. The system consists of 80% of a slow component, with a "relaxation time" of 90 channels. The remainder is a fast component with a relaxation time of 30 channels.

The lower half (Set II) tests an extreme set of parameters. Here we have only 7% of the fast component. The increased uncertainty in the determination of the parameter representing the rate constant for the fast component is evident. Exclusion of the data from the first few correlator channels would make the estimation of parameter 4 nearly impossible.

It is concluded that:

1. parameter estimates obtained from the fitting procedure constitute "true" values without significant bias and
2. the confidence intervals for the parameter estimates are valid intervals.

Table 1

Each of the two sets of simulations involves 24 "experiments". The parameters are fixed to their ideal values and normally distributed pseudo-random numbers, with a standard deviation of 1×10^3 are added to construct 120 values. Results of the analysis are summarized as average estimate, standard deviation of the estimate, and average estimated standard deviation for each parameter.

	Parameters			
Set I	1	2	3	4
10^3 ideal value	500	11.1..	800	33.3..
10^3 av. estimate	500	11.1	794	33
10^3 std.dev.est.	0.54	0.21	23	1.9
10^3 av.est.std.dev.	0.62	0.25	27	2.3
Set II				
(as above)	500	11.1..	930	33.3
	500	11.1	922	33
	0.59	0.18	21	5.0
	0.59	0.21	28	5.9

References

1. Koppel, E.D., J. Chem. Phys. 57, 4814 (1972).
2. Lancsos, C., Applied Analysis, Prentice-Hall, Englewood Cliffs, N.J., (1956).
3. Marquardt, D.W., SIAM J. Appl. Math., 11, 431 (1963).

APPLICATION OF SPECKLE TECHNIQUES IN MACROMOLECULAR PHYSICS

Jaroslav Holoubek, Josef Mikeš*, Blahoslav Sedláček
Institute of Macromolecular Chemistry, Czechoslovak Academy of Sciences, 162 06 Prague 6, Czechoslovakia

*Institute of Chemical Process Fundamentals, Czechoslovak Academy of Sciences, 165 02 Prague 6, Czechoslovakia

Introduction

The random intensity distribution called speckle pattern (SP) is formed when fairly coherent light propagates through a medium with random refractive index fluctuations. At present, the interest in SP lies in many different areas (1-4) of which we mention 1) applications in the study of object deformations, speckle decorrelation and some dynamic properties of SP and 2) structural aspects of non-Gaussian SP. The deterministic relation between the object and speckle movement is the basis of different methods of displacement measurements (both in spatial and temporal meaning). Roughly speaking, methods that generally analyze the temporal aspects of SP are dealing mainly with "boiling" (B) speckles and utilize the spatial "fixation" of speckles at a single spatial point, whereas those engaged in spatial characteristics mostly analyze the "translational" (T) speckles.

1) Object Deformations, Speckle Decorrelation and Dynamic Properties of Speckle Patterns

a) Speckle photography, T speckles

The displacement vector and Poisson´s ratio have been determined with an accuracy of ± 5% in the range of small deformations by the optical processing of double-exposed negatives (5,6). To deal with larger deformations, the method has been

Physical Optics of Dynamic Phenomena and Processes in Macromolecular Systems

extended by using an optical analysis of double-exposed scattered light SP in small consecutive deformations (deformation steps) covering the range under investigation (Fig.1).

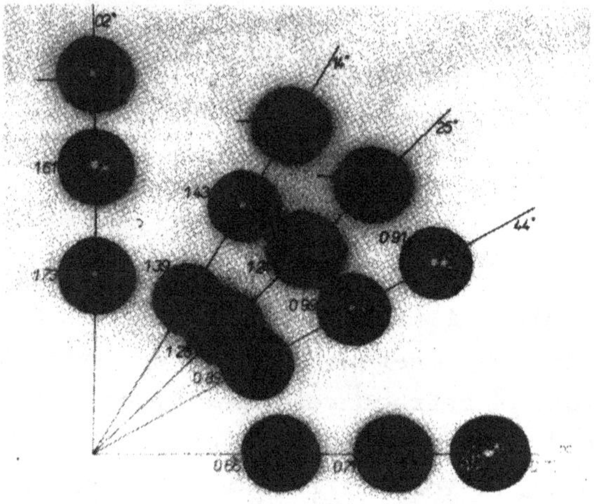

Fig.1. Actual appearance of fringes obtained by point-by-point filtration of double-exposure speckle pattern corresponding to uniaxial deformation of polyethylene foil of about 1.7%. The displacement is inversely proportional to the fringe spacing, and the direction of displacement is perpendicular to the fringes. Only results of processing of one quadrant of SP is shown. The vertical axis corresponds to elongation, the horizontal axis corresponds to contraction. Values of displacement are recalculated to the object plane and are given in per cent of relative deformation, orientation of fringes is given with respect to the horizontal coordinate axis.

b) Speckle photography, B speckles

The results of measurements of speckle decorrelation (7) and determination of slow correlation times (8) are based on the application of the random phase screen theory (9-11). Both methods use a relation between the contrast C of SP and the parameter M (giving the number of the degrees of freedom of light) (Fig.2).

c) Focussed image SP

In the analysis of both enlarged images of two-dimensional and three-dimensional structures, it is necessary to correct

Fig.2. Reduction of speckle contrast by increasing the number of temporal degrees of freedom M
(a) Small angle light scattering speckle pattern of polymer film (H_V type, sample between crossed polarizers), static sample, geometry of B speckles, time of exposure T = 1 s, contrast C = 1.0; (b) T_C = 0.34 s, T = 1/5 s, C = 1.00; (c) T_C = 0.34 s, T = 1 s, C = 0.57; (d) T_C = 0.34 s, T = 3 s, C = 0.33.

the displacement of field slewing. We obtained the corrected field of displacement of a commercial high-density polyethylene under uniaxial tensile deformation by using correction according to Stetson (12,13). Such a study was performed in order to understand changes of the structure and orientation of supermolecular regions during the drawing process of polymer films.

2) Structural Aspects of Non-Gaussian Speckle Patterns

For the spatial coherence properties of SP to carry direct structural information, measurements must be made under experimental conditions where the scattered field does not follow the Gaussian distribution. This situation can be achieved by suitably choosing and varying the coherence pro-

perties of incident light or by limiting the number of randomly phased contributions (14). The relevant structural parameters available by an analysis of scattered light SP are then the numbers of scatterers N, their correlation length (a measure of the size of scatterer) and the value of phase variations introduced by the scatterers $<\phi^2>$.
Thus, speckles are no longer regarded merely as a noise element, but as a carrier of information useful in many different applications.

References

1. Dainty, J.C., Editor: Laser Speckle and Related Phenomena, Plenum Press, Berlin 1975.
2. Erf, R.K., Editor: Speckle Metrology, Academic Press, N.Y. 1978.
3. Françon, M.: Laser Speckle and Applications in Optics, Academic Press, N.Y. 1979.
4. Applications of Speckle Phenomena, SPIE Vol. 243, 1980.
5. Holoubek, J., Růžek, J.: Opt.Acta 26, 43-54 (1979).
6. Holoubek, J., Sedláček, B.: J.Polym.Sci.,Polym.Phys.Ed. 18, 265-275 (1980).
7. Holoubek, J., Sedláček, B.: Makromol.Chem. 185 (1984) in press.
8. Holoubek, J.: this volume, P55.
9. Asakura, T., Takai, N.: Appl.Phys. 25, 179-194 (1981).
10. Baltes, H.P., Editor: Inverse Source Problem in Optics, Springer-Verlag, Berlin 1978.
11. Crosignani, B., Di Porto, P., Bertolotti, M.: Statistical Properties of Scattered Light, Academic Press, N.Y. 1975.
12. Holoubek, J., Sedláček, B.: J.Macromol.Sci.-Phys. B23, 143-152 (1984).
13. Stetson, K.A.: Miscellaneous Topics in Speckle Metrology, in Speckle Metrology, Erf, R.K., Editor, Academic Press, 1978, p.295.
14. Jakeman, E., McWhirter, J.G., Pusey, P.N.: J.Opt.Soc.Amer. 66, 1175-1182 (1976).

LIGHT SCATTERING SPECKLE PHOTOGRAPHY: DETERMINATION OF SLOW CORRELATION TIMES

Jaroslav Holoubek

Institute of Macromolecular Chemistry, Czechoslovak Academy of Sciences, 162 06 Prague 6, Czechoslovakia

Introduction

The basic features of scattering are shared by a number of closely related scatterer systems. The most elementary model for these systems is the random phase screen (RPS) (1-4). In this contribution predictions of the RPS theory are compared with experimental data obtained by analyzing the dynamical behaviour of scattered-light speckle patterns from a film made of a semicrystalline polymer. It is known that with such a material one can easily achieve correlation times within a time scale of 10^{-2} to 10^{2} s by a simple sample movement (5). Irrespective of another physical background of the correlation time used in model experiments, the reported method allows one to measure the average correlation times of slowly relaxing fluctuations in polymers, e.g., very near the glass transition, where photon correlation spectroscopy is not so effective as for shorter correlation times.

Theory

The method is based on a relation between the contrast of the speckle pattern C_I and the parameter M (specifying the number of the degrees of freedom of light) (3,6,7). Combining the expressions for spatial and temporal degrees of freedom, an explicit expression for M can be found in the form (8)

Physical Optics of Dynamic Phenomena and Processes in Macromolecular Systems

$$M = [2/(1+P^2)]\{(S/S_c)^{1/2}\mathrm{erf}(\pi S/S_c)^{1/2} - (S_c/\pi S)[1-\exp(-\pi S/S_c)]\}^{-2}$$

$$\{1/\gamma + (1/2\gamma^2)[\exp(-2\gamma)-1]\}^{-1} \quad (1)$$

where $\mathrm{erf}(x) = 2/(\pi)^{1/2} \int_0^x \exp(-z^2)dz$, and $\gamma = T/T_c$.
T is the observation (exposure) time, S is the physical area of aperture, T_c and S_c denote the coherence time and coherence area (speckle size) at the detector surface, respectively, and P is the degree of polarization. The validity of eq.(1) is limited by the assumption of validity of gamma density distribution for the probability density function of the time-averaged intensity and by the assumption of cross-spectral and polarizational purity of scattered radiation. The contrast C_I and parameter M are related through the equation (3,6,7)

$$C_I = M^{-1/2} \quad (2)$$

where C_I can be obtained at the same time from an experimental analysis as

$$C_I = [\langle(I-\langle I\rangle)^2\rangle]^{1/2}/\langle I\rangle \quad (3)$$

The RPS theory offers a simple relation for T_c (for "boiling" speckle), namely (4,5,9),

$$T_c = W/|\vec{v}| \quad (4)$$

where W is the radius of the laser beam at the object plane and v is the linear speed of motion (circumferential speed at the plane of illumination for rotational motion with constant velocity).
Equations (1)-(3) are basis of determination of the correlation time T_c, which consists in finding the contrast values of scattered light speckle patterns.

Experimental

Our interest was focussed on space-time statistical properties of dynamic speckle due to an in-plane moving diffuse object (polymer film) with the axis of rotation of the object parallel to the illuminating beam. Different circumferential speed at the place of illumination gives a possibility to vary suitably the coherence time T_c. The following elements were used in the experiment: He-Ne laser Spectra Physics 125 A (single transverse mode TEM_{00}), a polymer film made of polyethylene Liten 380 μm thick. The exposure were recorded on a Polaroid 665 P/N film and further processed with the Joyce-Loebl double beam microdensitometer for density fluctuation measurements (5). The typical T_c values used in experiment were T_c = 0.013 s, 0.195 s, 0.340 s.

Results

The comparison of experimental and theoretical values of the contrast is given in Table I for T_c = 0.195 s, S/S_c = 82.6 and indicated ratios of T/T_c values.

Table I: Comparison of theoretical values of C_{th} (eqs.(1),(2)) and experimental values of C_{exp} (eq.(3))

T/T_c	M	C_{th}	C_{exp}
-	82.6	0.110	0.110 ± 0.009
0.103	88.4	0.106	0.093 ± 0.010
0.205	94.2	0.103	0.098 ± 0.010
0.513	113.2	0.094	0.094 ± 0.013
1.026	147.9	0.082	0.084 ± 0.008
2.564	262.7	0.062	0.071 ± 0.010
5.128	470	0.046	0.047 ± 0.005
15.385	1313.3	0.028	0.029 ± 0.004
51.282	4278.6	0.015	0.019 ± 0.002

It follows from Table I that the experimental and theoretical values of contrast are in good accord. Even if the accuracy of the determination of experimental values is about 10%, one could expect that by exploitation of the whole information content of detected SP one should reach the accuracy of the QELS method for the same value of correlation times (at the long time limit).

Conclusions

We have shown that the concept of RPS is useful in describing dynamical effects accompanying the scattering of coherent light by polymer systems. The relation between the spatial and temporal properties of SP was analyzed.
The method of determination of slow correlation times in the time scale 10^{-2}-10^{2} s was suggested and tested for dynamic processes for which T_c does not depend significantly on the scattering vector with good accord between experimental and theoretical values. The applicability of the suggested method to another scattering geometry (e.g. to 90° scattering arrangement (10)) is expected on the basis of known relation between objective and subjective (image) speckle patterns (4,11).

References

1. Crosignani, B., DiPorto, P., Bertolotti, M.: Statistical Properties of Scattered Light, Academic Press, N.Y. 1975.
2. Cummins, H.Z., Pike, E.R., Editors: Photon Correlation and Light Beating Spectroscopy, Plenum Press, N.Y. 1974.
3. Baltes, H.P., Editor: Inverse Source Problem in Optics, Springer-Verlag, Berlin-Heidelberg 1978.
4. Asakura, T., Takai, N.: Appl.Phys. 25, 179-194 (1981).
5. Holoubek, J., Sedláček, B.: Makromol.Chem. 185 (1984), in press.
6. Dainty, J.C., Editor: Laser Speckle and Related Phenomena, Springer-Verlag, Berlin-Heidelberg 1975.
7. Peřina, J.: Progress in Optics, Vol.18 (1980), p.127-203.
8. Holoubek, J.: Opt.Acta, to be published.
9. Jakeman, E., McWhirter, J.G., Pusey, P.N.: J.Opt.Soc.Amer. 66, 1175-1182 (1976).
10. Carroll, P.J., Patterson, G.D., Cullerton, S.A.: J.Polym. Sci.,Polym.Phys.Ed. 21, 1889-1896 (1983).
11. Asakura, T., Takai, T.: in Application of Speckle Phenomena, SPIE, Vol. 243 (1980) p.114-123.

STUDY OF POLYMER SOLUTION/SOLID INTERFACES BY EVANESCENT WAVE SPECTROSCOPY

Hubert Hervet, Dominique Ausserré and Francis Rondelez
Physique de la Matière Condensée, Collège de France,
11 Place Marcelin-Berthelot, 75231 Paris Cedex 05 - France

Introduction

Interfacial properties of polymer solutions play an important role in many practical problems such as colloid stability, wetting, chromatography, biocompatibility, enhanced oil recovery and lubrication. Depending on the system one has to deal with an adsorption or "a depletion layer". In both cases the surface excess Γ is defined as :

$$\Gamma = \int_0^\infty (C(z) - C_v)dz \qquad (1)$$

where C(z) is the monomer concentration at a distance z from the interface, and C_v the bulk concentration. If the substrate is adsorbing Γ is positive and generally of the order of a few mg/m^2. For dilute solutions the surface concentration can be 100 fold the bulk concentration. If the surface is repulsive or neutral for the polymer, Γ is negative and can be written as :

$$\Gamma = (C_s - C_v)\xi \qquad (2)$$

where C_s is the monomer concentration on the solid surface,much lower than C_v. ξ is an average thickness of the depletion layer, of the order of the molecular length L. Thus the absolute value of Γ in the depletion case is much smaller than in the adsorption one.

A quantitative knowledge of the interfacial layer requires the determination of Γ, ξ and if possible, C(z). Conventional techniques such as ellipsometry(1) or hydrodynamic flow(2) provide respectively an average optical thickness ξ_e or a hydrodynamic thickness ξ_H. While supernatant

Physical Optics of Dynamic Phenomena and Processes in Macromolecular Systems

analysis[3] gives the surface excess Γ. On the other hand infrared measurements[4] can measure the number of adsorption sites of the polymer chains on the surface. Up-to-now, only small angle neutron scattering technique[5] provides an access to the concentration profile in the interfacial layer. It is applicable but to systems with vary large specific area.

We have developed a new method : E.W.I.F. (Evanescent Wave Induced Fluorescence) based on an extension of the Total Internal Reflection (T.I.R.) technique. E.W.I.F. enables investigation of fluorescent polymer solution from 500 Å to 10000 Å above the solid surface. A determination of the concentration profile can be obtained if the interfacial layer thickness is comparable to the penetration depth of the evanescent wave. Thanks to its high sensitivity E.W.I.F. is suitable to study even the weak concentrations expected in the depletion case. In the following, we shall first describe the experimental method and then report preliminary data in the case of a depletion layer.

Experimental method

1) Generalities. The cornerstone of the technique is to use the evanescent wave which appears when an incident light wave, coming from an optically dense material (refractive index n_1) impinges on an optically rare material (refractive index $n_2 < n_1$), at an incidence angle θ higher than the critical angle θ_c for total reflection. θ_c is related to the refractive index mismatch by : $\theta_c = \sin^{-1}(n_2/n_1)$. This wave penetrates in medium two over a characteristic distance Λ defined by :

$$\Lambda = \frac{\Lambda_0}{4\pi n_1}\left[\sin^2\theta - \left(\frac{n_2}{n_1}\right)^2\right]^{-1/2} \qquad (3)$$

Λ_0 is the wavelength of the laser beam.

This energy excites fluorescent probes which are attached to the polymer molecules in solution. Measuring the intensity of the emitted fluorescence versus the angle of incidence θ, gives a determination of the profile concentration of polymer near the interface.

2) Experimental procedure. The intensity of the emitted fluorescence by the tagged polymer in the solution for an angle of incidence θ ($\theta > \theta_c$) is :

$$I_p(\Lambda) = A(\Lambda)\ U_0(\Lambda) \int_0^\infty C(z)\ e^{-z/\Lambda}\ dz \qquad (4)$$

where $U_0(\Lambda)$ is the energy density at $z = 0$, and $A(\Lambda)$ is a factor containing the quantum efficiency, geometrical factors and the solid angle of collection. In principle $A(\Lambda)$ and $U_0(\Lambda)$ could be determined experimentally but that would require absolute light intensity measurements which are both tedious and inaccurate. To overcome this problem, a second experiment is performed with a solution of free fluorescent molecules in the same solvent, and in the same optical conditions. In that case one can assume that the concentration of fluorescent molecules is uniform throughout the solution. Therefore the intensity writes :

$$I_F(\Lambda) = A'(\Lambda)\ U_0(\Lambda)\ C_v\ \Lambda \qquad (5)$$

where $A'(\Lambda)$ is the equivalent of $A(\Lambda)$ in (3). The ratio $K = A(\Lambda)/A'(\Lambda)$ is assumed to be a constant equal to the change in quantum efficiency and in concentration of fluorescent probes. In (5) the integral has been replaced by $C_v\ \Lambda$, the result obtained for an isotropic concentration C_v. Performing the ratio of (4) to (5) at corresponding Λ yields :

$$R(\Lambda) = K \frac{1}{C_v \Lambda} \int_0^\infty C(z) e^{-z/\Lambda}\ dz \qquad (6)$$

$\Lambda R(\Lambda)$ is the Laplace Transform of the concentration profile.
K is just an amplitude factor, which can be, by renormalization sets equal to 1. If the penetration depth is much larger than the interfacial layer in thickness, i.e. θ close to θ_c, or the minimum penetration depth, Λmin, much greater than ξ, $R(\Lambda)$ can be expanded into a Taylor series and written as :

$$R(\Lambda) = 1 + \frac{(C_s - C_v)\xi}{\Lambda C_v} + O\left(\frac{\xi}{\Lambda}\right)^2 \qquad (7)$$

$$\text{or } R(\Lambda) = 1 + \frac{\Gamma}{\Lambda C_v} \tag{8}$$

Therefore in this case the initial slope of $R(\Lambda)$ vs $\frac{1}{\Lambda}$ yields the surface excess Γ. On the contrary if the minimum penetration depth is much smaller than ξ, the situation is more favorable and an accurate determination of the entire profile of the polymer concentration at the interface can be expected. Fig. 1a and 1b shows ideal $R(\Lambda)$ curves for adsorption and depletion cases :

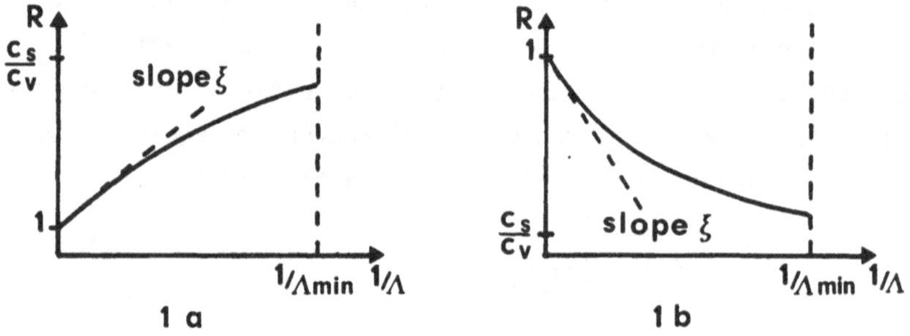

Fig. 1. Ideal curves for $R(\Lambda)$ vs $1/\Lambda$
a) adsorption case ; b) depletion case.

All the above discussion rests on the assumption of the invariance of n_2 with z, which makes possible a correct evaluation of Λ. This is quite well fulfilled in the depletion case where C_s is small. However it does not hold for the adsorption case,and this problem has to be overcome by matching the optical indices of the solute (polymer) and of the solvent.

3) Experimental set-up and materials.

a) The experimental set-up is schematically presented in Fig. 2. The important part is the automatic optical system which ensures that the laser beam impinges on a fixed point of the optical interface when θ is varied. This optical system is computer driven and the incidence angle is controlled with an accuracy of 1/100 of a degree. A glass plate inserted on the incoming beam is used to divert a small fraction of the laser beam and monitor the laser power. A complete curve $R(\Lambda)$ can be recorded in a minimum of two minutes depending on the intensity of the fluorescent signal.

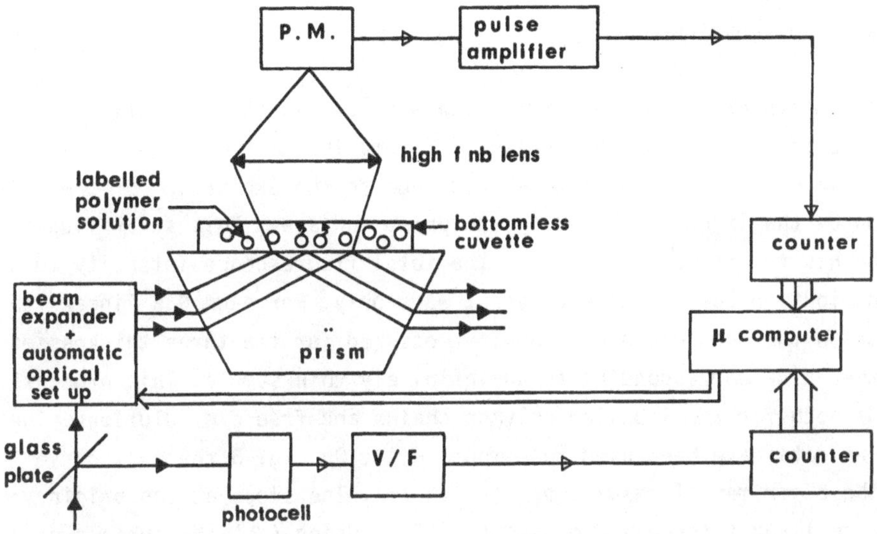

Fig. 2. Experimental set-up.

b) The prism is made of fused silica, with a carefully polished upper surface. Its roughness being of the order of 50 Å r.m.s. the following cleaning procedure has been used : first the prism was soaked in a detergent solution under sonication for two hours, then it was carefully rinsed and put in an oxygen flow under U.V. irradiation for two hours to remove all traces of organic contaminants. This procedure turned out to be very efficient and to give reproducible results. The polymer was a polysaccharide , xanthan , of high molecular weight , $M_W = 2 \times 10^6$, with a rather large polydispersity : $M_w/M_n = 1.4$. All chains were labelled with 0.1 % of fluorescein markers, the fluorescence of which could be excited by the 488 nm wavelength line of an Ar^+ laser. The xanthan was dissolved in water at a concentration of 96 ppm with 400 ppm of NaN_3 and 0.1 M of NaCl, the sodium azide prevents bacterial growth in the solution. The clean silica surface behaves as a neutral wall for the xanthan molecules in the saline solution. This behavior is due to the short range electrostatic repulsive forces between the polymer molecules and the silica surface. Therefore the concentration of polymer molecules right on the solid surface vanishes.

Results and discussion

The experiments have been performed with a 96 ppm xanthane solution with three cuvettes of various thicknesses : 50 μm, 100 μm and 200 μm. This was done to evaluate the amount fluorescence due to the excitation of the bulk solution by the light scattered by the prism surfaces. This stray fluorescence has to be substracted from the total fluorescence intensity to get the contribution due to the evanescent wave only. For each θ a linear regression on the fluorescence intensity measured for the three thicknesses gives the value corresponding to an infinitely thin sample. This processing was made both for the labelled polymer chains and free dye solutions. The resulting data have been used to compute $R(\Lambda)$. On Fig. 3 the full curve shows the experimental results of $R(\Lambda)$ vs $1/\Lambda$. The slope at the origin yield an average layer thickness ξ of about 1250 Å. Using (2) the surface excess Γ is then calculated to be -1.2×10^{-2} mg/m^2.

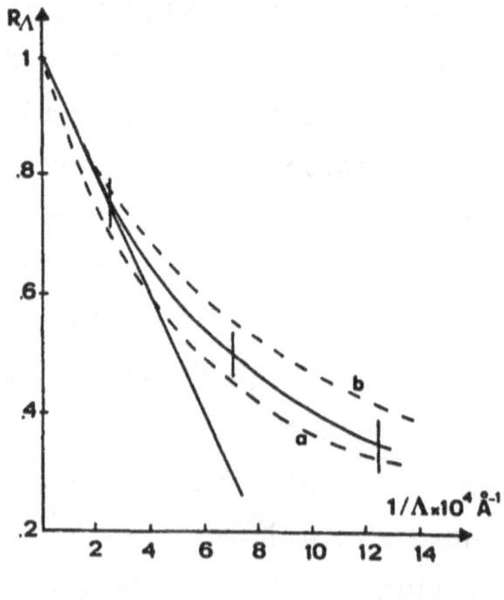

Fig. 3. $R(\Lambda)$ vs $1/\Lambda$. Full line represents the experimental data for a 96 ppm xanthane solution on a silica surface. Dashed lines a & b represent theoretical predictions of the Auvray's model for ξ = 1650 Å and 1100 Å.

This experimental result has to be compared to the theoretical predictions of L. Auvray[6] for an ideal gas of rigid rods of length L at a low concentration C_v :

$$C(z) = C_v(z/L)(1 - \ln(z/L)) \tag{8}$$

This model leads to an average thickness ξ equal to L/4. On Fig. 3, dashed curves a and b represent the theoretical variations of $R(\Lambda)$ computed from (8) for L respectively equal to 6600 Å and 4400 Å. The data points lie just in between the two theoretical curves. Therefore one can deduce that the average layer thickness in our experimental case is included between 1650 Å and 1100 Å. The model can provide a fair, although coarse, approximation of the concentration profile in the depletion layer. We can review some of the reasons which explain why this agreement cannot be made better. First, the model applies to perfect rigid rods while the xanthane molecule is known to be semi-flexible with a persistence length of 450 Å. Second, the theoretical computation is done in the regime where the rods are not interacting ($C < C_v^* = M/\mathcal{N}L^3$, where $\mathcal{N}$ is Avogadro's number and M the molecular weight). Assuming a molecular length of 7000 Å, $C_v^* \sim 10^{-5}$ g/g, which is smaller than the concentration under study. Last, the large sample polydispersity can blur the concentration profile and contribute to the misfit of the data with the model. More experiments are underway to check these points.

Conclusion

The E.W.I.F. technique is a very powerful tool, which can give detailed information on the concentration profile of polymer solutions near a solid wall, when the minimum penetration depth is smaller than the interfacial layer thickness. E.W.I.F. can be used both for depletion and adsorption studies. Its sensitivity makes it well suited to the detection of the small surface excesses encountered in depletion or weak adsorption cases. We have reported preliminary data on a depletion case formed by a solution of long semi-rigid molecules in front of a quasi-neutral solid surface. In this favorable case, the experimental data display a definite concentration profile which extends over a typical distance of 1200 Å. Comparison with the theoretical predictions of L. Auvray[6] show at least a qualitative agreement with the model for rigid rods. A more precise processing of the datas is currently underway.

References

1. Killmann, E., Eisenbauer, J., and Korn, M. : J. Polym. Sci. Part C, 61, 413 (1977).
2. Dejardin, P., and Varoqui, R. : J. Chem. Phys. 75, 4115 (1981).
3. Cohen-Stuart, M.A., Fleer, G.J., and Bijsterbosh, B.H. : J. Colloid Interface Sci. 90, 310 (1982).
4. Fontana, B., and Thomas, J. : J. Phys. Chem. 65, 480 (1961).
5. Cosgrove, T., Crowley, T.L., and Vincent, B. : Symposium on adsorption from solution, Bristol, Academic Press, sept. 1982.
6. Auvray, L. : Journal de Physique 42, 79 (1981).

SINGULAR VALUE DECOMPOSITION IN THE ANALYSIS OF SPECTRA

W. Curtis Johnson, Jr.

Department of Biochemistry and Biophysics
Oregon State University
Corvallis, Oregon 97331 USA

The singular value decomposition (SVD) theorem (1) provides a powerful method for analyzing a set of related vectors. Spectroscopists will find SVD particularly useful (2) for 1) finding the number of independent components in a set of spectra, and 2) averaging the set of spectra to remove both high frequency noise and low frequency random error.

Fig. 1 shows seven circular dichroism (CD) spectra from a set of 15 for hyaluronic acid as a function of solvent composition (3). The least intense spectrum corresponds to aqueous

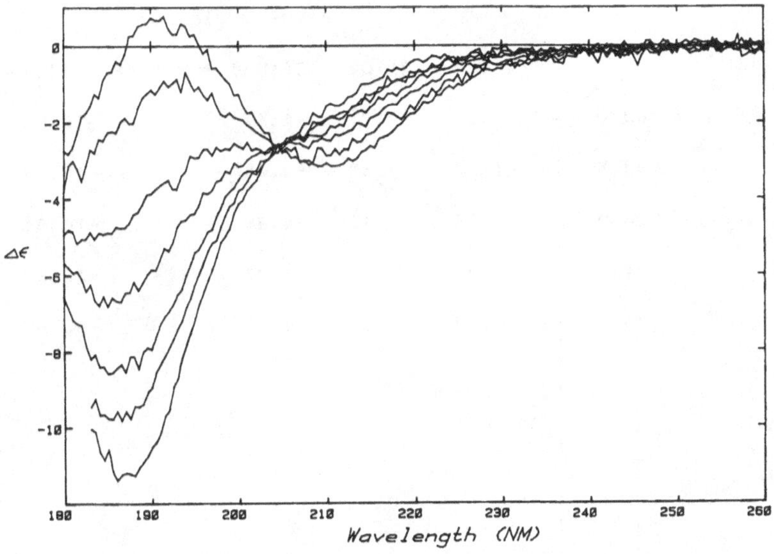

Fig. 1. The CD of hyaluronic acid as a function of alcohol concentration.

Physical Optics of Dynamic Phenomena and Processes in Macromolecular Systems

solution, and as alcohol is added the polysaccharide undergoes a cooperative transition which is complete at 48% vol/vol alcohol. The spectra were collected on a computer and digitized as the average value covering every one-half nm, as is obvious from the sawtooth type noise.

Each digitized CD spectrum is a vector, and the entire set of 15 related CD spectra measured as a function of alcohol concentration can become the columns of a data matrix, **A**, with the values recorded every one-half nm as the rows (Fig. 2). The SVD theorem states (1) that any matrix can be decomposed into the product of three matrices: 1) a column matrix of orthogonal basis vectors, U; 2) a matrix with singular values on the main diagonal and zeros elsewhere, S; and 3) a matrix of coefficients in which the columns fit the basis vectors to the original vectors and the rows fit the original vectors to

Percent alcohol

		1(0%)	2(10%)	3(15%)	→	15(48%)
	λ = 183.0	-1.57	-2.71	-4.87		-10.03
	183.5	-1.24	-2.80	-5.05		-10.12
	184.0	-1.35	-1.43	-4.88		-10.65
	184.5	-0.88	-2.47	-4.93		-10.87
	185.0	-0.69	-2.26	-4.89		-10.57
A =	↓					
	260.0	-0.05	-0.13	+0.10		- 0.19

Fig. 2. The CD of hyaluronic acid at various alcohol concentrations as a column matrix, **A**.

the basis vectors, V^T. This is shown in Fig. 3. Programs are available on most central computers for this type of decomposition and a program is given in Forsythe, Malcolm and Moler (4).

The three matrices from SVD have important properties for the spectroscopist. The singular values on the main diagonal of S are the positive square roots of the nonzero eigenvalues for the square matrix, AA^T, or equivalently A^TA. The eigenvectors

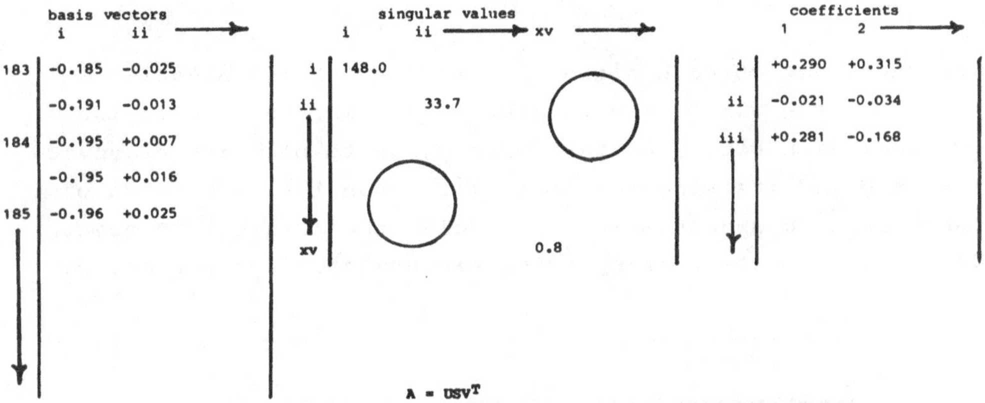

Fig. 3. Decomposition of the data matrix containing the CD of hyaluronic acid at various alcohol concentrations.

$$I = +0.290 \times 1 + 0.315 \times 2 + \qquad +0.106 \times 15$$

$$II = -0.021 \times 1 - 0.034 \times 2 + \qquad +0.381 \times 15$$

↓

$$XV = +0.098 \times 1 + 0.064 \times 2 \qquad +0.048 \times 15$$

$$US = AV$$

Fig. 4. Basis CD spectra expressed as a linear combination of the original CD spectra for hyaluronic acid at various alcohol concentrations.

of AA^T are the columns of U. They are all orthogonal and form a unitary matrix. We weigh each eigenvector in U with its corresponding singular value to form a column matrix, US, of basis CD spectra that are orthogonal (Fig. 4). These basis CD spectra are linear combinations of the original CD spectra. The rows of V^T are the eigenvectors of A^TA and thus each one also corresponds to a singular value on the main diagonal of S (Fig. 3). These row vectors are orthogonal and V^T is unitary. The rows of V^T are the coefficients in Fig. 4 that form the basis CD spectra from the original spectra.

The magnitude of each eigenvalue (square of the singular value) is the sum of the squares of the intensities in the original data matrix A that corresponds to both its eigenvector in U and its eigenvector in V^T. Thus the importance of each basis CD spectrum in reproducing the original CD spectra is denoted by its corresponding eigenvalue. For perfect data

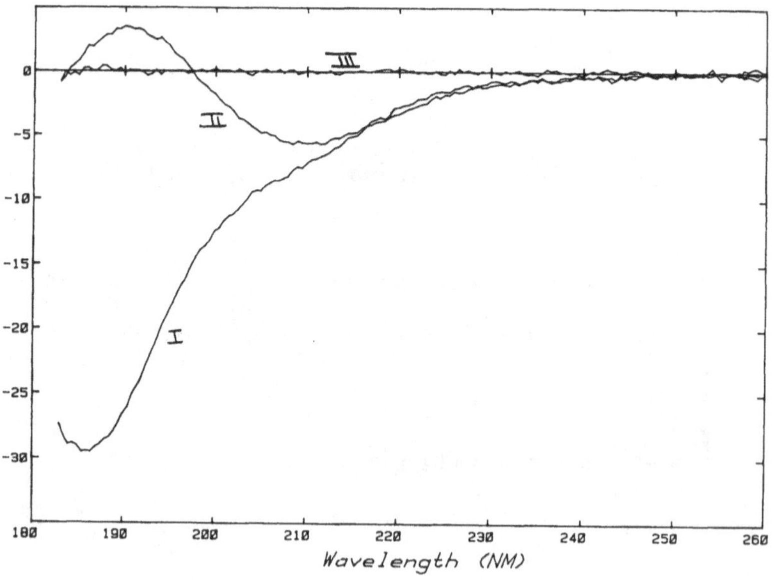

Fig. 5. The three most important basis CD spectra for hyaluronic acid at various alcohol concentrations.

the number of nonzero eigenvalues is the number of independent components in the set of data. Real data are noisy, so there is usually no eigenvalue that is identically zero. The problem then is to find the basis CD spectra that have significance above the noise in the data. This is easily done by inspection. The three basis CD spectra with the largest eigenvalues for the CD measurements on hyaluronic acid (Fig. 1) are given in Fig. 5. They have corresponding eigenvalues 21,900, 1,133, and 4.31. We easily see from inspection of Fig. 5 that only the first two basis CD spectra are significant above the noise. The other 12 basis CD spectra have eigenvalues ranging from 3.22 down to 0.65. Since the original data (Fig. 1) show an isodichroic point at 204 nm, we already know there are two independent components in this particular case.

The original data matrix, A, is a linear combination of the basis CD vectors with the columns of V^T as coefficients, as shown in Fig. 6. If we reconstruct the original data using only the two basis CD spectra that have significance above the noise, then we have effectively averaged the original CD spectra among themselves even though each has a different

1 = +0.290 x I -0.021 x II + +0.098 x XV

2 = +0.315 x I -0.034 x II + +0.064 x XV

↓

15 = +0.106 x I +0.381 x II + +0.048 x XV

$$A = USV^T$$

Fig. 6. The original CD spectra for hyaluronic acid at various alcohol concentrations, expressed as a linear combination of the basis CD spectra.

shape and intensity. Reconstructed CD spectra for hyaluronic acid as a function of solvent are shown in Fig. 7. They average the random errors in measurement as well as the high frequency noise.

Although we have discussed SVD in terms of eigenvectors and eigenvalues for AA^T and A^TA, it is important <u>not</u> to use a matrix diagonalization subroutine on these square matrices in order to find the basis CD spectra and their singular values. This method is extremely susceptible to round-off error and thus is not a good algorithm for computer work. SVD programs are specially developed to avoid this problem.

The CD spectrum of a protein is extremely sensitive to its component secondary structures as shown in Figs. 8-11. However, workers have not been particularly successful in ana-

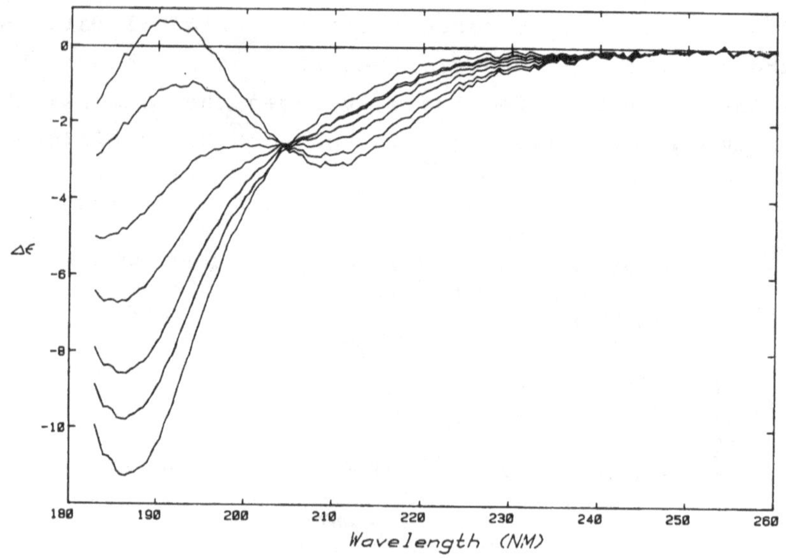

Fig. 7. The CD spectra of hyaluronic acid at various alcohol concentrations reconstructed from the two most important basis CD spectra.

lyzing such CD spectra for secondary structure. We have attacked this problem by measuring spectra further into the vacuum UV to 178 nm in order to obtain more information, and by making use of SVD (5). Figs. 8-10 show CD spectra measured in our laboratory for 11 proteins with known secondary structure, while Fig. 11 shows spectra for five more proteins with known secondary structure measured by Brahms and Brahms (6). These 16 CD spectra can be analyzed with SVD to generate 16 orthogonal basis CD spectra. Here the corresponding eigenvalues do not decrease significantly at any particular point, but inspection shows that the first five basis CD spectra are

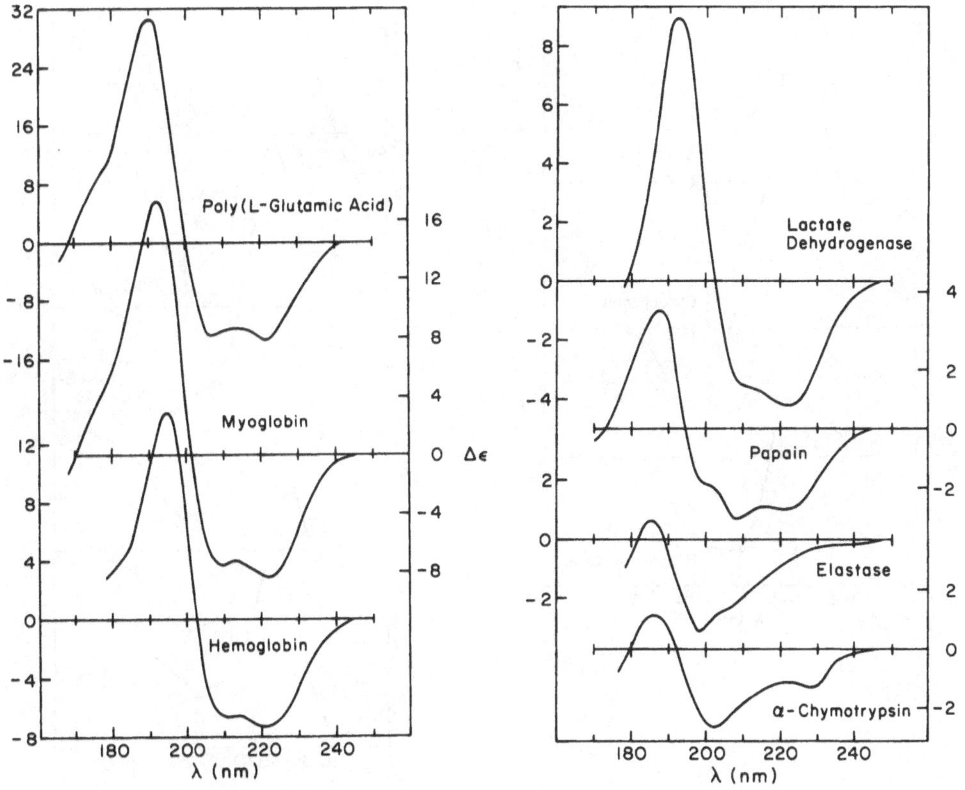

Figs. 8 and 9. CD spectra of proteins with known secondary structure redrawn from reference 5.

sufficient to fit the original protein CD spectra. In the case of papain (Fig. 12), four basis CD spectra do not mimic the papain spectrum particularly well while five basis CD spectra do a good job of reproducing the shape. A sixth basis CD spectrum does not improve the fit significantly. Thus the CD spectrum of a protein measured to 178 nm contains five significantly different components above the noise. Similarly,

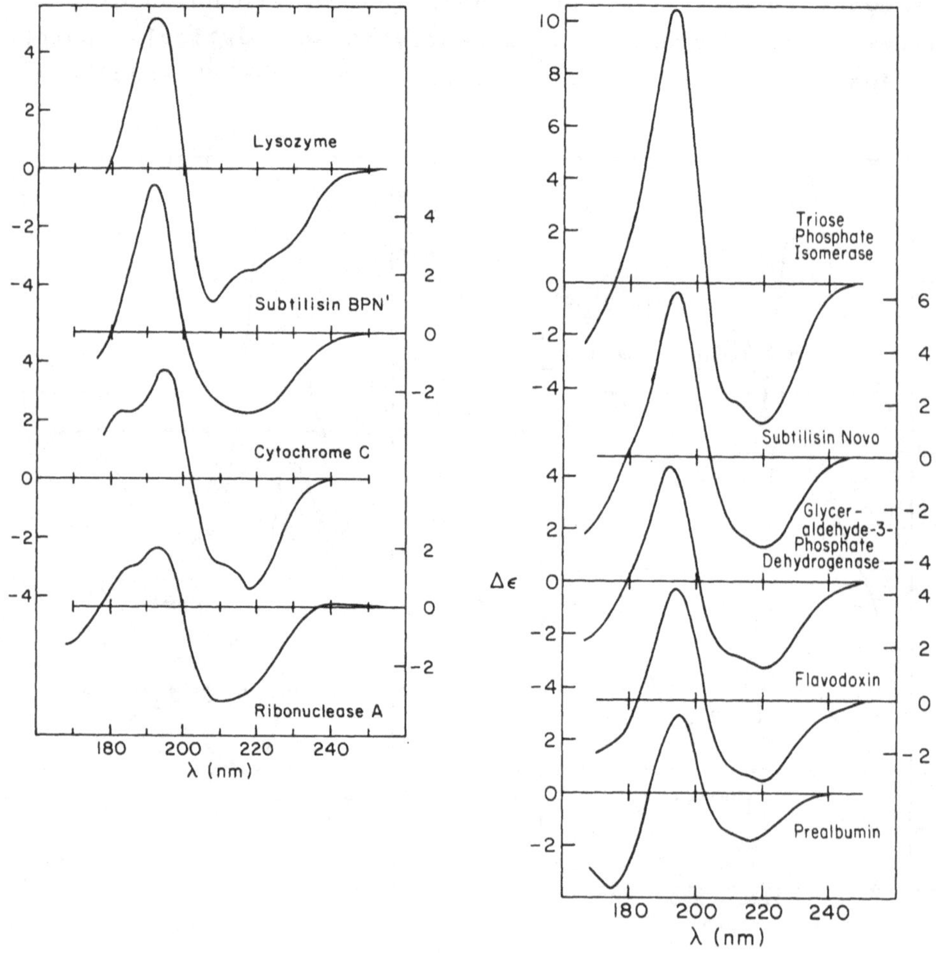

Figs. 10 and 11. CD spectra of proteins with known secondary structure redrawn from references 5 and 6.

spectra measured to 190 nm show three significantly different components while spectra measured to 200 nm show only two significantly different components. One can add another equation to the problem by requiring that the sum of secondary structures for a protein equal 100%. However, even with this constraint, spectra measured to 190 nm do not contain enough information to solve for the secondary structure of a protein. If one lumps all of the β-turns together, there is still a minimum of five variables that contribute to the CD spectrum of a protein (α-helix, H; antiparallel β-sheet, A; parallel β-sheet, P; β-turns, T; and "other" structures, O, that do not fit into the first four categories). Thus it is important to measure the CD spectrum of a protein into the vacuum UV region if it is to be analyzed for secondary structure.

The five significant basis CD spectra for proteins (Fig. 13) are linear combinations of the CD spectra for the 16

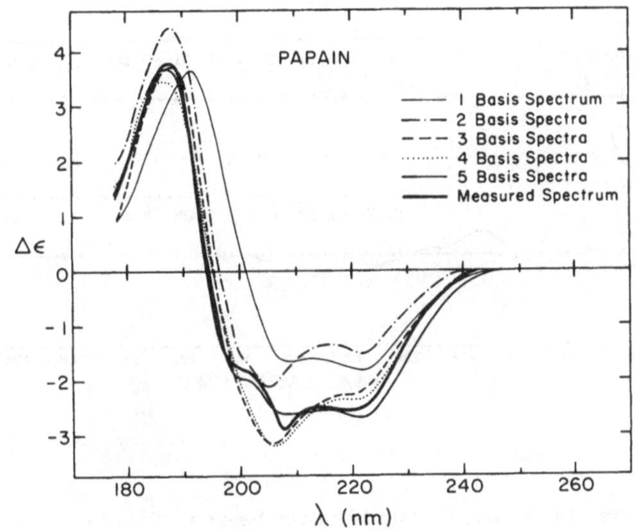

Fig. 12. Reconstructing the CD spectrum of papain using various numbers of basis CD spectra for proteins, redrawn from reference 5.

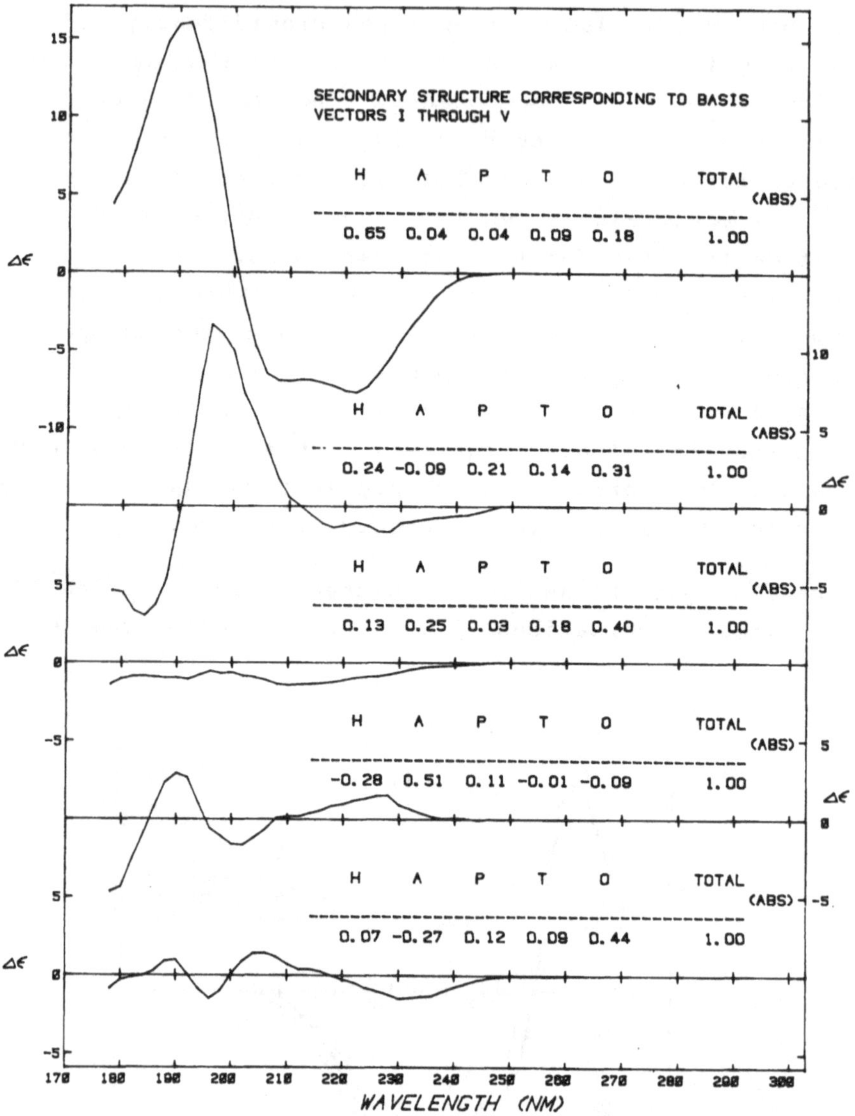

Fig. 13. The five most important basis CD spectra for proteins and the secondary structures corresponding to these spectra.

original proteins (Fig. 4). Since these 16 proteins have known secondary structure, we can use the same set of coefficients to compute the amount of each type of secondary structure that corresponds to each of the five basis spectra (Fig. 14). Thus each basis CD spectrum corresponds to a mixture of secondary structures just as the CD spectrum of a protein corresponds to a mixture of secondary structures

$$H_I = C_{i1} H_1 + C_{i2} H_2 + \cdots + C_{i16} H_{16}$$

$$H_{II} = C_{ii1} H_1 + C_{ii2} H_2 + \cdots + C_{i16} H_{16}$$

$$\downarrow$$

$$A_I = C_{i1} A_1 + C_{i2} A_2 + \cdots + C_{i16} A_{16}$$

$$\downarrow$$

$$O_V = C_{v1} O_1 + C_{v2} O_2 + \cdots + C_{v16} O_{16}$$

Fig. 14. Computation of secondary structures corresponding to each of the five basis CD spectra for proteins.

$$CD = c_I\, I + c_{II}\, II + c_{III}\, III + c_{IV}\, IV + c_V\, V$$

$$H = c_I H_I + c_{II} H_{II} + c_{III} H_{III} + c_{IV} H_{IV} + c_V H_V$$

$$A = c_I A_I + c_{II} A_{II} + \cdots$$

$$\downarrow$$

$$O = c_I O_I + c_{II} O_{II} + \cdots$$

Fig. 15. The CD spectrum of a protein with unknown structure is a linear combination of five basis CD spectra and the coefficients can be used to predict its secondary structure.

(Fig. 13). The five significant basis CD spectra for proteins are orthogonal and can be used to fit the CD spectrum of a protein with unknown structure (Fig. 15). This fit can be used to predict the secondary structure of the protein using

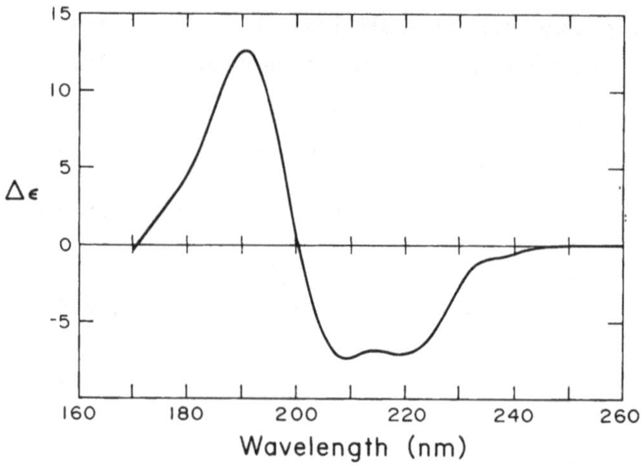

Fig. 16. The CD spectrum for interferon (7).

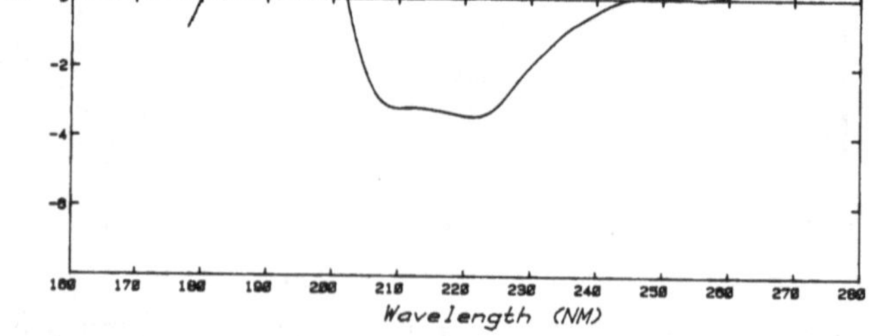

Fig. 17. The CD spectrum for acetyl cholinesterase (8).

the secondary structures corresponding to each basis CD spectrum. When one applies this scheme to 15 of the original 16 proteins with known secondary structure, and uses five basis CD spectra to predict the secondary structure of the 16th, the results are very good (5). Here we show the predictions for two proteins whose secondary structures have yet to be determined from X-ray crystallography. Fig. 16 shows the CD spectrum for interferon which is predicted to have 59% α-helix, 16% antiparallel β-sheet, no parallel β-sheet, 18% β-turn, and 13% other structures (7). The CD spectrum for acetyl cholinesterase is shown in Fig. 17 and the prediction is 34% α-helix, 13% antiparallel β-sheet, 8% parallel β-sheet, 16% β-turn, and 26% other structures (8). No constraint is used and the accuracy with which the predicted secondary structures sum to 100% is a measure of the success of the method.

Acknowledgment

This work was supported by National Science Foundation grant PCM80-21210 from the Biophysics Program and National Institutes of Health grant GM21479 from the Institute of General Medical Sciences.

References

1. Noble, B., Daniel, J. W.: Applied Linear Algebra, 2nd Edition, Prentice-Hall, Englewood Cliffs, NJ, 1977, pp. 323-330.

2. Lloyd, D.: Ph.D. thesis, University of California, Berkeley, 1969.

3. Staskus, P. W., Johnson, W. C. Jr.: Unpublished research.

4. Forsythe, G. E., Malcolm, M. A., Moler, C. B.: Computer Methods for Mathematical Computations, Prentice-Hall, Englewood Cliffs, NJ, 1977, pp. 229-235.

5. Hennessey, J. P. Jr., Johnson, W. C. Jr.: Biochemistry 20, 1085-1094 (1981).

6. Brahms, S., Brahms, J.: J. Mol. Biol. 138, 149-178 (1980).

7. Manavalan, P., Johnson, W. C. Jr., Johnston, P. D.: Unpublished research.

8. Manavalan, P., Johnson, W. C. Jr., Taylor, P.: Unpublished research.

FLOW BIREFRINGENCE OF MACROMOLECULAR SOLUTIONS IN VARIOUS ELONGATIONAL FLOWS

Robert Cressely, Roger Hocquart, Jean-Paul Decruppe and Tomasz Wydro

Groupe Physique des Polymères
Faculté des Sciences, Ile du Saulcy, 57045 METZ CEDEX 1 (France)

Introduction

The Flow Birefringence (F.B.) or Maxwell effect (for review see ref. 1-5) is an optical anisotropy resulting from the action of an hydrodynamical field on the particles. It can be visualized in the case of pure liquids, solutions of polymers, melts and even in gases submitted to different flows. Usually this effect is studied when the fluid is submitted to the action of shear stresses, the observations being made in a direction perpendicular to the plane of flow. The simpliest but most widely used device, the Couette cell consists of two concentric cylinders : one at rest, the other turning. When annular space between the cylinders is small enough and in laminar flow, the liquid layer is submitted to a shear rate G roughly constant in all the annular space. When one of the cylinders rotates, shearing stress appears and consequently the liquid behaves like a biaxial medium. More recently theoretical as well as experimental studies of the behaviour of macromolecules in elongational flow have been under consideration, and F.B. has also been used (see for example ref. 6-15).

Results

1) Rigid particles. For a rigid particle, as we show in our experiments, the orientation effect of the elongational gradient G_L is stronger than in the case of shear stresses, the only one reason for desorientation being the Brownian motion. In a cell, called "cross cell", represented

Physical Optics of Dynamic Phenomena and Processes in Macromolecular Systems

schematically on Fig. 1 we can produce a plane elongational field of hyperbolic type defined by V (the velocity) : $V = G_L(x, -y, o)$. On the photograph of Fig. 2 we can see this cell placed between crossed polarizers, oriented at 45° of the axes of the cell. In this case, the particles are rigid. When the flow rate is higher (after the transition laminar turbulent), a curious figure is obtained (photograph of Fig. 3)

2) Flexible particles. In the case of transverse gradient, the deformation of the flexible macromolecules remains quite small in consequence of their continual rotation. On the other hand in the case of elongational flow (in the direction of the flow), we can expect more important deformations showing in some cases the coil-stretch transition of the dissolved macromolecules. This transition can be obtained in our cross cell for flexible macromolecules of quite high mass (for example, solution of polyethylene oxide (POE) in water with masses higher than 10^6) when two conditions a) and b) are simultaneously realized :

a) $G_L \tau > 1$ where τ represents the longest relaxation time of the particle. This relation implies for the elongational gradient G_L to have a minimum value.

b) $t.G_L \gg 1$ shows that the macromolecules have to stay long enough in the elongational field to stretch.

The latter condition explains the strong localization of the F.B. obtained under certain conditions in our experiments and clearly visible on the pictures presented (Fig. 4).

It can be quoted that these strong deformations can be obtained with composite flows where a rotational part $\vec{\Omega}(\vec{\Omega} = 1/2\vec{\nabla} \wedge \vec{V})$and an elongational component exist as long as the latter prevails over the first one ($G_L > \Omega$). For example, we have been able to realize the transition coil-stretch in a device with two rolls rotating in the same way for $\Omega = 0.94 G_L$ with solutions of POE in water at the concentration of 0.6 % ($M = 5 \times 10^6$).

Conclusion

With these new manifestations and possibilities in elongational flow (mea-

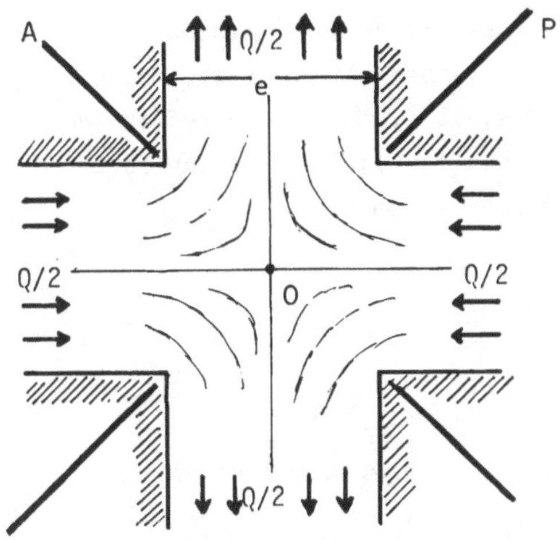

Q : flow rate   e = 2mm
A : analyseur   P : polariseur
O : stagnation point

Fig. 1 : The cross cell
(plane hyperbolic flow)

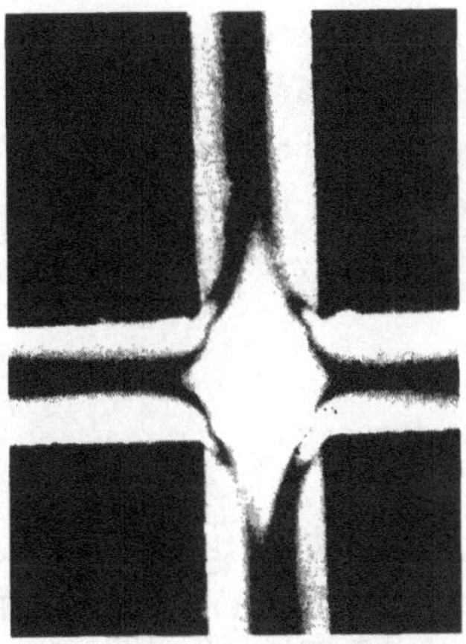

Fig. 2 : Laminar flow of rigid particles (bentonites) in the cross cell.

Fig. 3 : Same situation as in Fig. 2, but at higher flow rate Q

Fig. 4 : Case of flexible particles (polyethylene oxide)

surements of the optical anisotropies of stretched and oriented particles, determination of the different relaxation times related to the molecular masses and conformation of macromolecules, ...), the Flow Birefringence is a non perturbating method which should see its applications developed in the study of the dynamics of macromolecular systems.

References

1. Jerrard, H.G. : Chem. Rev. 59, 345 (1959).
2. Harrington, R.E. : "Flow Birefringence", Encyclopedia of Polymer Science, 7, 100-179 (1967).
3. Tsvetkov, V.N.: "Flow Birefringence", Newer Methods of Polymer Characterization, 536 (1964), Bacon, K., Ed., Interscience, N.Y.
4. Peterlin, A.: "Streaming and Stress Birefringence, Rheology, 1, 615 (1956), Eirich, F.R., Ed., Academic Press, N.Y.
5. Cerf, R., Scheraga, H.A. : Chem. Rev., 51, 185 (1952).
6. Peterlin A. : Polymer, 2, 257 (1961)
7. Mackley, M.R., Keller, A. : Phyl. Trans. Roy. Soc. Lond. A. 278, 29 (1975).
8. Crowley, D.G., Frank, F.C., Mackley, M.R., Stephenson, R.G. : J. Polym. Sci., 14, 1111 (1976).
9. Fuller, G.G. : Ph. D. Thesis, Calif. Inst. Technol. (1980).
10. Cressely,R., Hocquart R. : 8th Int. Congt. on Rheology, Naples, 1980, "Rheology", 2, 377 (1980) Plenum Press.
11. Cressely, R.: Doctorat d'Etat Sc. Phys. Metz (1982).
12. Cressely,R., Hocquart, R., Scrivener, O. : Optica Acta, 25, 559 (1978)
13. Cressely, R., Hocquart, R., Scrivener, O. : Optica Acta, 26, 1173 (1979)
14. Cressely, R. , Hocquart, R. : Optica Acta, 27, 699 (1980)
15. Cressely, R., Hocquart, R. : ACS Polymer Preprints, 22, 120 (1981)

USE OF LIGHT SCATTERING METHODS FOR THE STUDY OF DYNAMIC PROCESSES OF MACROMOLECULES COAGULATION IN HYDROCARBON MEDIUM

G.F. Bolshakov
Institute of Petroleum Chemistry, Siberian Branch of the USSR Academy of Sciences, Tomsk, USSR

In the period from 1962 up to 1984 noncontact methods of defining kinetics of microparticle coagulation from 0.1 to 10 μm and from 2 up to 100 μm in hydrocarbon medium were developed. An experimental installation for defining particle dispersity in the range from 2 to 100 μm by the method of small-angle light scattering comprises a light source, an optical system, a cell and a recorder (1). The source of light is a laser. Intensity gradient of scattered light in the interval of $\beta =$ $= 0 - 10^\circ$ is $I_{max}/I_{min} \approx 10^6 - 10^5$. This intensity range was covered by means of an attenuator with light passage 46.3; 21; 10; 0.64; 0.75; 0.0016%.

A rectangular cell with plane parallel walls was used for changing indicatrix of small-angle light scattering. A light scattering receptor is assembled in a case that provides the required writing speed at the angle β.

The order of processing indicatrixes is (cf. List of symbols):

1. making up of tables (particle size against the light scattering intensity and the angle β); 2. definition of the regions of angles β_i corresponding to optical data; 3. obtaining tables and diagrams $\beta_i^3\ \Phi(\beta)$; 4. analysis of the course of the function $\beta_i^3\ \Phi(\beta)$ and determining the limiting angle of the integration, β^*; 5. graphical or numerical differentiation of the function $\beta_i^3 \Phi(\beta)$; 6. making up tables of products $\rho_j\ \beta_i$ and finding out tabular values of the function $F(\rho_j,\ \beta_i)$; 7. calculation of the sum of the products

$$\Pi_{ij} = \frac{d}{d\beta}\{\beta_i^3\ \Phi(\beta)\}\ F(\rho_j, \beta_i)\Delta\beta$$

Physical Optics of Dynamic Phenomena and Processes in Macromolecular Systems

8. calculation of the spectrum $f(\rho_j,\beta_i)$ and making a plot of particle size distribution.

Relative spectrum of particle sizes $\Phi(\beta_i)$ is calculated as follows:

$$f(\rho_j) = \frac{1}{\rho^2} \sum_{\beta=0}^{\beta^*} \frac{d}{d\beta}\{\beta_i^3 \Phi(\beta)\} F(\rho_j,\beta_i)\Delta\beta$$

Number of particles of a definite size and the total number of particles in the volume studied was determined from the equation

$$n(\rho_j) = -\frac{16\pi^4}{x^3 \Phi_0} \frac{D_0^2}{D_3} \sum_{0}^{\beta^*} \frac{d}{d\beta}\{\beta^3\Phi(\beta)\} F(\rho,\beta)\Delta\beta$$

A program has been composed and a special calculation has been made: it enabled us to obtain automatically the needed results in 7-10 minutes. The small-angle method can be successfully used for particles of irregular and spherical forms (2). Using large-angle scattering, it is possible to obtain information on particle size distribution from 0.1 to 10 μm. The experimental device contains a laser, an optical system and a recording apparatus (1).

The receiving optical system and the receiver of radiation energy are moved horizontally, their velocity being in the range from 1.25'/s to 360°/s. A cylindrical cell was especially developed for the device.

After alignment of the installation and reading calibration indicatrix, the experimental indicatrixes are read out at a velocity from 0.1 to 0.2°/s in the range from 5 to 165°. To find out the true indicatrix, the indicatrix read out without the sample is subtracted from the experimental one. The scattering volume correction is inserted into the obtained difference curve. The corrected difference curve $\Delta I(\beta) = [I(\beta)-I_0(\beta)]\Phi(\beta)$ was processed using the method of statistical regularization, by solving the following simultaneous equations:

$$\sum_{j=1}^{n} F(\rho_j,\beta_i) f(\rho_j) = \Delta I(\beta_i), \quad i = 1,2,3,\ldots,m$$

found after algebraizing the initial integral equation

$$I(\beta) = \int_0^\infty F(\rho,\beta)f(\rho)d\rho$$

in one or another statistical combination. The statistical regularization algorithm is taken as the basis, where prior information on the function looked for is given by using the definition of some probability distribution, and solution is sought in a statistical combination of smooth functions with the function parameter α. Prior requirements imposed on the solution to the problem are reduced to positiveness of the sought function $f(\rho)$ calculated by the computer.

The developed apparatus may be used for scientific and practical purposes.

The use of light scattering methods is considered in detail in works (1-4). The applicability of light scattering methods are illustrated by the following experimental data.

Definition of the contamination level and particle dispersity of fuel contamination is especially necessary before fuelling aircrafts. At present, only the common level of fuel contamination is defined, and its filtration is obligatory. Filtration is considered to provide the necessary purity and, potentially, also flight safety. This conception may be wrong, for, firstly, there is no guarantee that the filter "cleans" fuel from contamination particles which are mostly larger than the openings of the filter and, secondly, a great danger consists in the passing of smaller contamination particles through the filter. The presence of small particles in fuel promotes formation of a solid phase in fuel at the expense of its unstable **partly** hetero-organic compounds, unsaturated and unsaturated aromatic hydrocarbons. These processes are intensified by rising temperature. Danger of this phenomenon is experimentally confirmed by the author.

0.00014% contamination thoroughly ground to 0.3 - 1.0 μm was added into a prepared fuel T-7 sample. The mixture was then heated up to 140°C at the rate of 1°C/min, maintained at this temperature for 4 hours, and cooled to room temperature. The number of forming particles and their average effective radii were measured by the light scattering method. The same fuel T-7 sample, but without added contamination, was heated simultaneously. It can be seen from the experimental results (4) that in presence of contamination the number and particle size of the forming solid phase increase considerably. Apparently, in fuelling aircrafts the content of contamination with particle size less than 10 μm should be limited. Such limitation is unreal without an operative control of the contamination content. Purity control is also important for benzenes and diesel fuels. This problem can be successfully solved using light scattering methods. The methods developed can be used to investigate the kinetics of olefine polymerization in a hydrocarbon medium.

List of symbols

- D diameter (mm): D_0 diameter of light beam image in the focal plane of receiving lens; D_3 diameter of receiving diaphragm
- F, f, Φ_0 function symbols
- I scattered light intensity; I_{max}, I_{min} intensity extremes; $\Delta I(\beta)$ excess scattering curve (ΔI of measurements with and without sample)
- n total number of particles in a unit volume; $n(\rho_i)$ number of particles of definite size in a unit volume (cm^{-3})
- α parameter in the statistical combination of smooth functions
- β scattering angle; β_i region of angles, $\Delta\beta$ their step; β^* limiting angle of integration
- ρ dimensionless radius
- Π_{ij} sum of products

References

1. Bolshakov, G.F., Timofeev, V.F., Novichkov, M.N.: Opticheskie metody opredeleniya zagryaznennosti zhidkih sred, Nauka (1984).

2. Bolshakov, G.F., Timofeev, V.F., Sibarova, I.I.: Express--metody opredeleniya zagryaznennosti nefteproduktov, Khimiya (1977).

3. Bolshakov, G.F: Vosstanovleniye i kontrol kachestva nefteproduktov, Nedra (1982).

4. Bolshakov, G.F.: Fizikokhimicheskiye osnovy obrazovaniya osadkov v reaktivnykh toplivakh, Khimiya (1972).

ASSOCIATION OF ALCOHOLS INVESTIGATED BY RAYLEIGH LIGHT SCATTERING

Barbara M. Fechner

Institute of Physics, A.Mickiewicz University,
Grunwaldzka 6, 60-780 Poznań, Poland

Abstract: The solutions of butanol-1, pentanol-1, hexanol-1 in n-hexane have been investigated by the method of Rayleigh light scattering. From the analysis of the isotropic and anisotropic components of the scattered light the dependence of the effective optical anisotropy of alcohol molecules on the mean number of molecules in the associate has been determined.

Introduction

Isotropic component R_i of the scattered light allows us to determine chemical potentials μ_1 and activity coefficients f_1 of the solution and, in consequence, the mean number of molecules in the associate χ versus the concentration of the associating component. The calculation of the chemical potential μ_1 as function of concentration is reduced to integration of equation(1,2)

$$\partial\mu_1/\partial x_2 = (\pi^2/2\lambda^4)\ kTV\ (\partial\varepsilon/\partial x_2)^2\ x_2/R_c$$

where: $V=x_1V_1+x_2V_2$ stands for molar volume of solution, V_1,V_2 - those of pure components, x_1,x_2 - concentrations of solvent and solute in molar fractions, λ - wavelength of scattered light, k - Boltzmann's constant, ε - optical dielectric constant and $R_c=R_i-R_d-R^*$ the therm of isotropic Rayleigh constant due to fluctuation in concentration;

Physical Optics of Dynamic Phenomena and Processes in Macromolecular Systems

R_d is the intensity scattered on density fluctuations and R^* is the intensity scattered on mixed (density-concentration) fluctuations.

In the case of an ideally associating solution, when but one of the components undergoes association, once the activity coefficients f_1 are available, the mean number of molecules in the associate can be evaluated from (3,4)

$$\chi = x_2 f_1/(1 - x_1 f_1)$$

The anisotropic component R_a of the scattered light allows us to determine the effective optical anisotropy Γ_2^2 per one molecule of the associating component (5,6,7)

$$\Gamma_2^2 = \frac{R_a}{(104/45)(\pi^2/\lambda^4)(N/x_2)[(x^2 + 2)/3]^2} - (x_1/x_2)\,\Gamma_1^2$$

where Γ_1^2 is the effective optical anisotropy of the solvent, N - the number of molecules per unity volume and n - the refractive index of the solution.

Thus, the analysis of both components of the scattered light makes it possible to present Γ_2^2 versus the mean number of molecules in the associate χ . This dependence evidences whether the associates have a tendency to form closed or open structures.

Results and discussion

The studies were performed for butanol-1, pentanol-1 and hexanol-1 in n-hexane.The results are presented in diagrams. Fig.1 presents the dependence of the chemical potential μ_1 of the solvent upon the alcohol concentration x_2. Fig.2 shows the dependence of Γ_2^2 on x_2 while the relation between Γ_2^2 and the mean number of molecules in the associate χ is presented in Fig.3.

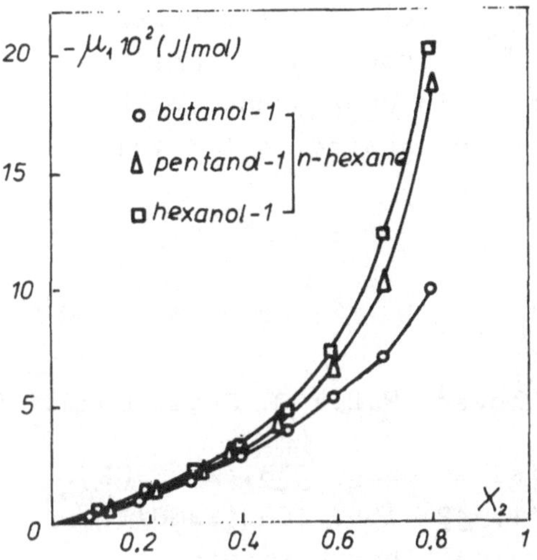

Fig.1. Chemical potentials versus concentration x_2 of the alcohols

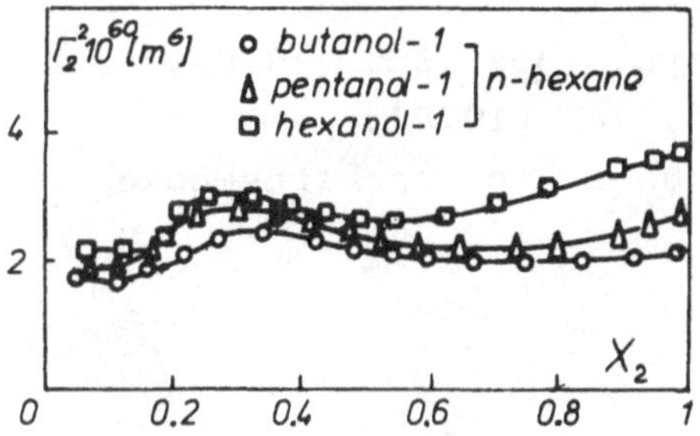

Fig.2. Effective optical anisotropy Γ_2^2 versus concentration x_2 of the alcohols

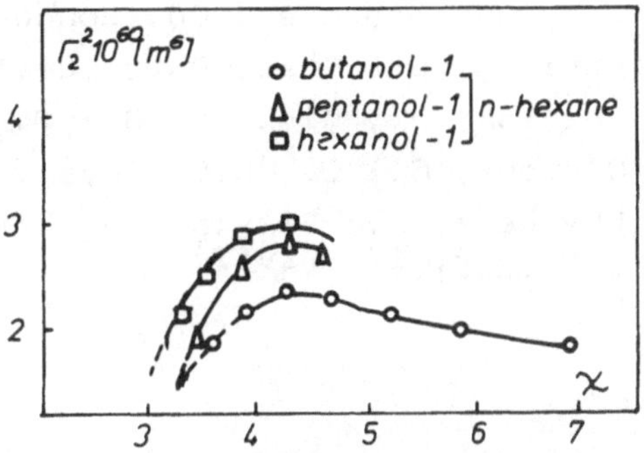

Fig.3. Effective optical anisotropy Γ_2^2 versus the mean number of molecules in the associate

The last dependence is similar for the three alcohols studied. For all of them Γ_2^2 is the highest when χ is close to 4.2 and when χ approaching 3 considerably

decreases. This suggests a cyclic form of trimers which becomes transformed in an open structure with increasing average dimension of an associate.
The course of the dependence of average square of dipole moment on χ , reported by Huyskens (8) for alcohols solutions is similar to that of Γ_2^2 versus χ . The interpretation given by Huyskens is in consistence with the conclusions given in the paper presented, as far the formation of associates in concerned.

References

1. Lewis, H.H., Schmidt, R.L., Clever, H.L.: J. Phys. Chem., 74, 4377 (1970).
2. Fechner, B.M.: Z. phys. Chemie, Leipzig, 263, 2 (1982).
3. Buchowski, H.: Roczniki Chemii, 42, 165, 168 (1968).
4. Prigogine, I., Defay, R.: Chemical Thermodynamics, Longmans Green and Co, London, New York, Toronto 1954.
5. Kielich, S.: Chem. Phys. Letters, 10, 516 (1971).
6. Wozniak, S., Kielich, S.: J. Phys., 36, 1305 (1975).
7. Clement, C.; J. Chim. Phys., 75, 747 (1978).
8. Huyskens, P.: Fizyka Dielektryków i radiospektroskopia 9 part 2, 143 (1977).

REFRACTOMETRIC STUDY OF PRESSURE EFFECT ON PREFERENTIAL SORPTION OF MIXED LIQUIDS ON GELS UNDER DYNAMIC CONDITIONS

Tibor Macko, Dušan Berek
Polymer Institute CCR, Slovak Academy of Sciences,
842 36 Bratislava, Czechoslovakia

Mária Chalányová
Chemical Institute, Comenius University,
842 15 Bratislava, Czechoslovakia

Introduction

In the system containing a porous gel and a mixture of liquids, the composition of liquid in the domain of gel as a rule differs from the composition of bulk liquid. This phenomenon is called preferential sorption or solvation and is caused by differences in the interactions between gel and particular components of mixed liquid. The extent of the preferential sorption is characterized by the coefficient λ that gives the excess of preferentially sorbed liquid component in millilitres per mass or surface area unit of the dry gel. Many authors reported that the extent of the preferential sorption fairly depends on the nature of all components of the system as well as on temperature.

We have recently revealed that the preferential sorption of gels varies also with the pressure changes within the system (1, 2). The present contribution brings additional experimental evidence of this effect.

Experimental

Both original and modified silica gels Separon Six and LiChrosorb packed into pressure resistant glass columns (Laboratory Instruments Works, Prague, Czechoslovakia) were used. The following chemical groups were grafted onto the silica surface: octadecyl-, cyanoethyl-, aminopropyl-, (Separon), glycerylpropyl-, (LiChrosorb). The mixture of benzene (BE) and monomethylether ethylene glycol (MMEEG) 90/10 vol/vol was pumped at the flow rate about 0.3 ml/min through the column according to the scheme in Fig. 1. The switching of the valve V caused a pronounced pressure change within the column C due to reversing the flow: In one part of the column, the pressure suddenly increased while in the other part it decreased. Approximately in the centre of the column, however, the pressure did not change (Fig. 2). The pressure variations produced changes in preferential sorption that resulted in the alterations of the composition of mixed liquid within the column. The latter were monitored by a differential refractometer (Model RIDK 101, Lab. Instruments Works, Prague, Czechoslovakia) in the column effluent. The characteristic shapes of the refractometric records are shown in Fig. 3.

Fig. 1

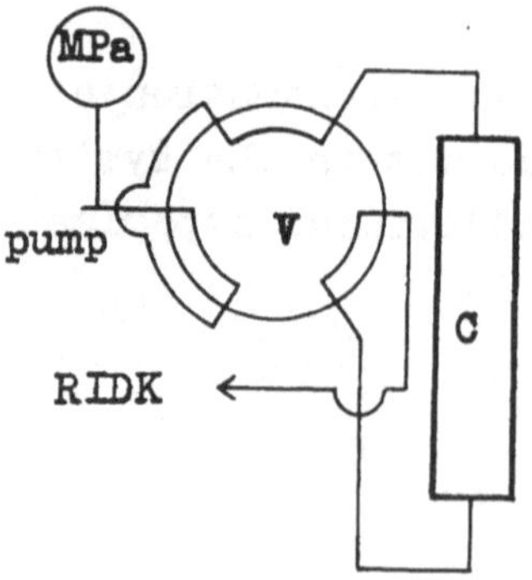

Fig. 2

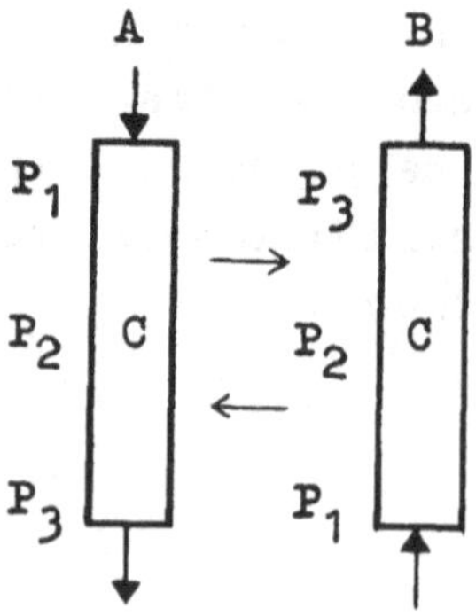

Fig. 3

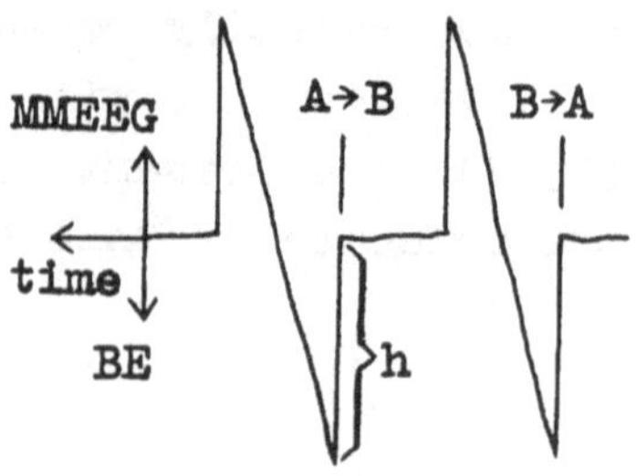

The maximum heights of the observed zones h were used for making the material balance of the system after corresponding calibrations and the changes of λ parameter with pressure variations could be approximately calculated (2).
Dynamic measurements were completed by static experiments at the atmospheric pressure: 30-150 mg of dry gel was poured by 3 ml of mixture BE plus MMEEG. After 4 hours, the composition of the supernatant was determined using a differential refractometer (Brice Phoenix, Model BP-2000-V, Virtis, Cardiner, NY, USA) and the corresponding λ parameters were calculated.

Results and discussion

The values of λ coefficients at atmospheric pressure as well as the data showing the changes of λ with pressure are collected in Table 1.

Table 1

Gel	λ (ml/g) at 0.1 MPa	$\Delta\lambda$ (%) at 9 MPa (λ at 0.1 MPa = 100%)
Separon Six non modified	- 0.187	3.9
Six C_{18}	(+ 0.005)	(130.0)
Six CN	- 0.018	38.9
Six NH_2	- 0.084	8.6
LiChrosorb Diol	- 0.073	10.3

Note: Minus sign means the preferential sorption of MMEEG.

The results obtained can be summarized as follows:

1) The extent of preferential sorption of gels varies with pressure changes at dynamic conditions. Relative changes in λ with pressure decrease with the increasing extent of preferential sorption under atmospheric pressure.
2) Generally, the increase in pressure causes further sorption of the component which has already been preferentially

sorbed at atmospheric pressure.

3) The changes of preferential sorption with pressure are fully reversible and repeatable.

The observed phenomenon can be brought about for example by:

1) The influence of pressure on the liquid - gel interactions.
2) The temperature variations within the column caused by friction that are again pressure dependent (3).
3) The pressure effects on the capillary phenomena -

- for its detailed explanation, however, it is necessary to collect experimental data also in the static system under higher pressures.

The preferential solvation of solute molecules in mixed solvents is expected to be also pressure dependent. This is indicated e. g. by the variations of second virial coefficients and chain dimensions of macromolecules with pressure observed by several authors (4, 5, 6).

The described effects may play a role in all processes where pressure variations take place in the multicomponent systems, viz. high pressure liquid chromatography, centrifugation, filtrations, high pressure phase separations and polymerizations, etc.

References

1. Berek, D., Chalányová, M., Macko, T.: J. Chromatogr. 286, 185-192 (1984).
2. Macko, T., Chalányová, M., Berek, D.: Conference on Modified Polymers, Bratislava, Czechoslovakia, Preprints of contributions, 256-259 (1984).
3. Poppe, H., Kraak, J.C., Huber, J.F.K., van der Berg, J. H.M.: Chromatographia 14, 515-524 (1981).
4. Gaeckle, D., Patterson, D.: Macromolecules 5, 136-141 (1972).
5. Schulz, G.V., Lechner, M.: J. Polym. Sci. A2, 8, 1855-1860 (1970).
6. Hammel, G.L., Schulz, G.V., Lechner, M.D.: Eur. Polym. J. 15, 209-222 (1979).

COMPATIBILIZATION THROUGH HYDROGEN BONDING OF PAIRS OF COPOLYMERS

Omar Aouadj, Abdelghani Lassoued and Said Djadoun
Laboratoire de Chimie Macromoleculaire, U.S.T.H.B.,
BP N°9 Dar El Beida, Alger, Algeria.

Introduction

It is well known that it is thermodynamically unfavorable for most pairs of polymers to form homogeneous mixtures with each other (1,2): the combinatorial entropy of mixing two polymers is small as compared to that of two low molecular weight analogs and cannot compensate even small positive value of the enthalpy of mixing. Yet several pairs of compatible polymers have been discovered over the past twenty years (3,4) , and because of the resulting properties of the mixture, the study of polymer compatibility remains an important area of applied and fundamental research. Most of the compatible polymer mixtures owe their compatibility to the presence of specific interactions between the two polymers. Indeed, we have previously showed that the introduction of interacting groups into the polymeric chains by free radical copolymerization, led to compatible pairs of polymers as evidenced from light scattering (5) , or viscosimetric studies (6,7). We have also showed (8) that such phenomena depended on the nature of the solvent.
In this communication, we will examine a generalization of this method of compatibilization, as applied to several pairs of copolymers containing varying densities and nature of interacting groups. Informations on the polymer-polymer interactions were obtained by viscosimetry and gel permeation chromatography for solutions and by differential scanning calorimetry after removal of the solvent. Temperature effect will also be discussed.

Physical Optics of Dynamic Phenomena and Processes in Macromolecular Systems

Results

Homopolymers of styrene,methyl or ethyl or butyl methacrylates and several copolymers of these monomers with acrylic or methacrylic acids, 4-vinylpyridine,N,N dimethyl amino ethyl methacrylate and N,N dimethyl acrylamide were prepared by free radical polymerisation. The copolymers were characterized by either U.V. spectroscopy or elemental analysis.
Qualitative tests showed for the several pairs of copolymers studied that indeed homogenous phases were observed ,and that the compatibilization of such systems depends on the densities of interacting groups,the nature of the solvent,and on the temperature and composition.
Viscosity measurements were carried out with an automatic Ubbelohde viscosimeter,type Fica.Using both approaches(9,10),positive and increasing values of Δb_{23} (difference between experimental and theoretical interaction parameters) from the most incompatible system to the totally compatible one, were obtained; this is an evidence of both hydrodynamic and thermodynamic interactions. Moreover,the non linear relations observed for the Huggins constant of the mixtures of copolymers versus temperature and for the intrinsic viscosity versus composition show that classical thermodynamic theories cannot interpret these results.
The results obtained by gel permeation chromatography (using a Water Associates model ALC/GPC liquid chromatograph) showed that while for incompatible mixtures, two chromatogram peaks, appearing at elution volumes of the components, only one peak at smaller elution volume has appeared with compatible mixtures. Transparent films and only one glass transition temperature (as determined with the Perkin Elmer DSC II) were obtained with the homogenous phases observed qualitatively, while opaque films and two glass transition temperatures are characteristic of incompatible polymer pairs.Some of the results obtained in this study are shown in the Figures below.

Fig. 1

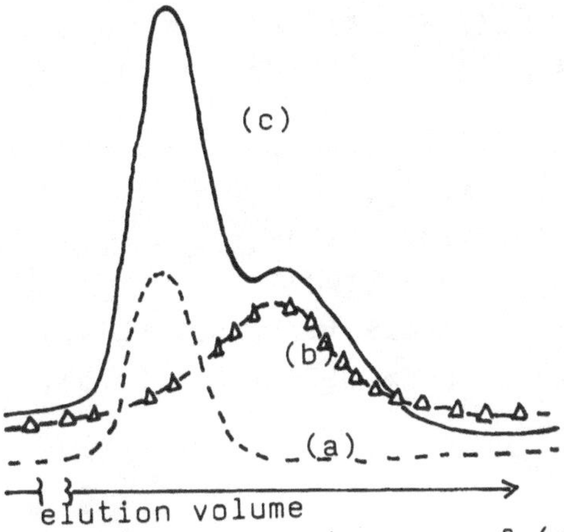

Fig. 2

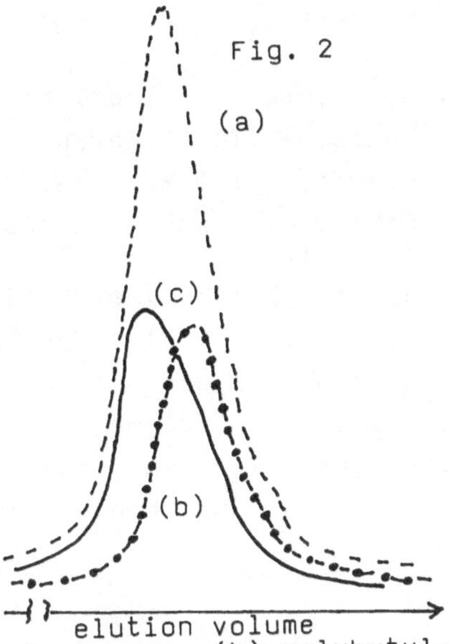

Figure 1. GPC chromatograms of (a) polystyrene, (b) polybutylmethacrylate and (c) their 1:1 mixture.

Figure 2. GPC chromatograms of (a) SA-33.6 (styrene copolymer containing 33.6 % of methacrylic acid), (b) MBDA-5.25 (copolymer of butyl methacrylate containing 5.25 % of N,N dimethylamino ethyl methacrylate) and their 1:1 mixture (c).

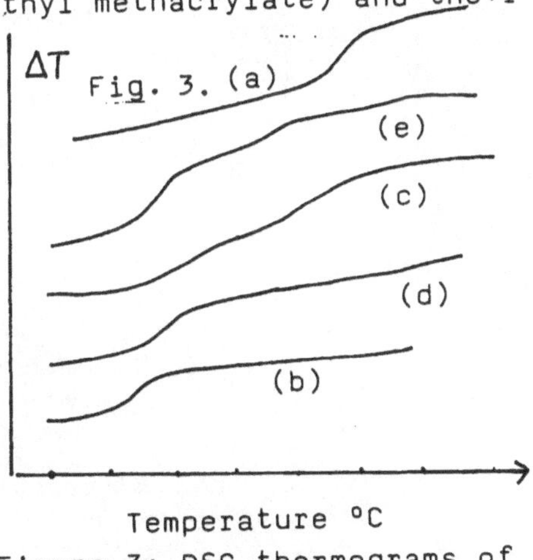

Fig. 4

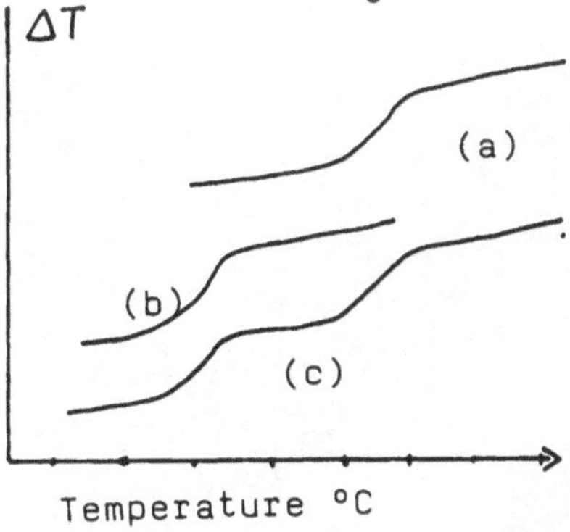

Figure 3: DSC thermograms of SA-33.6 (a), MBDA-5.25 (b) and their mixtures: (c) 1:1 ratio, (d) 1:3 and (e) 2:1 ratios.

Figure 4: DSC thermograms of Polystyrene (a), polybutyl methacrylate (b) and their 1:1 mixture.

References

1. Scott,R.L. : J.Chem.Phys.,17,279 (1949).
2. Tompa,H. :Trans.Faraday Soc.,45, 1142 (1949).
3. Krause,S. : J.Macromol.Chem.,C-7, 251 (1972).
4. Paul,D.R.,Newman,S. : "Polymer Blends" Academic Press, New-York, 1978.
5. Djadoun,S.: Polymer Bulletin, 7, 607 (1982).
6. Djadoun,S. et al.: Macromolecules, 10, 1015 (1977).
7. Lassoued A.,and Djadoun S.: (Submitted 1984).
8. Djadoun S.: Polymer Bulletin, 9, 313 (1983).
9. Krigbaum,W.R. and Wall,F.W.: J.Polym.Sci.,5, 505 (1950).
10. Dondos,A. and Benoit H.:Makromol.Chem.,176,3441 (1975).

CONTRIBUTORS

M. Adam (59)
Service de Physique du Solide et de Résonance Magnétique, CEN-Saclay, 91191 Gif-sur-Yvette Cedex, France

C. Andries (371)
University of Antwerp, Department of Biochemistry, Biophysics Research Group, Universiteitsplein 1, B-2610 Antwerpen, Belgium

M. Ankrim (145)
Laboratoire de Spectrométrie et d'Imagerie Ultrasonores (formely Laboratoire d'Acoustique Moléculaire), ERA au CNRS, Université Louis Pasteur, 4, rue Blaise Pascal, 67070 Strasbourg Cedex, France

M. Antonietti (191)
Institut für Physikalische Chemie der Universität Mainz, Jakob-Welder-Weg 15, D-6500 Mainz, BRD

E. Anufrieva (251)
Institute of Macromolecular Compounds of the Academy of Sciences of the USSR, Bolshoi Prosp. 31, 199004 Leningrad, USSR

O. Aouadj (525)
Laboratoire de Chimie Macromoléculaire, USTHB, BP N° 9, Dar El Beida, Alger, Algeria

D. Ausserré (485)
Physique de la Matière Condensée, Collège de France, 11, place Marcelin-Berthelot, 75231 Paris Cedex 05, France

N. K. Balabaev (413)
Computer Research Centre, USSR Academy of Sciences, Pushchino, 142292 USSR

D. Berek (521)
Polymer Institute CCR, Slovak Academy of Sciences, 84236 Bratislava, Czechoslovakia

Z. Błaszczak (325)
Institute of Physics, Adam Mickiewicz University, Grunwaldzka 6, 60-780 Poznań, Poland

J. Blok (321)
Department of Biophysics, Physics Laboratory of the Free University, De Boelelaan 1081, 1081 HV Amsterdam, The Netherlands

H. Bohidar (67)
Laboratoire de Spectrométrie Physique, Université de Grenoble I, B.P. 68, 38402 St Martin d'Hères Cedex, France; present address: Department of Physics, Indian Institute of Technology, New Delhi 110016, India

C. Bojarski (275)

Institute of Physics, Technical University of Gdańsk, 80-952 Gdańsk, Poland

G. F. Bolshakov (511)
Institute of Petroleum Chemistry, Siberian Branch of the USSR Academy of Sciences, Tomsk, USSR

S. Borbély (317)
Joint Institute for Nuclear Research, Dubna, USSR

W. Boyer (43)
Department of Nuclear Engineering, University of Michigan, Ann Arbor, Michigan 48109, USA

P. Brak (301)
Laboratory of Chemical and Biological Dynamics, Celestijnenlaan 200 D, B-3030 Leuven, Belgium

W. Brown (71)
Institute of Physical Chemistry, University of Uppsala, Sweden

F. Candau (397)
Centre de Recherches sur les Macromolécules, CNRS, 6, rue Boussingault, 67083 Strasbourg Cedex, France

S. Candau (145, 397)
Laboratoire d'Acoustique Moléculaire, Université Louis Pasteur, 4, rue Blaise Pascal, 67070 Strasbourg Cedex, France

M. Chalányiová (521)
Chemical Institute, Comenius University, 84215 Bratislava, Czechoslovakia

B. Chu (363)
Department of Chemistry, State University of New York at Stony Brook, Stony Brook, N.Y. 11794, USA

J. Clauwaert (371)
University of Antwerp, Department of Biochemistry, Biophysics Research Group, Universiteitsplein 1, B-2610 Antwerpen, Belgium

C. Cohen (49)
School of Chemical Engineering, Cornell University, Ithaca, New York 14853, USA

J. Coutandin (191)
Institute für Physikalische Chemie der Universität Mainz, Jakob-Welder-Weg 15, D-6500 Mainz, BRD

R. Cressely (507)
Groupe Physique des Polymères, Faculté des Sciences, Ile du Saulcy, 57045 Metz Cedex 1, France

L. Cser (317)
Central Research Institute for Physics, Hungarian Academy of Sciences, Budapest, Hungary

C. Cuniberti (267)
Istituto di Chimica Industriale dell'Università di Genova, Centro di Studi ChimicoFisici di Macromolecole Sintetiche e Naturali del CNR, Corso Europa 30, 16132 Genova, Italia

G. Damaschun (367)
Department of Molecular Biophysics, Central Institute of Molecular Biology of the Academy of Sciences of the GDR, 1115 Berlin, GDR

J.-P. Decruppe (507)
Groupe Physique des Polymères, Faculté des Sciences, Ile du Saulcy, 57045 Metz Cedex 1, France

C. De Jonghe (359)
Laboratorium voor Macromoleculaire Chemie, KULeuven, B-3030 Heverlee, Belgium

M. Delsanti (59)
Service de Physique du Solide et de Résonance Magnétique, CEN-Saclay, 91191 Gif-sur-Yvette Cedex, France

S. Djadoun (525)
Laboratoire de Chimie Macromoléculaire, USTHB, BP N° 9, Dar El Beida, Alger, Algeria

C. J. C. Edwards (1)
Department of Polymer Science and Technology, University of Manchester, Institute of Science and Technology, Manchester M60 1QD, UK

D. Ehlich (191)
Institut für Physikalische Chemie der Universität Mainz, Jakob-Welder-Weg 15, D-6500 Mainz, BRD

B. Faragó (317)
Central Institute for Physics, Hungarian Academy of Sciences, Budapest, Hungary

B. M. Fechner (517)
Institute of Physics, A. Mickiewicz University, Grunwaldzka 6, 60-780 Poznań, Poland

F. Franěk (317)
Institute for Molecular Genetics, Czechoslovak Academy of Sciences, Prague, Czechoslovakia

S. Fujime (17, 33)
Mitsubishi-Kasei Institute of Life Sciences, Machida, Tokyo 194, Japan

Y. Fukuda (349)
Department of Chemistry, College of Arts and Sciences, University of Tokyo, Komaba, Meguroku, Tokyo, Japan

G. Fytas (205)
Max-Planck-Institut für Polymerforschung Mainz, c/o Institut für Physikalische Chemie der Universität Mainz, Postfach 3980, D-6500 Mainz, BRD

K. Gast (367)
Department of Molecular Biophysics, Central Institute of Molecular Biology of the Academy of Sciences of the GDR, 1115 Berlin, GDR

E. Geissler (67, 157)
Laboratoire de Spectrométrie Physique, Université de Grenoble I, B.P. 68, 38402 St Martin D'Hères Cedex, France

A. G. Grivtsov (413)
Institute of Physical Chemistry, USSR Academy of Sciences, Leninskii Prosp. 31, Moscow, 117312, USSR

T. Hashimoto (233)
Department of Polymer Chemistry, Kyoto University, Kyoto 606, Japan

P. Hauptmann (329)
Technische Hochschule Leuna-Merseburg, Sektion Physik, 42 Merseburg, DDR

A. Havránek (245)
Department of Polymer Physics, Faculty of Mathematics and Physics, Charles University, 18000 Prague, Czechoslovakia

A.-M. Hecht (67, 157)
Laboratoire de Spectrométrie Physique, Université de Grenoble I, B.P. 68, 38402 St Martin D'Hères Cedex, France

H. Hervet (83, 485)
Physique de la Matière Condensée, Collège de France, 11, place Marcelin-Berthelot, 75231 Paris Cedex 05, France

W. Hess (335)
Fakultät für Physik der Universität Konstanz, D-7750 Konstanz, BRD

G. Hild (145)
Centre de Recherches sur les Macromolécules, CNRS, 6, rue Boussingault, 67083 Strasbourg Cedex, France

Y. Hirokawa (197)
Research and Development Center, Nippon Zeon Company, Kawasaki, Japan

R. Hocquart (507)
Groupe Physique des Polymères, Faculté des Sciences, Ile du Saulcy, 57045 Metz Cedex 1, France

J. Holoubek (477, 481)
Institute of Macromolecular Chemistry, Czechoslovak Academy of Sciences, 16206 Prague 6, Czechoslovakia

C. Holtzcherer (397)
Centre de Recherches sur les Macromolécules, CNRS, 6, rue Boussingault, 67083 Strasbourg Cedex, France

D. H. Hwang (49)
School of Chemical Engineering, Cornell University, Ithaca, New York 14853, USA

T. Igarashi (165)
Research Institute of Composite Materials, Gunma University
1-5-1, Tenjin-cho, Kiryu 376, Japan

M. Irie (287)
Institute of Scientific and Industrial Research, Osaka University, Osaka, Japan

R. Johnsen (473)
Institute of Physical Chemistry, University of Uppsala, Uppsala, Sweden

W. C. Johnson, Jr. (493)
Department of Biochemistry and Biophysics, Oregon State University, Corvallis, Oregon 97331, USA

E. Kato (229)
Kobe University of Mercantile Marine, Fukae-Minami, Higashinada, Kobe 658, Japan

J. S. King (43)
Department of Nuclear Engineering, University of Michigan, Ann Arbor, Michigan 48109

R. Klein (335)
Fakultät für Physik der Universität Konstanz, D-7750 Konstanz, BRD

B. Kołodziej (37)
Institute of Experimental Physics, University of Gdańsk, Wita Stwosza 57, 80-952 Gdańsk, Poland

Č. Koňák (271, 353, 401, 405, 457, 461)
Institute of Macromolecular Chemistry, Czechoslovak Academy of Sciences, 16206 Prague 6, Czechoslovakia

S. Kondo (165)
Research Institute of Composite Materials, Gunma University
1-5-1, Tenjin-cho, Kiryu 376, Japan

R. Koyama (279)
College of General Education, Kyoto University, Kyoto 606, Japan

M. Krakovyak (251)
Institute of Macromolecular Compounds of the Academy of Sciences of the USSR, Leningrad, USSR

J. Křepelka (457)
Laboratory of Optics, Palacký University, 77146 Olomouc, Czechoslovakia

E. Kröner (131)
Institute für Theoretische und Angewandte Physik, Universität Stuttgart und Max-Planck-Institut für Metallforschung, Stuttgart, D-7000 Stuttgart, BRD

J. Kucera (197)
Massachusetts Institute of Technology, Cambridge, MA 02139, USA

K. Kubota (33)
Department of Physics, Ochanomizu University, Bunkyo-ku, Tokyo 112, Japan

M. E. Kuil (321)
Department of Biophysics, Physics Laboratory of the Free University, De Boelelaan 1081, 1081 HV Amsterdam, The Netherlands

K. H. Langley (107)
Departments of Physics and Astronomy, University of Massachusetts, Amherst, Massachusetts 01003, USA

A. Lassoued (525)
Laboratoire de Chimie Macromoléculaire, USTHB, BP N° 9, Dar El Beida, Alger, Algeria

L. Léger (83)
Laboratoire de Physique de la Matière Condensée, Collège de France, 11, place Marcelin-Berthelot, 75231 Paris Cedex 05, France

L. V. Lunevskaya (413)
Computer Research Centre, USSR Academy of Sciences, Pushchino 142292, USSR

T. Macko (521)
Polymer Institute CCR, Slovak Academy of Sciences, 84236 Bratislava, Czechoslovakia

T. Maeda (17, 33)
Mitsubishi-Kasei Institute of Life Sciences, Machida, Tokyo 194, Japan

M. Mandel (87)
Department of Physical and Macromolecular Chemistry, Gorlaeus Laboratories, University of Leiden, P.O. Box 9502, 2300 RA Leiden, The Netherlands

M. A. Mazo (413)
Institute of Chemical Physics, USSR Academy of Sciences, ul. Kosygin, 4, Moscow 117977, USSR

R. V. Mehta (377)
Sardar Vallabhbhai Regional College of Engineering and Technology, Surat-395007, India

F. Mezei (317)
Institute Laue-Langevin, Grenoble, France

M. F. Millet (83)
Laboratoire de Physique de la Matière Condensée, Collège de France, 11, place Marcelin-Berthelot, 75231 Paris Cedex 05, France

J. Mikeš (477)
Institute of Chemical Process Fundamentals, Czechoslovak Academy of Sciences, 16502 Prague 6, Czechoslovakia

A. Morita (427, 443)

Department of Chemistry, College of Arts and Sciences, University of Tokyo, Komaba, Meguroku, Tokyo 153, Japan

L. Mrkvičková (353)
Institute of Macromolecular Chemistry, Czechoslovak Academy of Sciences, 16206 Prague 6, Czechoslovakia

G. Muller (117)
Faculté des Sciences de l'Université de Rouen, Haute-Normandie, 76130 Mont-Saint-Aignan, France

J. J. Müller (367)
Department of Molecular Biophysics, Central Institute of Molecular Biology of the Academy of Sciences of the GDR, 1115 Berlin, GDR

J. P. Munch (145)
Laboratoire de Spectrométrie et d'Imagerie Ultrasonores (formerly Laboratoire d'Acoustique Moléculaire), ERA au CNRS, Université Louis Pasteur, 4, rue Blaise Pascal, 67070 Strasbourg Cedex, France

G. Nägele (335)
Fakultät der Physik der Universität Konstanz, D-7750 Konstanz, BRD

T. Nakano (349)
Department of Chemistry, College of Arts and Sciences, University of Tokyo, Komaba, Meguroku, Tokyo, Japan

B. Nordén (325)
Physical Chemistry Department, Chalmers University of Technology, S-41296 Göteborg, Sweden

B. Nyström (305, 469)
Department of Chemistry, University of Oslo, P.O. Box 1033, Blindern, 0315 Oslo 3, Norway

R. Okasha (145)
Centre de Recherches sur les Macromolécules, CNRS, 6, rue Boussingault, 67083 Strasbourg Cedex, France

E. F. Oleinik (413)
Institute of Chemical Physics, USSR Academy of Sciences, ul. Kosygin, 4, Moscow 117977, USSR

Y. M. Ostanevich (317)
Joint Institute for Nuclear Research, Dubna, USSR

A. Patkowski (363)
Quantum Electronics Laboratory, Institute of Physics, A. Mickiewicz University, Grunwaldzka 6, 60-780 Poznań, Poland

A. Perico (267)
Istituto di Chimica Industriale dell'Università di Genova, Centro Studi ChimicoFisici di Macromolecole Sintetiche e Naturali del CNR, Corso Europa 30, 16132 Genova, Italia

J. Peřina (447, 457)

Laboratory of Optics, Palacký University, 77146 Olomouc, Czechoslovakia

A. Persoons (301, 409, 465)
Laboratory of Chemical and Biological Dynamics, Celestijnenlaan 200 D, B-3030 Leuven, Belgium

P. Plietz (367)
Department of Molecular Biophysics, Central Institute of Molecular Biology of the Academy of Sciences of the GDR, 1115 Berlin, GDR

J. Pospíšil (245)
Department of Polymer Physics, Faculty of Mathematics and Physics, Charles University, 18000 Prague, Czechoslovakia

P. Rempp (145)
Centre de Recherches sur les Macromolécules, CNRS, 6, rue Boussingault, 67083 Strasbourg Cedex, France

H. Reynaers (359)
Laboratorium voor Macromoleculaire Chemie, KULeuven, B-3030 Heverlee, Belgium

F. Rondelez (485)
Physique de la Matière Condensée, Collège de France, 11, place Marcelin-Berthelot, 75231 Paris Cedex 05, France

J. Roots (305, 469)
Department of Chemistry, University of Oslo, P.O. Box 1033, Blindern, 0315 Oslo 3, Norway

P. S. Russo (107)
Polymer Science and Engineering, University of Massachusetts, Amherst, Massachusetts 01003, USA; present address: Department of Chemistry, Louisiana State University, Baton Rouge, Louisiana 70803, USA

Y. Saji (229)
Kobe University of Mercantile Marine, Fukae-Minami, Higashinada, Kobe 658, Japan

M. A. Scheerhagen (321)
Department of Biophysics, Physics Laboratory of Free University, De Boelelaan 1081, 1081 HV Amsterdam, The Netherlands

W. Schnabel (287)
Hahn-Meitner-Institut für Kernforschung Berlin, Bereich Strahlenchemie, D-1000 Berlin 39, BRD

B. Sedláček (271, 353, 477)
Institute of Macromolecular Chemistry, Czechoslovak Academy of Sciences, 16206 Prague 6, Czechoslovakia

D. B. Sellen (177)
Astbury Department of Biophysics, University of Leeds, Leeds LS2 9JT, UK

K. Sienicki (275)
Institute of Physics, Technical University of Gdańsk,

80-952 Gdańsk, Poland

H. Sillescu (191)
Institute für Physikalische Chemie der Universität Mainz, Jakob-Welder-Weg 15, D-6500 Mainz, BRD

P. Štěpánek (271, 401, 405, 457, 461)
Institute of Macromolecular Chemistry, Czechoslovak Academy of Sciences, 16206 Prague 6, Czechoslovakia

R. F. T. Stepto (1)
Department of Polymer Science and Technology, University of Manchester, Institute of Science and Technology, Manchester M60 1QD, UK

B. Stuhn (217)
Institute für Physikalische Chemie der Universität Mainz, 6500 Mainz, BRD; present address: Chemische Werke Hüls, Marl, BRD

S.-T. Sun (197)
Hercules Company, Wilmington, Delaware, USA

I. A. Tale (245)
Institute of Solid State Physics, Latvian State University, 226063 Riga, Latvia, USSR

T. Tanaka (20)
Department of Physics, Massachusetts Institute of Technology, Cambridge, MA 02139, USA

R. Taxerman-Krozer (131)
Institut für Theoretische und Angewandte Physik, Universität Stuttgart und Max-Plansk-Institut für Metallforschung, Stuttgart, D-7000 Stuttgart, BRD

Y. Tominaga (33)
Department of Physics, Ochanomizu University, Bunkyo-ku, Tokyo 112, Japan

Z. Tuzar (401, 405, 461)
Institute of Macromolecular Chemistry, Czechoslovak Academy of Sciences, 16206 Prague 6, Czechoslovakia

R. Ullman (43)
Research Staff, Ford Motor Co., Dearborn, Michigan 48121, USA

H. Urabe (33)
Department of Physics, Ochanomizu University, Bunkyo-ku, Tokyo 112, Japan

R. van Amerongen (321)
Department of Biophysics, Physics Laboratory of Free University, De Boelelaan 1081, 1081 HV Amsterdam, The Netherlands

F. van der Touw (87)
Department of Physical and Macromolecular Chemistry, Gorlaeus Laboratories, University of Leiden, P.O. Box 9502,

2300 RA Leiden, The Netherlands

W. Van de Sande (409)
Laboratory for Chemical and Biological Dynamics, University of Leuven, Celestijnenlaan 200 d, B-3030 Leuven, Belgium

R. van Grondelle (321)
Department of Biophysics, Physics Laboratory of the Free University, De Boelelaan 1081, 1081 HV Amsterdam, The Netherlands

M. Van Laethem (371)
University of Antwerp, Department of Biochemistry, Biophysics Research Group, Universiteitsplein 1, B-2610 Antwerpen, Belgium

B. Van Wonterghem (465)
Laboratory for Chemical and Biological Dynamics, University of Leuven, Celestijnenlaan 200 d, B-3030 Leuven, Belgium

K. B. van Treslong (359)
Gorlaeus Laboratoria der Rijkuniversiteit Leiden, Leiden, The Netherlands

F. Varkevisser (359)
Gorlaeus Laboratoria der Rijkuniversiteit Leiden, Leiden, The Netherlands

C. H. Wang (217)
Department of Chemistry, University of Utah, Salt Lake City, Utah 84112, USA

H. Watanabe (349, 427, 443)
Department of Chemistry, College of Arts and Sciences, University of Tokyo, Komaba, Meguroku, Tokyo, Japan

H. Welfle (367)
Department of Molecular Biophysics, Central Institute of Molecular Biology of the Academy of Sciences of the GDR, 1115 Berlin, GDR

G. D. Wignal (43)
Oak Ridge National Laboratory, P.O. Box X, Oak Ridge, Tennessee 37831, USA

S. Wijmenga (87)
Department of Physical and Macromolecular Chemistry, Gorlaeus Laboratories, University of Leiden, P.O. Box 9502, 2300 RA Leiden, The Netherlands

K. Witkowski (37)
Institute of Experimental Physics, University of Gdańsk, Wita Stwosza 57, 80-952 Gdańsk, Poland

M. Witrh (325)
Physical Chemistry Department, Chalmers University of Technology, S-41296 Göteborg, Sweden

L. Woliński (37)
Institute of Experimental Physics, University of Gdańsk,

Wita Stwosza 57, 80-952 Gdańsk, Poland

T. Wydro (507)
Groupe Physique des Polymères, Faculté des Sciences, Ile du Soulcy, 57045 Metz Cedex 1, France

D. Zirwer (367)
Department of Molecular Biophysics, Central Institute of Molecular Biology of the Academy of Sciences of the GDR, 1115 Berlin, GDR

INDEX

Acetylcholinesterase, 504

Activation energy
  diffusivity, 239
  excimer formation, 272
  relaxation process, 247-249

Aggregation
  ι-carrageenan, 361, 362
  magnetic colloid, 389
  poly(γ-benzyl-L-glutamate), 349-352

n-Alkanes, 413

Amorphous polymer
  Rayleigh-Brillouin scattering, 217

Analysis of DLS (ICS, PCS, QELS) data
  bimodal, 51, 52, 55, 69, 72-74, 473
  cumulant method, 7, 29, 31, 34, 35, 68, 71, 301, 304
  non-linear regression procedure, 474

Anisotropy
  diamagnetic, 380
  optical, 518
  paramagnetic, 380

Antibunching, 448, 453

Association, 101
  alcohols, 517
    optical anisotropy, 518
  by static light scattering (SLS), 517-520

Attenuation constant of coherent field, 457

Babinet's teorem, 170

Bending motion, 17, 18, 20-22, 24, 30, 31

Bimodal analysis of autocorrelation function, 51, 52, 55, 69, 72-81, 473

Biological macromolecules, 305, 306; see Nucleic acids, Polysaccharides, Proteins, Viruses

Birefringence, 29, 87, 91, 94, 97, 98, 131-133, 139, 141, 349-352, 380, 507-510; see also Electric birefringence, Streaming birefringence, Transient electric birefringence

Blob, 101, 102

Block copolymers, 233-243, 401-404, 405-408
  micelles, 401-408
    DLS (ICS, PCS, QELS), 401-403, 405-408
    SLS, 401-404, 405-408
    Time-resolved SAXS, 234

Bovine serum albumin, 185,187, 188, 306, 308, 311, 312, 314, 363-366

Branched polymers, 145

Brillouin scattering
  amorphous polymers, 217
  broadening of lines, 167
  gelation, 165-175
  half-width, 229
  low temperature, 229, 231
  segmental motion, 229
  shift, 166, 167, 170, 229-231
  spectral intensity, 166

Brownian motion, 12, 22, 85,

234, 313

Bulk polymers (regime, region), 44-46, 205-215, 217-226, 233-242, 245-250, 477-480

Butadiene-styrene copolymers, 233-243, 401-404, 405-408

Carrageenan

κ- and ι-carrageenan, 359, 360

aggregation (ι), 361, 362

segmented (κ, ι), 360, 362

Charged dispersions

binary mixtures of macroions, 340

DLVO potential, 336

interaction potentials, 337

Mean-spherical-approximation (MSA), 336

One-component macrofluid (OCM), 336

polydispersity effects, 336-338

radial distribution function, 335, 336

rescaled MSA, 336

static structure factor, 335, 336

systems with small ions, 338

Chemical processes

by luminescence, 259

Chromophores, 268

α-chymotrypsin, 499

Cluster formation, 375

Coherency matrix, 380

Coil-globule transition, 255, 267, 271

Collective diffusion, see Diffusion

Colloids, 335, 377, 397, 401, 405, 409, 485

magneto-optics, 377

magnetic, 378, 379, 390, 391

Compatibilization, 525-528

Concentration fluctuations, 68

Concentrated solutions (regime, region), 44, 46, 50, 51, 54-56

Condensed polymer systems, 132

Conformation

flexible, 43, 49, 59, 318, 443, 507

rigid, 1, 33, 507

semiflexible, 17, 18, 21, 67

Conformation od polymers, 251, 254, 255

coil contraction, 289-293

copolymer glutamic acid-leucin, 255

copolymer styrene-4-(methylacryloylamino)azobenzene, 290

polyamide, azobenzene groups, 288, 289

poly(1,2-dimethoxy ethylene), 255

poly(glutamic acid), 255

poly(methacryloyl-ω-oxyundecane acid), cholesterol ester, 256, 257

polystyrene, azobenzene groups, 289, 291

coil expansion, 294-296

polyamide, azobenzene groups, 288

poly(methacrylic acid), 254, 255, 294-296

flash photolysis, 287, 291-298

SLS, 287-298

polarized luminescence (PL), 251-265

Conformation of proteins, 306, 308-311

bovine serum albumin, 308, 313, 363, 366
DNA-protein complex, 321
circular dichroism (CD), 323, 324
DLS (ICS, PCS, QELS), 322, 323
electric birefringence, 321, 322
hemoglobin, 325
immunoglobulin G, 317-320
lysozyme, 310, 313, 472
tRNA-BSA complex, 363, 366
under pressure, 310-313, 472
Contrast variation method, 162
Cooperative diffusion, see Diffusion
Copolymerization, styrene-biunsaturated comonomers, 145-155
Correlation hole, 237
Correlation length, 70, 172
Correlator improvement, 301
Counterions, 336-347
static properties, 336-343
dynamic properties, 343-347
Critical micelle concentration (CMC), 405-408
Critical micelle temperature (CMT), 405-408
Crosslinked polymers, 262, 263
Cross-over concentration, 67
α-Crystallin, 371-375
cluster formation, 375
DLS (ICS, PCS, QELS), 372
SLS, 374
Cumulant method and analysis, 7, 29, 31, 34, 35, 51, 52, 55, 68, 69, 71, 72-74, 301, 304, 307, 473
Cytochrome C, 500

Denaturants of proteins, 305, 309, 311
Denaturation, 305, 308
Deswelling, 50
Dichroism
circular (CD), 321, 323, 324
magnetically induced, 377, 386
Dielectric relaxation
Cole-Cole plots, 443-446
Debye semicircle, 443-445
flexible molecule, 443-446
Diffusion, see also Diffusion coefficients
anisotropy in, 17, 32, 33
collective, cooperative, 50, 54, 68-70, 75, 197, 401
gradient diffusion, 77, 78, 80
lateral, 409
mutual, 50, 56, 412
rotational, 17-19, 26, 32, 33, 35, 37, 103, 107, 318-320, 321-323
self-diffusion, 53, 56, 83, 235, 236, 239, 240
through gels, 177-190
tracer diffusion, 412
translational, 1, 17-19, 26, 32, 33, 35, 37-41, 49, 51, 52-54, 56, 59-62, 67-70, 71, 97, 98, 109, 114, 115, 121, 123-128, 146, 148, 151, 153, 154, 177-190, 191, 301-303, 307, 312, 313, 317-320, 321, 323, 335, 343-347, 364, 367-370, 371, 372, 374, 397-400,

401, 403, 405-408, 409, 411, 412

Diffusion coefficients, l, 13, 18-20, 23, 28-35, 37-41, 49, 52, 54, 56, 59, 61, 62, 71, 67-70, 71, 73, 75, 77, 78, 80, 93, 97, 99, 103, 121, 123-128, 146, 148, 151, 153, 154, 177-190, 191, 192, 194, 235, 301-303, 305-307, 312, 313, 318-320, 321, 364, 367-370, 371, 372, 397-400, 401, 403, 408, 411, 412

concentration effect on, 37-41

α-crystallin, 371

D_0, D_1, D_3, 18-21, 23, 28-35, 52, 54

D_c, D_g, 62

Fluorescence recovery after photobleaching (FRAP), 411

globulins, 367-370

good solvents, 73

interacting micelles, particles, 401

inverse latices, 397

macromolecules in melt, 192

'marginal' solvent, 75

model computation, 318-320

molecular weight effect on, 59, 62

molten polystyrene, 191

poly(acrylamide), 125-128, 397

poly(butyl methacrylate), 54

polydisperse polymers, 41

polymer blends, 41

poly(methyl methacrylate), 49

polystyrene, 37, 59, 61, 62, 67-70, 71, 97, 98

Diffusion equation, 19, 133, 135

Diffusion-relaxation coupling, 54

Diffusion through gels, 177-190

Dilute solutions (regime, region), 1, 21, 29-32, 49, 88, 112, 117, 145, 251-265, 267-270, 271-274, 279-286, 287-298, 301-304, 305-314, 317-320, 321-324, 325-328, 329-333, 335-347, 353-357, 359-362, 363-366, 367-370, 371-375, 397-400, 409-412

Disorder-order transition, 234, 241

Dispersion, see Charged dispersions

DLS, Dynamic light scattering

Dynamic electro-optical effect (DEOE), 427-441

extremely high field, 439-440

static field, 428-432

varying field, 428-438

Dynamic light scattering (DLS); also Intensity correlation spectroscopy (ICS), Photon correlation spectroscopy (PCS), and Quasi-elastic light scattering (QELS) 15, 17, 29-33, 35, 37-39, 49-52, 59, 67-70, 71-81, 97, 98, 121, 123-128, 145-154, 177-190, 205-215, 301, 305, 321, 343, 347, 356, 363, 364, 367-369, 371, 372, 374, 375, 397, 401-403, 405-408, 461, 464, 465, 469, 473

Elastase, 499

Electric birefringence, see also Transient electric birefringence

decay of, 87, 91, 94, 97

Na-poly(styrene sulfonate), poly(γ-benzyl-L-glutamete), 349-352

Electro-optical relaxation
  dynamics, 87
  polyelectrolytes, 87
Elongational flow, 507, 508
Entanglements, 49, 67, 83-86, 195, 196
Epoxy resins, 174
Evanescent wawe induced fluorescence (EWIF), 485-491
  absorption or 'depletion' layer, 485, 487
  experimental set-up, 488-490
  interfacial properties, 485, 487, 491
  penetration depth, 491
  xanthan solution on silica surface, 490, 491
Excimer, 267-270, 271-274, 275
  activation energy, 272
  coil-globule transition, 267, 270, 271, 274
  formation, 268, 271, 275
  influence of M_w, 272
  polystyrene, 271, 277
  poly(vinyl acetate), 269
  poly(1-vinylnaphthalene), 278
  poly(2-vinylnaphthalene), 278
Excluded volume effect, 43, 134
Extinction of light scattering, 383-385
Eye lens proteins (calf and bovine), 371-375
  DLS (ICS, PCS, QELS), 372
  SLS, 374
Factorial moments, 452
Ferrofluids, 379
Fickian diffusion, 235
Flavodoxin, 500
Flexibility parameter, 20, 21, 24, 28, 29, 31
Flexible macromolecules, 43, 49, 59, 318, 443-446, 508
Flory equation, swelling, 158
Flory-Huggins parameter, 158, 160, 233-243
Flow birefringence, 507-510
Fluorescence, see also Luminescence
  chromophoric groups, 275, 277
  coil-globule transition, 267, 270, 271, 274
  decay of, 275-278
  excimer formation, 268, 271, 275
  polystyrene, 271, 277
  poly(vinyl acetate), 269
  poly(1-vinylnaphthalene), 278
  poly(2-vinylnaphthalene), 278
  spectra, 269, 271
Fluorescence recovery after photobleaching (FRAP), 409
  lateral diffusion of micelles, 409
  mutual diffusion of micelles, 412
  tracer diffusion of micelles, 412
Fokker-Planck equation, 344, 447, 451
Forced Rayleigh light scattering (FRS), 56, 83, 191-193
Friction coefficient (molecular, segment), 1, 11, 63, 68, 114, 132, 135, 367-370

globulins, 369
poly(γ-benzyl-L-glutamate),

Gels
agarose, 182-185
calcium alginate, 182-184
copolymer styrene-comonomer, 145-155
dextran, 183
diffusion of compact macromolecules, 177
gelation, 145, 152, 165
poly(acrylamide), 157, 158, 180-182, 185-189
poly(dimethylsiloxane), 168
polysaccharides, 182-185
polystyrene, 194
scattering of incoherent light, 182, 187
swelling, 158, 160
Generation functions, 451, 452
Glass transition temperature
amorphous poly(ethyl methacrylate), 205
polystyrene, 195
Globulins
axial ratio, 369
DLS (ICS, PCS, QELS), 367-370
frictional coefficient, 369
small-angle X-ray scattering (SAXS), 367-370
shape model, 369, 370
solvation, 367-370
Glow curve, 246, 249
Glyceraldehyde-3-phosphate dehydrogenase, 500
Gobrecht's method, 246
Gradient diffusion, 77, 78, 80
Green's function method, 174

Heisenberg-Langevin equations, 447, 449
Hemoglobin, human, 325, 328
deoxy-Hb, 327, 328
met-Hb, 327, 328
oxy-Hb, 327, 328
High pressure effect on dynamical properties, 214, 305, 469
Holographic grating techniques, 191
Hydrodynamic interactions, 1, 2, 12, 23, 26-29, 43, 54-56, 134, 344
Hydrodynamic radius, 37, 306-311, 397, 402, 404, 406

ICS, Intensity correlation spectroscopy
Immunoglobulin G, pig, 317, 319
Inhomogeneities, static, 173
Integral light scattering, see Static light scattering
Intensity of saturation magnetization, 391
Intensity correlation spectroscopy (ICS), see Dynamic light scattering
Interacting Brownian particles, 313
Interactions
intermolecular reactions, 1, 23, 26-29, 37, 41, 54-56, 134, 257, 258, 264, 310, 312-314, 349-352, 361, 362, 389
intramolecular processes, 18, 37, 252, 254, 267, 289-293, 294-296, 306, 308-311, 313, 317-320, 321-323, 325, 363,

366, 472

Junction
  decay, 131, 132, 137
  formation, 131, 132, 137

Kerner's equation, 162
Kerr effect
  hemoglobin, 427
  Kerr constant, 87
  relaxation in high electric fields, 427
Kirkwood equation, 1, 12, 14, 15
Kirkwood-Riseman radii, 9-15

Labelled luminescent polymers, 191, 251
Lactate dehydrogenase, 499
Langevin equation, 21, 134
Lateral diffusion of micelles, 409
Light scattering 'anomaly' 353
  aged poly(vinyl alcohol), 353
  external interference, 357
  interparticle ordering, 356
Light scattering speckle photography
  'boiling' speckles, 477
  contrast of speckle pattern, 481, 483
  parameter M, 481-483
  random phase screen, 481-484
  slow correlation times, 481, 484
  translational speckles, 477
Longitudinal compliance, 225
Longitudinal configuration, 382
Longitudinal modulus, 218, 219, 225
Lorenz-Mie light scattering, see Static light scattering
LS, light scattering
Luminescence, see also Fluorescence
  polarized luminescence, 251-265
    chemical processes, 259-265
    coil-globule transition, 255
    coil-helix transition, 255
    coil-isotropic liquid globule-liquid-crystalline globule transition, 256
    conformational transition, 255
    intermolecular interactions, 253, 254
    intermolecular complexes, 253, 257, 258
    intramolecular mobility, 253, 258
    nanosecond relaxation process, 252, 254
    relaxation time, 252, 254
    rotational motion of macromolecules, 253
    stereocomplexes, 258
  thermostimulated luminescence, 245-250
    activation energy, 246, 249
    anomalous behaviour, 245, 246, 249
    Gobrecht's method, 246
    relaxation region, 245-248
Lysozyme, 258, 306, 308-313, 472, 500
  DLS (ICS, PCS, QELS), at high pressure, 472

Magnetic colloids, 378, 379, 390, 391
Magneto-optical effect

experimental set-up, 387
reversal in, 385
Magneto-optics of colloids, 377-395
Micelles
styrene-butadiene copolymers, 401
DLS (ICS, PCS, QELS), 401, 405
fluorescence, 409
hydrodynamic radius, 402-404
interaction between micelles, 401, 406-408
interaction parameter, 402-404
lateral diffusion, 409
micellar mass, 402-404, 406
Microgel, polystyrene, 170, 194, 195
Microdomain structure, 233
Microphase
dissolution, 233-243
separation, 157, 233, 234, 240
Microsyneresis, 157, 233, 234, 240
Mobility, molecular
acoustical methods, 329
liquid polymers, 329
polymerization process, 329
Molecular dynamic simulation, 413
C_{10}-n-alkane, 413
crystal structure, 417
introduction of chemical deffects, 415, 416
molecular conformations, 419
polyethylene single crystal model, 413
rotational mobility of molecules, 421
structural model (cell), 414-416
valence angles, 420
Molecular scattering factor, 280, 283
worm-like chain polymer, 283
Monte Carlo technique, 2-4, 10, 12
Motility, intramolecular, 317
immunoglobulin G (pig), 317
neutron spin echo technique, 317
Multiple scattering, 31
Mutual diffusion of micelles, 50, 56, 412
Myoglobin, 499

Network
collective diffusion, 197
permanent, 134
temporary, 131, 132, 141
Neutron scattering, see Small-angle neutron scattering
Neutron spin echo technique, 56, 317
Newtonian flow, 240

Order-disorder transition
order parameter, 233-243
Ornstein-Zernike equation, 337
Oseen tensor, 12-15
preaveraging, 13
Osmotic
coefficient, 302, 303
compressibility 286
polyelectrolyte solutions, 286

modulus, 59, 63, 68

Papain, 499, 500
Partition coefficients
  agarose gels, 182
  alginate gels, 182
  bovine plasma albumin, 185, 187
  dextrane, 183, 185
  diffusion of macromolecules within gels, 183
Percolation theory, 145
PCS, Photon correlation spectroscopy
Phase separation
  poly(acrylamide) gels, 157, 158
Phonons, 171, 172, 174, 175
  damping, 171, 172, 174
  scattering, 173
Photocount
  distribution, 452
    broadening of, 455
  statistics, 447
    in light scattering, 447
Photodetection equation, 447, 454
Photon correlation spectroscopy (PCS); also Intensity correlation spectroscopy (ICS), Quasi-elastic light scattering (QELS); see Dynamic light scattering (DLS)
Photostationary ratio, 268
Picosecond laser pulses, 465-468
  polystyrene, semidilute, 71
Polarization of scattered light, 380, 381, 383
Polarized luminescence, 251-265, 275
Poly(acrylamide)
  diffusion coefficient, 125-128
  DLS (ICS, PCS, QELS), 125-128, 397
  effect of charge density, 126
  effect of molecular weight, 125
  gels, 157
  inverse latices, 397
  microemulsion polymerization, 397, 398
  scaling relations, 127, 128
  seeding process, 399
  semidilute solutions, 117
Poly(acrylic acid), 264
Poly(alkyl methacrylates), 229
  Brillouin scattering, 229
Polyamides
  with azobenzene groups, 292
Poly(γ-benzyl-L-glutamate), 107, 349
  aggregation, 349
  diffusion coefficient, 109, 114, 115
  electro-optical methods, 349-352
  friction coefficient, 114
  semiconcentrated solution, 349
Poly(n-butyl methacrylate), 49-53, 209, 229
  Brillouin scattering, 229
  diffusion coefficient, 53
Poly(1,2-dimethoxy ethylene), 255
Poly(dimethylsiloxane), 2, 8, 56

cyclic, 8
gels, 168
linear, 8
Polydispersity index, 461
DLS (ICS, PCS, QELS), 461, 462
Gaussian and log-normal distribution, 462
influence of dissymmetry and internal modes, 463
α,ω-Divinyl-poly(dimethylsiloxane), 168
Polyelectrolytes
chains, 280, 285
dilute and semidilute solution, 88, 101
ionic strength, 101
persistence length, 117
small-angle neutron scattering (SANS), 279
small-angle X-ray scattering (SAXS), 279
Poly(ethylene glycol), 258
Poly(ethyl methacrylate), 205, 209-212
Brillouin scattering, 229
molecular weight effect, 210
pressure effect, 211-214
temperature effect, 211, 212
Poly(L-glutamic acid), 499
Polymer dynamics
below Θ, 267
Polymer network, see Network
Poly(methacrylic acid), 254, 255, 258, 304
Poly(methacryloyl-ω-oxyundecane acid), 256, 257
Polymethylene, 2, 3
Poly(methyl methacrylate), 15, 50-53, 229, 257-261
Brillouin scattering, 229
diffusion coefficient, 49, 53
Polyoxyethylene, 2, 7, 8
Poly(propylene glycol), 223
DLS (ICS, PCS, QELS), 223
Polysaccharides
Agarose, 182-185
Calcium alginate, 182-184
Dextran, 183
Hyaluronic acid, 498
Xanthan, 117, 119-123
Polystyrene, 15, 37, 43, 44, 59, 67, 68, 71, 83, 88, 257, 289, 306, 312, 313
with azobenzen groups, 291
crosslinked, 194
diffusion coefficient D(c,M), 61, 62, 67-70, 71, 97, 98
DLS (ICS, PCS, QELS), 59, 71
fluorescein labeled, 192, 194, 195
fluorescence, 269, 271-275
Forced Rayleigh light scattering (FRS), 83
friction coefficient, 63, 68
in good solvent, 73
scaling, 63, 65, 69, 70, 100-103
screening length, 43
semidilute solutions, 59, 71
self-diffusion, 83
shear modulus, 67
small-angle neutron scattering (SANS), 43-47
solvent quality effect, 71-81
stilben labeled, 192

in theta-solvent, 59, 67, 76-80
Poly(vinyl acetate)
excimer technique, 270
fluorescence, 268
polymerization, by acoustical methods, 329
pyrenyl-labeled, 269
Poly(vinyl alcohol), aged
DLS (ICS, PCS, QELS), 353, 356
Lorenz-Mie light scattering, 354-356
'liquid-like type' structure, 353
Rayleigh-Debye light scattering, 353-356
Poly(1-vinylnaphthalene)
fluorescence, 278
Poly(2-vinylnaphthalene)
fluorescence, 278
Poly(vinylpyrrolidone), 264
Prealbumin, 500
Preferential solvation, 524
Preferential sorption, 521
mixed liquids on silica gels, 521-524
pressure effect, 521-524
Pregel state, 145
Pressure denaturation, 308, 310, 311
Pressure effect, 214, 305, 469
Pretransition contraction of protein, 311, 312
Primitive model, 336
Processability, 234
Proteins
acetylcholinesterase, 504
bovine plasma albumin, 185, 187
bovine serum albumin, 188, 306, 308, 311, 312, 314, 363-366
α-chymotrypsin, 499
circular dichroism, 493-505
α-crystallin, 371-375
cytochrome C, 500
DLS (ICS, PCS, QELS), 367, 368, 371
elastase, 499
eye lens proteins, 371
flavodoxin, 500
glyceraldehyde-3-phosphate dehydrogenase, 500
hemoglobin, 325-328, 499
immunoglobulin G, 317, 319
interferon, 504
lactate dehydrogenase, 499
lysozyme, 306, 308-313, 472, 500
myoglobin, 499
papain, 499, 500
poly(γ-benzyl-L-glutamate), 107, 349-352
poly(L-glutamic acid), 499
prealbumin, 500
pretransition contraction, 311, 312
ribonuclease A, 500
small-angle X-ray scattering (SAXS), 368-370
solvation of, 367-370
subtilisin BPN', 500
subtilisin Novo, 500
triose phosphate isomerase, 500
Pseudo-gel, 59, 64

QELS, Quasi-elastic light scattering

Quantum characteristic function, 449, 450

Quantum effects, 455

Quasi-elastic light scatering (QELS); also Intensity correlation spectroscopy (ICS) and Photon correlation spectroscopy (PCS); see Dynamic light scattering (DLS)

Quasi-probability function, 450

Radial distribution function

monomers, 279, 280

charged macromolecules, 339, 340

Radius of gyration, 43-45, 47, 49, 146, 160, 161

concentration effect on, 45

Random phase approximation (RPA), 237

Rayleigh-Debye light scattering, see Static light scattering

Rheology, 240

Relaxation time, 49, 60, 61, 64, 87, 131, 133, 138, 139, 206, 211-214, 245-249, 270

Reptation model, 83, 85, 86, 100, 192, 194-196, 234, 235, 242

Ribonuclease A, 500

t-RNA-BSA interaction

aggregation rate, 365

DLS (ICS, PCS, QELS), 363-366

ionic strength, 365

nature of binding, 365, 366

Rigid particles, 10, 33, 507

Rod-like polymer, 107

Rotational diffusion, see Diffusion

Rouse dynamics, 54-56

SALS, Small-angle light scattering

SANS, Small-angle neutron scattering

SAXS, Small-angle X-ray scattering

Scaling relations, 50, 55, 63, 65, 68, 69, 100-103, 152, 154

Scattering, electromagnetic

conducting spheroids, 383

dipole, 381-383

large particles, 384

magnetic sphere, 384

Rayleigh-Gans particles, 383

small number of particles, 455

Screening length, 43-47

concentration effect on, 46

Segmental motion

anharmonicity, 230

Brillouin scattering, 229

Self-diffusion, 53, 83, 235, 236, 239, 240

diffusion coefficient, 83

entangled chains, 83

Self-radiation, 454

Semidilute solutions (regime, region), 21, 29, 31, 32, 45, 46, 49-54, 59-65, 67, 71-81, 85, 88, 94, 96, 103, 107, 117, 145, 312, 317-320, 349-352, 353-357

Semiflexible chains, filaments, 17, 18, 20, 21, 31, 67

Serum albumin, bovine, 185, 187, 188, 306, 308, 311, 312, 314, 363-366

Shape factor, 388, 390

Shear modulus, 64, 67, 68, 158, 160, 162, 240

Shear stress, 507

Shift factor, 240

Shrinking of gels, 197

Side chain

- length, 229, 230
- motion, 229, 230

Size and size distribution, 388, 390, 461-463, 511-515

Small-angle light scattering (SALS), 511

Small-angle neutron scattering (SANS), 43-47, 162, 279-284

- flexible chains, 43-47
- gels, 157-162
- polyelectrolytes, 279
- polystyrene, 43-47
- radius of gyration, 43, 45, 47
- screening length, 43, 46, 47
- worm-like chain polymer, 279, 283

Small-angle X-ray scattering (SAXS), 233-243, 279-286, 367-370

- globulins, 367-370
- disorder-order transition, 234, 241, 242
- order-disorder transition, 234, 237-241
- polyelectrolytes, 279-286
- styrene-butadiene block copolymers, 233-243
- time resolved SAXS, 233

Sodium poly(styrene sulfonate), 284, 306-308, 310

- small-angle neutron scattering (SANS), 284

Sol-gel transition, 145

Speckle technique

- 'boiling' speckles, 477, 478
- dynamic properties, 477
- focused image speckle pattern, 478
- in macromolecular science, 477
- object deformations, 477, 478
- speckle contrast, 479
- speckle decorrelation, 477, 478
- speckle pattern, 477
- speckle photography, 477, 478
- structural aspects of non-Gaussian speckle pattern, 477, 479, 480
- 'translational' speckles, 477, 478

Spherical macromolecules

- static structure factor, 339-341
- dynamic structure factor, 344

Spinodal decomposition, 242

Spring-bead model, 131, 133, 134, 136

Squeezing, 448, 453

Static light scattering (SLS); also Integral light scattering, Rayleigh-Debye light scattering, and Lorenz-Mie light scattering 38, 40, 120, 122, 125, 146, 160-162, 287, 288, 290-293, 296, 297, 338-343, 345, 346, 353-356, 359-362, 371, 374, 380-384, 401-404, 405-408, 457, 460, 511, 512, 517

Statistics
sub-Poissonian, 448, 453
super-Poissonian, 453
Stokes-Einstein equation, 1, 4, 6, 307, 318, 368, 375, 402
Streaming birefringence, 131-142
Subtilisin BPN', 500
Subtilisin Novo, 500
Super-chaotic light, 455
Superposition of coherent and chaotic light, 451
Suspension
bentonite, 390
graphite, 378, 390
charged macroparticles, 335
Suspension polymerization, 50
Swelling
of gels
buckling of surface, 198
kinetics of, 197
mechanical instability, 197
pattern formation, 199-204
poly(acrylamide), 198
of protein molecules, 308, 311

Temperature-jump method, 233-243
Temporary network, 131, 132, 141
generalized diffusion equation, 133-135
relaxation time approach, 138
transition probability, 136
Thermoplastic elastomer, 234
Theta solutions, 67, 68
Three-block copolymer (SBS), 405-408
Time-resolved small-angle X-ray scattering, 233-243
Tobacco mosaic virus (TMV), 17, 33-35
Torsional vibration method, 165
Torsional braid analysis, 165
Tracer diffusion of micelles, 412
Transient electric birefringence (TEB), 29
of virus fd, 31
Translational diffusion, see Diffusion
anisotropy in, 17, 32, 33
Translational motion, 49
Translation-rotation coupling, 112, 115
Transition probability, 131-137, 139, 141
Transition temperatures
glass-glass phase transition, 229, 232
glass-rubber transition, 229, 232
T_1, T_2, T_3, 230
Transverse configuration, 381
Triose phosphate isomerase, 500
Turbidity, 157, 160, 161, 354-356

Ultrashort laser pulses
estimation of local motions, 465
on picosecond time scale, 465
Ultrasonic absorption
excess absorption, 329-332
relaxation process, 330-332
velocity process, 330-332

Upper critical solution temperature (UCST), 234, 241

Verwey-Overbeek potential, 336

Viruses

fd virus, 28-30, 32

tobacco mosaic virus (TMV), 17, 33-35

Viscoelasticity, 346

measurements, 246, 249

relaxation, 60, 61

Viscosity of colloidal systems, 346

Williams-Watts function, 206, 209, 211

WLF equation, 195, 206, 211

Worm-like chain, 122, 123, 283

Xanthan

diffusion coefficient, 122

Evanescent wave-induced fluorescence (EWIF), 485-491

ordered conformation, 119

persistance length, 122

SLS, 121

Zimm dynamics, 54-56

FSC
www.fsc.org
MIX
Papier | Fördert gute Waldnutzung
FSC® C083411

Kolibri 360 GmbH
Ferdinand-Jühlke-Straße 7
99095 Erfurt, Deutschland
produktsicherheit@kolibri360.de